CLIMATES OF THE BRITISH ISLES

Climates of the British Isles commemorates the twenty-fifth anniversary of the founding of the internationally acclaimed Climatic Research Unit at the University of East Anglia. Written by present and recent members of the Unit, the sixteen chapters of this book distil much of the work and expertise for which the Climatic Research Unit is famous, presenting to the reader through the geographical lens of the British Isles an integrated synopsis of what we know about climate at the end of the twentieth century. This domain is deliberately wider than just the United Kingdom. Yet climate knows no boundaries other than those wrought by mountains and oceans and while climate change has become a political issue in recent years the climate system itself continues to function oblivious to political boundaries.

Climates of the British Isles combines the historical and geographical dimensions of climate to provide a more comprehensive account of the changing climate than has previously been attempted. The climates of past ages in this region are reconstructed, from the great Quaternary Ice Ages through to the Little Ice Age of the seventeenth century. A full description of the climate of the present century is provided, illustrated with a wealth of graphs, photographs and colour maps. Some important climate datasets are also listed. The book also addresses the prospects for climate change in the British Isles over the next hundred years and further into the future.

CLIMATES OF THE BRITISH ISLES

Present, Past and Future

Edited by
Mike Hulme and Elaine Barrow

London and New York

First published 1997
by Routledge
11 New Fetter Lane, London EC4P 4EE

Simultaneously published in the USA and Canada
by Routledge
29 West 35th Street, New York, NY 10001

© 1997 Climatic Research Unit
Typeset in Garamond by Florencetype Ltd, Stoodleigh, Devon

Printed and bound in Great Britain by
Butler and Tanner Ltd, Frome and London

All rights reserved. No part of this book may be reprinted or
reproduced or utilized in any form or by any electronic,
mechanical, or other means, now known or hereafter
invented, including photocopying and recording, or in any
information storage or retrieval system, without permission in
writing from the publishers.

British Library Cataloguing in Publication Data
A catalogue record for this book is available from the British Library

Library of Congress Cataloging in Publication Data
Climate of the British Isles: present, past and future / edited by Mike
Hulme & Elaine Barrow.
    p. cm.
    Includes bibliographical references and index.
    1. British Isles – Climate. 2. Paleoclimatology – British Isles.
I. Hulme, M.   II. Barrow, E.
QC989.G69C58   96-31980
551.6941–dc20   CIP

ISBN 0–415–13016–6
ISBN 0–415–13017–4 (pbk)

This book is dedicated to Professor H.H. Lamb,
Founder and Director of the Climatic Research Unit, 1972–1978

CONTENTS

PLATES

FIGURES

TABLES

CONTRIBUTORS

Tim Atkinson (B.Sc. Geology, Ph.D. Hydrology) is a Reader in the School of Environmental Sciences and a former Senior Research Associate in the Climatic Research Unit. His research interests vary widely in hydrogeology and hydrology, Quaternary geology and palaeoclimatology. He has a special interest in the methodology of inferring palaeoclimate from geological data.

Elaine Barrow (B.Sc. Environmental Sciences, M.Sc. Atmospheric Sciences) is a Senior Research Associate in the Climatic Research Unit specialising in the construction of climate change scenarios for agricultural applications and is also involved in the Climate Impacts LINK Project.

Graham Bentham (MA Geography) is Lecturer in the School of Environmental Sciences and specialises in research into the effects of environmental conditions on health.

Keith Briffa (B.Sc. Biological Sciences, Ph.D. Dendroclimatology) is a Reader in the Climatic Research Unit specialising in tree-ring-related studies. He also works with instrumental and other proxy climate data, mostly in the context of late glacial and Holocene climates.

Peter Brimblecombe (B.Sc., M.Sc., Ph.D. Chemistry) is a Professor in the School of Environmental Sciences. He is interested in the effects of air pollution on material and the history of atmospheric composition. He also works on the chemistry of cloud droplets both in the troposphere and in the stratosphere.

Keith Clayton (CBE, M.Sc., Ph.D. Geomorphology, Hon. D.Sc.) was the founding Dean of the School of Environmental Sciences between 1967 and 1971 and since 1993 has been Emeritus Professor. While locally known for his radical views on coastal management, his dominant research interest is the long-term evolution of the British landform.

Declan Conway (B.Sc., M.Sc. Geography, Ph.D. Climatology) is a Senior Research Associate in the

	Climatic Research Unit specialising in climate change and water resources in Africa. He is currently working on the generation of daily rainfall time-series from weather types and circulation patterns.
Trevor Davies	(B.Sc. Physical Geography, Ph.D. Air Pollution and Atmospheric Circulation) has been Director of the Climatic Research Unit since 1993 and is Professor in the School of Environmental Sciences. He has worked on links between atmospheric circulation and atmospheric composition and pollutant deposition. He is interested in the nature of changing atmospheric circulations.
Michael Dukes	(B.Sc. Geography, M.Sc. Meteorology and Climatology) has worked as a Research Associate in the Climatic Research Unit but he currently runs his own weather consultancy. His main interests are in extreme weather events and climate change.
Philip Eden	(BA Geography, M.Sc. Meteorology and Climatology) runs his own commercial weather consultancy, providing services for a variety of media outlets, and is the chief weather presenter on BBC Radio 5 Live. His main interest is in the synoptic meteorology of major weather events.
Graham Farmer	(B.Sc., Ph.D. Geography) is currently Director of the FAO/SADC Early Warning System based in Harare, Zimbabwe. He was a Senior Research Associate in the Climatic Research Unit from 1981 to 1989 where his research interests were the historical climate of England, African rainfall change and the global instrumental temperature record.
Brian Funnell	(BA, Ph.D. Geology) is an Emeritus Professor in the School of Environmental Sciences, and a Visiting Professor in the Climatic Research Unit. His research has involved investigations of both deep-sea and coastal sediments, covering a broad spectrum of palaeoenvironmental processes and changes, from the very long term (the past 250 million years) to the shorter term (the past 10,000 years).
Clare Goodess	(B.Sc. Environmental Sciences) is a Senior Research Associate in the Climatic Research Unit specialising in the study of long-term climate change and the implications for underground radioactive waste disposal in the UK. She has also worked on the construction of climate change scenarios for the study of desertification processes in the Mediterranean region.
Tom Holt	(B.Sc., Ph.D. Environmental Sciences) is a Senior Research Associate in the Climatic Research Unit with a particular interest in mid-latitude storms and severe tropical cyclones. He has also worked on carbon-cycle modelling and has produced several reports on climate variations for the insurance industry.

Mike Hulme   (B.Sc., Ph.D. Geography) is a Senior Research Associate in the Climatic Research Unit specialising in the construction of observed climatologies and in the validation of global climate models. He has also worked extensively on trends in African rainfall and their relationship to desertification.

Phil Jones   (BA Environmental Sciences, M.Sc., Ph.D. Hydrology) is a Reader in the Climatic Research Unit. He is involved in four principal research areas: monitoring climate on a global scale, palaeoclimatology, bringing the instrumental and palaeoclimatic data together in the context of the climate change detection issue and riverflow reconstruction in the UK.

P. Mick Kelly   (B.Sc. Physics with Meteorology, Ph.D. Environmental Sciences–Climatic Change) is a Reader with the Climatic Research Unit and the Centre for Social and Economic Research on the Global Environment. An atmospheric scientist by training, he has worked extensively on instrumental data analysis, Arctic climate variability and causes of climate change, and is currently involved in a number of interdisciplinary studies of climate and development issues.

John Kington   (B.Sc. Geography, M.Sc. Meteorology) is a Visiting Fellow in the Climatic Research Unit with special interests in historical climatology and synoptic meteorology.

Hubert Lamb   (MA, D.Sc., Hon. D.Sc.) is an Emeritus Professor in the Climatic Research Unit and was the founding Director of the Climatic Research Unit from 1972 to 1978. His interests lie in the history of climate and in its interactions with human society.

Astrid Ogilvie   (BA European History, Ph.D. Environmental Sciences) is a Senior Research Associate in the Climatic Research Unit. However, she is currently based in the USA where she is Associate Director of the Institute of Arctic and Alpine Research at the University of Colorado in Boulder. Her main areas of interest are the use of historical records to reconstruct past climate, the impact of climate on societies and the comparison of different proxy climate records.

Tim Osborn   (B.Sc. Geophysical Sciences, Ph.D. Environmental Sciences) is a Senior Research Associate in the Climatic Research Unit specialising in the analysis, simulation and validation of natural climate variability in numerical models and in observations.

Jean Palutikof   (B.Sc., Ph.D. Geography) is a Reader in the Climatic Research Unit specialising in climate change impacts, particularly related to global warming, and the

	application of climate data to economic and planning issues. For a number of years she has worked on the analysis of wind data for the renewable energy and insurance industries in the UK.
Graham Parker	(B.Sc. Mathematics and Physics) is a retired government meteorologist who founded the Norwich Weather Centre. He is an ex-BBC national and regional weatherman and radio broadcaster.
Clive Pierce	(B.Sc. Environmental Sciences, M.Sc. Applied Meteorology and Climatology) worked as a Research Associate in the Climatic Research Unit and the School of Environmental Sciences before pursuing a career in weather forecasting. He is currently a research scientist in the Met. Office working on the conceptual modelling of convective precipitation.
Sarah Raper	(B.Sc., Ph.D. Environmental Sciences) is a Senior Research Associate in the Climatic Research Unit specialising in simple models for simulating past and future global-mean temperature change and sea-level rise.
Andrew Skellern	(B.Sc. Environmental Sciences, M.Sc. Information Technology) has worked as a Research Associate in the Climatic Research Unit, but is currently completing his Ph.D. on modelling global peatland distributions utilising both climate and topography in the School of Geography, University of Leeds.
David Viner	(B.Sc. Physical Geography, Ph.D. Civil Engineering) is a Senior Research Associate in the Climatic Research Unit specialising in the construction of datasets for climate change scenario construction. He runs the UK Department of the Environment's Climate Impacts LINK Project which acts as the interface between the Hadley Centre and the international climate change research community.
Tom Wigley	(B.Sc., Ph.D. Mathematical Physics) is a former Director of the Climatic Research Unit and is currently a Senior Scientist at the National Center for Atmospheric Research, Boulder, Colorado. His main interests are in carbon cycle modelling, projections of future climate and sea-level change and interpretation of past climate change particularly with a view to detecting anthropogenic influences.

PREFACE

In 1997 the Climatic Research Unit will have existed for twenty-five years. Although the first member of staff was appointed by the University of East Anglia in October 1971, it was in January 1972 that Professor H.H. Lamb arrived at the University, becoming the founding Director of the Unit. The original vision was to establish a research centre devoted to the study of past, present and future climates, a vision inspired very largely by Professor Lamb. Twenty-five years later his foresight has proved remarkably prescient. The Unit has long been established as one of the world's outstanding centres for research into climate change, a research field in which there has been escalating interest, especially in recent years. As a mark of his contribution, not only to the Unit but to the endeavour of climate change research world-wide, we dedicate this book to Professor Lamb.

The book commemorates this twenty-fifth anniversary of the Climatic Research Unit by providing an integrated synopsis of what we know at the end of the twentieth century about the climates of the British Isles. Written by present and recent members of the Unit, the sixteen chapters of this book distil much of the work and expertise for which the Climatic Research Unit is famous and present it to the reader through the geographical lens of the British Isles. This domain is deliberately wider than just the United Kingdom. Climate knows no boundaries other than those wrought by mountains and oceans and while climate change has become a political issue in recent years the climate system itself continues to function oblivious to political bound-

aries. The British Isles are a coherent climatic region; the United Kingdom is not. If there is a bias in the book towards the United Kingdom in some of the data and maps shown, it has been forced on us by the restriction of certain datasets or studies to purely national or sub-national domains.

The rationale and organisation of the book draws upon two earlier studies published during the 1970s: *Climate: Present, Past and Future* written by Hubert Lamb, and published as two volumes in 1972 and 1977, and *The Climate of the British Isles* written by Tony Chandler and Stan Gregory and published in 1976. The respective historical and geographical dimensions of these two books have been combined in the present volume to provide a more comprehensive account of the changing climate of the British Isles than has previously been attempted. Our understanding of climate and its role in human affairs has changed markedly since the 1970s, as have our climate observation systems and modelling capabilities. It is timely, therefore, that this book should appear when the prospect of human-induced climate change is registering increasingly in the minds of the public. Some of the understanding of our climate, and its interpretation, which is offered within these pages will doubtless be superseded as science advances; but we believe that this book, as a whole, will provide a lasting, authoritative and accessible view of the changing climate resource of the British Isles as we approach the millennium.

The text is written not as a research document, but as one intended for the non-specialist. Where necessary, we separate more technical details from the flow of the text by placing such material in boxes. Terms that have specific climatic or environmental definitions appear in bold the first time they occur in a chapter and are formally explained in the Glossary at the back of the book. Each chapter is fully referenced through endnotes and provides suggestions for general reading on the subject matter by listing a small selection of key texts. We also provide a series of climate maps, graphs and data listings relating to the climate of the British Isles in four appendices, information that has never before been published in single volume.

No book comes into being lightly or without pain. This one is no exception. As editors we wish to thank our colleagues in the Climatic Research Unit, and our ex-colleagues outside it, both those whose names appear alongside one or more of the chapters and those whose names do not, but who nevertheless contributed to the overall effort in many ways.

Mike Hulme and Elaine Barrow
Climatic Research Unit, Norwich, June 1996

FOREWORD

The Climatic Research Unit at Twenty-five Years

It is tempting to suggest that a research unit as original, as timely and as successful as the Climatic Research Unit was designed and implemented in a co-ordinated way, that a fine vision was realised through careful planning and much hard work. There was of course a founding vision and there have undoubtedly been many years of hard work; good research does not just happen and funds for research, however innovative, are never easy to come by. But chance, serendipity, and simple good luck have also played a role alongside the more premeditated events, the strong appointments made, and the successful bids for funding.

Hubert Lamb had for years been trying to convince the meteorological establishment that the climate system in its natural state is highly variable on time-scales of decades to centuries and longer. The founding of the Climatic Research Unit in 1971–2 gave him the chance to concentrate on this battle, one that he and the climatological community eventually won so convincingly that ideas of climate constancy have faded almost completely from scientific memory. The creation of the Unit was due to initial grants from Shell International and the Nuffield Foundation, followed in the early years by the beneficence of the Rockefeller Foundation and repeatedly from the Wolfson Foundation which, in 1986, gave the Unit its current building. The initial establishment of the Unit also owes much to the advocacy of the late Sir Graham Sutton (Director-General of the UK Met. Office from 1953 to 1965 and subsequently of the Natural Environment Research Council) and its location in Norwich at the

University of East Anglia to the support of the late Lord Solly Zuckerman, an adviser to the University, and of Professors Keith Clayton and Brian Funnell, Deans of the School of Environmental Sciences in 1971 and 1972 and within whose School the Unit was based.

The Climatic Research Unit set itself four aims, laid out in its first annual report. It is worth reproducing those aims here since, twenty-five years on, they still effectively describe the accomplishments and mission of the Climatic Research Unit as we reach the end of the twentieth century. Indeed, these aims are even more pertinent today than they were in 1972:

- To establish firmer knowledge of the history of climate in the recent and distant past.
- To monitor and report on current climatic developments on a global scale.
- To identify the processes (natural and man-made/anthropogenic) at work in climatic fluctuations and the characteristic time-scales of their evolution.
- To investigate the possibilities of making advisory statements about future trends of weather and climate from a season to many years ahead, based on acceptable scientific methods and in a form likely to be useful for long-term planning purposes.[1]

That these aims remain valid is testimony not only to an early appreciation by the founder of the Unit of why climate research is such an important scientific endeavour, but also to the successes of the 200

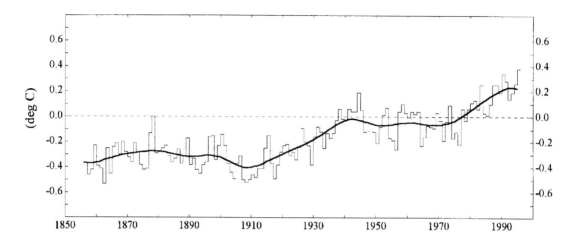

Figure 0.1 The global record of annual near-surface temperature from 1856 to 1995 expressed as anomalies, in degrees Celsius, from the 1961 to 1990 average. The bold curve is the result of applying a filter to the yearly values which emphasises variations on time-scales longer than thirty years. The land component of this record is compiled by the Climatic Research Unit and the marine component by the Hadley Centre. The merged record, as shown here, is the one used by the Intergovernmental Panel on Climate Change in all their publications.

or so individuals who have been members of the Unit over the past twenty-five years and who have researched these objectives and communicated their results to the outside world. All of these scientists have made contributions to the intellectual assets maintained by the Unit and should be proud of their achievements.

The global near-surface temperature record is perhaps the best, but by no means the only, example of such work. Painstakingly compiled during the early 1980s by the Unit, then under the directorship of Professor Tom Wigley, the record is now routinely updated in conjunction with the Hadley Centre (see Figure 0.1). This authoritative record of global temperature fluctuations over the last 140 years has provided the science of climate change with an invaluable resource and has acted as a powerful symbol to the public at large of the reality of climate change. The initial funding for this work provides a good example of where serendipity underwrites enterprise. It became clear, in the second half of the 1970s, that there was a need for an improved data-bank of

climate observations from around the world. The demand came initially from those involved in climate reconstruction, such as Hal Fritts of the Tree Ring Laboratory at the University of Arizona in Tucson, and then to support efforts to detect the potential effects on climate of greenhouse gas emissions. The Unit was well-placed to meet this demand – and acted on the opportunity. Strong contacts in the United States, such as the late J. Murray Mitchell, the doyen of American climatologists, Ray Bradley at the University of Massachusetts and Henry Diaz at NOAA, led to collaboration and funding through a series of contracts with the US Department of Energy that have continued to be renewed to the present.

This book is a commemoration of the first quarter century of the existence of the Climatic Research Unit. We are very happy to see it and proud that the Unit has prospered during this time and established a world-wide reputation. With all who find interest in the climate of the British Isles, in its variability and change over time, and in its relationship to the wider global climate system, we join in thanking the authors and editors of the book. We also express our wishes for a continuing successful future for the Unit under its current Director, Professor Trevor Davies. The Unit, now with other institutions in Britain and abroad, maintains its critical watch on the changing climate, both in the British Isles and world-wide. Some of these changes are shaped by human actions; some of them are quite natural in origin. Either way, climate change fundamentally affects the conditions of life in our world. Seeking to understand and predict such change remains a critical endeavour deserving our strongest efforts, a worthy and worthwhile challenge to human creativity, ingenuity and discipline.

EMERITUS PROFESSOR H.H. LAMB
Holt, Norfolk
(Founding Director, CRU, 1972–8)

EMERITUS PROFESSOR K.M. CLAYTON
Norwich, Norfolk
(Dean, Environmental Sciences, UEA, 1967–71 and 1987–93)

PROFESSOR T.M.L. WIGLEY
Boulder, Colorado
(Director, CRU, 1978–93)

NOTE

1   *Climatic Research Unit Monthly Bulletin*, 1972, vol. 1, p. 9.

ACKNOWLEDGEMENTS

Many people whose names do not appear as chapter authors have contributed to the production of this book. They have all provided invaluable assistance and it is no exaggeration to say that this book would not have appeared without their help. In particular we thank Julie Burgess, Christine Jeffery, and Susan Boland for secretarial support; Sheila Davies, Phillip Judge and Adam James for their photographic, cartographic and computer graphic skills, respectively; Mick Kelly for allowing us to use his nifty graphics package, Doodler for Windows; Mike Bristow for help with computing problems; Mary Spence of the Royal Meteorological Society, Frankie Pullinger and Graham Bartlett of the National Meteorological Library, Bracknell, Andrew Joyce of the University of Durham, Don McKinlay of the Climatic Research Unit and Martin Ingram of Brasenose College, Oxford, for their help in tracking down photos which we have used in this book. We would also like to thank the Hadley Centre and the UK Met. Office for providing the global climate model data and the station normal data which have been used in various chapters. These data were provided through the Department of the Environment's Climate Impacts LINK Project. The daily precipitation data from the England and Wales precipitation series used in Chapter 10 were made available by agreement of Tom Wigley and Phil Jones of the Climatic Research Unit at the University of East Anglia and John Cole of the Water Research Centre. The assistance of the UK Met. Office in updating the various national series is acknowledged. Precipitation data for the fourteen stations in the Republic of Ireland were supplied by Dennis Fitzgerald of the Irish Meteorological Service.

We have made every effort to trace the copyright holders of photographic material used in this book and acknowledgement is given to them below. Unfortunately, we have not been able to trace the sources of some of the figures. We would be grateful to readers for any further information they may be able to provide.

The quotation from Manley, p. 9, appears courtesy of Chapman and Hall who, as Allen and Unwin, published this passage in 1985 in the book *The Climatic Scene* (eds Tooley, M.J. and Sheail, G.M.). It is translated from an article which originally appeared in Russian in 1963 in the magazine *Anglia*. All other quotations used as epigraphs were extracted from the Penguin Dictionaries of Quotations.

Chapter 2: The University of Dundee for Figures 2.3 and 2.13.

Chapter 3: Ronald F. Saunders for Figure 3.4 (previously published in *Weather*, December 1993); University of Dundee for Figure 3.6; Kevin J. Richardson – Photography for Figure 3.8 (previously published in *Weather*, February 1995).

Chapter 4: Figure 4.5 is reproduced by permission of Science Reviews Ltd from D.Q. Bowen, 'The Pleistocene of North West Europe', *Science Progress*, 1992, vol. 76, pp. 209–23; The National Museum of Wales for Figure 4.6; Figure 4.9 is reproduced by permission of Oxford University Press (from F.W. Shotton (ed.), *British Quaternary Studies: Recent Advances*, 1977).

Chapter 5: Barbara Maher for Figure 5.3.

Chapter 6: Stofnun Árna Magnússonar (The Árni Magnússon Manuscript Institute at the University of

Iceland), in particular Sigurgeir Steingrímsson, for Figure 6.2.

Chapter 7: The Mary Evans Picture Library for Figure 7.1; the Royal Meteorological Society for Figure 7.2 (previously published in *Weather*, 27, p. 495, from Symons 1891); J. Kington, *The Weather of the 1780s over Europe*, 1988, Cambridge University Press, for Figures 7.3, 7.4, 7.5; Derek Hudspeth for Figure 7.7; the Royal Meteorological Society for Figure 7.8; Figures 7.9, 7.10 and 7.11 are from Fitzroy 1863; Graham Bartlett for Figure 7.12; the Tate Gallery, London for Figure 7.13.

Chapter 8: Penny Tranter and the BBC Weather Centre for Figure 8.1; University of Dundee for Figures 8.2 and 8.4; J.M. Cook for information about Figure 8.5 (previously published in *Weather*, January 1992), although the photographer is unknown; Longmans, Green and Co. for Figure 8.6; the Royal Geographical Society with the Institute of British Geographers for Figure 8.7; Figure 8.8 is reprinted from *Atmospheric Environment*, 24A, T.D. Davies, G. Farmer and R.J. Barthelmie, 'Use of simple daily atmospheric circulation types for the interpretation of precipitation composition at a site (Eskdalemuir) in Scotland, 1978–1984', pp. 63–72, copyright (1990), with kind permission of Elsevier Science Ltd, The Boulevard, Langford Lane, Kidlington, OX5 1GB, UK.

Chapter 9: The Royal Meteorological Society for Figure 9.1; the Bodleian Library, University of Oxford for Figure 9.2; the UK Met. Office for Figure 9.4; J.F.P. Galvin, by permission of the Royal Meteorological Society, for Figure 9.10.

Chapter 10: The Royal Meteorological Society for Figure 10.2; the Water Services Association for Figure 10.17.

Chapter 11: Surrey Herald, *News* and *Leader* Series for Figure 11.1; Betty Bosworth for Figure 11.4 (previously published in *Weather*, April 1996); the Royal Meteorological Society for Figure 11.7 which is reproduced courtesy of The Met. Office.

Chapter 12: the Department of the Environment for Figures 12.6, 12.9, 12.10 and Plate 8(a); David Ball for Figures 12.1 and 12.8.

Chapter 13: Figure 13.1 is reproduced courtesy of the *Daily Mirror* and Mirror Syndication International; Mark Davison/ Frosted Earth for Figure 13.4 (previously published in *Weather*, February 1990); Figure 13.8 appears courtesy of the *Western Morning News* Company Ltd.; George D. Anderson for Figure 13.15; S.D. Burt for Figure 13.19.

Chapter 14: Malcolm Walker for Figure 14.1; the Royal Meteorological Society and Kenneth Woodley for Figure 14.2; the University of Dundee for Figure 14.3; the UK Met. Office for Plate 8(b) and Figures 14.5 and 14.6.

Chapter 15: Julian Williams for Figure 15.9.

Chapter 16: Figure 16.2 is reprinted from the *Journal of Glaciology* with the permission of the International Glaciological Society; André Berger for Figures 16.3 and 16.9; Figure 16.4 is reprinted from *Palaeogeography, Palaeoclimatology, Palaeoecology* (Global and Planetary Change Section), vol. 97, Walter and Kasting, 'Effects of fuel and forest conservation on future levels of atmospheric carbon dioxide', pp. 151–89, copyright (1992) with kind permission of the authors and Elsevier Science – NL, Sara Burgerhartstraat 25, 1055 KV Amsterdam, The Netherlands; Figure 16.6 is reprinted with permission from the authors and *Nature* (Manabe and Stouffer, 1993), copyright (1993), Macmillan Magazines Ltd; Figure 16.7 Kim and Crowley, *Geophysical Research Letters*, 21(8), pp. 681–4, 1994, published by the American Geophysical Union; Figure 16.8 Crowley and Kim, *Geophysical Research Letters*, 22(8), pp. 933–6, 1995, published by the American Geophysical Union.

Appendix B: Professor H.H. Lamb.

Appendix C: The Hadley Centre of the UK Met. Office for the daily CET data.

Appendices D1 and D5: The Hadley Centre.

Appendices D2, D3 and D4: Phil Jones of the Climatic Research Unit, University of East Anglia.

1

INTRODUCING CLIMATE CHANGE

Mike Hulme and Elaine Barrow

The more things change, the more they are the same.
Alphonse Karr

CHANGING VIEWS OF CLIMATE

With hindsight, the twentieth-century view of climate was dominated by the perspective that climate is constant. Climate was effectively stationary for the purposes of human decision-making and only varied in any significant way over geological time. This view of climate was partly conditioned by the dominance of the developments in weather forecasting which took place during the first half of the century. The excitement of scientific discoveries which led to improvements in the understanding and prediction of the weather of the next day marginalised work which was more concerned with variations in climate over decades and centuries. The adoption of 'normal' periods by the fledgling International Meteorological Organisation reinforced this rather static view of climate. Weather statistics collected over thirty or thirty-five years were thought of as adequate to define the climate of a region, statistics which could then be safely used in future design and planning applications. Climate change was largely irrelevant.

This was in contrast to much thinking in the nineteenth century. The evidence of glaciation discovered during the early decades of that century and the emergence of evolutionary ideas were more consistent with a dynamic view of nature and of climate than one in which all things remained constant. Concern amongst colonialist conservationists about deforestation in the tropics causing climate change – in this case loss of rainfall[1] – was also consistent with this view. Thus a major English national newspaper could observe in 1818 that,

> a prospect far more gloomy than the mere loss of wine had begun to present itself by the increased chilliness of our summer months. It is too well known that there was not sufficient warmth in the summer of 1816 to ripen the grain; and it is generally thought that if the ten or twelve days of hot weather at the end of June last had not occurred, most of the corn must have perished. The warm and settled appearance of the weather at this early period of the season, leads us to hope that an agreeable change is about to take place in our planet; and that we shall not, as for many past years, have to deplore the deficiency of solar heat which is so necessary to ripen the productions of the earth.[2]

Only during the last quarter of the twentieth century – from the 1970s onwards – has this more dynamic view of climate been rediscovered. The twentieth-century view that climate is constant was first seriously challenged by a few pioneering scholars, of whom Hubert Lamb must rank in this country as perhaps the most important. They were followed by the growing body of climate scientists and, by the time of our present decade, by an increasing constituency of decision-makers. A strong sense of history was characteristic of those originally challenging the twentieth-century orthodoxy. Indeed, Hubert Lamb and others were almost as much historians as they were climatologists. More recently, events in the climate system itself reinforced the challenge and have now led to a re-writing of the orthodoxy.

The prospect of significant global climate change induced by human pollution of the atmosphere has acted as a powerful agent in consolidating the revisionist view of climate as non-stationary. This process of re-thinking has been underpinned by the twin developments of more abundant global climate observations and rapid increases in computer modelling capability. It is now possible to describe truly global changes in climate using observational data and to explore future changes in climate using credible climate models. The changed attitude towards climate has also been institutionalised in recent years. In 1988, for example, the World Meteorological Organisation and the United Nations Environment Programme established an Intergovernmental Panel on Climate Change to assess the evidence for the enhanced greenhouse effect, or so-called 'global warming'. This Panel continues to produce reports for the world community on the prospect of climate change[3] and they have also considered the consequences of global climate change for individuals, ecosystems and nation-states.[4] The concern about changing global climate was sufficient to yield a United Nations Framework Convention on Climate Change. This Convention was signed by 155 nations at Rio de Janeiro in June 1992 and subsequently came into force in March 1994. The British and Irish governments ratified the Convention in December 1993 and April 1994, respectively, and both diplomatic delegations have played their part in the ongoing negotiations to establish a legally binding climate protocol.

These developments have taken place against a background of a warming climate. Since the 1970s, both the British Isles and the world have warmed by about 0.3°C. The reality of this warming, and the prospect of accelerated warming over the next few decades, has focused more attention on the interactions and interdependencies of climate and society than was hitherto the case. Thus the United Kingdom government, through its Department of the Environment, commissioned national reviews in 1991 and again in 1996 of the potential impacts of climate change for the country.[5] This type of national review of the importance of climate change is required of many countries under the Framework Convention and is a mode of reporting that has been adopted around the world.

There is a danger that this recent political concern about climate change and its impacts bestows on climate an unwarranted importance as an agent that shapes our lives. Such thinking has led, perhaps rather curiously, to a return in some quarters to a variant of the climatic determinism prevalent at the start of the century. Determinism is a reductionist philosophy that sees events and behaviour as controlled by a very limited set of physical factors. Ellsworth Huntington, the Yale geographer, is the most well-known proponent of such a role for climate. He argued in 1915 that, 'The climate of many countries seems to be one of the great reasons why idleness, dishonesty, immorality, stupidity, and weakness of will prevail.'[6] Although not always as strident or doctrinaire as Huntington, the importance of the climatic influence on our lives has been stressed by numerous thinkers, starting with the Ancient Greeks and their supposedly uninhabitable, torrid and frigid 'climata'. The influence of climate has also been interpreted psychologically. In the middle of this century, for example, Gordon Manley stated that, 'Appreciation of the British climate depends largely on temperament. That it has not been conducive to idleness has been reflected in the characteristics of the people',[7] and, more recently, Richard Beck argues that, 'the historical record is highly suggestive . . . that a mild climate in mid-latitudes helps to foster a tolerant society or that an extreme climate may predispose people towards intolerance.'[8] These psychological interpretations of climatic determinism may seem hard to defend. Nevertheless, the prospect of global warming, and the study of the impacts of such climate change, introduces a new variant to the climatic determinists' repertoire of arguments.

Many studies of the possible impact of future climate change seem, implicitly, to elevate climate to being the major factor that will influence future human activity and welfare. Thus the conventional climate change impact study would attempt to simulate the effect of climate change by, say, 2050 on a particular aspect of the environment, say cropping patterns or forest distribution. Little attention is usually paid to whether or not climate is the main driving factor behind observed changes in such distributions. Even if it is recognised explicitly that other factors are involved (e.g., changes in technology, consumer behaviour, work and leisure patterns), these are so unpredictable that climate often retains the appearance of being the main controlling factor. Climate change determinism thus re-appears. Some studies have shown, however, that factors such as the future of the Common Agricultural Policy of the European Union will have a much larger impact on the future British landscape than climate change.[9] And it only takes a simple thought experiment to realise that other considerations, too, will swamp the effects of climate change on future human and animal welfare. For example, civil conflict, technological and demographic change and global epidemics, are all likely to influence welfare to a greater extent than will climate change. This is not to say that climate change is unimportant or does not matter. We merely stress that to assess the true significance of climate change it must be evaluated against changes that will occur due to other environmental constraints and social constructs.

Climatologists talking about climate change always run the risk, therefore, of being seen to be slanted in their views. They may be interpreted as being unnecessarily alarmist by those who reckon that human ingenuity and technical change will minimise the effects of climate change,[10] or overly complacent by those who see climate as a dominant control on human choices and action. Seeing climate, and therefore climate change, as a resource to manage and to benefit from, and as essentially neutral, is surely a more constructive view to take. The notion of 'good' and 'bad' climates is a hard one to defend in any absolute sense. Temporal changes in climate, seen in this way, present societies with challenges to cope with and opportunities to exploit. These challenges and opportunities presented by climate change are very much those that every colonising community through history has realised are posed by geographical differences in climate. This view of climate, whether implicit or explicit, has been true of, for example, Mongols in Europe, Vikings in Greenland or Europeans in Africa. For example, one may view the nineteenth-century history of the interaction between climate and society as one about the ability of the European colonising powers to exploit geographical differences in climate in the Tropics – rubber in Malaysia, cocoa in West Africa or bananas in the Caribbean – and to manage the regional climate impacts that such exploitation might bring with it.[11] A twenty-first-century history of such interaction may well be about the ability of different communities or regions to exploit and manage the forthcoming temporal changes in climate brought about by human pollution of the atmosphere.

Taking this view, the United Nations Framework Convention on Climate Change is concerned primarily with the regulation of this exploitation and management process as implied in Article 2,

> The ultimate objective of this Convention . . . is to achieve . . . stabilisation of greenhouse gas concentrations in the atmosphere at a level that would prevent dangerous anthropogenic interference with the climate system. Such a level should be achieved within a time frame sufficient to allow ecosystems to adapt naturally to climate change, to ensure that food production is not threatened and to enable economic development to proceed in a sustainable manner.[12]

It is unlikely, however, that such regulation can ensure that communities and nation-states benefit or suffer equally from climate change. The partitioning of these benefits or costs between nations will depend on two things: serendipity and access to human and technological capital. The impact of climate change on the world is likely to be dictated largely by the existing inequalities in human vulnerability, with some luck and institutional regulation thrown in.

The view of climate prevailing at the end of the twentieth century is, therefore, as follows. Climate is no longer regarded as a constant, but is continually subject to change. These changes are increasingly being caused, inadvertently, by human behaviour. The changes in climate remain largely unpredictable, but will have important consequences for human welfare, decision-making and planning. In addition to the prospect of climate change, new developments in daily, monthly and seasonal weather forecasting provide an even greater impetus to take climate variability seriously. Organisations, charged with investment decisions, environmental management and future planning strategies, need to include climate in their decision-making structures as a key variable rather than as an assumed constant. If this book contributes to such an awareness with respect to the British Isles, then it will have achieved one of its purposes. It will, in the process, have also contributed to one of the four original aims of the Climatic Research Unit cited when it was established in 1972, namely, 'To investigate the possibilities of making advisory statements about future trends of weather and climate from a season to many years ahead, based on acceptable scientific methods and in a form likely to be useful for long-term planning purposes.'[13]

OUTLINE OF THE BOOK

This book unashamedly exploits the title and structure of one of Professor Lamb's most important works – *Climate: present, past and future* – published in two volumes in 1972 and 1977. This title so well captures the essential temporal dimension of the study of climate and also embraces so fully the scope of research for which the Unit is renowned, that we cannot formulate a better description of the subject matter with which we are concerned. We have therefore followed his template in this anniversary volume.

Part 1 of the book is concerned with an overview of the causes and character of the climate of the British Isles as we know it at the end of the twentieth century. Trevor Davies, Tim Osborn and Mick Kelly in Chapter 2 put the climate of the region into a global context by describing the factors that shape and control the climate of these islands. These include the proximity of the British Isles to the warm ocean currents of the North Atlantic and the latitudinal position of the region in the main path of the mid-latitude westerlies. The outcome of these broad-scale influences is an ensemble of daily weather events that, when averaged over a suitable length of time (conventionally thirty years), yield a set of statistics that provide a short-hand description of the climate of the region. Such a statistical creation is termed 'a climatology'. The 1961 to 1990 climatology for the British Isles is described by Elaine Barrow and Mike Hulme in Chapter 3. Ranges and patterns of a large number of climate variables are described in this chapter with the help of maps and tables. A supplementary set of colour maps are also provided in Appendix A.

The reconstruction of past climates, prior to the commencement of formal meteorological observation during the seventeenth century, is addressed in Part 2 of the book, 'Reconstructing the Past'. The landscapes of the British Isles have been shaped over millennia by a continually changing climate as well as, more recently, by the activities of our own distant ancestors. The response of the physical and biotic environment to these climate changes has been preserved with differing levels of detail and reliability in various depositories both on land and in the ocean. These landscapes and depositories enable the forensic work of reconstructing past climates stretching back over thousands and millions of years. Brian Funnell takes the longest view of all in Chapter 4 in his assessment of the climates of the British Isles in past geologic ages, using a mixture of landscape and ocean core evidence from the region and from the wider North Atlantic basin. The glacial and interglacial pulses of the **Quaternary** era dominate the record on these time-scales, pulses which now seem to be orchestrated largely by changes in the orbital characteristics of the Earth around the Sun. In Chapter 5, Keith Briffa and Tim Atkinson tackle the reconstruction of climates during the period – called the **Holocene** – since the last glaciation of the British Isles, which came to an end about 15,000 years ago. Although closer in time to the present, the physical and biotic evidence of climate fluctuations in the region through the Holocene do not always allow a robust picture of climate change to be drawn during this period of increasing human acculturation. The Roman occupation of the British Isles for the four centuries following the birth of Christ coincides with the earliest documentary evidence of the British environment. The Dark Ages also left a limited legacy of written descriptions of climate to future generations, but the centuries following the Norman invasion in 1066 possess a great abundance of written documents that contain direct and indirect references to weather and climate conditions in the British Isles. Some of these documents, dating from the medieval period of the twelfth to fifteenth centuries, are evaluated by Astrid Ogilvie and Graham Farmer in Chapter 6 and compared with comparable documentary sources from elsewhere in the North Atlantic basin – Iceland and Greenland.

The British Isles is a region endowed with some of the richest sources of instrumental data for the study of historical and contemporary climate variation and change. In Part 3 of the book, 'Monitoring the Present', this vast store of quantitative climate information is used to present a selective evaluation of the observed trends and characteristics of British and Irish climate in the present and recent past. The origins of regular meteorological observations in the British Isles are explored by John Kington in Chapter 7. The stimulus given to science in general, and careful meteorological measurements in particular, by the seventeenth-century Enlightenment is quite clear and the first dedicated meteorological observatories in the British Isles soon followed in the eighteenth century. By the middle of the nineteenth century, a more regional view of weather was achievable and, prompted by the demands of the British Navy, daily synoptic weather charts were a regular feature of the meteorological enterprise. An impressive archive of a synoptic-scale weather classification for the British Isles is provided by the Lamb Catalogue of circulation types. This Catalogue was originally developed by Hubert Lamb in the 1940s and the ubiquity of historical daily synoptic charts for the Isles enabled him to reconstruct the Catalogue

back to 1 January 1861. Professor Lamb still updates the Catalogue each month and the full 135-year record is used by Mick Kelly, Phil Jones and Keith Briffa in Chapter 8 to describe the influence of seasonal and decadal variations in atmospheric circulation on the British and Irish climate. Hubert Lamb last published a complete listing of his Catalogue in 1972 and we therefore include in Appendix B an update of the Catalogue from 1972 to 1995. The 'Central England' temperature record, originally compiled by the late Gordon Manley, is the longest continuous instrumental climate record in the world. This record commenced in 1659 – the year following the death of Oliver Cromwell and the year before the creation of the Royal Society – and is now updated routinely by the United Kingdom Met. Office. It provides a unique opportunity to evaluate climate change in the region and beyond on century time-scales and is discussed by Phil Jones and Mike Hulme in Chapter 9. This 337-year temperature record provides the clearest indication of a warming climate for the British Isles, and the last fifty years have been the warmest such period in the entire record. The daily values of this record for the period since 1961 are plotted in Appendix C.

In addition to temperature, rainfall (or more correctly precipitation) is the other primary climate variable. Records of precipitation in the British Isles extend back almost as far as they do for temperature and an analysis of over two hundred years of measurements is undertaken in Chapter 10 by Phil Jones, Declan Conway and Keith Briffa. Precipitation is much more variable than temperature and it is hard to discern any significant trends over time in British and Irish precipitation. Droughts and floods are the manifestation of extreme precipitation variability and the frequencies of these environmental hazards are also described. Extreme wind storms are also a major hazard in the British Isles, but equally important for the region is the potential offered by winds for energy generation and recreational pursuits. These two contrasting roles for wind are explored in Chapter 11 by Jean Palutikof, Tom Holt and Andrew Skellern as they examine changes over time in wind as both resource and hazard.

The atmosphere provides the medium by which weather and climate are delivered to us who live at the Earth's surface. The composition of the atmosphere in terms of gaseous compounds and particulates is important, not only for climate but also because it affects the quality of the air that we and other animals and plants breathe each day. Not only does the British Isles possess some of the longest climate records in the world, this region also has one of the longest histories of air pollution and air pollution legislation extending back over several centuries. These interactions between climate and air quality in the British Isles are discussed by Peter Brimblecombe and Graham Bentham in Chapter 12. One of the perennial popular interests afforded by the weather is the establishment of new records of climatic extremity, whether hot or cold, wet or dry, windy, sunny or cloudy. Michael Dukes and Philip Eden examine this fascination with weather records in the British Isles in Chapter 13 and provide examples of some of the more popular and the more unusual. Of course, the more indices one has which describe the weather measured at more and more places, the more common it will be on purely statistical grounds for new records to be established on a given day, month or year. One must be cautious about interpreting a preponderance of new records as necessarily indicative of a changed or more extreme climate.

What of the future? Can we expect our weather and climate to continue to provide us with the same environmental conditions that we have experienced in our own lifetimes or that were experienced in previous generations or eras? In Part 4 of the book, 'Forecasting the Future', our ability to envision future weather and climate is discussed on three different time-scales: short-term weather forecasting, climate prediction over the next one hundred years and, finally, climate prediction many thousands of years into the future. Weather forecasting takes on many guises and over time has reflected the cultural changes and diversity that have characterised the history of the human species. Thus, superstition, legend and folklore, intuition, deterministic and chaotic science, have all generated their own fore-

casting methodologies at different times and in different cultures. Clive Pierce, Michael Dukes and Graham Parker provide a short summary of some of this history of weather forecasting in Chapter 14, before explaining the basis of modern weather forecasting techniques in the British Isles. Of course computers are now central to this enterprise, and some of the most powerful computers in the world, and in the British Isles, are dedicated to this activity. Similarly, for predicting climate many decades or centuries hence, very complex models operated on very powerful machines are necessary. Model complexity and computer power are no guarantees of acceptable predictive capacity, however, and Sarah Raper, David Viner, Mike Hulme and Elaine Barrow discuss the problems of understanding and modelling climate change for a region like the British Isles in Chapter 15. Although current predictions suggest that the climate of the twenty-first century will be dominated by continued warming, exactly how this will be manifest in the ensemble of weather elements and time-scales that comprise climate remains unclear. Planning for change, however, now seems a more necessary approach for public and private sector organisations than assuming stationarity of climate.

There are a small number of planning activities, however, that require information about the future climate on much longer time-scales, time-scales that extend over thousands of years. Clare Goodess and Jean Palutikof close the book with their analysis in Chapter 16 of the very long-term outlook for climate in the British Isles. To achieve this goal requires a re-examination of some of the evidence for past climate change presented in Chapters 4 and 5, as well as the use of the computer models described in Chapters 14 and 15, which simulate regional and global climate. And here, of course, lies the paradox. The increasing human imprint on the world in which we live is leading to a general warming of the climate, a warming which we expect will become increasingly dominant during the lifetime of our children and grandchildren. From the perspective of the planet and solar system, however, and recognising that their time metric is very different from ours, the climate of the Earth seems set to cool over the next 10,000, 50,000 or 100,000 years. Whether global warming, and associated climate change in regions like the British Isles, will appear as a minor blip on this longer-term trend, or whether the ability of the human species to modify its own environment is now so great as to be able to offset these planetary and cosmic forces, is a question that we will probably have to leave to our evolved descendants several millennia hence . . . or else to science fiction.

NOTES

1 See R.H. Grove, 'A historical review of early institutional and conservationist responses to fears of artificially induced global climate change: the deforestation–desiccation discourse 1500–1860', *Chemosphere*, vol. 29, pp. 1001–13.

2 *The Observer*, 18 June 1818.

3 IPCC, J.T. Houghton, L.G. Meiro Filho, B.A. Callendar, N. Harris, A. Kattenburg and K. Maskell (eds), *Climate Change 1995: the Science of Climate Change*, 1996, Cambridge, Cambridge University Press, 572 pp.

4 IPCC, R.T. Watson, M.C. Zinyowera and R.H. Moss (eds), *Climate Change 1995: Impacts, Adaptations and Mitigation of Climate Change: Scientific–technical Analyses*, 1996, Cambridge, Cambridge University Press, 878 pp.

5 See CCIRG, *The potential effects of climate change in the United Kingdom*, Department of the Environment, London, HMSO, 1991, 123 pp.; also CCIRG, *Potential impacts and adaptations of climate change in the United Kingdom*, Department of the Environment, London, HMSO, 1996, 248 pp.

6 E. Huntington, *Civilization and climate*, Hamden, Conn. USA, Shoe String Press, 1915, p. 411.

7 G. Manley, *Climate and the British scene*, London, Collins, 1952.

8 R.A. Beck, 'Climate, liberalism and intolerance', *Weather*, 1993, vol. 48, pp. 63–4.

9 M.L. Parry, J.E. Hossell, P.J. Jones, T. Rehman, R.B. Tranter, J.S. Marsh, C. Rosenzweig, G. Fischer, I.G. Carson and R.G.H. Bunce, 'Integrating global and regional analyses of the effects of climate change: a case study of land use in England and Wales', *Climatic Change*, 1996, vol. 32, pp. 185–98.

10 J. Ausubel, 'Technical progress and climate change', *Energy Policy*, 1995, vol. 23, pp. 411–16.

11 R.H. Grove, op. cit.

12 Article 2, *The United Nations Framework Convention on Climate Change*, New York, United Nations, 1992.

13 Quoted in *Climatic Research Unit Monthly Bulletin*, 1972, vol. 1, p .9.

Part 1:

THE BRITISH ISLES CLIMATE

So we can claim that these islands of frequent changes, of the terrible
Atlantic gales whose endless roar besets our coasts in winter, of the exquisite long June
days celebrated by our poets throughout the centuries, of the harsh biting north-easter in April,
the wind-driven rain day after day if there comes a wet autumn, the occasional spell of three
weeks of snow and frost, the persistent dryness that quite frequently leads to shortage of water
in early summer – all these give us much cause to grumble, but even more cause to enjoy the
march of the seasons and the opportunities for such a variety of flowers that the poorest
man can still grow them in his garden.
Gordon Manley, 'The Weather in Britain', Anglia, 1963

2

EXPLAINING THE CLIMATE OF THE BRITISH ISLES

Trevor Davies, P. Mick Kelly and Tim Osborn

It's a warm wind, the west wind, full of birds' cries.
John Masefield, 'The West Wind'

INTRODUCTION

The climate of the British Isles is the cumulative result of each day's weather. The weather on a particular day depends on the character of the atmospheric circulation over the Islands and on the **synoptic system – depression** or **anticyclone** – affecting the region: in other words, the overall 'weather type' (see Chapter 8). The nature of the winds and the character of the synoptic system depend, in turn, on the interactions between the atmosphere, the oceans and the land surface at every other point on the globe.

The weather experienced at a particular location in the British Isles is a function of the precise part of the synoptic system which is overhead at the time and of local influences. The weather associated with a particular synoptic system (which has a typical horizontal scale of 1–2,000 km) can have important variations on scales of 10–100 km. The weather which is 'delivered' to the British Isles by the synoptic systems is, in turn, modified by the underlying surface: high land may enhance the precipitation process (see Chapter 3); different land surfaces will react to incoming solar radiation in different ways, affecting the near-surface air temperature; local circulations (on the scale of one to a few tens of km) can be affected by topography or by the discontinuity across a coastal zone. Local effects such as these can exert a strong influence on the climate experienced by particular places.

The processes which produce the British climate are therefore enormously complex. There are interactions on all space- and time-scales. Consequently, it is over-simplistic to attempt to define a 'beginning' and an 'end' to the various interplays of factors. The picture gets even more complex when one considers that the nature of the interactions in the land–atmosphere–ocean system varies over time. This is either because of natural variability internal to the system, or because of outside 'forcings', such as variations in the receipt of solar radiation at the top of the atmosphere, or changes in the composition of the atmosphere because of volcanic activity or **anthropogenic** gas emissions.

Having made these cautionary remarks, we will consider in this chapter the most important processes affecting the climate of the British Isles sequentially, from the global to the local scale – and thereby run the risk of over-simplification. We start by considering the manner in which global climate is shaped by the planetary-scale atmospheric circulation, or 'general circulation'. The general circulation of the atmosphere takes the form it does because of the way energy from the Sun is distributed and utilised, because the Earth rotates, and because of the particular geographical pattern and orography of the land and ocean basins. We first look at the radiation and consequent heat budgets of the surface-atmosphere system, the driving force behind the wind systems

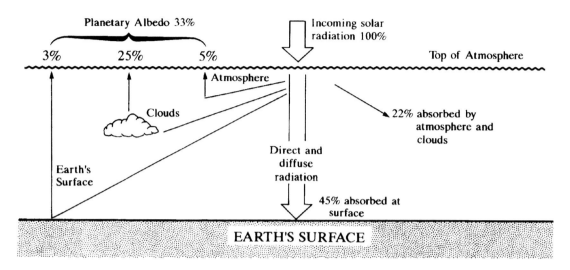

Figure 2.1 A representation of what happens to the incoming solar radiation.

of the Earth.[1] Later sections look at links between climate and the oceans and at local-scale influences on climate.

GLOBAL-SCALE RADIATION AND HEAT BALANCES

The main source of energy available to the planet is the Sun. Radiation emitted by the Sun is short in wavelength. Around 33 per cent of the solar radiation which is received by the planet from the Sun is scattered back to space (see Figure 2.1). The fraction lost in this way is called the planetary **albedo**. Most of the scattering occurs in the atmosphere, predominantly by clouds. The Earth's surface contribution to the planetary albedo is small. There are, however, considerable geographical differences in surface albedo – desert albedos are generally greater than those of forests and snow has a high albedo, as does water when the Sun angle is low. The atmosphere (gases, clouds and dust) absorbs about 22 per cent of the incoming solar radiation, leaving 45 per cent of the original received solar radiation to arrive at the Earth's surface, where it is absorbed.

The Earth's surface also receives radiation from the atmosphere, where the solar radiation has been absorbed and re-emitted as longwave radiation.[2] This longwave radiation is also absorbed at the Earth's surface. At the surface and in the atmosphere there is therefore a complex pattern of longwave radiation absorption, emission, re-absorption and re-emission.

An issue of considerable current concern is the change in concentration of the so-called 'greenhouse gases' in the atmosphere because of human activities. These greenhouse gases are the constituents of the atmosphere which play the important role of absorbing longwave terrestrial radiation (while not interfering significantly with the incoming solar radiation) thereby trapping heat near the Earth's surface. This **greenhouse effect** maintains temperatures which make the planet habitable; without it the Earth's surface would be much colder than it is. An increase in the concentrations of greenhouse gases should lead to significant global warming. This enhanced greenhouse effect is likely to change global climate, and the climate of the British Isles, in important ways. This issue is addressed at some length in Chapter 15, but we will neglect it for now

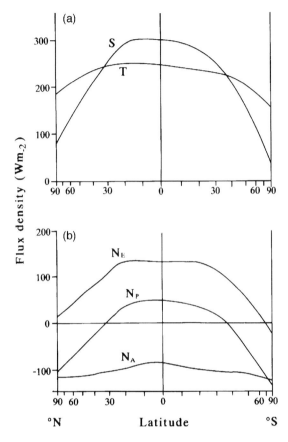

Figure 2.2 (a) Latitudinal averages of solar radiation (S) and longwave (terrestrial) radiation (T), and (b) latitudinal averages of net planetary radiation (N_P) net radiation at the Earth's surface (N_E), and net radiation to the atmosphere (N_A). Adapted from McIlveen (1992) (see p. 32 note 13).

as we attempt to explain the atmospheric processes which give the Islands' 'normal' climate its essential character.

Although the net radiation (the difference between the incoming radiation and the outgoing radiation) for the planet as a whole is zero, this is not the case at all latitudes; neither is it the case for the atmosphere and the Earth's surface separately. For now, we shall ignore geographical variations, and only consider the radiation budget for the planet averaged over all lines of longitude from 90°N to 90°S. This gives us a globally averaged latitudinal distribution. Since the Sun's overhead position traverses between the Tropics of Capricorn (23.45°S) and Cancer (23.45°N), the input of solar radiation is at a maximum in equatorial latitudes. The low Sun angle towards the Poles, and the greater albedo of snow and ice, means that the input of solar radiation at the surface is much smaller at these high latitudes (Figure 2.2a). The emission of longwave radiation shows a smaller latitudinal variation and exceeds the solar energy input outside tropical latitudes. The resulting planetary net radiation distribution (Figure 2.2b) is positive between 30°N and 40°S and negative at higher latitudes. The general circulation of the atmosphere and the oceans is driven by this latitudinal energy imbalance and it serves to transport heat polewards.

When we consider the net radiation distribution for the atmosphere (Figure 2.2b), we see that the values are negative at all latitudes. In terms of radiant energy, the atmosphere is cooling at a rate equivalent to about 0.8°C per day. This radiational cooling is offset by the transfer of energy from the Earth's surface which experiences positive net radiation at practically all latitudes (Figure 2.2b). This transfer is effected through the **convection** of **sensible heat** and the release of **latent heat** through the condensation of water vapour which was evaporated from the surface. Globally, latent heat transfer from the surface to the atmosphere is around four times more important than the convection of sensible heat, although there are large latitudinal variations. For example, near the Equator sensible heat accounts for only around 5 per cent of the total vertical heat transfer from surface to atmosphere, and yet at 70°N it accounts for about one-half of the total transfer.

In terms of polewards heat **advection** in the atmosphere, however, sensible heat transport is more important than that of latent heat. This indicates that most water vapour re-condenses in the atmosphere in much the same latitudinal zone as it was evaporated from the Earth's surface. More than half of the atmosphere's sensible heat originates in the global atmospheric 'engine': the tropical rain belt (between around 0–10°N). Polewards sensible heat

Figure 2.3 Two examples of infra-red satellite images of the British Isles showing the contrast between (above) winter, 27 February 1986, when the cold land shows light against a warm sea, and (p.15) summer, 16 May 1980, when the warm land shows dark against the colder sea. The land surface cools and warms much more rapidly than the ocean surface because of its lower heat capacity. Images courtesy of the University of Dundee.

transfer exhibits a double maximum in each hemisphere; polewards of the tropical rain belt at around 20°N and S, and again at around 50–60°N and S in response to condensation of water vapour in the mid-latitude cyclonic storm belts. Ocean currents account for around one-third of the polewards sensible heat transport.

Thus far, we have considered annual energy distributions averaged over latitude. There are important seasonal and geographical variations. We shall come back to these later but, at this point, some important features will be introduced. The **specific heat capacity** of land surfaces is much less than that of the oceans (Figure 2.3), so the adjustment time of the oceans to

Figure 2.3 Continued

any change in energy input is greater than of the land. Moreover, unlike the land, the oceans can transport heat, so **sea-surface temperatures** are not related in a simple way to the energy balance. Figure 2.4, which illustrates the annual range in temperature at the Earth's surface, clearly reflects the large-scale land/sea distribution. The annual temperature range in the interior of the high northern continents approaches, or exceeds, 50°C. The sea-surface temperature range is much smaller than the range over land and reflects the ocean circulations. The moderating influence of the North Atlantic Drift stretching over to the British Isles, for example, is most pronounced. The greater adjustment time of the oceans means that maximum and minimum sea-surface temperatures lag the solstices by around six weeks. Bathing in the sea around the British Isles is most pleasurable late in the summer.

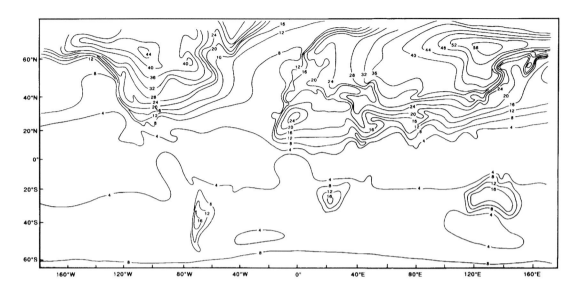

Figure 2.4 Average annual temperature range at the Earth's surface (°C). The annual temperature range is the difference in mean surface air temperature between the warmest and coldest months. Adapted from Wallace and Hobbs (1977).

THE GENERAL CIRCULATION

Early conceptualisations of the general circulation of the atmosphere took the energy imbalance between equator and pole as a starting point. They then focused on variants of a single, large-scale thermally driven circulation cell in both hemispheres – with a rising limb in tropical latitudes, descent in higher latitudes, and a return flow near the surface – as a means of redressing the imbalance. Such a large convective cell, with one dimension much more extensive than the other, is not stable. Improved observations, coupled with better understanding of the workings of a very thin atmospheric skin on a rotating Earth, produced a latitudinal conceptualisation which looks something like the scheme shown on the right-hand side of the hemisphere in Figure 2.5.

In tropical latitudes there is the so-called Hadley Cell, named after the eighteenth-century English scientist, George Hadley. The Hadley Cell can be viewed as a thermally direct cell, although much of the upwards energy transport near the Equator is concentrated over a relatively small area in vertically extensive convective cloud clusters. A zone of rapidly moving air, the subtropical jet stream, travelling west-to-east, is located near the **tropopause** at around 30°. This results from a need to conserve **angular momentum** as air moves from the Equator to regions closer to the Earth's rotational axis. Just beneath the tropopause in mid-latitudes is another zone of air moving rapidly from west-to-east – the polar front jet stream. This mid-latitude jet stream is a consequence of the strong thermal gradient in the vicinity of the **polar front**, a transitional zone between the relatively cold tropospheric **air masses** of high latitudes and the relatively warm air masses of subtropical latitudes. The latitudes where the polar front occurs represent an area of **slantwise convection**, interleaves of descending cold air moving equatorwards and ascending warm air moving polewards. These are the latitudes of travelling cyclones and anticyclones, synoptic systems which play a crucial part in the maintenance of the atmospheric general circulation, and in shaping the climate of the British Isles.

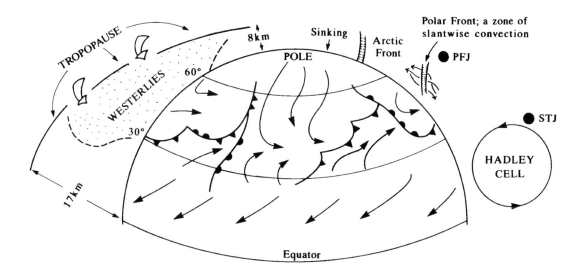

Figure 2.5 A latitudinal cross-section of the general circulation of the atmosphere (at right-hand side). STJ and PFJ are the Sub Tropical and Polar Front Jet streams embedded in the broad zone of westerly flow (see left-hand side). The tropopause (left-hand side) is the top of that part of the atmosphere where weather systems occur, representing a 'lid' on the troposphere, although there is some exchange with the **stratosphere** above. A schematic of the airflow at the Earth's surface is also shown, indicating that the Polar Front is heavily perturbed on a day-to-day basis, and that even in the broad band of westerlies, airflow with an easterly component does occur. The surface easterlies in tropical latitudes are the trade winds.

The average annual latitudinal distribution of evaporation and precipitation (Figure 2.6) confirms the role of the tropical Hadley Cell and the mid-latitude cyclones, in both hemispheres, in global atmospheric energy transport. The subtropical dry zones (where evaporation is greater than precipitation) correspond with the subsiding and equatorward moving parts of the Hadley Cell circulations. These provide atmospheric moisture transport into the tropical 'engine', where uplift produces condensation (releasing latent heat) and precipitation. The relatively wet mid-latitudes are the zones of high cyclone frequency, where frontal uplift leads to condensation of water vapour evaporated from the ocean surfaces.

The Earth rotates west-to-east and, because the atmosphere clings to the Earth, it rotates with it. When viewed from the Earth's surface, however, some winds travel from east-to-west (i.e., they are easterly relative to an observer at the surface). In absolute terms, they are rotating in a west-to-east sense at a rate which is slower than that of the Earth's surface. In broad terms easterlies occur in the latitudes of the lower limbs of the tropical Hadley Cells and in a narrower latitudinal band of their upper limbs; and restricted regions of high latitudes – see Figure 2.5. On the other hand, of course, the westerlies are rotating west-to-east at a rate which is faster than that of the Earth's surface. The drag of the Earth's surface extracts angular momentum from the westerlies. In order to keep blowing, there must be a reliable mechanism to inject angular momentum into the westerlies from the easterlies, which have angular momentum fed into them from the Earth's surface.

In tropical latitudes, the Hadley Cell plays an important role in this poleward angular momentum transport. In mid-latitudes, however, it is the cyclones, and the waves in the westerlies associated with them, which play the important role in the

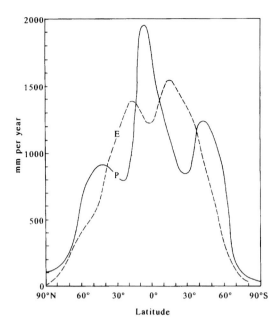

Figure 2.6 The average latitudinal distribution of precipitation (P) and evaporation (E) in mm per year.

poleward transport of angular momentum. They accomplish this by exchanging air with a stronger westerly component (moving polewards) with air with a weaker westerly component (equatorwards). It has been calculated that five simultaneous well-developed mid-latitude cyclones can provide all the necessary angular momentum transport required in winter. Since similar calculations lead to similar conclusions about poleward heat transport, it is clear that the mid-latitude cyclone belt is a crucial feature of the general circulation of the atmosphere. Since it also determines the character of the climate of the British Isles, it is appropriate that we examine the mid-latitude westerlies a little more closely.

The westerlies

In the following discussion, we shall focus on the Northern Hemisphere. Much is also pertinent to the Southern Hemisphere, although the contrasting land and sea distributions of the two hemispheres do

result in some differences. In the Northern Hemisphere, the subtropical jet stream and the polar front jet stream represent the zones of strongest flow within the broad band of mid-latitude westerlies. The westerlies are more vigorous and extensive in winter; their kinetic energy (that energy due to motion) is three times greater in winter than in summer. The flow in the upper westerlies can be particularly strong, with wind speeds up to 140 ms^{-1} near the top of the **troposphere**. This represents a jet stream in the strict sense of the term which says that jet streams should be characterised by wind speeds of 30 ms^{-1} or above. The term is also used more loosely to describe the locally stronger flow of the subtropical and polar front jet streams.

A prominent feature of the westerlies is their wave-like form. The wave pattern is more noticeable, and simpler, away from the surface. The **500 hPa pressure level** (at a height of around 5.5 km at the latitude of the British Isles) is commonly used to describe the free atmosphere westerlies, although wind speeds are greater at higher levels. On a day-to-day basis the flow can be very complex, but if the flow is averaged over a few days there tends to be five waves in the westerlies encircling the Northern Hemisphere. Waves with this sort of wavelength are called planetary waves. If the averaging period is increased to seasonal time-scales, then the further smoothing produces three waves in the Northern Hemisphere winter and four in the summer (Figure 2.7). The changeover from the winter to summer westerly pattern occurs relatively abruptly around June and back again, equally abruptly, around October (see Chapter 8). There is a general relationship between westerly vigour and the number of planetary waves, even over shorter periods. The stronger the westerly (or zonal) flow, the smaller the number of planetary waves.

In reality, the planetary waves drift slowly eastwards, but statistical smoothing highlights the major regions of occurrence. The main planetary wave **troughs** are situated over eastern North America and over the east coast of Asia. These troughs are especially pronounced in winter. A weaker planetary wave trough is located over Europe, between about

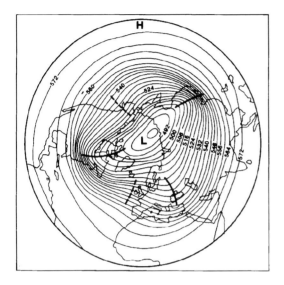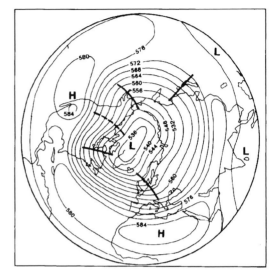

Figure 2.7 Average height (tens of metres) of the 500 hPa surface in January (left) and July (right). The mid-latitude westerlies blow parallel to the contours – the closer the contour spacing the stronger the wind. Dashed bold lines indicate the average position of ridges, continuous bold lines the position of troughs.

10–60°E. The position and strength of this European trough are much more variable than are those of the North American and East Asian troughs, which have more anchored positions. The European trough is very sensitive to changes in the westerly circulation. This is an important characteristic for the European and British climate since, as we shall see later, the overlying westerly waves exert a strong control on surface weather patterns. It is of some interest that the average position of the European trough axis, when averaged by decade, has varied by as much as 20° longitude over the last 200 years. A planetary wave crest (**ridge**) is located to the west of the European trough, with a position somewhere between 10°E and 30°W. This ridge is less discernible in summer.

The preferred positions of the planetary waves owe much to the continental distributions of land masses and to the presence of major mountain ranges. Experiments using dishpans to model the Earth–atmosphere system have replicated some of the major features of the general circulation without an underlying topography (see Box 2.1). The nature of the underlying surface, however, strongly conditions the location and character of the planetary waves.

BOX 2.1 DISHPAN EXPERIMENTS

Dishpans are physical models of the Earth–atmosphere system, so named because the Earth is modelled by a dishpan and the atmosphere is modelled by a fluid contained therein. A temperature gradient is applied across the dishpan (the outside representing the 'equator' and the centre the 'pole') and the dishpan is rotated. The resulting circulation of the fluid reproduces many features observed in the atmosphere, including the westerly planetary waves (as well as circulations which resemble cyclones and anticyclones). The dishpan models do not have the equivalent of mountain ranges or land masses, so the planetary waves and other circulation features form as a consequence of a pole–equator temperature gradient impressed on a rotating fluid system, irrespective of the underlying topography. Nevertheless, in the real Earth–atmosphere system, it is obvious that the location and precise character of the planetary waves are strongly conditioned by the nature of the underlying surface. Before we return to the real atmosphere, it is worth pointing out another aspect of dishpan circulation behaviour that is analogous to observed behaviour in the atmosphere. Some dishpan experiments exhibit semi-regular fluctuations in the westerly wavelength and amplitude. The real atmosphere also tends to demonstrate this type of behaviour; the so-called (Zonal) Index Cycle which, typically, spans several weeks. One extreme is strong westerly flow (high Zonal Index) with small amplitude waves; the other extreme is weaker westerly flow with large amplitude waves producing more north-to-south and south-to-north (meridional) flow.

An important property of atmospheric circulations is **vorticity**, which is a measure of spin within a fluid. If we assume that an airstream, flowing west-to-east, is approaching a large mountain barrier, such as the North American Rockies, then as it is forced to rise over the mountain range and since the tropopause acts as a sort of 'lid', it becomes vertically squeezed as it passes over the peaks. Meteorology textbooks show mathematically that there are important links between vertical squeezing (and stretching) and vorticity. In our case, it leads to a decrease in the vorticity of the airstream which in the Northern Hemisphere represents a clockwise turning – anticyclonic curvature. As it passes over the mountain barrier, therefore, the air starts to turn equatorwards. As the air flows beyond the mountain barrier, however, it is allowed to stretch vertically. This now leads to an increase in vorticity, which in the Northern Hemisphere leads to anticlockwise turning – cyclonic curvature. The result is a (cyclonic) trough in the westerlies in the lee of the mountains. This explains the anchoring of a pronounced wave trough in the westerlies over eastern North America (Figure 2.7).

Once such a large-scale wave has been initiated, there are good reasons why a wave form should be maintained downstream. In large-scale motion, at mid-troposphere levels (around 500–600 hPa), there is a need to conserve absolute vorticity. This is the sum of atmospheric vorticity and the local vorticity of the Earth's surface (the Earth is 'spinning' and therefore has vorticity). The local value of the vorticity of the Earth's surface is known as the **Coriolis parameter** (named after Gustave-Gaspard Coriolis, see Chapter 14) and is proportional to the sine of the latitude. It is therefore at a minimum at the Equator and a maximum at the poles. The large-scale westerly motion downstream of the North American trough has a poleward component; hence the Coriolis parameter is increasing. To maintain absolute vorticity, the atmospheric vorticity must decrease. The airstream starts to take on increasing anticyclonic curvature, eventually turning through a ridge, and so eventually taking on an equatorward component. Now the Coriolis parameter is decreasing and absolute vorticity must be conserved by the atmosphere adopting more cyclonic curvature. This downstream oscillation would continue indefinitely in the absence of any other factors and explains the point made above that the North American trough largely controls the behaviour of the European trough. In reality, other factors do come into play and the character of downstream waves is influenced by, for example, surface temperature patterns.

Cyclone waves in the westerlies

On a day-to-day basis, the smoothed planetary wave pattern is obscured by the superimposition of smaller wavy perturbations which appear, grow and decay

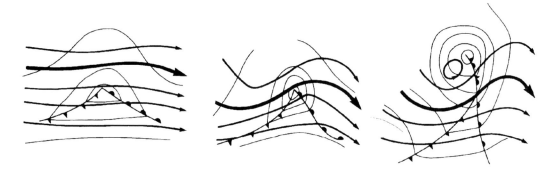

Figure 2.8 Schematic representation of the development of a cyclone wave. The heavier lines represent the flow at upper levels, with the heaviest line showing the most rapidly moving air. The frontal depression at the surface is denoted by the warm (semi-circles) and cold (triangles) fronts and by the lightest lines which represent the surface isobars (lines of constant atmospheric pressure). Adapted from McIlveen (1992) (see p. 32 Note 13).

over a few days. These waves are the result of instabilities in the smooth, planetary flow and they generally propagate rapidly eastwards, apparently steered by the flow in the planetary waves, at a rate of about 10–12° longitude per day. Typically, their amplitude increases by a factor of two to three over a couple of days. Their wavelength is of the order of 3–6,000 km; so the number of waves around the hemisphere is between 6 and 10. These unstable perturbations are known as **baroclinic waves**, or cyclone waves. One of the most important destabilising factors responsible for these waves is the marked increase of wind speed with height in the polar front jet stream zone. This, combined with the strong horizontal temperature gradients concentrated into this zone, produces the sort of instability which leads to cyclone wave development.

Figure 2.8 shows the development of a cyclone wave at the level of the polar front jet stream and its associated frontal depression at the Earth's surface. In order for these cyclone waves to grow, the kinetic energy of the wave must increase. The kinetic energy is made available by warm air rising and cold air sinking. The upsliding of warm air, exchanging with the downsliding of cold air across the polar front zone, takes place along a slope of only around 1°. The horizontal and vertical temperature gradients in this zone are such, however, that they make the

conversion of potential energy (dependent on the relative positions of warm and cold air) to kinetic energy more effective than the very shallow slope of the exchange might suggest. As far as energy conversion is concerned, what is happening is convection (although we generally understand convection to be a process which involves greater vertical exchange); hence the term 'slantwise convection'. The cyclone waves pass kinetic energy into the westerly wind belt, playing the crucial role in the maintenance of the general circulation which has been mentioned before. Early notions of the mid-latitude cyclones were that they were akin to turbulent eddies, which are maintained by the energy of the mean flow. The opposite is in fact the case.

Frontal depressions often form in families. Figure 2.9 is an idealised picture of four planetary waves in the 500 hPa flow and associated frontal depressions (or cyclone waves). Four families of frontal depressions are shown with new depressions forming on the trailing cold front of the 'parent', or **occluding**, depression. The depression families are seen to lie on the poleward-moving limbs of the planetary waves, downstream of the troughs. The reason for this is as follows.

The planetary wave troughs have cyclonic curvature and the planetary wave ridges anticyclonic curvature. For reasons we will not explain here, this

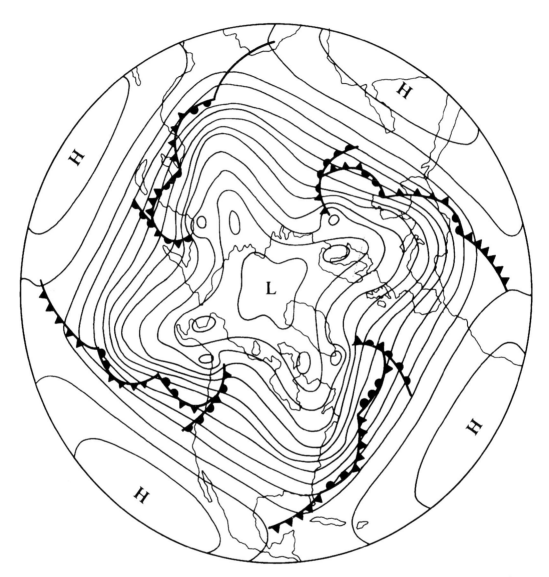

Figure 2.9 Four idealised planetary waves, showing how the formation of families of frontal depressions is favoured under the poleward-moving limb of the waves (see text and Figure 2.10). In the real atmosphere on a day-to-day basis, the westerly wave pattern is more complex than this. The isolines represent heights of the 500 hPa pressure surface.

has the effect of horizontally stretching volumes of air passing through the poleward-moving limb of the planetary waves at upper levels, whereas the effect is one of horizontal squeezing on the equatorward-moving limb. This upper stretching (**divergence**) and upper squeezing (**convergence**) produces compensatory patterns of convergence and divergence near the surface (Figure 2.10). The surface convergence leads to an increase in cyclonic vorticity, and the surface divergence leads to an increase in anti-

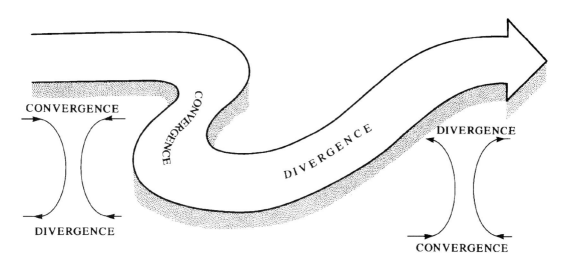

Figure 2.10 Convergence at upper levels in the equatorward-moving limb of a planetary wave is compensated by divergence closer to the Earth's surface. Under the poleward-moving limb, the upper divergence is compensated by surface convergence.

cyclonic vorticity. Consequently, cyclone development is encouraged downstream of a planetary wave trough, and is inhibited upstream of the trough, where anticyclone development is more favoured. The upper flow pattern is, therefore, seen to be an important control on the development of cyclones and travelling anticyclones.

From our discussion it is probably quite easy to get the impression that the polar front is a continuous band snaking round the hemisphere, coinciding with the path of the polar front jet stream (although Figures 2.5 and 2.9 have indicated that this is a misleading simplicity). In reality, the geography of the Earth's surface, including the influence of ocean currents, favours certain zones of 'frontogenesis'. One such zone stretches from the south-eastern United States, across the North Atlantic, towards the British Isles. It is weaker, and less tilted, in summer compared to winter (in particular, its eastern end shifts northwards). The zone off North America is the strongest part of the North Atlantic Polar Front. In winter there is a large temperature contrast between a land mass with an extensive snowcover and warm offshore currents. The positioning of

the poleward-moving limb of the planetary wave anchored over North America (and, hence, upper divergence) over one of the strongest frontal zones in the Northern Hemisphere favours the development of cyclonic waves. These are then steered by the upper flow, growing as they move towards Iceland and northern Scandinavia. This depression track is much weaker in summer and tends to take a more northerly course over the eastern North Atlantic. Something like 170 depressions a year follow this track, although there are significant variations from year to year and from decade to decade. There are also variations in the typical depression tracks, depending on, amongst other factors, variations in the westerly wave pattern. Some of these variations will be discussed below.

We have spent some time discussing the westerlies and cyclonic waves and their role in the general circulation of the atmosphere. We have concentrated on the area of the North Atlantic, although there are other regions, in both hemispheres, of important frontogenesis and cyclone wave development. We will return to some features of the global-scale circulation later, but the reason for the present emphasis is that the travelling cyclones, and their intervening high pressure ridges, create the essential character of the British weather. Seasonal fluctuations in their behaviour control much of the regional-scale climate, and long-term variations in behaviour – responding to some other feature of the circulation of the atmosphere and oceans – can help explain much of the observed longer-term fluctuations in the climate of the British Isles.

We do not intend to describe in detail the particular pattern of weather associated with frontal depressions. Most readers will be familiar with the precipitation bands and temperature changes associated with the passage of the **warm** and **cold fronts** in a frontal depression, which can swing over the British Isles even when the depression centre is passing far to the north. The mobility and precise path of a particular system are important characteristics for daily weather. The severe windstorms which afflicted Western Europe in the early 1990s are a graphic example (see Chapter 11). The passage of the intervening high pressure ridge, with its different weather (weaker winds, clearer skies, cool winter nights, etc.; although high pressure can also produce persistently cloudy skies in winter) generally provides a clear contrast to their fellow travelling depressions. Together, they can produce a characteristic 2–3 day sequence of weather – although this sequence is regularly disturbed.

Blocking

When we were discussing the westerlies, we noted that the wave pattern exhibits fluctuations in wavelength and amplitude. This characteristic is known as the Index Cycle.[3] The term is derived from the parameter, the Zonal Index, which is a measure of the strength of the middle-latitude westerly winds.

At low values of the Zonal Index, the westerly flow is weak and the wave pattern becomes so exaggerated that there are large areas of higher-than-average and lower-than-average atmospheric pressure (positive and negative pressure anomalies, respectively) encircling mid-latitudes. Once the wave pattern starts to become strongly perturbed in one

longitudinal band, there is a tendency for the large-scale amplitude pattern to spread throughout the mid-latitudes within days. A chain of four positive and four negative pressure anomalies encircling the globe is a common pattern. When the positive anomalies become well-established and remain quasi-stationary, they are known as 'blocks'. The longitudes of the planetary wave ridges at 150°W and 15°W (the latter position being off the west European coastline, see Figure 2.7) are particularly favoured for the development of blocks.

When the mid-latitude flow is zonal (high Zonal Index), the vigorous westerly flow over extensive regions means that fast-developing cyclonic waves move quickly eastwards. When the flow is meridional (low Zonal Index), then the development and passage of the cyclonic waves is 'blocked' over extensive regions. **Blocking** is most common in spring/early summer, although it can occur at any time of year. A typical position for a block to be centred is about 15°W. In line with the seasonal shifts in the planetary wave pattern, however, there is a tendency for the preferred position of the block axis to move from east of the British Isles in winter to the west of Ireland in May, and continue out into the Atlantic to its most westerly position in summer, whence it starts its slow progress eastwards again. The high frequency of blocking in spring/early summer when averaged over 100 years is quite pronounced. This common spring/early summer block allows more airflow from the north, south and east, at the expense of westerly, progressive, conditions. By the end of June, with the block declining or starting to shift eastwards, more westerly flow has been resumed with generally more precipitation. Some climatologists have described this late June period as heralding the '**European Monsoon**'. Chapter 8 further discusses this characteristic of the annual cycle of weather over the British Isles.

Fluctuations in the Zonal Index, and the associated hemispheric-scale adjustments to the mid-latitude flow, operate on a time-scale such that a particular region's weather over significant parts of whole seasons may be strongly influenced by them. That being so, at least part of the year-to-year variations

(a) Mean pressure (hPa): Summer (1961-90)

(b) Mean pressure (hPa): Winter (1961-90)

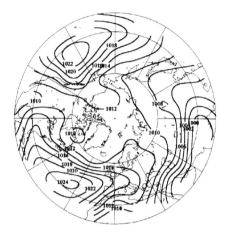
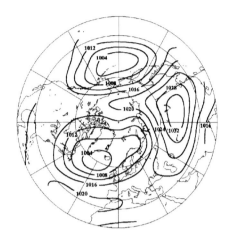

Figure 2.11 Average mean sea-level pressure (hPa) for the Northern Hemisphere (1961–90), summer (a) and winter (b). Note the change in contour interval between the seasons. Pressure gradients are stronger in winter and, as a result, the atmospheric circulation is more vigorous.

which are such a strong feature of short-term climatic variations can be explained by the 'Index Cycle'.

SURFACE PRESSURE PATTERNS

Average surface airflow is parallel to the isobars and its vigour is proportional to the pressure gradient – the tighter the isobars, the stronger the flow. The most noticeable difference between the maps of summer and winter surface pressure over the Northern Hemisphere (Figure 2.11) is the replacement of winter high pressure by summer low pressure over Eurasia. This produces the very pronounced seasonal reversal of flow which is the south-east Asian Monsoon. As far as the most immediate features of relevance for the British Isles are concerned, the dominant centres of action are the area of low pressure near Iceland (the Icelandic Low), most marked in winter, and high pressure to the west of Spain, which extends over western Europe in summer as the Icelandic Low migrates westwards.

The high pressure area is known as the Azores High and is an extension of the permanent subtropical high pressure over the Atlantic. This is matched by another over the Pacific and which are broad zones of subsiding air reflecting descent in the Hadley Cell (Figure 2.5). The Icelandic Low is really a statistical manifestation of the passage of travelling cyclones over this part of the Atlantic. The eastern part of Europe is influenced by the western extremity of the Siberian High in winter; this is a shallow high pressure caused by radiational cooling of the Eurasian land mass. During periods of low Zonal Index, there can be outflows of very cold air from the Siberian High which lead to cold weather over the British Isles. In summer, the high pressure is replaced by low pressure, caused by heating of the Eurasian land mass. The surface pressure patterns confirm what we already know; winter flow over the Atlantic and into the British Isles is much stronger than summer flow. The summer reduction in the Icelandic Low reflects the decline in cyclone vigour and the northwards expansion of the Azores High reflects the tendency of all components of the global climate to 'follow the Sun'.

Since the two centres of action – the Icelandic Low and the Azores High – dominate the pattern of surface pressure over the eastern Atlantic, a useful index to describe conditions upwind of the British Isles is the difference in pressure between the Azores region and over Iceland. This is, in effect, a 'local' Zonal Index. Besides describing part of the annual variation in surface pressure over the Atlantic, this Azores/Iceland pressure index also characterises changes in the strengths and positions of the Icelandic Low and the Azores High. The behaviour which the index characterises is known as the **North Atlantic Oscillation** (NAO). The oscillation is the link between the two centres of action – when the Azores High is more intense (higher pressure), the Icelandic Low also tends to be more intense (lower pressure). This NAO signal, when averaged over several years, is present for all seasons, although it changes its precise character with the seasons. The NAO is an important component of the interannual variability of the whole Northern Hemisphere circulation.

It will come as no surprise that there are links between the behaviour of the NAO index and the weather experienced over the British Isles during a particular year (see Chapter 9 for its relationship with temperature). Changes in the circulation patterns over the Atlantic are associated with shifts in storm tracks – high values of the NAO index push storms further into Northern Europe, accompanied by higher temperatures than usual. Stronger Atlantic westerlies increase the atmospheric transport of moisture into northern Europe. This leads to heavier precipitation over the northern half of the British Isles, although indications are that precipitation may be reduced over the southern half. So, even though the initial control is the North Atlantic large-scale circulation, we still have to consider the sub-regional scale response. This reflects the similarity of scale between the British Isles and the synoptic systems which produce the Islands' day-to-day weather.

On occasions the NAO index is negative; that is, the normal south-to-north pressure gradient is reversed. This is an extreme circulation mode, reflecting a strong pattern of blocking and leads to flow with an easterly component over the North Atlantic European sector. As we indicated earlier, periods of weeks or, occasionally whole seasons, can be dominated by such conditions (see Box 2.2).

LINKS WITH THE OCEAN

Variations in the NAO have been linked to sea-surface temperature (SST) changes in the North Atlantic. From year to year, the SST patterns are probably caused (or forced) by the atmospheric circulation, with the surface wind influencing the ocean circulation and hence the distribution of SST anomalies. The picture is not clear, however, and there is evidence that SST patterns in the western part of the North Atlantic influence the British weather on time-scales of months.[4] Warm SST anomalies in this part of the ocean tend to precede a greater incidence of cyclonic circulations over the British Isles in the following months, whereas a cold SST anomaly is frequently followed by months which are more anti-cyclonic in character.

The precise linking mechanism between the ocean and the atmosphere appears to be related to the shift in the position of the maximum surface temperature gradient, affecting the formation and path of cyclone waves. Over time-scales of several years to decades, although two-way interactions between the atmosphere and ocean still operate, there are indications that the SST anomalies (this time over a larger area of the North Atlantic Ocean) are playing an important role in forcing the circulation of the overlying atmosphere, and in influencing the climate of Europe. Figure 2.12 shows that SSTs over the North Atlantic were relatively low up to the 1920s, higher up to the 1960s, then lower thereafter. There are indications that the high SSTs in the 1940s and 1950s were associated with the production of more cyclones over the mid-North Atlantic Ocean at around 45°N. There are also some hints of links between the SST anomalies and the frequency of different types of circulation over the British Isles. Robert Ratcliffe and Roy Murray amongst others[5] have emphasised, however, that it is likely to be the precise pattern of SSTs which is important for

BOX 2.2 BLOCKING AND EXTREME SEASONAL WEATHER

One of the lowest values of the NAO index occurred in 1963. This winter (January to March) was one of the coldest in the last 250 years in the British Isles, when the temperature in parts of England did not rise above 0°C for three months, because of the persistent easterly flow from the cold European mainland. The first three months of 1996 were also rather cold, as a result of blocking highs in the Scandinavia to east North Atlantic region, leading to persistent easterly or northerly flow over the British Isles. This recent cold winter was a timely reminder to a public, which had been fed an oversimplified diet of global warming by many parts of the media, that interannual variability is still a strong characteristic of the British Isles climate. Understanding the regional response of climate to the enhanced greenhouse effect demands rather more sophisticated consideration (see Chapter 15).

The 1963 block was centred over Iceland. Blocks do not have to occur over the Atlantic to have a dominating control on the British weather over a season. Another extreme winter, in 1947, was caused by a blocking high over Scandinavia. This was less noticeable in the NAO index, but had the effect of steering depressions further south

than usual, over the southern half of the British Isles, producing much snowfall (snow fell somewhere over the British Isles every day from 22 January to 17 March in 1947). In this case, the Atlantic depressions provided the moisture source for a deep-white winter, whereas the extremely cold 1963 winter was relatively deficient in snowfall because of the easterly flow from the dry European mainland. So, although the prime control on the British climate comes from the atmospheric circulation over the Atlantic, and part of this control can be represented by simple indices such as the NAO index, we need to remember that the precise configuration of anomalous circulations is important.

Blocks can also produce anomalously hot or dry summers. The prolonged drought of 1975–6 (one of the driest 18 month periods on record over England and Wales, see Chapter 10) resulted from blocking summer highs over, or close to, the British Isles. The clear settled conditions resulted in high temperatures and the rain-bearing depressions were steered to the north and to the south, around the blocks. Persistent high pressure also dominated in the very hot summer of 1995; a particularly pronounced ridge in the westerlies occupied a position which stretched from the British Isles to as far east as 25°E (cf. Figure 2.7).

the British climate, in particular the position/orientation of the zone of maximum SST gradient across the North Atlantic.

SST patterns in parts of the North Atlantic are intimately linked with sea-ice distributions and there are clearly established relationships between sea-ice extent around Iceland and the European climate.[6] Periods of extensive northerly winds over the northeast Atlantic and western Europe, for example, bring cooling to the continental landmass and ice to the shores of Iceland. The warmth of the 1920s and 1930s round the North Atlantic sector was associated with strong westerly winds and a period when ice was a relatively infrequent visitor to Iceland. These relationships arise because, on this geographical scale, the atmospheric circulation is the primary cause of the fluctuations in ice and climate conditions. Locally, however, the advance and retreat of the ice edge is associated with a marked change in surface heating and albedo and can exert a strong influence on the overlying atmosphere.

The interactions between the ocean surface and the atmosphere over the North Atlantic, and the consequent 'downstream' impact on the weather or

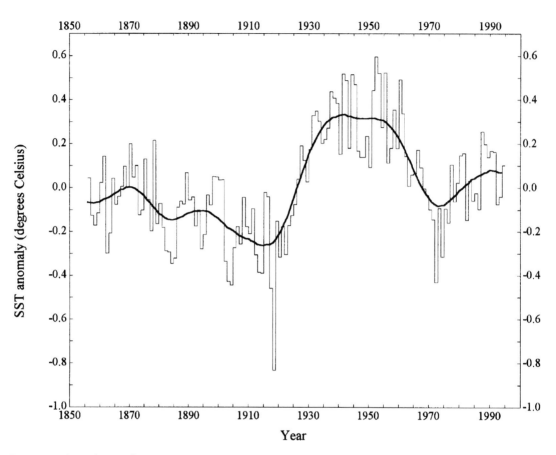

Figure 2.12 Annual sea-surface temperature anomalies, with respect to the 1961–90 mean, in the North Atlantic Ocean from 1856 to 1995. The smooth line is the result of applying a filter which emphasises variations on time-scales greater than 30 years. The region is defined as 20° to 70°N and from 0° to 80°W.

climate of the British Isles, are complex and two-way. The nature of the interaction depends on time-scale and geographical scale. We must also consider the atmosphere–ocean interactions on a wider, indeed global, scale. Similarly, we have to extend our interest to deep ocean circulations, not only in the northern part of the Atlantic Basin, but further south and, as for the atmosphere, to other ocean basins.

First, we examine the deep overturning of water which is particularly vigorous in the Atlantic Ocean due to the formation of North Atlantic Deep Water (NADW). The sinking of this water mass in the northernmost parts of the ocean causes the Gulf Stream and its extension – the **North Atlantic Drift** – to turn more northward to replace the sinking water. This does not happen in the Pacific Ocean which has no deep water formation. The sinking of the NADW results from its high density, which is a consequence of its high salinity as much as its low temperature. It is self-sustaining, to some degree, since its high salinity is due to the northward transport of saline water from more tropical latitudes. Its warmth is also important, since it induces evaporation, further increasing salinity.

Since the sinking is sensitive to changes in the input of freshwater to the North Atlantic, the suggestion has been made that the overturning in the Atlantic could vary in strength, stop, or even reverse. Such switches could occur over very short time-scales. There is evidence that the warming from the last **Ice Age** was interrupted around 10,000 years ago by a dramatic return towards Ice Age temperatures, sometimes called the **Younger Dryas** period, only to be followed by a rapid recovery, all within 1,000 years (see Chapter 5). This rapid climate deterioration was probably caused by the Atlantic overturning being disturbed by a large input of freshwater of low density into the North Atlantic from the melting of ice. The mechanisms which are likely to be involved in these fluctuations are complex.[7]

There are some tantalising indications from computer models that the sinking in the North Atlantic may fluctuate over shorter time-scales – possible oscillations of around 40–60 years have been reported.[8] The reason for such oscillatory behaviour is uncertain, but some aspects may be triggered by a short-term change in the input of freshwater in the sinking region north of 60°N (more or less ice melt, or even heavy precipitation). Other aspects may be more self-sustaining. The implication for the British climate is that these multi-decadal oscillations may be reflected in changes in the North Atlantic SST patterns which, we know, have an important downstream influence. The reason why these computer experiments are tantalising is that climate reconstructions from tree-rings from parts of Europe also exhibit oscillatory-type behaviour on the same time-scale.[9]

We have mentioned the **teleconnection** character of the North Atlantic Oscillation. The most pronounced teleconnections globally are those associated with the El Niño Southern Oscillation (ENSO). This is a surface pressure oscillation across the tropical Pacific, related to Pacific Ocean currents.[10] The ENSO dominates interannual climate variability in tropical latitudes and there is a pronounced teleconnection between it and surface pressure over the North Pacific Ocean and North America. Although strong ENSO signals have not been detected in the atmospheric circulation over the North Atlantic/European sector, there are some indications of possible linkages with European-scale weather patterns.[11] There appear to be weak links with some aspects of the British climate, particularly the frequency of anticyclonic and cyclonic weather types in winter, and winter precipitation over England and Wales. The strongest links appear to be in January and February.[12] It is possible that more pronounced ENSO signals in British climate will emerge as research progresses.

AIR MASSES

We have already discussed how air flowing over the North Atlantic to the British Isles has a different character to that flowing out of Siberia in winter – relatively warm and moist versus cold and dry. The concept of air masses is a useful one and underpins the usefulness of the weather type classifications (see Chapter 8), since direction of airflow is one of the bases of identification of many types. When air resides in a source region for weeks it starts to develop a homogeneous character, a few kilometres deep, which it acquires from its source region. The source regions are geographically distinct, cover hundreds of thousands of square kilometres and differ between summer and winter.

Two of the winter land source regions are Canada and (approximately) the former Soviet Union, the origin of 'continental polar' air masses. The character of air masses from these source regions will clearly differ considerably between summer and winter. There are also seasonally differing source regions for 'continental arctic' and 'continental tropical' air masses, as well as 'maritime arctic', 'maritime polar' and 'maritime tropical' air masses. The path the air mass takes from its source region to the British Isles is important. For example, a relatively cool air mass flowing over a warmer surface will often bring convective clouds, good visibility and gusty winds; warmer air flowing over a similarly warm surface may produce stratus cloud, fog and poor visibility. Many weather features result from such modification of an air mass along its path; the air mass type may change from such modification – 'continental polar' air flowing out of Canada over the warm North Atlantic Drift may have developed into a cool moist 'maritime polar' air mass by mid-Atlantic, producing bright periods and some showers.

SMALLER WEATHER SYSTEMS AND LOCAL INFLUENCES

A number of circulations on a smaller scale than the features we have been describing make contributions to the climatic character of the British Isles. This short section is not all-inclusive. We will not, for example, look at thunderstorms which can provide a significant proportion of summer rainfall (see Chapter 3), nor at mountain and valley wind systems which can influence local wind climatology, or cold nocturnal drainage flow which may fill 'frost hollows'.

Midway in scale between these circulations and the cyclones and travelling depressions are **polar lows**. We shall refer briefly to these circulations since, although they are relatively small (often several hundreds of kilometres) and shallow (5 km), they can produce severe weather over parts of the British Isles, with strong gusts and contributing to much of the heavier snowfalls. They represent a special case of cyclone formation in that they are generally non-frontal. They usually develop over the ocean in the northerly 'maritime polar' or 'maritime arctic' airflow to the rear of a cold front, often between Iceland and the British Isles (see Figure 2.13). The cold airflow across zones of relatively high SST gradients provides the mechanism for their formation.

Another type of low, but of entirely different origin and type, is the so-called heat low. Localised heating of land in summer can produce such features, usually in the afternoon. There are occasions, however, where they may survive night-time cooling and persist for some days. They can be relatively small-scale (for example, over East Anglia), or they may cover an area such as most of England. Thunderstorms may develop in the heat lows.[13]

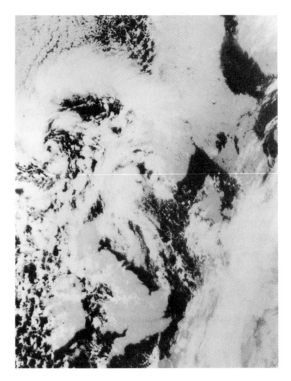

Figure 2.13 An infra-red image of a polar low near the Faroes, 25 November 1978, observed from the NOAA 5 satellite. This day is classified as NW in the Lamb Catalogue (see Appendix B).

Within tens of kilometres of the coastline, the day-to-day weather may be modified by sea-breezes.[14] They are also caused by the daytime heating of the land, which produces a pressure gradient between the sea (high pressure) and the land (low pressure). During late morning a sea-breeze starts to blow inshore and penetrates inland causing moister, cooler conditions, frequently accompanied by cloud. Typically, during summer at some British locations, sea-breezes will blow on 20–30 per cent of days, but there are periods when they are much more frequent. Consequently, they may affect the character of a whole summer at some near-coastal locations.

Another important factor for local climate is the local orography. A glance at a long-term precipitation map of the British Isles provides clear confirmation of this (see Plate 4 or Appendix A). When moist air is forced to rise over high land, the air can be cooled to a point where condensation occurs and the precipitation process starts. There is a distinct west-to-east gradient in precipitation over the British Isles which largely reflects this orographic effect on moist air blowing in from the Atlantic. Another reason for enhanced precipitation over the highest land is that the passage of fronts can be slowed down. Detailed precipitation maps show a dependency of precipitation on elevation, even where the orography is not pronounced and even in eastern locations. Modest orographic enhancement of precipitation is apparent, for example, even in the far-eastern and relatively flat Norfolk. It goes without saying that cloud also is more common over higher land.

We have noted the importance of the land surface during our discussion of global-scale radiation and heat balances. It is not only albedo (which can change seasonally with vegetation changes) which is important, but also heat conductivity. The thermal conductivity of the soil is an important factor in the response of the surface to changes in net radiation. Soil conductivity is strongly influenced by water content. So, if the soil is coarse, sandy and dry (thus containing a lot of air; a good insulator), night-time radiational cooling will not be offset by the conduction of heat from lower levels in the soil. An example of this, again from East Anglia, is the sandy soil area of the Breckland area of Norfolk, where night-time minimum temperatures can be 3–4°C lower than in the surrounding areas where the soil is less freely drained (e.g., Santon Downham – see Chapter 3).

Other important surface differences occur in urban areas. The city fabric acts as a 'storage heater' maintaining night-time temperatures above those of the surrounding regions. The urban heat island tends to be most pronounced on calm, clear nights after sunny days but, for many towns and cities, is apparent even in yearly averages. Other climate variables, such as humidity and wind speed, are also modified to such an extent by some urban areas that differences are apparent in the long-term statistics for urban and adjacent rural locations.

THE SCENE IS SET

The next chapter describes the surface climatology of the British Isles and it is necessary for the reader to bear in mind that local influences, such as those we have introduced in the previous section, will be very important in modifying the climates of specific locations. Just as it is necessary to adapt a weather forecast for a region to take account of local conditions, so the broad climatology presented here must be modified to suit the reader's neighbourhood. It is possible to focus on very small scales – whole books have been written on the climates of a single city[15] – but the aim of this chapter, and of much of this book, has been to look outwards from the climate of the British Isles to the large-scale processes which deliver our average weather. When one considers the myriad factors which shape the climate of a particular area, it is clear why modelling weather and the climate system presents such a challenge. It is necessary to take account of all the processes discussed in this chapter, and many more, on spatial scales ranging from the global to the local. Computer power is limited, so compromises have to be made, often in sacrificing local detail. It is testament to the skills of those developing the models and forecasts that, despite the difficulties, they manage such a degree of accuracy.

NOTES

1   The reader wishing to learn more of the global climate system and the general circulation of the atmosphere is directed to R.G. Barry and R.J. Chorley, *Atmosphere, Weather and Climate* (6th edn) London, Routledge, 1995. For more technical accounts, see R. McIlveen, *Fundamentals of Weather and Climate*, London, Chapman and Hall, 1992, 497 pp., or J.T. Houghton, *The Physics of Atmospheres* (2nd edn), Cambridge, Cambridge University Press, 1986, 271 pp.

2   This radiation is longer in wavelength than that emitted by the Sun since the temperature of the Earth's atmosphere is much lower than that of the Sun – hotter bodies emit shorter wavelength radiation

3   The term 'Index Cycle' is something of a misnomer, since 'cycle' does imply some regularity. In reality, the

cycle has a characteristic time-scale of several weeks or so, but the period is very variable.

4   R.A.S. Ratcliffe and R. Murray, 'New lag associations between North Atlantic sea temperature and European pressure applied to long-range weather forecasting', *Quarterly Journal of the Royal Meteorological Society*, 1970, vol. 96, pp. 226–46.

5   For example, A.H. Perry, 'Eastern North Atlantic sea-surface temperature anomalies and concurrent temperature and weather patterns over the British Isles', *Weather*, 1975, vol. 30, pp. 258–61.

6   P.M. Kelly, C.M. Goodess and B.S.G. Cherry, 'The interpretation of the Icelandic sea-ice record', *Journal of Geophysical Research*, 1987, vol. 92, pp. 10835–43.

7   I.M. Held, 'Large-Scale Dynamics and Global Warming', *Bulletin of the American Meteorological Society*, 1993, vol. 74(2), pp. 228–41.

8   T. Delworth, S. Manabe and R.J. Stouffer, 'Inter-decadal variations of the thermohaline circulation in a coupled ocean-atmosphere model', *Journal of Climate*, 1993, vol. 6, pp. 1993–2011; R.J. Greatbach and S. Zhang, 'An interdecadal oscillation in an idealised ocean basin forced by constant heat flux', *Journal of Climate*, 1995, vol. 8, pp. 81–91.

9   T.F. Stocker, 'The variable ocean', *Nature*, 1994, vol. 367, pp. 221–2.

10   G.R. Bigg, 'El Niño and the Southern Oscillation', *Weather*, 1990, vol. 45, pp. 2–8.

11   K. Fraedrich and K. Müller, 'Climate anomalies in Europe associated with ENSO extremes', *International Journal of Climatology*, 1992, vol. 12, pp. 25–31.

12   R. Wilby, 'Evidence of ENSO in the synoptic climate of the British Isles', *Weather*, 1993, vol. 48(8), pp. 234–9.

13   See R. McIlveen, op. cit., for interesting descriptions of polar lows and heat lows.

14   J.E. Simpson, *Sea Breezes and Local Wind*, Cambridge, Cambridge University Press, 1994, 234 pp.

15   T.J. Chandler, *The Climate of London*, London, Hutchinson and Co. Ltd, 1965, 292 pp.

GENERAL READING

R.G. Barry and R.J. Chorley, '*Atmosphere, Weather and Climate*', London and New York, Routledge 1995, 6th edn.

R. McIlveen, '*Fundamentals of Weather and Climate*', London, Chapman and Hall, 1992, 497 pp.

J.M. Wallace and P.V. Hobbs, '*Atmospheric Science, an Introductory Survey*', New York, Academic Press Inc., 1977, 467 pp.

3

DESCRIBING THE SURFACE CLIMATE OF THE BRITISH ISLES

Elaine Barrow and Mike Hulme

Wherever you go, the weather is, without exception, exceptional.
Kingsley Martin

INTRODUCTION

The British Isles have a more equable climate than would be expected at a latitude of between 49° and 61°N. Their maritime location, their position within the main flow of the mid-latitude westerlies and their proximity to the mild waters of the north-east Atlantic Ocean all contribute to a climate which knows little of the extremes of winter and summer typical of Moscow or the Hudson Bay, places at equivalent latitude to the British Isles. Whereas Chapter 2 examined some of the important reasons for this temperate climate to prevail, the present chapter describes the main features of the British climate. To do this we use maps, diagrams and tables containing data averaged over the most recent thirty-year climate 'normal' period,[1] namely 1961 to 1990.

When averaged over three decades, climate statistics for variables such as temperature, precipitation and sunshine smooth out the year-to-year fluctuations in weather and give a better description of the climate experienced over a human lifetime. The 1961 to 1990 period is the most relevant for contemporary applications of climate data, although there will clearly be differences between the climate defined by these data and climates described by earlier normal periods, such as 1941 to 1970 and 1951 to 1980.[2] How representative the climate of 1961 to 1990 is of past and future climates of the British Isles is

a question examined in later chapters, especially Chapters 9, 10, 15 and 16.

The surface features of the climate of the British Isles are described using maps constructed on a regular 10' latitude by 10' longitude grid using 1961 to 1990 station data supplied by the United Kingdom and Irish Met. Offices (see Box 3.1). In addition to presenting these maps, data from a number of individual sites are used to illustrate more specific aspects of British climate and the locations of these sites are shown in Figure 3.1. We also examine some of the climatic classifications which have been used to define British climate in relation to world climates.

SURFACE AIR TEMPERATURE

The location of the British Isles plays a central role in governing our climate (see Chapter 2). Our situation means that we experience the combined influences of the mid-latitude westerly winds and the **North Atlantic Drift** which comprises warm water of tropical origin. The British Isles are surrounded by this comparatively warm oceanic water, the temperature of which varies only slowly from month to month because of the high **thermal inertia** of the oceans. This means that in coastal areas average temperatures are usually similar to those of the sea surface, whereas areas farther inland and away from

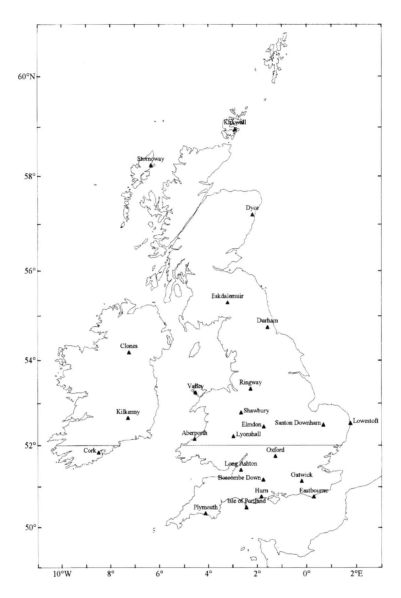

Figure 3.1 Location of the stations mentioned in this chapter. The latitude transect at 52°N is also indicated.

the maritime influence experience larger temperature extremes. In these inland areas of the British Isles the climate is more 'continental' in nature. Interruption of the predominantly westerly circulation may result in extremes of temperature. For example, cold winter spells are usually a result of northerly or easterly airflow, whilst mildness in a British winter is typically a result of airflow from the west or south. Cool summers tend to be produced by westerly or northerly airflows whilst southerly or easterly airflow over the British Isles results in warm summer episodes (see Chapter 8).

BOX 3.1 CONSTRUCTION OF THE 1961 TO 1990 GRIDDED CLIMATOLOGY

A number of attempts have been made to construct gridded climate surfaces for parts of the British Isles from scattered station observations. These have used both multiple regression and spatial interpolation techniques, but have almost always been confined to either a limited number of climate variables, a smaller domain than the whole of the British Isles, or have been based on short or non-standard records of climate. **Partial thin-plate splines**, a technique developed for climate applications by Mike Hutchinson, were used to construct the climatology shown here.[a] This technique included elevation as an independent predictor variable, in addition to the more usual latitude and longitude. As a result of using this approach, the climatology is available at three different elevations corresponding to the maximum, minimum and modal elevations of each 10' grid cell. This grid resolution corresponds to an area approximately 19 km by 19 km, roughly equivalent in size to the city of Birmingham. The number of stations used to construct the climate surfaces was dependent on the climate variable in question. The number of stations ranged from eighty, in the case of wind speed, to 750 for precipitation. The precipitation dataset contained almost 2,500 sites, but only 750 of these were used in the interpolation. The climatology described in this chapter represents, to the best of our knowledge, the most contemporary, comprehensive and widely available climate dataset for the British Isles presently in use.[b] Maps extracted from this climatology are shown in this chapter and also in Appendix A.

[a] E.M. Barrow, M. Hulme and T. Jiang, *A 1961–90 Baseline Climatology and Future Climate Change Scenarios for Great Britain and Europe. Part I: 1961–90 Great Britain Baseline Climatology*, a report accompanying the datasets prepared for the 'Landscape Dynamics and Climate Change' TIGER IV Consortium, Norwich, Climatic Research Unit, 1993.

[b] The 1961 to 1990 climatology may be obtained through the Climate Impacts LINK Project (contact David Viner at the Climatic Research Unit). Monthly average values for all the variables described here are available at the three elevations (minimum, maximum and modal).

Victor Conrad[3] devised a 'continentality index' based on the annual range of average temperature and the sine of the latitude of the site in question. Under this convention the oceanic regime of Thorshavn in the Faeroes has an index value of zero whereas the extreme continental climate of Verkhoyansk in Siberia has an index of 100. In North America, where values rise to more than sixty, the area with values less than twelve is restricted to the tip of the Florida peninsula and the Pacific coast. When this index is applied to the British Isles the extreme oceanicity of stations exposed to the Atlantic Ocean, especially those of the Hebrides, Orkneys and the north-western tip of Scotland, is apparent (see Figure 3.2). Values of the index in these areas are between only two and three. The climate of Ireland is relatively oceanic, with the index ranging from three in coastal areas to seven inland, whereas the most continental areas of the British Isles are in eastern and south-eastern England where the index has values of between eleven and twelve.

Table 3.1 illustrates the monthly mean[4] temperatures of a number of sites in the British Isles. Lowest mean temperatures are at the more northerly and higher altitude sites, and the annual mean temperatures range from approximately 7°C in northern Scotland to in excess of 10°C in south-west England and Wales. The annual range of monthly mean temperature, defined as the temperature in the warmest month minus the temperature in the coldest month, is less at coastal sites compared to those inland. At Stornoway, for example, the range is

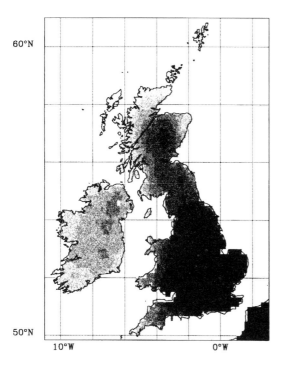

Figure 3.2 Conrad's continentality index, based on the annual range of average temperature and latitude. The larger the value (and the darker the shading) the more 'continental' the site. Index ranges from more than eleven around Greater London to less than three in the northwest of Scotland.

approximately 9°C compared to nearly 13°C at Elmdon near Birmingham. The year-to-year variability of temperature is discussed in Chapter 9.

The seasonal[5] patterns of mean temperature at the average elevation within each grid cell are illustrated in Plate 1. In winter, the maritime influence dominates temperatures with coastal stations tending to be warmer than inland areas, mainly as a result of higher minimum temperatures in the coastal areas. **Sea-surface temperatures** around the British Isles do not reach their lowest values until February or early March with the result that the lowest mean temperatures of those stations exposed to the full influence of the Atlantic occur in February rather than in January (Table 3.1). At Plymouth, for example, the mean temperature in February is 0.2°C lower than in January. The east coast of the British Isles tends to be the leeward side of the country and so stations here are not affected by the influence of the cool February maritime air to the same extent. At Durham, for example, the mean temperature in February is 0.2°C higher than in January.

Highest mean temperatures in winter are between 6°C and 8°C and are experienced along south-western coasts, whilst in central and eastern England winter temperatures are of the order 2°C to 4°C. In upland and highland areas mean temperatures range from less than −2°C to 2°C, whilst in Ireland they are warmer in the south (4°C to 6°C) compared to the north (2°C to 4°C).

Mean maximum temperatures in winter show a similar pattern to the mean temperature (see Plate 2), with highest values in south-west and southern areas (8°C to 10°C) and lowest values in the Scottish Highlands (less than 2°C). Mean minimum temperatures in winter range from between 2°C and 6°C on south-western coasts, to between 0°C and 2°C in central areas and to below −2°C in the Scottish Highlands (Plate 3). Topographic features, vegetation cover and soil type probably have a larger influence on mean minimum temperatures than they do on mean maximum values. These **sub-grid scale** variations in temperature are not captured by the gridded climatology and are probably more important for minimum than for maximum temperature.

In summer, coastal areas are generally cooler than inland areas at similar latitudes. Sea-surface temperatures around the British Isles reach their highest values in August or early September and this affects the month of highest mean temperatures at coastal stations. The impact of this oceanic effect extends farther around the coast than in January, with stations on the North Sea coast being affected as far north as Scarborough. Table 3.2 lists the mean maximum temperatures for a number of coastal stations. Equivalent temperatures for a nearby inland station are shown for comparison. Thus in winter, Eastbourne, for example, experiences daytime maxima about 0.5°C higher than Gatwick, whereas in summer Eastbourne is on average 2°C cooler.

Table 3.1 Average 1960–90 monthly mean temperature (°C). The elevation of each site above mean sea level (m) is indicated in parentheses next to the station name.

Site	J	F	M	A	M	J	J	A	S	O	N	D	Annual	Range
Kirkwall (26)	3.7	3.6	4.5	6.0	8.4	11.0	12.3	12.5	10.9	9.0	5.7	4.4	7.7	8.9
Stornoway (15)	4.2	4.1	5.1	6.5	9.0	11.4	12.7	12.8	11.2	9.2	5.9	4.9	8.1	8.7
Dyce (65)	2.7	2.9	4.6	6.4	9.1	12.2	13.8	13.7	11.7	9.0	5.1	3.6	7.9	11.1
Durham (102)	3.0	3.2	5.0	7.0	10.0	13.0	14.8	14.7	12.7	9.7	5.7	3.9	8.6	11.8
Santon Downham (24)	3.0	3.2	5.1	7.3	10.7	13.8	15.6	15.4	13.2	10.0	5.9	3.8	8.9	12.6
Elmdon (96)	3.2	3.2	5.2	7.6	10.7	14.0	15.9	15.5	13.2	10.0	6.1	4.3	9.1	12.7
Oxford (63)	4.1	4.2	6.2	8.5	11.9	15.0	17.1	16.7	14.4	11.1	6.9	4.9	10.1	13.0
Shawbury (72)	3.4	3.4	5.4	7.5	10.7	13.7	15.6	15.3	13.1	10.0	6.0	4.2	9.0	12.2
Gatwick (59)	3.6	3.8	5.7	8.0	11.3	14.5	16.5	16.2	13.8	10.7	6.5	4.5	9.6	12.9
Eskdalemuir (242)	1.8	1.8	3.5	5.7	8.7	11.7	13.2	13.0	10.8	8.2	4.1	2.6	7.1	11.4
Ringway (75)	3.9	3.9	5.7	8.0	11.3	14.2	15.8	15.7	13.5	10.6	6.4	4.6	9.5	11.9
Valley (10)	5.5	5.1	6.5	8.3	11.1	13.6	15.3	15.4	13.9	11.6	8.1	6.4	10.1	10.3
Long Ashton (51)	4.5	4.5	6.3	8.4	11.5	14.5	16.5	16.2	14.1	11.0	7.2	5.1	10.0	12.0
Plymouth (50)	6.0	5.8	7.0	8.8	11.6	14.3	16.2	16.1	14.4	12.0	8.5	7.0	10.6	10.4
Cork (154)	5.2	5.1	6.2	7.8	10.1	12.9	14.8	14.5	12.7	10.3	7.2	6.0	9.4	9.7
Kilkenny (66)	4.6	4.8	6.2	7.9	10.4	13.3	15.2	14.8	12.6	10.0	6.5	5.3	9.3	10.6
Clones (89)	4.0	4.2	5.7	7.5	10.1	12.9	14.5	14.2	12.2	9.8	5.9	4.8	8.8	10.5

Table 3.2 Comparison of coastal and inland mean monthly maximum temperatures (°C). (Bold indicates coastal sites – the corresponding inland site is immediately below. The approximate distance (km) between the coastal and corresponding inland site is also given.

Distance (km)	Site	Jan	Feb	Mar	Apr	May	Jun	Jul	Aug	Sep	Oct	Nov	Dec
75	**Lowestoft**	**6.1**	**6.2**	**8.6**	**10.6**	**14.2**	**17.6**	**19.8**	**20.0**	**18.2**	**14.4**	**9.7**	**7.2**
	Santon Downham	6.3	6.8	9.7	12.5	16.6	19.7	21.5	21.4	18.8	14.8	9.6	7.1
55	**Eastbourne**	**7.3**	**7.1**	**9.1**	**11.5**	**14.9**	**17.8**	**19.7**	**20.0**	**18.1**	**15.0**	**10.8**	**8.5**
	Gatwick	6.7	7.1	9.9	12.6	16.3	19.6	21.7	21.4	18.8	15.0	10.1	7.7
80	**Isle of Portland**	**7.7**	**7.3**	**8.8**	**11.0**	**13.8**	**16.5**	**18.5**	**18.8**	**17.3**	**14.7**	**11.0**	**9.0**
	Boscombe Down	6.4	6.8	9.4	12.3	15.8	19.1	21.2	20.8	18.2	14.3	9.7	7.4
120	**Aberporth**	**7.0**	**6.9**	**8.4**	**10.5**	**13.4**	**16.0**	**17.6**	**17.7**	**16.0**	**13.5**	**9.8**	**8.0**
	Lyonshall	6.0	6.0	8.8	11.7	15.3	18.4	20.5	19.9	17.1	13.4	9.0	6.9

Mean temperatures in summer are between 14°C and 16°C in southern and central lowland Britain and between 12°C and 14°C in the north, although in higher altitude areas such as North Wales, the Lake District, Exmoor and Dartmoor, mean temperatures are slightly lower (Plate 1). London is relatively warm due to the **urban warming** effect with mean temperatures approximately 2°C higher than those in the surrounding area. Mean summer temperatures are lowest in the Scottish Highlands (8°C to

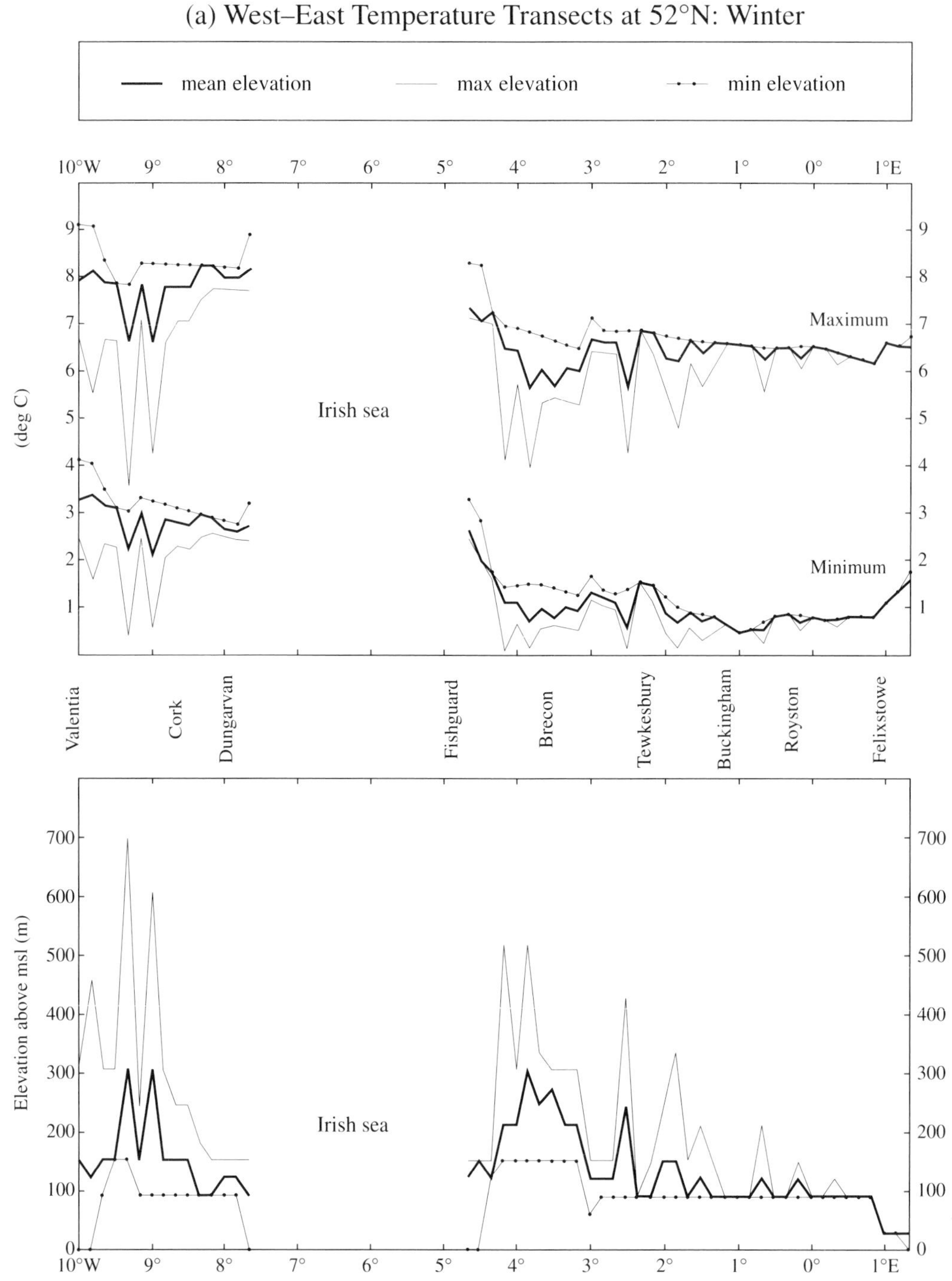

(a) West–East Temperature Transects at 52°N: Winter

Figure 3.3 West–east transect of (a) average winter and (b) average summer maximum and minimum temperature (°C) at 52°N. The bold line refers to the temperature at the average elevation of each 10 km grid cell, whereas the thin

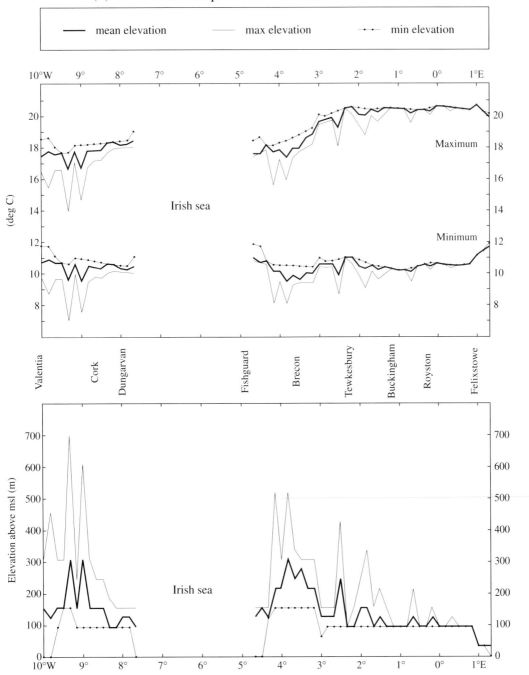

(b) West–East Temperature Transects at 52°N: Summer

lines indicate the minimum and maximum elevations and temperatures of each cell. The lower plots show the average, minimum and maximum elevation of each 10' cell.

12°C). Mean maximum summer temperatures in central areas exceed 20°C, while coastal areas are between 2°C and 4°C cooler (see Plate 2). Mean minimum summer temperatures are highest in southern and eastern areas (10°C to 12°C).

Elevation has a strong influence on temperature and this is illustrated by west–east transects of mean maximum and minimum temperatures at a latitude of 52°N (see Figure 3.3). Both maximum and minimum temperatures follow the same general pattern in winter with higher temperatures in the west than in the east and decreasing temperature with increasing elevation. In the west of Wales (for example, at Fishguard) mean minimum temperatures are, in general, about 2.5°C higher than those in more eastern areas (for example, at Royston). Closer to the east coast (for example, at Felixstowe), however, this difference lessens because of the ameliorating maritime influence on mean minimum temperatures. For mean maximum temperatures, however, this longitudinal difference is only about 1.5°C. The **diurnal temperature range**, therefore, is about 1°C less in the west compared to the east.

In summer, the situation is reversed. This is particularly the case for mean maximum temperatures where west–east differences are of the order of 2°C to 3°C with a cooler west and warmer east. The difference between west–east mean minimum temperatures in summer is small, meaning again that the diurnal temperature range in summer is less in the west than in the east. The summer diurnal range in temperature is between 3°C and 4°C larger in the central areas of England compared to western coastal regions of the British Isles, where the proximity to the slow-changing sea-surface temperature reduces the daily range of temperatures. On the east coast the diurnal range is approximately 2°C to 3°C less than in central areas.

The seasonal patterns of mean minimum temperature tend to be reflected in the average number of frostdays over a season. For our purposes, a frostday is defined when grass minimum temperature falls below 0°C, i.e., when a ground frost occurs. An air frost, on the other hand, occurs when the air temperature recorded in the **Stevenson screen** (at a height

Figure 3.4 Heavy rime deposits (opaque, white ice crystals) around Lincoln Cathedral at 1130 (GMT) on 24 December 1992. This photograph was taken in the middle of a spell of ten consecutive anticyclonic days (according to the Lamb Catalogue) with calm conditions and very cold polar air.

of 1.4 m above ground level) falls below 0°C. Ground frosts occur more frequently than air frosts. Many people assume that frost on rooftops and cars means that an air frost has occurred, when actually this is not the case. Such surfaces are good conductors of heat and therefore radiate away their heat freely thus causing their temperature, and that of the air in contact with them, to fall and frost to form. Calm, cloud-free conditions are ideal for frosts to occur (Figure 3.4), especially if the air is of polar origin. In winter, frostdays are most frequent – fifty days or more – in central and eastern England and Scotland,

Table 3.3 Average 1961–90 seasonal number of ground frostdays

	Winter	Spring	Summer	Autumn
Kirkwall	42.1	28.4	1.1	13.5
Stornoway	41.0	27.3	2.9	17.9
Dyce	60.8	41.3	5.0	28.1
Elmdon	51.2	36.6	3.4	24.4
Oxford	52.1	29.2	0.6	18.3
Shawbury	52.4	40.8	5.6	26.7
Gatwick	52.4	37.3	1.6	23.7
Eskdalemuir	59.6	44.7	6.2	29.9
Ringway	40.3	22.9	0.2	12.7
Valley	31.5	18.2	0.2	8.4
Long Ashton	50.6	31.1	0.8	19.4
Plymouth	31.9	21.4	0.1	10.8
Cork	42.0	27.9	0.0	15.0
Kilkenny	51.0	39.9	3.0	27.9
Clones	47.1	35.1	3.0	21.9

whereas in more coastal areas, such as Plymouth and Valley, their number is less than forty (see Table 3.3).

PRECIPITATION: AMOUNT AND FREQUENCY

Precipitation[6] over the British Isles is produced from three main sources: frontal systems, local atmospheric **static instability** (thunderstorms and thundershowers) and atmospheric uplift by hills and mountains (orographic precipitation). Average seasonal precipitation depends on the frequency, intensity and tracks of rain-bearing systems near the British Isles. More active and more frequent frontal systems cross Scotland from west to east and this, combined with the influence of the mountains, leads to high precipitation totals in this region. The west coast of Scotland receives between four and five times as much precipitation as the east coast.

The highest precipitation totals over the British Isles are usually produced by cyclonic, southerly or westerly circulations (see Chapter 8). Westerly circulations drive moist maritime **air masses** originating from the Atlantic over the British Isles. When these

air masses are forced to rise, either by hill and mountain barriers or in frontal systems, large quantities of cloud and precipitation result. The influence of elevation is very marked over the hills and mountains of southern Ireland, south-western England, north and south Wales, the Lake District and the Highlands of Scotland and there is therefore a strong west–east contrast in precipitation. This is illustrated in Figure 3.5 which shows winter and summer precipitation totals along the same west–east transect at latitude of 52°N as used for temperature. The effect of the hills and mountains is clear. In winter, seasonal precipitation totals over the highest areas of Wales are about 600 mm compared to about 150 mm in the eastern **rain-shadow** areas. In summer, seasonal precipitation totals in the east are similar to those in winter, but in western areas they are only about half their winter values.

The seasonal pattern of precipitation distribution over the British Isles is illustrated in Plate 4. Large precipitation totals are obtained in the higher areas of Ireland, Wales, south-west England, Scotland and the Lake District, but these gradually decrease towards the south and east. The relative pattern of precipitation tends to be similar over all seasons, although the absolute magnitude of the geographical differences varies.

The monthly contribution of precipitation to the annual total for a number of sites is illustrated in Table 3.4. For many sites precipitation in the autumn and winter months makes the largest contribution to the annual precipitation total. This is especially true of the northern and western areas of the British Isles and is caused by the most frequent and intense **depressions** being experienced in these months. In much of central and eastern England, however, the annual cycle is much less marked and summer precipitation can make the largest contribution to the annual total. At Santon Downham and Elmdon, for example, more than 25 per cent of precipitation falls in summer. In this region a higher proportion of summer precipitation is likely to be of convective origin than elsewhere. In no part of the country is spring the wettest season. The variability of precipitation from year to year is discussed in Chapter 10.

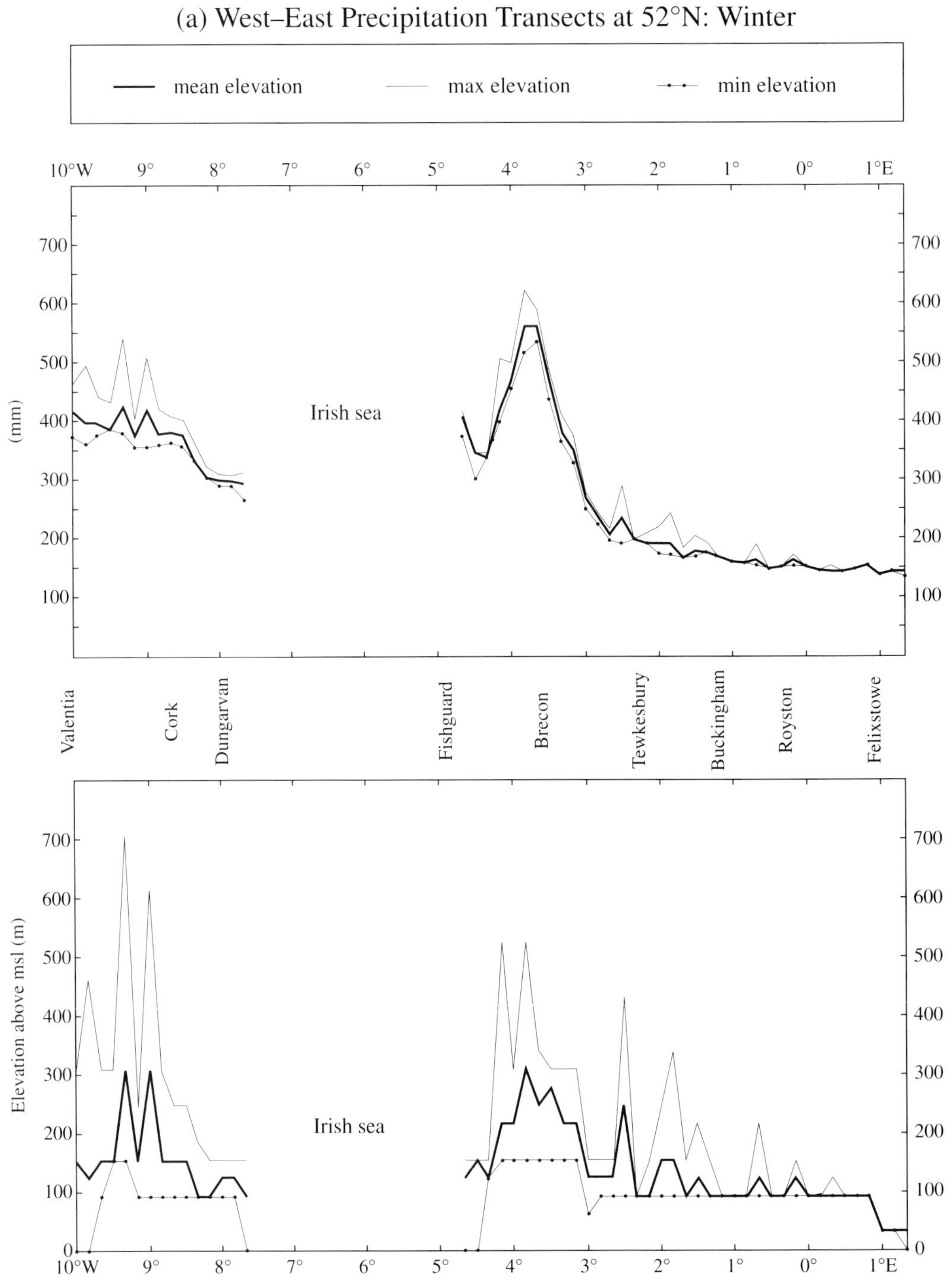

(a) West–East Precipitation Transects at 52°N: Winter

Figure 3.5 West–east transect of (a) average winter and (b) summer precipitation totals (mm) at 52°N. The bold line refers to the precipitation at the average elevation of each 10 km grid cell, whereas the thin lines indicate the minimum

(b) West–East Precipitation Transects at 52°N: Summer

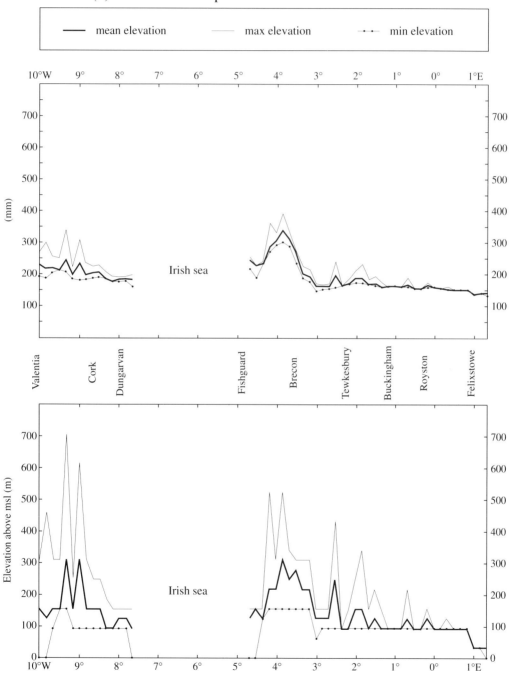

and maximum elevations and precipitation of each grid cell. The lower plots show the average, minimum and maximum elevation of each 10' cell.

Table 3.4 The contribution, as per cent, of average monthly precipitation to the average annual total, 1961–90 period. (Bold type indicates the wettest season.)

	Winter			Spring			Summer			Autumn		
	D	J	F	M	A	M	J	J	A	S	O	N
Kirkwall	11	11	7	8	6	5	5	6	8	**10**	**11**	**12**
Stornoway	11	10	7	9	6	5	5	6	7	**10**	**12**	**11**
Dyce	9	10	7	7	7	8	7	8	10	9	10	10
Durham	9	9	6	8	7	8	8	8	10	9	8	10
Santon Downham	9	9	6	7	8	8	**9**	**9**	**9**	8	9	10
Elmdon	**10**	**9**	**7**	8	7	8	9	7	10	8	8	9
Oxford	10	9	6	8	7	9	9	7	9	9	9	9
Shawbury	10	8	6	8	7	9	8	8	9	9	9	10
Gatwick	10	10	7	8	7	7	8	6	7	9	10	10
Eskdalemuir	11	11	7	9	5	6	6	6	8	**10**	**11**	**10**
Ringway	10	9	6	8	6	8	8	8	10	9	10	10
Valley	11	10	7	8	6	6	6	6	9	9	**11**	**12**
Long Ashton	**11**	**10**	**7**	8	6	7	7	7	8	9	9	10
Plymouth	**12**	**12**	**9**	9	6	6	6	6	7	8	10	10
Cork	**11**	**12**	**10**	8	6	7	6	5	7	8	11	9
Kilkenny	**11**	**11**	**8**	8	6	7	6	6	9	9	10	9
Clones	10	10	7	8	6	7	7	6	9	**9**	**10**	**9**

Table 3.5 Average 1961–90 monthly number of days of thunder. (Bold values indicate the highest monthly frequency for each site.)

	J	F	M	A	M	J	J	A	S	O	N	D
Kirkwall	**0.7**	0.5	0.3	0.1	0.3	0.3	0.6	0.4	0.1	0.4	**0.7**	0.4
Stornoway	**0.6**	0.4	0.4	0.0	0.2	0.2	0.2	0.3	0.2	0.4	0.5	0.5
Dyce	0.1	0.0	0.0	0.1	0.6	0.8	0.6	**1.0**	0.3	0.2	0.1	0.1
Elmdon	0.3	0.3	0.6	1.1	**3.0**	2.5	2.6	2.2	1.1	0.3	0.2	0.2
Shawbury	0.1	0.1	0.2	0.4	**2.3**	1.7	2.1	1.7	0.9	0.3	0.1	0.1
Eskdalemuir	0.3	0.1	0.3	0.5	**2.1**	1.7	1.3	1.4	1.1	0.5	0.5	0.1
Ringway	0.4	0.4	0.6	0.9	**2.5**	2.1	1.8	1.7	1.3	0.7	0.3	0.4
Valley	0.2	0.2	0.2	0.1	0.8	0.8	**0.9**	0.8	0.6	0.7	0.6	0.5
Plymouth	1.0	0.3	0.4	0.3	1.0	1.3	**1.7**	1.0	0.7	0.8	0.4	0.5

Although thunderstorms tend to be localised and are usually of short duration they can produce high precipitation totals which may lead to local flooding (see Chapter 13). For thunderstorms to occur a deep, moist, unstable layer of air is needed to allow the growth of large cumulonimbus clouds. In eastern and central England the number of days of thunder tends to be highest between May and August, when favourable conditions occur as a result of local heating (see Table 3.5). Further north, winter thunderstorms, associated with the marked temperature contrast in a **cold front**, are more frequent. Hail tends to be

Table 3.6 Average 1961–90 number of 'raindays' per season and average precipitation intensity (mm/day) on raindays. (Bold type indicates the season in which precipitation intensity is highest.)

	Winter		Spring		Summer		Autumn	
	Rain days	Intensity (mm/day)	Rain days	Intensity (mm/day)	Rain days	Intensity (mm/day)	Rain days	Intensity (mm/day)
Kirkwall	22.9	4.2	18.5	3.4	17.4	3.5	22.9	**5.0**
Stornoway	22.4	4.9	20.0	3.8	18.7	3.9	23.9	**5.4**
Dyce	17.0	4.0	15.8	3.6	15.2	4.0	16.8	**4.4**
Elmdon	15.1	3.8	15.1	3.5	12.6	**4.6**	13.8	4.0
Oxford	14.9	3.5	14.5	3.5	11.6	**4.6**	13.5	4.0
Shawbury	15.8	3.4	15.3	3.3	13.1	**4.1**	15.0	4.0
Cork	20.8	**6.4**	17.1	4.9	14.8	4.9	19.2	5.8
Kilkenny	18.8	4.3	16.4	3.5	14.4	3.9	17.6	**4.4**
Clones	20.3	4.0	18.4	3.6	17.7	3.8	20.8	**4.2**

associated with summer thunderstorms in southern and eastern areas of the British Isles, whilst over the hills and coasts of western and northern areas it is most frequent in winter and tends to be associated with maritime polar and arctic airstreams.

Another index of wetness is the 'rainday'.[7] The seasonal average number of raindays and the corresponding values of precipitation intensity on these days are shown in Table 3.6 for a number of sites. The smallest number of raindays is recorded in the summer at all nine sites, but for central and eastern sites precipitation intensity is highest in this season; convective precipitation is dominant in these areas in summer. Elsewhere, precipitation intensity is generally highest in autumn associated with the arrival of moisture-laden frontal systems from the Atlantic, perhaps enhanced by orographic uplift.

Frontal systems, **polar lows** or **troughs** and instability showers may result in snowfall over the British Isles. The role which each of these factors plays in producing snowfall varies throughout the country, but in general the frequency of snow falling increases towards the north and east and with altitude. Areas which are exposed to northerly, north-westerly or easterly winds suffer most from instability snow showers especially where high ground is close to the coast, for example, the North York Moors (Figure 3.6). Arctic air may bring sudden snowfalls from

Figure 3.6 An early morning (0950 GMT) visible satellite image (from NOAA 5) of the North Sea showing extensive snow cover over eastern England and Scotland on 17 February 1978. The day was classified as 'easterly' in the Lamb Catalogue. Courtesy of the University of Dundee.

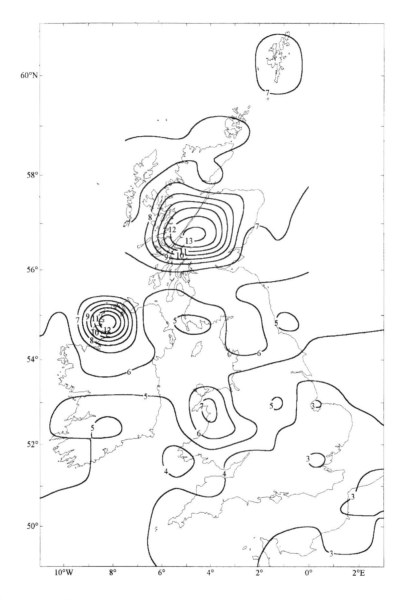

Figure 3.7 Average number of 'snowdays' in winter, 1961 to 1990. A snowday is defined as 'snow lying at 0900 GMT'. The values shown in this map are derived from observing stations mostly at low elevations; the actual number of snow-days on the hills and mountains are much greater than these.

small depressions, known as polar lows, which develop unexpectedly.

The average number of 'snowdays'[8] in winter is illustrated in Figure 3.7. The greatest frequency of snowdays is experienced in the Scottish Highlands, north-west Ireland and North Wales (Figure 3.8). In central and southern England, an average winter would have between three and six snowdays

Figure 3.8 Snow lying on the south face of Cader Idris, Snowdonia National Park. The uplands of the British Isles often experience several weeks of snow cover in contrast to lowland areas where, on average, less than ten days of snow cover occur.

compared to between only one and three snowdays in the lower elevation areas of the south-west. Similar patterns exist in spring and autumn, although totals are obviously not as great as in winter months. The variation in snowdays from year to year is discussed in Chapter 10 and unusually snowy winters in Chapter 13.

OTHER CLIMATE VARIABLES: SUNSHINE, HUMIDITY AND WIND SPEED

Sunshine

The annual variation in daylength associated with the changes in solar declination means that there is a marked annual variation in average daily sunshine duration. In general, there is a decrease in sunshine hours from south to north, from the coast to inland and with altitude. The south coast tends to be the sunniest area of the British Isles because it is most sheltered from the cloud-bearing westerly and easterly winds and is also influenced by continental high pressure systems. Plate 5 illustrates the seasonal variation in sunshine receipt over the British Isles. In winter, maximum sunshine totals (between 1.5 and 2.5 hours per day) are experienced along the south coast, whereas in spring many coastal areas of mainland Britain and Ireland have their highest sunshine totals (between 5 and 5.5 hours per day). In summer, areas of maximum sunshine (more than 6 hours per day) are in the south and east with totals almost double those in Scotland. In autumn, sunshine totals

Table 3.7 Average 1961–90 monthly relative humidity (per cent). (Range is the difference between the least and most humid months of the year.)

	J	F	M	A	M	J	J	A	S	O	N	D	Range
Kirkwall	89	87	87	86	85	85	88	90	90	90	90	88	5
Stornoway	87	88	88	87	85	85	87	89	90	91	91	90	6
Dyce	86	84	80	80	81	80	79	82	83	85	85	86	7
Durham	89	87	83	82	81	82	81	84	85	88	89	89	8
Santon Downham	95	94	89	86	83	82	84	86	88	95	97	97	15
Elmdon	90	87	83	78	77	75	75	79	83	87	88	89	15
Oxford	89	86	81	78	75	74	73	77	81	86	89	91	18
Shawbury	89	87	83	81	79	78	79	81	84	88	89	90	12
Gatwick	91	87	84	80	79	78	78	81	86	90	91	93	15
Eskdalemuir	92	88	85	83	81	82	84	86	89	90	92	92	11
Ringway	86	83	78	74	70	72	74	76	79	82	84	86	16
Valley	85	83	83	80	78	80	81	82	84	84	84	86	8
Long Ashton	89	87	82	79	77	77	77	79	83	89	89	92	15
Plymouth	89	86	84	80	80	81	81	83	85	87	88	89	9
Cork	92	89	86	82	82	82	82	84	87	90	92	93	11
Kilkenny	90	86	82	79	78	79	79	81	84	89	90	92	14
Clones	92	87	83	80	79	80	81	83	86	88	92	93	14

are again largest in southern and eastern England (3.5 to 4 hours per day). The sunniest month for much of the British Isles tends to be May even though the average daylength is shorter than in June.

Relative humidity

The water-holding capacity of the atmosphere is dependent on temperature; as temperature increases the amount of water vapour the air can hold also increases. Saturated air at 20°C, for example, holds 3.6 times more water vapour than air at 0°C. Relative humidity is a measure of the amount of water vapour in the air compared to the maximum amount the air can hold at that particular temperature, expressed as a percentage. Table 3.7 illustrates average monthly relative humidity for selected sites. Coastal sites have a smaller range of relative humidity over the year (for example, Kirkwall 5 per cent) than sites in inland areas (for example, Oxford 18 per cent). This is partly because of the constant supply of moisture around the coast, but also because of the less vari-able temperature regime in coastal areas. Plate 6 illustrates the seasonal variation in relative humidity over the British Isles.

Relative humidity also affects how comfortable people feel in a particular place and at a particular time. People are generally accustomed to the range of temperatures of the country where they live, and whether or not they feel comfortable at a given temperature depends largely on the relative humidity. If the relative humidity is different from usual then similar temperatures do not produce the same feeling of comfort or discomfort. In order to maintain a constant body temperature in warm conditions perspiration – evaporation of water from the skin – occurs. As this process proceeds, energy is required to evaporate the water and the body is cooled. If relative humidity is low and temperatures are high then evaporation is rapid and perspiration evaporates from the skin easily. Replacement of fluid is required to avoid dehydration. If relative humidity is high, however, then body sweat does not evaporate readily and people feel sticky and hot. As temperatures

Mean Temperature

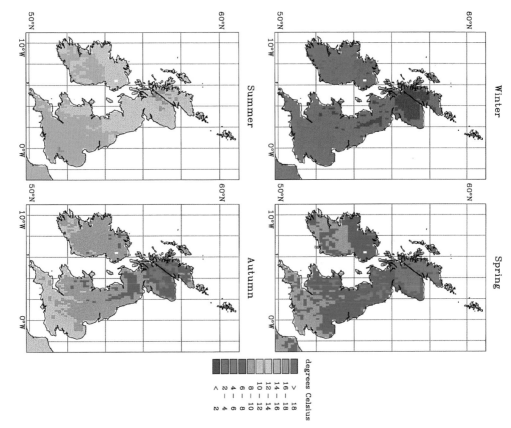

Winter

Spring

Summer

Autumn

degrees Celsius

> 18
16 — 18
14 — 16
12 — 14
10 — 12
8 — 10
6 — 8
4 — 6
2 — 4
< 2

Plate 1 Seasonal average mean temperature, °C, 1961 to 1990 period. On this, and all other maps, the data shown are for the average elevation in each 10 km grid cell. Winter = DJF; spring = MAM; summer = JJA; autumn = SON.

Mean Maximum Temperature

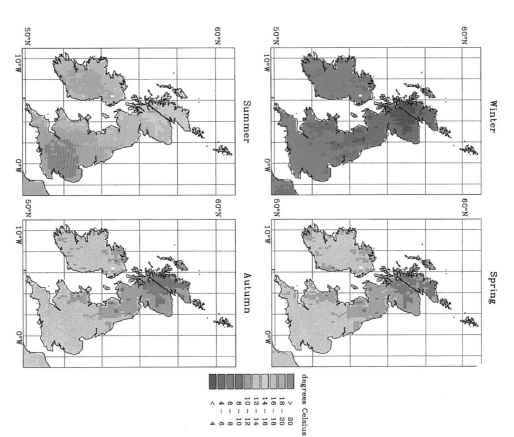

Winter

Spring

Summer

Autumn

degrees Celsius

> 20
18 — 20
16 — 18
14 — 16
12 — 14
10 — 12
8 — 10
6 — 8
4 — 6
< 4

Plate 2 Seasonal average maximum temperature, °C, 1961 to 1990 period.

Mean Minimum Temperature

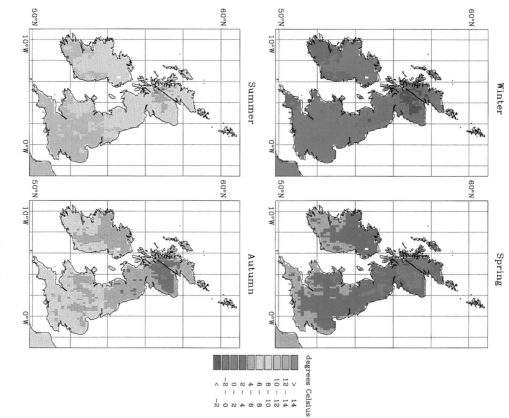

Winter

Spring

Summer

Autumn

degrees Celsius

> 14
12 — 14
10 — 12
8 — 10
6 — 8
4 — 6
2 — 4
0 — 2
-2 — 0
< -2

Plate 3 Seasonal average minimum temperature, °C, 1961 to 1990 period.

Mean Precipitation

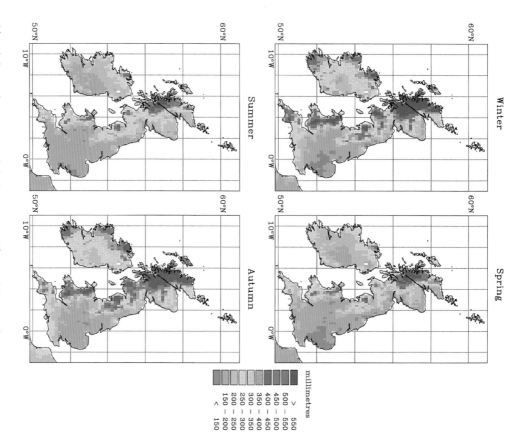

Winter

Spring

Summer

Autumn

millimetres

> 550
500 — 550
450 — 500
400 — 450
350 — 400
300 — 350
250 — 300
200 — 250
150 — 200
< 150

Plate 4 Seasonal average precipitation, mm, 1961 to 1990 period.

Mean Sunshine

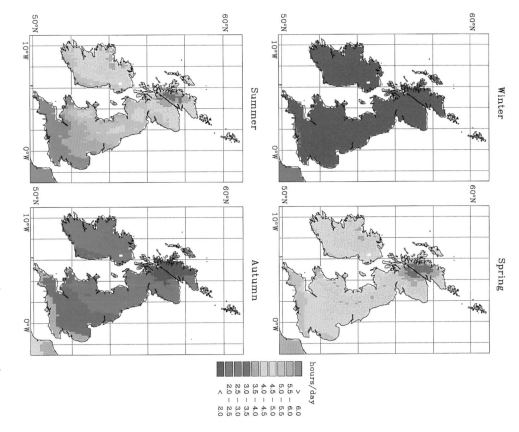

Plate 5 Seasonal average sunshine rate, hours per day, 1961 to 1990 period.

hours/day

> 6.0
5.5 — 6.0
5.0 — 5.5
4.5 — 5.0
4.0 — 4.5
3.5 — 4.0
3.0 — 3.5
2.5 — 3.0
2.0 — 2.5
< 2.0

Mean Relative Humidity

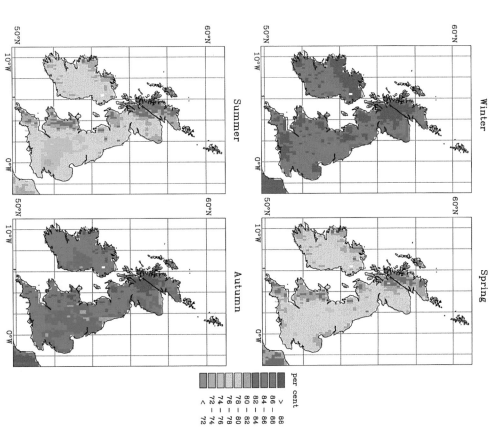

Plate 6 Seasonal average relative humidity, per cent, 1961 to 1990 period.

per cent

> 88
86 — 88
84 — 86
82 — 84
80 — 82
78 — 80
76 — 78
74 — 76
72 — 74
< 72

Mean Wind Speed

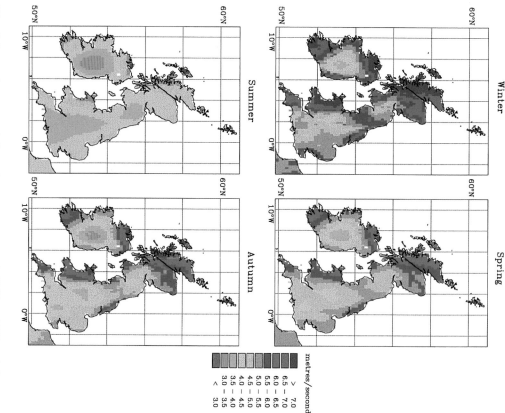

Winter

Spring

Summer

Autumn

metres/second

> 7.0
6.5 — 7.0
6.0 — 6.5
5.5 — 6.0
5.0 — 5.5
4.5 — 5.0
4.0 — 4.5
3.5 — 4.0
3.0 — 3.5
< 3.0

Plate 7 Seasonal average wind speed, metres per second, 1961 to 1990 period.

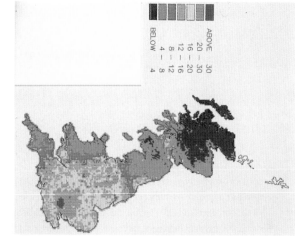

ABOVE 30
20 — 30
16 — 20
12 — 16
8 — 12
4 — 8
BELOW 4

ABOVE 700
600 — 700
400 — 600
200 — 400
100 — 200
50 — 100
BELOW 50

Plate 8(a) Estimated nitrogen dioxide concentrations (ppb) in Great Britain, July to December 1991[18] (above left) and estimated carbon monoxide concentrations (ppb) in Great Britain in 1991 (above right).[18]

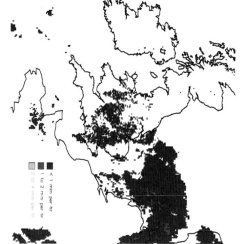

< 1 mm per hr
1 to 2 mm per hr
2 to 4 mm per hr

Plate 8(b) A Nimrod radar network image for 1200 GMT on 10 May 1996. The radar image shows a crop of showers moving south-westwards across England and Wales. An area of more persistent rain can be seen over the North Sea. This picture is supplied courtesy of the UK Met. Office.

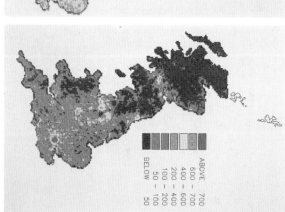

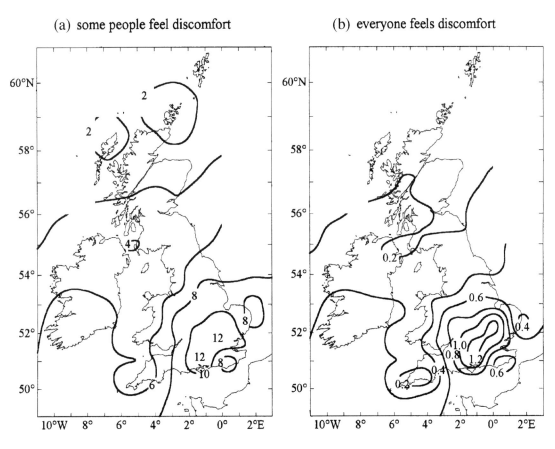

Figure 3.9 Average number of days in each discomfort class in July, 1961–90: (a) up to 50 per cent of people feel discomfort; (b) everyone feels discomfort and some may suffer distinct stress.

increase, the more people perspire and the more uncomfortable they feel. This discomfort is eased somewhat if there is a breeze to assist evaporation.

A discomfort index[9] can be calculated by combining maximum temperature with relative humidity. In this instance the index refers to still air conditions. High night-time minimum temperatures can also make people feel uncomfortable. Based on this index six discomfort classes can be identified: no discomfort; some people are uncomfortable; over 50 per cent of people are uncomfortable; everyone feels uncomfortable; distinct stress; and great discomfort – danger of heat-stroke. Figure 3.9 illustrates the average number of days in particular discomfort classes for

July. The highest number of days when everyone feels uncomfortable occurs in central and south-east England with more than one such day occurring in an average July. In Scotland, there are very few occasions when everyone feels discomfort. These are average values and in individual years the picture will be different. Figures 3.10a–c illustrate the number of days in different discomfort classes for individual years over the period 1961 to 1987 for three sites around the British Isles. Central England (in this case Oxford) experiences the highest number of days with discomfort, whereas Lowestoft due to its coastal location experiences significantly fewer days when everyone feels stress. The most uncomfortable years in

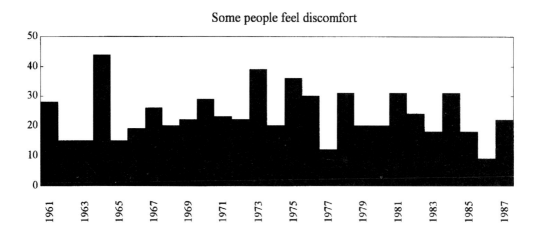

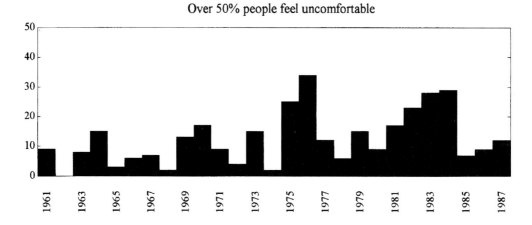

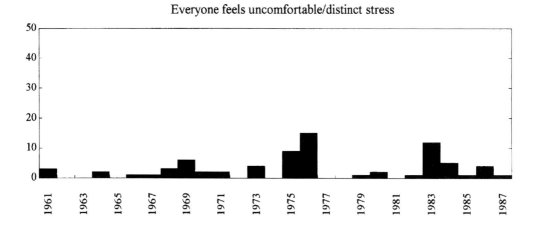

Figure 3.10a The number of days in different discomfort classes over the period 1961 to 1987 for Oxford.

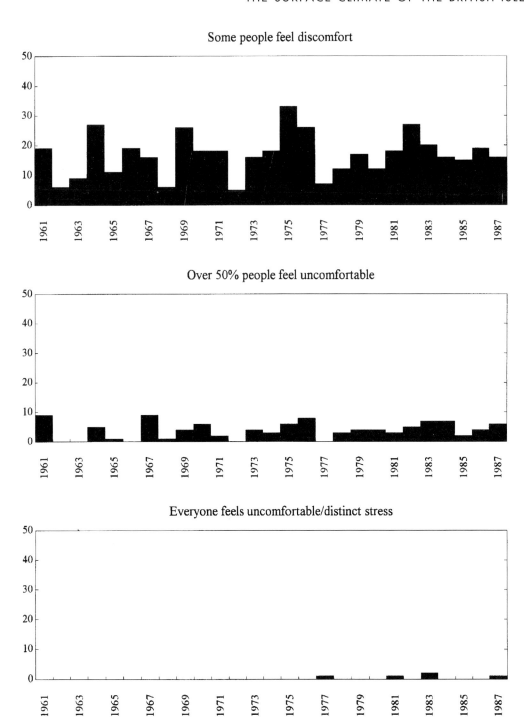

Figure 3.10b The number of days in different discomfort classes over the period 1961 to 1987 for Lowestoft.

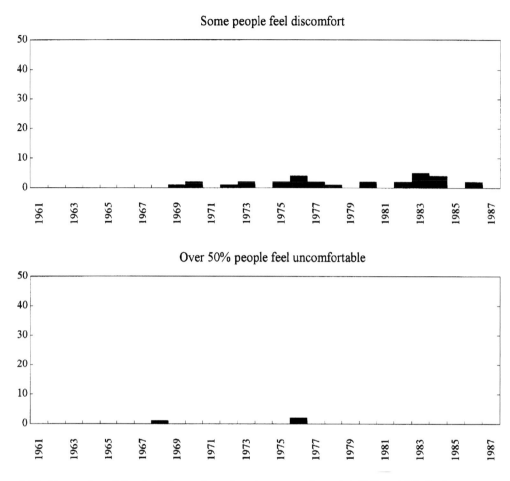

Figure 3.10c The number of days in different discomfort classes over the period 1961 to 1987 for Stornoway.

central England during this period are clearly identified as 1976 and 1983 (these were both very hot summers; see Chapter 9). Discomfort was even experienced by over 50 per cent of people in the Western Isles (Stornoway) on two days in 1976.

Wind speed

The prevailing wind direction over the British Isles is generally from between the south and west, whilst wind with an easterly and northerly component is less frequent (see Chapter 11). In some areas topog-

raphy has an important effect on local winds resulting in different speeds and directions to the prevailing airflow. In the north and west of the British Isles wind speed tends to be greater and periods of higher sustained wind speed are more frequent. Lower wind speeds are experienced in central and eastern areas. This is indicated in Table 3.8. Highest wind speeds occur in January or March and lowest values are obtained in the summer. In all seasons wind speeds are greatest in coastal areas compared to the more sheltered inland regions (see Plate 7). Chapters 11 and 13 contain more discussion about winds in the British Isles.

Table 3.8 Average 1961–90 monthly wind speed (metres per second). (Bold type indicates the month with highest average wind speed.)

	J	F	M	A	M	J	J	A	S	O	N	D
Kirkwall	**8.3**	7.6	8.1	6.6	6.3	5.7	5.7	5.5	6.6	7.4	7.7	8.1
Stornoway	**7.7**	7.2	7.3	6.1	5.9	5.6	5.4	5.1	6.3	6.7	6.7	7.5
Dyce	5.3	5.1	**5.6**	4.8	4.6	4.2	4.1	4.1	4.5	4.8	5.1	5.3
Elmdon	4.7	4.5	**4.8**	4.5	4.3	4.0	4.0	3.9	4.0	4.0	4.5	4.7
Oxford	**5.1**	4.9	5.0	4.6	4.3	4.0	3.9	4.0	4.1	4.2	4.6	4.8
Shawbury	4.4	4.4	**4.8**	4.3	4.0	3.9	3.8	3.8	3.8	3.8	4.1	4.3
Eskdalemuir	4.7	4.4	**4.8**	4.2	3.9	3.8	3.4	3.4	4.0	4.2	4.5	4.6
Ringway	4.9	4.9	**5.0**	4.5	4.4	4.0	3.8	3.9	4.0	4.2	4.5	4.7
Valley	**8.0**	7.1	7.3	6.0	6.0	5.8	5.6	5.8	6.7	7.3	7.6	7.8
Plymouth	**6.4**	6.2	5.9	5.3	5.1	4.7	4.5	4.5	4.9	5.4	5.8	6.3
Cork	**6.6**	**6.6**	6.3	5.6	5.5	4.9	4.7	4.8	5.2	5.8	5.9	6.4
Kilkenny	3.8	3.8	**4.0**	3.4	3.3	3.0	2.9	2.9	3.0	3.3	3.3	3.7
Clones	5.0	5.0	**5.2**	4.4	4.1	3.7	3.6	3.6	3.9	4.4	4.4	4.8

As with relative humidity, wind speed also plays an important role in determining how comfortable a person feels at a particular **ambient temperature**. Wind decreases the apparent temperature experienced by a person; at high temperatures this is beneficial, but at low ones it can be dangerous. Wind speed and average temperature can be combined to define a **wind chill equivalent temperature**.[10] These equivalent temperatures have been calculated at a number of sites around the British Isles and the results are shown in Table 3.9. The number of days below particular temperatures have been summed over the winter season both with and without the effect of wind speed being taken into account. Wind chill has a dramatic effect on the number of days when the 'felt' temperature is below different thresholds.

REGIONAL CLIMATES

With respect to the climates of the world, the British Isles are usually defined as falling within one major climatic type. It is only when studies are undertaken at a finer resolution that different regional climates within the Isles may be identified. Differences within

the British Isles which are sometimes defined using a global classification include areas of milder winters in the far south-west, areas of summer precipitation maximum in parts of eastern England and contrasts between areas of moisture deficit in the east and moisture surplus in the west. All of these differences have been noted in the appropriate sections earlier in this chapter.

A number of studies have been undertaken over the years to identify regional climates in the British Isles using globally based classifications. For example, as a result of Charles Thornthwaite's[11] precipitation minus evapotranspiration index only three of the global categories – wet, humid and sub-humid – were identified in the British Isles. Hartmut Walter and Helmut Lieth[12] developed Thornthwaite's approach and identified sixteen climatic types over the British Isles, but there were only two major categories which equated with global-scale conditions – namely, warm temperate humid climate with occasional frost and humid temperate climate with a cold season. According to the classification of Carl Troll and Karl Paffen[13] the climate of the British Isles may be described as either 'oceanic' or 'sub-oceanic'. Oceanic climates have an annual fluctuation in temperature of

Table 3.9 Average 1961–90 annual number of days with mean temperature below 0°C and –10°C for actual and wind-chill equivalent temperature

| | Actual mean temperature | | Wind-chill equivalent temperature | |
	Days below 0°C	Days below –10°C	Days below 0°C	Days below –10°C
Kirkwall	4	0	42	17
Stornoway	2	0	38	9
Dyce	7	0	42	12
Durham	9	0	34	6
Elmdon	12	0	37	5
Oxford	11	0	32	9
Shawbury	9	0	35	6
Gatwick	11	0	33	6
Eskdalemuir	13	0	45	8
Ringway	8	0	41	6
Valley	5	0	41	8
Plymouth	5	0	34	5

less than 16°C, mild winters, autumn and winter maxima of precipitation and moderately warm summers, whilst sub-oceanic climates have an annual temperature fluctuation of between 16°C and 25°C, mild to moderately cold winters, autumn to summer maxima of precipitation and moderately warm summers.

Earlier studies of regional climates of the British Isles include that of Sir Arthur Tansley[14] which distinguished between coastal and inland climates. For example, extreme Atlantic coastal areas were defined as having low summer and high winter temperatures, moderate precipitation and below-average sunshine for their latitude, whilst the east coasts have lower winter temperatures and more snow, less precipitation (especially in the south) and rather more sunshine than in the west. The nearest approach to a continental-type climate was in the English midlands, especially in the east (cf. Conrad's index shown in Figure 3.2). Stan Gregory[15] determined regional climates in the British Isles by considering three basic climatic characteristics and their interaction – namely, the length of the growing season, a precipitation magnitude factor and precipitation seasonality – while Tom Wigley and colleagues[16] determined regions of coher-

ent precipitation variability in England and Wales based on time-series analyses. This latter work has since been expanded to cover the whole of the British Isles and is described in Chapter 10.

Identification of regional climates within the British Isles obviously depends to a large extent on the subject of interest. For example, whether one was interested in water resources, agriculture or tourism would change the nature of the classification employed and the size of the regions would also vary depending on the subject under consideration. Instead of attempting to define formal regional climates in the British Isles[17] we have selected a number of sites to illustrate some of the points made above. Figure 3.11 shows average maximum and minimum temperature, precipitation and sunshine distributions for six sites around the British Isles (see Figure 3.1 for their location). Values of some of the global classifications mentioned above have been calculated for each of these sites and are also indicated.

The three sites which are most influenced by the Atlantic Ocean are Stornoway, Plymouth and Kilkenny and this is reflected in the mild winter and cool summer temperatures at these sites. According to Conrad's continentality index, Stornoway is the

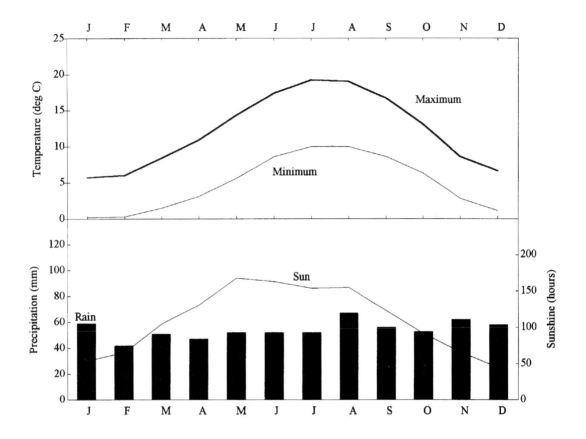

Global Classification

Continentality index	8.2
Walter/Lieth	humid, temperate with a cold season
Troll/Paffen	sub-oceanic
Thornwaite	humid
Gregory	B – growing season of 7 or 8 months
	D – receives 750 mm rain or less with at least 30% probability
	2 – rainfall maximum during second half of year

Figure 3.11a The average 1961 to 1990 climate of Durham and its climatic classification according to five different schemes.

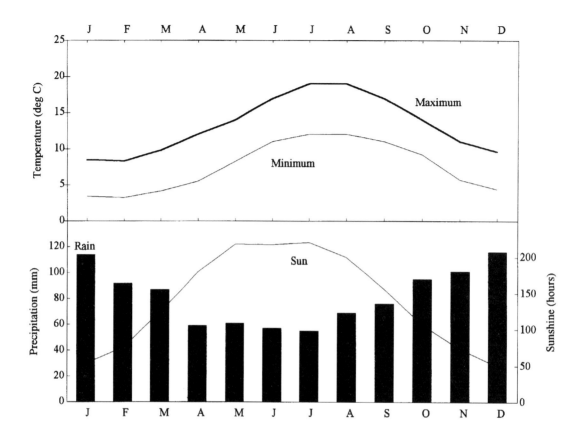

Global Classification

Continentality index	6.3
Walter/Lieth	warm, temperate humid with only occasional frost
Troll/Paffen	oceanic
Thornwaite	humid
Gregory	A – growing season of 9 or more months
	M – receives 750 mm rain or less with at least 30% probability
	W – rainfall maximum in winter half of year

Figure 3.11b The average 1961 to 1990 climate of Plymouth and its climatic classification according to five different schemes.

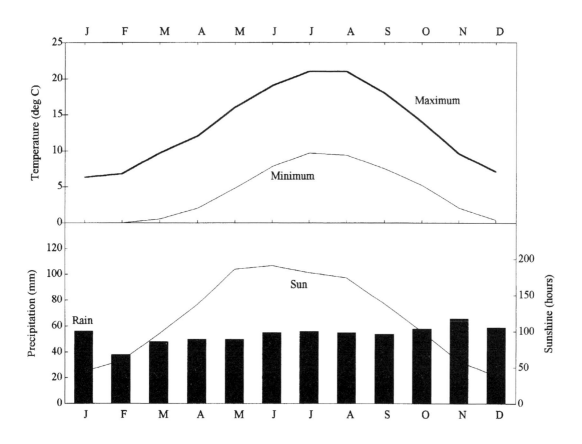

Global Classification

Continentality index	10.2
Walter/Lieth	humid, temperate with a cold season
Troll/Paffen	sub-oceanic
Thornwaite	humid
Gregory	B – growing season of 7 or 8 months
	D – receives 750 mm rain or less with at least 30% probability
	S – rainfall maximum tends to be in summer

Figure 3.11c The average 1961 to 1990 climate of Santon Downham and its climatic classification according to five different schemes.

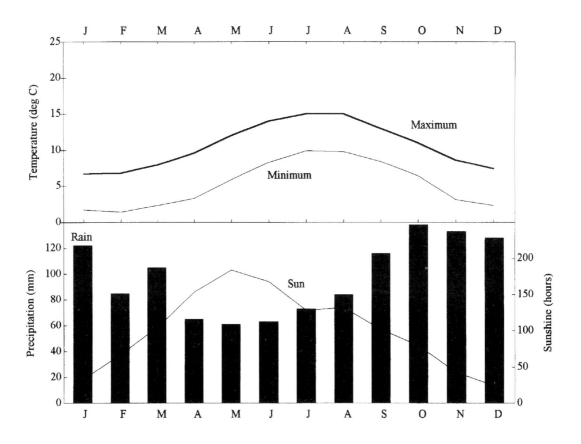

Global Classification

Continentality index	1.9
Walter/Lieth	humid, temperate with a cold season
Troll/Paffen	sub-oceanic
Thornwaite	humid
Gregory	B – growing season of 7 or 8 months
	M – area of moderate rainfall
	W – rainfall maximum in winter half of year

Figure 3.11d The average 1961 to 1990 climate of Stornoway and its climatic classification according to five different schemes.

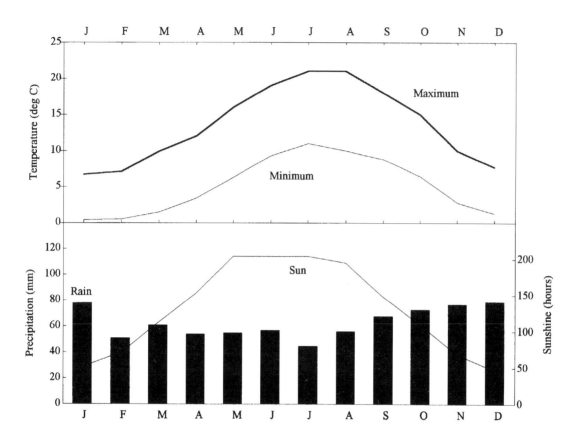

Global Classification

Continentality index	11.0
Walter/Lieth	humid, temperate with a cold season
Troll/Paffen	sub-oceanic
Thornwaite	humid
Gregory	B – growing season of 7 or 8 months
	D – receives 750 mm rain or less with at least 30% probability
	W – rainfall maximum in winter half of year

Figure 3.11e The average 1961 to 1990 climate of Gatwick and its climatic classification according to five different schemes.

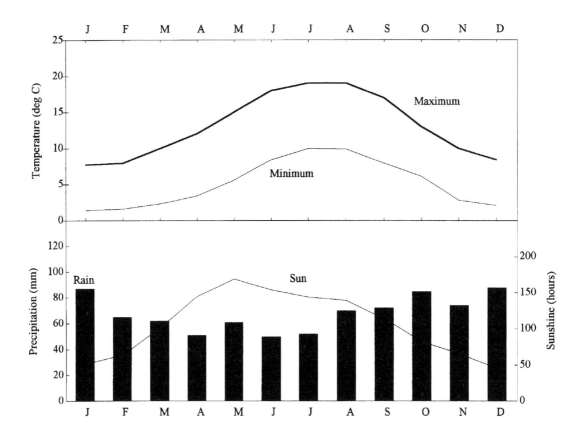

Global Classification

Continentality index	6.3
Walter/Lieth	transitional between warm, temperate humid with only occasional frost and humid, temperate with a cold season
Troll/Paffen	sub-oceanic
Thornwaite	humid
Gregory	B – growing season of 7 or 8 months
	M – area of moderate rainfall
	W – rainfall maximum in winter half of year

Figure 3.11f The average 1961 to 1990 climate of Kilkenny and its climatic classification according to five different schemes.

most oceanic of all the six sites with an index of 1.9. The index at both Plymouth and Kilkenny has a value of 6.3, in spite of Plymouth's more coastal location. The most continental climates are experienced at Santon Downham and Gatwick; the continentality indices at these two sites are amongst the highest in the British Isles (10.2 and 11.0, respectively). Both these sites experience lower winter and higher summer temperatures than the other four sites and also the largest diurnal temperature ranges. Precipitation maxima at Santon Downham, the most easterly site, occur in summer and autumn compared to the autumn and winter maxima of the other five sites. Troll and Paffen's classification resulted in five of the six sites being defined as 'sub-oceanic'; only Plymouth was classified as having an 'oceanic' climate. Thornthwaite's classification yields a definition of 'humid' for five out of the six sites; Santon Downham was the exception in this case and is defined as 'sub-humid'. Durham, Santon Downham, Gatwick and Stornoway were classified as having humid, temperate climates with a cold season according to Walter and Lieth, whilst Plymouth has a warm, temperate, humid climate with only occasional frost. Kilkenny has a climate which is transitional between these two types.

Gregory's more complicated classification defines all sites except Plymouth as having a growing season length of between seven and eight months; at Plymouth it is of nine or more months. Further classification of the climate using Gregory's approach classifies three of the sites (Durham, Gatwick and Santon Downham) as receiving 750 mm or less of precipitation in at least three years out of ten. Stornoway, Kilkenny and Plymouth are in areas where 'high rainfall does not occur frequently, but low rainfall is itself infrequent or absent'.[18] Further classification of the climate in Gregory's scheme includes a measure of precipitation seasonality.

SUMMARY

This chapter has described the climate of the British Isles as experienced during the period 1961 to 1990 and as represented by gridded climatologies constructed from individual station data. Since it is a description of the average climate only the general features have been discussed; more detail concerning the year-to-year variability of, for example, temperature, precipitation and wind speed, and also the frequency and severity of extreme weather events, can be found in the subsequent chapters of this book.

The climate of the British Isles is greatly influenced by the surrounding relatively warm, oceanic water and by their location in the mid-latitude westerly wind belt. This means that the Atlantic coastal areas have lower summer and higher winter temperatures compared to more inland and eastern regions. Precipitation is greater in the western half of the British Isles because of the combined effects of orography and exposure to the moisture-laden westerly winds. Eastern areas tend to be much drier and have colder winters but warmer summers. A higher proportion of summer precipitation in eastern and central regions is of convective, i.e., thunderstorm, origin. Southern England is the sunniest region in the British Isles and the south-west Midlands the least windy. Which region of the British Isles has the best climate? It of course depends on what your criteria are; whether you wish to maximise sunshine or temperature, minimise precipitation or humidity, eliminate discomfort due to wind-chill or avoid the worst storms. At least some of the data presented in this chapter should help make that decision an easier one.

NOTES

1   It is the convention of the World Meteorological Organisation that thirty-year periods are used to describe climatic averages.
2   For example, P.D. Jones, 'Hemispheric surface air temperature variations: a reanalysis and an update to 1993', *Journal of Climate*, 1994, vol. 7, pp. 1794–1802.
3   V. Conrad, 'Usual formulas of continentality and their limits of validity', *Transactions of the American Geophysical Union*, 1946, vol. 27, pp. 663–4.
4   Mean temperature is defined as half the sum of maximum and minimum temperature.
5   The standard climatological definitions of the seasons are used throughout this book, i.e. winter is December,

January and February; spring is March, April and May; summer is June, July and August and autumn is September, October and November. See Chapter 8 for a discussion about a different approach to defining 'seasons'.

6   Precipitation includes snowfall as a liquid equivalent.

7   A 'rainday' is defined when precipitation is greater then 0.2 mm in 24 hours.

8   A 'snowday' is defined if snow is lying at 9 a.m. (GMT).

9   This discomfort index is based on that in E.A. Pearce, and C.G. Smith, *The World Weather Guide*, London, Melbourne, Sydney, Auckland and Johannesburg, Hutchinson, 1984.

10   Wind-chill equivalent temperatures are based on A. Henderson-Sellers and P.J. Robinson, *Contemporary Climatology*, Longman Scientific and Technical, 1986.

11   C.W. Thornthwaite, 'The climates of the Earth', *Geographical Review*, 1933, vol. 23, pp. 433–40.

12   H. Walter and H. Lieth, *Klimadiagramm Weltatlas*, Jena, Fisher, 1967.

13   C. Troll and K.H. Paffen, 'Seasonal climates of the Earth', in E. Rodenwaldt and H. J. Jusatz (eds), *World Maps of Climatology*, Berlin, Springer, 1965, pp. 19–25.

14   A.G. Tansley, *The British Isles and Their Vegetation*, Cambridge, Cambridge University Press, 1939.

15   S. Gregory, 'Regional climates', in T.J. Chandler and S. Gregory (eds), *The Climate of the British Isles*, London and New York, Longman, 1976, pp. 330–42.

16   T.M.L. Wigley, J.M. Lough and P.D. Jones, 'Spatial patterns of precipitation in England and Wales and a revised, homogeneous England and Wales precipitation series', *Journal of Climatology*, 1984, vol. 4, pp. 1–25.

17   For a specifically regional approach to the climate of the British Isles see D. Wheeler and J.R. Mayes (eds), *The Regional Climates of the British Isles*, 1997, London, Routledge.

18   Gregory, 1976, op. cit., p. 339. There is an error on Figure 15.4 in this chapter which classifies Stornoway, incorrectly, as BDW instead of BMW.

GENERAL READING

T.J. Chandler, and S. Gregory (eds), *The Climate of the British Isles*, London and New York, Longman, 1976.

M. Hulme, D. Conway, P.D. Jones, T. Jiang, E.M. Barrow, and C. Turney, 'Construction of a 1961–1990 European climatology for climate change modelling and impact applications', *International Journal of Climatology*, 1995, vol. 15, pp. 1333–63.

R. Stirling, *The Weather of Britain*, London, Faber and Faber, 1982.

D. Wheeler and J.R. Mayes (eds), *The Regional Climates of the British Isles*, London and New York, Routledge, 1997.

Part 2

RECONSTRUCTING THE PAST

What's past is prologue.
William Shakespeare, *The Tempest*

4

THE CLIMATES OF PAST AGES

Brian Funnell

In Nature's infinite book of secrecy, a little can I read.
William Shakespeare, *Antony and Cleopatra*

INTRODUCTION

Sediments, and sedimentary rocks, provide us with a remarkable record of past conditions at the surface of the Earth. This record extends many hundreds of millions of years into the past (Table 4.1). Different sediments are produced under different climatic conditions. Evaporites – gypsum and rock salt – which are often associated with desert dune sands, are produced under hot arid conditions. Glacial boulder clays – also called tills and diamictons – are produced by glaciers and ice-sheets. There are many other examples. The fossil skeletal remains of organisms preserved in sediments also give evidence of past climates. Massive colonial coral reefs are characteristic of tropical seas, whereas less diverse faunas of cold-water molluscs typify high-latitude seas. Many sophisticated methods have been developed to infer more precise climatic parameters from sediments deposited during the last few million years. These methods include stable isotopes[1], statistical faunal[2] and floral[3] analyses, and organic compound 'biomarkers'.[4]

Sediments are destroyed over time, both by erosion and other processes, and the record for the more distant past is therefore less complete; certainly it is geographically less comprehensive. Also, the fossil record becomes much less easy to interpret (because of the evolution, creation and extinction of life forms with no living equivalents) and the geochemical methods become less reliable (because of changes in global biogeochemical equilibria, and gradual, **diagenetic changes** in the chemistry of the sediments themselves).

To obtain a history of past climate change the sedimentary record also has to be placed in a time sequence. There are various ways of doing this (see Box 4.1), each of which have very different accuracies and are applicable over widely different periods of time. In general, the rocks of the British Isles include a broad variety of sedimentary rocks covering a very wide time-span. Unfortunately, much of the sedimentary record of terrestrial climate change in the British Isles during the last few million years has been destroyed by erosion by the British ice-sheets during the **Quaternary** glaciations. An alternative, continuous record of climate change for that period is, however, fortunately available from the deep-sea floor.

In the present chapter we consider the climate of the British Isles during the following three ages: before the present (Late Cenozoic) '**Ice Age**'; leading up to the present Ice Age; and during the present Ice Age up to and including the Last (**Devensian**) Glacial Maximum and the onset of deglaciation. Chapter 5 continues the story into the post-glacial period of the **Holocene**. Table 4.1 provides a simple summary of the overall geological time-scale against which these changes have occurred.

Table 4.1 Key features of the geological time-scale relating to climatic change: >600–0 million years BP. (Quaternary Ice Age *Interglacial* periods are in italics.)

Era	Period	Epoch	Ice Ages[5]	Million years BP
Cenozoic	Quaternary	Holocene	*Flandrian*	0.011–0.000
		Pleistocene	Devensian	0.117–0.011
			Ipswichian	0.130–0.117
			Wolstonian	c. 0.14
			Hoxnian	c. 0.36
			Anglian	c. 0.42
			Cromerian	c. 0.56
	Tertiary	Pliocene	'Red Crag'	c. 2.60–c. 2.40
			'Coralline Crag'	c. 3.75
				65
Mesozoic	Cretaceous			
	Jurassic			
	Triassic			
				245
Palaeozoic	Permian		Permo-Carboniferous	c. 290
	Carboniferous			
	Devonian			
	Silurian			
	Ordovician		Ordovician	c. 440
	Cambrian			
				570
Precambrian	Vendian		Precambrian	c. 610
				c. 4560

Sources: W.B. Harland, R.L. Armstrong, A.V. Cox, L.A. Craig, A.G. Smith and D.G. Smith, *A Geologic Time Scale 1989*, Cambridge, Cambridge University Press, 1990, 263 pp. T.J. Crowley and G.R. North, *Paleoclimatology*, Oxford, Oxford University Press, 1991, 339 pp. A.E. Wright and F. Moseley (eds), *Ice Ages: Ancient and Modern*, Liverpool, Seel House Press, 1975, 320pp.

CLIMATE IN THE PRE-CENOZOIC ERA: >600 TO 65 MILLION YEARS BP

The climate of the British Isles has been continually changing as far back as we can read the record, i.e., for more than 600 million years. Some of these climate changes have been global and involved variations in the Earth's climate. These global climate changes have been caused by changes in the Earth's climate boundary conditions; namely the geographical configuration of the continents and oceans, the chemical composition of the atmosphere, biological productivity, and variations in solar output. Other climate changes, as far as the British Isles are concerned, have been produced by the geological movements of Britain relative to the surface of the Earth caused by ocean-floor spreading and plate tectonics. These processes have generated both

BOX 4.1 METHODS OF DATING

All proxy data from the sedimentary record of past climates need to be related to the *time* to which they refer. Originally, the Earth's geological time-scale, which has been progressively assembled during two centuries of study, was only a 'relative' time-scale. In recent years, however, more and more methods have been developed which enable the allocation of 'absolute' calendar dates to the sedimentary record. These dates are expressed in thousands or millions of years before present (BP).

- Because **radioactive decay** proceeds at a constant rate, measurement of its products can be used to establish the time elapsed since an original radioactive element was incorporated into a mineral or biogenic compound. Common targets for such investigations are: uranium series elements (for both very long time-scales and the last *c.* 500,000 years), potassium–argon decay (from *c.* 500,000 years BP to *c.* 500 million years BP), and ^{14}C radiocarbon (for the last *c.* 50,000 years). All of these methods involve critical assumptions about the conditions of formation and the subsequent history of the minerals or compounds containing the radioactive elements (see Chapter 5, for example, for a discussion about ^{14}C dating).
- The polarity of the Earth's geomagnetic field has reversed repeatedly over geological time. The present-day polarity is regarded as normal and the alternate polarity is referred to as reversed. The last major reversal occurred at *c.* 0.78 million years BP. Because magnetic minerals in sediments and volcanic rocks orientate themselves in relation to the Earth's magnetic field at the time of their formation, sediment cores and volcanic rocks can be 'dated' in relation to the pattern of reversals that they preserve.

- In the last few years it has proved possible to map the sedimentary, chemical and biological cycles observed in deep-sea sediment cores on to back calculations of the solar radiation variations produced by the Earth's orbital cycles (see Box 4.2). By using check-points provided by the palaeomagnetic reversal record and biological evolution events, astronomically calibrated dating has now been applied from the present-day to beyond 5 million years BP,[a,b] with a sustainable accuracy of ±*c.* 5,000 years.
- Nearer the present day, annual cycles preserved in tree-rings (see Chapter 5) or glacier ice have been used to create calendar time-scales extending back to *c.* 10,000 years and *c.* 150,000 years BP respectively. In both cases counting errors become more problematical in older sequences.
- Evolution, creation and extinction events, particularly of ocean plankton (**coccolithophores, diatoms, planktonic foraminifera** and **Radiolaria**) in deep-sea cores, were originally the basis of geological 'relative' dating methods. Many of these have now been calibrated to an 'absolute' time-scale, so that they can themselves be used as approximate 'absolute' age indicators, often to within *c.* 0.1 million years BP.

[a] N.J. Shackleton, A. Berger and W.R. Peltier, 'An alternative astronomical calibration of the lower Pleistocene timescale based on ODP Site 677', *Transactions of the Royal Society of Edinburgh, Earth Sciences*, 1990, vol. 81, pp. 251–61.
[b] N.J. Shackleton, M.A. Hall and D. Pate, 'Pliocene stable isotope stratigraphy of Site 846', in N.G. Pisias, L.A. Mayer, T.R. Palmer-Julson and T.H. van Andel (eds), *Proceedings of the Ocean Drilling Program, Scientific Results*, Ocean Drilling Program, College Station, Texas, 1995, vol. 138, pp. 337–55.

'**continental drift**' between continents and continental fragments and movements relative to the Earth's magnetic/rotational axis.

Just before multicellular organisms became common on the Earth, at *c*. 610 million years before present (BP) (Table 4.1), there was a period of widespread glaciation (the Vendian or Late Precambrian Ice Age);[5] 170 million years later, at around 440 million years BP, there was an Ordovician Ice Age. Ironically, some of the best evidence of glacial conditions at that time is found at the edges of the present-day Sahara desert! Another 150 million years on, for an extended period around 290 million years BP, there were extensive and repeated glaciations in the Southern Hemisphere (the Permo-Carboniferous Ice Age). Finally, in the last few million years, there has again been extensive high-latitude glaciation, initially of the South Polar region, and subsequently also of the North (the Late Cenozoic or Quaternary Ice Age).

Evidence of climate change from the British Isles starts with glacial deposits dating from the Late Precambrian Ice Age (approximately 610 million years BP), when the relevant parts of Britain were located at about 45°S (Figure 4.1[6]). Although globally there was an Ordovician Ice Age at *c*. 440 million years BP, coral reefs then grew in the shallow seas in and around the British Isles. During the next Ice Age – the Permo-Carboniferous, centred around 290 million years BP – because Britain was then situated on the Equator, extensive tropical rainforest growth contributed to massive coal accumulations. The global sea-level changes, associated with repeated glacial and interglacial episodes in the Southern Hemisphere at that time, are clearly reflected in the rhythmic pattern of coal seam development, as the rainforest peats sank and became entombed in thick piles of sediment. A period of increasing aridity followed as the British crustal rocks came to form part of the interior of a great North American-Eurasian continent. This continent was located under the '**Horse latitudes**', beneath which the main deserts of the world existed then as they do at the present day. By that time the Permo-Carboniferous Ice Age had come to an end. Hot desert conditions continued in the British Isles into

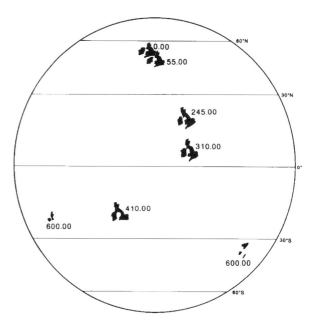

Figure 4.1 Global track of the British Isles terrain during the last 600 million years. Dates shown are in million years BP. Compiled from various sources.[6]

the early Mesozoic era, until 210 million years BP, but by that time a globally warm period had set in.

During the Mesozoic (245 to 65 million years BP) sea-level rose world-wide and the total global area of land above sea-level was significantly reduced (see Figure 4.2).[7] During this era of global warmth dinosaurs spread to become dominant throughout the world.

THE APPROACH TO THE PRESENT ICE AGE: 65 TO 2.6 MILLION YEARS BP

Whether or not the dinosaurs finally became extinct as the result of an asteroid impact, the global climate was already beginning to cool again by that time (65 million years BP). For some time afterwards, however, the British Isles continued to enjoy a relatively warm climate. When at about 45°N (which was about 55 million years BP; see Figure 4.1), palm

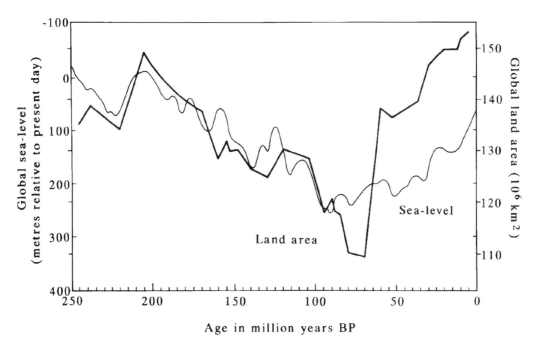

Figure 4.2 Global sea-level (thin line) and land areas (bold line) during the last 250 million years. In general, global temperatures were proportional to sea-level and inversely proportional to land area during the entire period. Adapted from Smith *et al.* 1994, Figure 2 (see p. 82 Note 6).

trees fringed the British Isles and tropical sea shells lived in the surrounding shallow seas. The association of subtropical plants and marine life suggest that average summer maximum temperatures reached about 25°C at that time, about 6°C warmer than at present.

Cooler conditions developed progressively from 30 to 5 million years BP as polar ice-sheets grew and global sea-level fell (see Figure 4.2). Very little evidence for this, however, is preserved in the British Isles. We know that glaciation of the Antarctic land mass had already begun by the Middle Tertiary period and that substantial growth of the Antarctic Ice-sheet, to near its present size, had been achieved by 15 million years BP. We also know that around 5 million years ago the Antarctic ice-sheet was probably even more extensive than it is at the present day. It was not until 2.52 million years BP, however, that the Northern Hemisphere ice-sheets first spread

southwards to the sort of limits that characterised their cyclical extension during the ensuing Late Cenozoic or Quaternary Ice Age.

Global sea-level was generally higher prior to the growth of these Northern Hemisphere ice-sheets and some marine deposits from that period are still preserved on land in the British Isles. These deposits are principally located in the south-east of England adjoining the southern North Sea. The earliest deposits of this type, preserving a wealth of marine fossils (see Figure 4.3), comprise the 'Coralline Crag' of Suffolk. In spite of their name these sediments contain very few true corals. Those that are present are occasional solitary corals, much as one might find off the coast of Cornwall at the present day. The overall composition of the marine fauna suggests, however, a climate of Mediterranean warmth with summer temperatures around 21°C and winter temperatures of 15°C.

Figure 4.3 Examples of Coralline Crag mollusc shells.

Resting on top of the 'Coralline Crag' in some places, and cut as cliffs into it in others, is another later deposit called the 'Red Crag'. By combining faunal and palaeomagnetic correlations (Box 4.1), we can deduce that these 'Red Crag' deposits were accumulated just before, during, and probably slightly after the first major expansion of glaciers into the North Atlantic area around 2.52 million years BP.[8] Interestingly, the earliest 'Red Crag' deposits indicate a climate that was not much cooler than the 'Coralline Crag', but the later faunas contain some distinctive species that had migrated into the North Sea from a North Pacific source, presumably via the Arctic seaway. During the course of the accumulation of the 'Red Crag', the proportion of Mediterranean, or **Lusitanian**, warmer water species decreases and the proportion of **Boreal** or even Arctic species increases. This indicates a climate cooling to conditions very similar to, or even cooler than, those

of the present day. The extremes of cooling may not always have been registered in the British Isles because of the sea-level reductions which accompanied them (Figure 4.4[9]). In the Netherlands, however, in the record of tree pollen preserved in the subsiding **graben**, which accumulated fluviatile sediments deposited by the Rivers Rhine and Meuse,[10] there is ample evidence of cyclical climate changes and notable overall cooling during this period.

THE PRESENT ICE AGE: 2.6 MILLION YEARS BP ONWARDS

As the present (Late Cenozoic or Quaternary) Ice Age came into being progressively, it is difficult to decide precisely which time or event should be chosen to define its start. Although the Antarctic had supported an extensive ice-sheet since at least 15 million

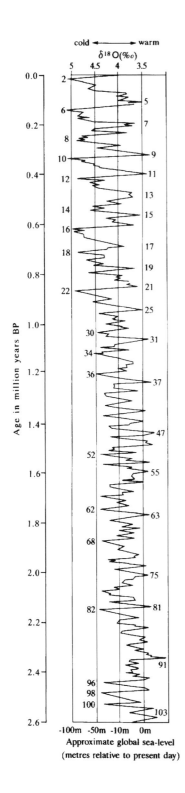

cold ◄————► warm

δ¹⁸O(‰)

Approximate global sea-level
(metres relative to present day)

years BP, the development of major Northern Hemisphere ice-sheets did not occur until after 2.6 million years BP. Strictly speaking this point falls within the later part of the Tertiary period – although some European geologists regard it as an event of such significance that they define the onset of the Quaternary period by it. The base of the Quaternary period is internationally defined, however, at a juncture dated to 1.8 million years BP, ostensibly at the first arrival of cool-water faunas in the Mediterranean Sea. Therefore, although it is probably more accurate to describe the present Ice Age as the Late Cenozoic Ice Age, it is commonly referred to as the Quaternary Ice Age. In practice, only the second half of the Quaternary (from about 900,000 years BP onwards) exhibits the strong alternating pattern of Glacials and Interglacials that is typified by the Last Glacial–Interglacial cycle.

The global record

The sedimentary record of Glacial and Interglacial periods, both in and around the British Isles, is generally very fragmentary and it is necessary to look to the global record for an overall chronology. Since the 1970s, that chronology has been provided by the δ¹⁸O record found in **benthic foraminifera** from deep-sea sediments. As global ice volume increases, ^{16}O is preferentially incorporated into ice-sheets and the δ¹⁸O of benthic foraminifera increases. Because the mixing time within the oceans is relatively short, the same pattern of benthic δ¹⁸O change can be found in all ocean basins. Figure 4.4 shows the record from Ocean Drilling Program Site 677, located in the Panama Basin of the eastern equatorial Pacific Ocean, back to the Gauss/Matuyama palaeomagnetic reversal (see Box 4.1) at 2.6 million years BP.

Odd-numbered stages starting from 1, corresponding to the present-day Interglacial, represent warm

Figure 4.4 Global sea-level and ice volume changes over the last 2.6 million years. In general, global temperatures during this period were proportional to sea-level and inversely proportional to the global ice volume implied by benthic δ¹⁸O. Adapted from Funnell 1995, Figure 1.[7]

or Interglacial periods. Even-numbered stages represent cold or Glacial periods. Because the creation of continental ice-sheets incorporates vast quantities of water, which is derived ultimately from the oceans, global sea-level is substantially lowered during Glacial (even-numbered) stages. Therefore the $\delta^{18}O$ stages also reflect global sea-level changes as well as climate changes.

It is impossible to extrapolate the absolute magnitude of either the climate or sea-level changes relative to the benthic $\delta^{18}O$ values back to 2.6 million years BP with total confidence. Nevertheless, a cyclical alternation of higher and lower global temperatures and sea-levels is an inescapable conclusion. What is more, the frequencies in the oscillations can be matched with those in the Earth's orbital cycles (see Box 4.2), and those correlations can be used to date the cycles and stages (Box 4.1).

A brief review of the global record shown in Figure 4.4 indicates that, initially, both sea-level and temperature (the inverse of global ice volume) were generally higher than those which characterise the present day. Then three cold stages – 100, 98 and 96 – are seen of which the first (now dated to 2.524 million years BP[11]) is known to represent the first major development of Northern Hemisphere ice-sheets.[12] These stages correspond to a period of major deforestation in north-west Europe and repeat at c. 41,000 year intervals. For the following 1.4 million years, spanning the Tertiary/Quaternary boundary at 1.8 million years BP and continuing up to about 0.9 million years BP, climate and sea-level oscillate between limits that are higher than those of the present day and no lower than stage 100. This oscillation occurs at a continuing frequency of c. 41,000 years, the frequency of the **obliquity cycle** of the Earth's orbit.

From about 1.2 million years BP onwards a c. 100,000-year cycle becomes more evident and from $\delta^{18}O$ stage 22, at about 900,000 years BP, the 100,000-year frequency becomes altogether dominant, with both sea-level and temperature falling during most cold stages to levels equivalent to those of the Last Glacial period. It is this 100,000-year cyclical pattern that is regarded as so characteristic of the Quaternary Ice Age and has been successfully simulated by ice-sheet modelling using the orbital cycles of **insolation** (see Box 4.2 and Chapter 16).

Causes of the Late Cenozoic Glacial–Interglacial Cycles

As already noted, the rhythm of the well-known major Quaternary Ice Age glaciations, when major continental glaciers developed in North America and Eurasia (including the British Isles) as well as persisting in Greenland, appears to have begun about 900,000 years ago. Since then, glaciation has peaked every 100,000 years or so. Interglacials – that is to say periods of maximum warmth such as the present day – have likewise occurred every 100,000 years or so. These interglacials, however, have each only lasted about 10,000 years. This periodicity is likely to be controlled by the interaction between the cycles of ellipticity (95,000 years), obliquity (41,000 years) and precession of the equinoxes (23,000 and 19,000 years) in the Earth's orbit (see Box 4.2), and the effect these combined variations have on the geographical and seasonal distribution of incoming radiation to the Earth's surface. These variations in received insolation are not, however, in themselves sufficient to induce the full range of temperature conditions that have been experienced. Some amplifying processes are therefore likely to be involved. Changes in atmospheric carbon dioxide levels may well be one of these processes. In addition, the relatively long response time and complexity of total ice-sheet changes may well introduce significant lags and amplifications in the longer-term climate response to insolation changes.

In the earlier part of the Quaternary and Late Cenozoic Ice Age – before 900,000 years BP when an obliquity modulated (41,000-year) rhythm is more evident in the glacial–interglacial sequence – climate change may have been more closely linked with changes in the Antarctic rather than Northern Hemisphere ice-sheets. Evidence from diverse sources – ocean and land-based sediments, pollen cores, ice-cores and coral reefs – all lend support to the primary role of orbital forcing of the Quaternary glacial–interglacial cycles.

BOX 4.2 THE EARTH'S ORBITAL CYCLES AND CLIMATE

Cyclical variations in the orbit of the Earth around the Sun cause cyclical changes in the seasonal and latitudinal distribution of incoming solar radiation (insolation) at the surface of the Earth. There are three main variables in the Earth's orbital geometry.

- Obliquity or tilt: the angle between the Earth's rotational axis and the plane of the ellipse in which it travels around the Sun varies over a 41,000-year cycle. The effect on insolation is greatest at high latitudes and is equal in both hemispheres.
- Precession of the equinoxes: the gravitational pull of the Sun and Moon on the Earth's equatorial bulge causes the Earth's axis of rotation to 'wobble' around its mean position. As a result, the shortest distance between the Earth and Sun progresses around the seasons on 19,000- and 23,000-year cycles. The effect on insolation is greatest at low latitudes and opposite in the Northern and Southern Hemispheres.
- Eccentricity: The ellipse of the Earth's orbit varies from near circular to more elliptical over a 95,000-year cycle. This modulates, alternately reducing and accentuating, the effects of the precessional cycle.

Orbital cycles, and the resultant changes in the seasonal and latitudinal distribution of insolation, can be calculated both backwards and forwards in time.[a] James Croll,[b] and later Milutin Milankovitch,[c] suggested that the Late Cenozoic glacial–interglacial cycles had been caused by these cyclical changes. Milankovitch argued that

glaciation was likely to have occurred when summer radiation was low (reducing snow and ice melt) and winter insolation was high (providing more evaporation and moisture for snowfall). A positive feedback mechanism to amplify the climate response to the insolation changes was also considered necessary.

Strong support for Milankovitch's hypothesis was obtained in the 1970s from spectral analysis of the periodicities found in deep-sea core benthic $\delta^{18}O$, which is a proxy of global ice-sheet volumes,[d] and subsequent direct comparisons of these inferred ice volume changes with later and more accurate calculations of the orbital and insolation cyclicity.[e]

[a] A. Berger, 'Long-term variations of caloric insolation resulting from the Earth's orbital elements', *Quaternary Research*, 1978, vol. 9, pp. 139–67.
[b] J. Croll, 'On the excentricity of the Earth's orbit, and its physical relations to the glacial epoch', *Philosophical Magazine*, 1867, vol. 33, pp. 119–31; on the change in the obliquity of the ecliptic, its influence on the climate of the polar regions and on the level of the sea, ibid., pp. 426–45.
[c] M. Milankovitch, 'Kanon der Erdbestrahlung und seine Anwendung auf Eiszeitenproblem', *Royal Serbian Academy, Special Publication*, 1941, vol. 133, pp. 1–633.
[d] J. Imbrie, J.D. Hays, D.G. Martinson, A. McIntyre, A.C. Mix, J.J. Morley, N.G. Pisias, W.L. Prell and N.J. Shackleton, 'The orbital theory of Pleistocene climate: support from a revised chronology of the marine ^{18}O record', in A. Berger, J. Imbrie, J. Hays, G. Kukla and B. Saltzman (eds), *Milankovitch and Climate*, Dordrecht, D. Reidel, 1984, pp. 269–306.
[e] N.J. Shackleton, A. Berger and W.R. Peltier, 'An alternative astronomical calibration of the lower Pleistocene timescale based on ODP Site 677', *Transactions of the Royal Society of Edinburgh, Earth Sciences*, 1990, vol. 81, pp. 252–61.

GLACIAL CYCLES IN THE BRITISH ISLES

Although the marine sequences in the British Isles, and marine and freshwater sequences in the Netherlands, both record responses to the more extreme

oscillations seen in the global record after 2.6 million years BP, no evidence of either glacial or **periglacial** conditions is found in those sequences. By about 1.9 million years BP, however, the British Isles were experiencing very cold conditions during the stronger

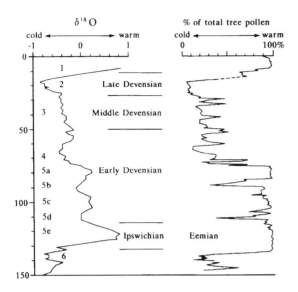

δ^{18}O

cold ◄——————► warm
-1   0   1

% of total tree pollen

cold ◄——————► warm
0   100%

1

2   Late Devensian

3   Middle Devensian

4
5a   Early Devensian
5b
5c
5d
5e   Ipswichian   Eemian
6

Figure 4.5 The Last Interglacial–Glacial cycle as revealed from deep-sea (North Atlantic) and land-pollen (France) records. Adapted from Bowen 1992, Figure 3.[14]

obliquity cycles. Marine faunas indicated climates more akin to those of Spitsbergen and periglacial conditions were occurring in the Netherlands, but there was still no evidence for a regional ice-sheet. The accompanying flora, as indicated by pollen, was bereft of trees and the terrestrial landscape converted to open steppe-like conditions.

In the British Isles, evidence for major climate fluctuations in the early Quaternary (1.8 to 0.78 million years BP) appears only to be present in the river terrace deposits of the Thames. These are very difficult to date in absolute terms. Any primary evidence for early glacial advances during this time has been destroyed by the devastation caused by the Anglian and the Last (Devensian) Glaciations.

Shortly after 0.6 million years BP, interglacial (Cromerian) peat deposits containing elephant remains on the coast of East Anglia are directly overlain by ground moraine deposits of the largest glaciation to affect southern Britain during the Quaternary. At the culmination of this Anglian Glaciation a continental glacier covered most of the British Isles and extended as far south as the outskirts of the

future site of London, displacing the Thames permanently southwards into a more southerly course. This Anglian Glacial may relate to δ^{18}O stage 12, at 420,000 years BP, but independent dating has so far proved elusive. Hoxnian Interglacial deposits post-dating the Anglian glacial deposits yield uranium series dates of 360,000 years BP. Although the Cromerian and Hoxnian Interglacials and the Anglian Glacial clearly form part of the sequence of major climate cycles that characterise the latter part of the Quaternary period, uncertainties about their dating still precludes their unequivocal placing in relation to the sequence of global cycles. In fact, there remains little if any primary evidence of many of the glacials that might be expected to have affected the British Isles over the last ten global glacial–interglacial cycles.

THE LAST INTERGLACIAL–GLACIAL CYCLE: 130,000 TO 11,000 YEARS BP

Evidence for earlier glacial–interglacial cycles in the British Isles is relatively scarce, so we turn our attention to the climates of the Last Interglacial and the Last Glacial for which inferences of age (see Box 4.1 and Table 4.2) and climate conditions are the most reliable. It should be remembered, however, that the cycles previous to this were also probably characterised by very similar conditions.

Recent research has produced some very significant advances in our understanding of the development of the Last Interglacial–Glacial cycle, particularly in relation to the role of ice-sheets[13] and deep-sea sediments in recording the detailed sequence of climate change. Figure 4.5[14] compares the δ^{18}O stages inferred from the deep-sea record with a pollen record obtained from south-east France. It shows a close correspondence between the global temperature (and sea-level), which can be inferred from the deep-sea record, and the development of forest in western Europe during the last 150,000 years. No such continuous record is available from the British Isles, but where British data are available they closely echo the global and regional record.

(a)

Figure 4.6 (a) *Hippopotamus amphibius* and (b) its distribution in the British Isles (shaded area) during the Last Interglacial.

The Last (Ipswichian) Interglacial: 130,000 to 117,000 years BP

The Last Interglacial, known in the British Isles as the **Ipswichian** and in Europe as the Eemian, peaked at about 124,000 years BP. It is chiefly remarkable for the occurrence of hippopotamus and other tropical or subtropical fauna in Britain. Remains of *Hippopotamus amphibius* (Figure 4.6)[15] are found as far north as Darlington and abundant skeletons have been recovered from river gravels near Cambridge and in London. These, together with remains of the European Pond Tortoise, *Emys orbicularis,* testify that temperatures during the Ipswichian were significantly higher – summer temperatures at least 2°C higher – than the present day. These higher temperatures were probably a global feature of the Last Interglacial, with more melting of ice-sheets than

(b)

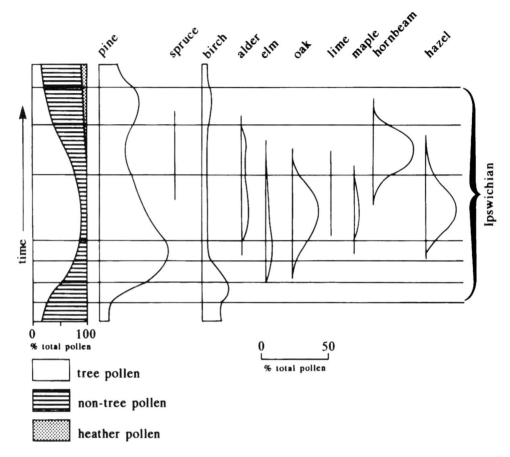

Figure 4.7 Changes in East Anglian flora during the last (Ipswichian) Interglacial. Adapted from West 1980.[16]

has occurred during the present Interglacial, because sea-levels appear to have been somewhat higher during the Ipswichian than at the present time.

Pollen from deposits in Britain laid down during the Last Interglacial display a characteristic interglacial sequence commencing with a low proportion of tree pollen, typical of an unforested landscape (Figure 4.7).[16] Trees arrive progressively and dominate the forest flora in succession. At first birch and pine are most important, then warmth-loving species such as elm, oak, maple and hazel become more important; finally, towards the end of the Ipswichian, hornbeam becomes very abundant. The overlap of hornbeam with the other **thermophilous** species coincides with

the occurrences of hippopotamus in the deposits and also indicates summer temperatures 1° to 2°C higher than the present day.[17] The end of the Ipswichian Interglacial is marked by a decline in the thermophilous forest trees, the entry of open heath non-tree species and the overall decline of tree pollen as colder conditions approached.

The Last (Devensian) Glacial: 117,000 to 11,000 years BP

At the end of the Last (Ipswichian) Interglacial the climate of the British Isles cooled sharply, leading to the deforestation seen in the pollen record at

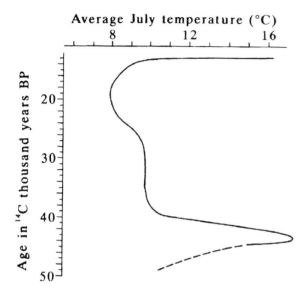

Average July temperature (°C)

Figure 4.8 Last Glacial (Middle and Late Devensian) July temperature record inferred from fossil beetle remains. Adapted from Coope 1977.[18]

the Ipswichian/Devensian transition (Figure 4.7). Although it is not yet possible to date this transition directly in the British record, it is widely accepted to correspond to the boundary between $\delta^{18}O$ sub-stages 5e and 5d of the oceanic record so that an age of about 117,000 years BP can be assigned (see Figure 4.5).

There are no continuous records of the ensuing glaciation in the British Isles. Once more it is necessary to use continuous records of climatic change from elsewhere (Table 4.2) to provide a template for the British record. Warm 'interstadial' floras and faunas are found preserved in river deposits at Chelford, in Cheshire, and at Upton Warren in Herefordshire. At Chelford (dated to *c.* 100,000–90,000 years BP by **thermoluminescence** and therefore probably equivalent to $\delta^{18}O$ stage 5c), average July temperatures of 15°C have been inferred from beetle faunas. At Upton Warren, (*c.* 43,000 ^{14}C years BP, equivalent to $\delta^{18}O$ stage 3), analysis of beetle faunas indicates that average July temperatures at that

time rose as high as 17.5°C, compared with values of much less than 10°C during most of the later Devensian (Figure 4.8)[18] and about 16.5°C at the present day.

Ice-sheet growth seems to have commenced in earnest from 74,000 to 59,000 years BP. After the partial amelioration of conditions in the Middle Devensian (59,000–24,000 years BP), which includes the Upton Warren Interstadial, ice-sheets then extended during the Late Devensian (24,000–11,000 years BP), obliterating most of the evidence of Early Devensian ice-sheet growth. The precise limits attained by the Late Devensian ice-sheet is disputed, especially where it extended into the area of the present North Sea, but an early reconstruction (Figure 4.9)[19] still provides a good indication of the likely extent and elevation of the ice-sheet at its maximum extent.

An alternative estimate of climate variability in the British isles during the Last Interglacial–Glacial cycle has been obtained from a study of the main growth periods shown by **speleothems** – stalagmites and stalactites.[20] Growth of speleothems may be totally inhibited under glacial or periglacial conditions for both hydrogeological and biological reasons, but usually proceeds normally under less cold and interglacial conditions. Uranium series dating of British Isles speleothems indicates that most growth has taken place over the intervals indicated in Table 4.2. These include the Ipswichian Interglacial – at 124,000 years BP – and later sub-stages of the $\delta^{18}O$ stage 5, including that corresponding to the Chelford Interstadial. They also clearly characterise the warmer $\delta^{18}O$ stage 3, including the Upton Warren Interstadial. Similar temporal variations in speleothem growth have been observed over a broader area in north-west Europe.

For the Middle and Late Devensian, corresponding to $\delta^{18}O$ stages 3 and 2 of the deep-sea record, increasing interest has recently been focused on evidence for periodic mass discharges of icebergs into the Atlantic Ocean. These originated mainly from the **Laurentian** (North American) ice-sheet and strongly influenced the palaeoceanography of the North Atlantic.[21] Initially, studies concentrated on

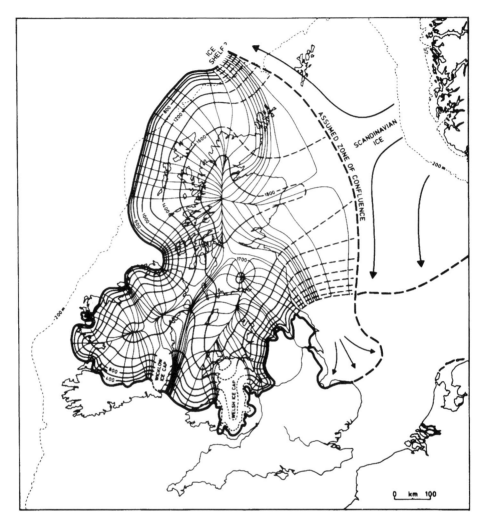

Figure 4.9 Modelled British Isles ice-sheet at the Last Glacial Maximum showing surface topography (ice depth in metres) and flow lines.[19]

so-called **Heinrich events** in deep-sea sediments and their relation to events in the ice-core records from the Greenland ice-sheet. Progressively more attention has been given to finer scale variation in ice-rafted materials in ocean sediments. Both the ice-rafted detritus and the ice-core records indicate continuous and sometimes rapid millennial scale variation in climate that is superimposed on the Glacial–Interglacial cycle.

The Heinrich events themselves are generally regarded as being correlated with episodes of colder climate, the penultimate event (H2, see Table 4.2) leading up to the **Last Glacial Maximum**. Other Heinrich events, however, seem to be very close in their timing to speleothem growth events (Table 4.2) and the last Heinrich event (H1) occurs during the onset of deglaciation. Clearly, bearing in mind the short wavelength of the climate cycles that they

Table 4.2 Selected events in the Last Interglacial–Glacial cycle, related to the global $\delta^{18}O$ stages. (All Uranium series and SPECMAP dates are essentially calendar dates, within the limits of error. Heinrich events ^{14}C dates are given in square brackets, except where calendar equivalents have been suggested. SPECMAP dates are based on stacked $\delta^{18}O$ records correlated with calculated insolation cycles. Colder $\delta^{18}O$ stage numbers are in **bold**.)

$\delta^{18}O$ stage	Speleothem events (Ur series ages; thousand years BP)[a,b]	Heinrich events [^{14}C based ages; thousand years BP][c,d]	SPECMAP ages (thousand years BP)[e,f]
1			12 (12.0)
2	(27–22)	H1 [14.3] 16.5 H2 [21.0] 23.0	**24 (24.1)**
3	D1 29 (31–28) D2 36 (42–35) D3 45 (46–44) D4 50 (56–49) D5 57 (62–56)	H3 [27.0] 29.0 H4 [35.5] 37 H5 [c. 51]	59 (59.0)
4	(71–63)	H6 [c. 70]	**71 (73.9)**
5a	D6 76 (78–72) (81–79)		(84.1)
5b	**I1 90.5 (98–87)**		(93.1)
5c	I2 105 (115–98)		(105.6)
5d			(117.3)
5e	I1 124 (133–115)		128 (129.8)
6			**186 (189.6)**

Sources:
[a] D. Gordon, P.L. Smart, D.C. Ford, J.N. Andrews, T.C. Atkinson, P.J. Rowe and N.S.J. Christopher, 'Dating of late Pleistocene interglacial and interstadial periods in the United Kingdom from speleothem growth frequency', *Quaternary Research*, 1989, vol. 31, pp. 14–26.
[b] A. Baker, P.L. Smart and D.C. Ford, 'Northwest European palaeoclimate as indicated by growth frequency variations of secondary calcite deposits', *Palaeogeography, Palaeoclimatology, Palaeoecology*, 1993, vol. 100, pp. 291–301.
[c] G. Bond, H. Heinrich, W. Broecker, L. Labeyrie, J. McManus, J. Andrews, S. Huon, R. Jantschik, S. Clasen, C. Simet, K. Tedesco, M. Klas, G. Bonani and S. Ivy, 'Evidence for massive discharges of icebergs into the North Atlantic ocean during the last glacial period', *Nature*, 1992, vol. 360, pp. 245–9; G. Bond, W. Broecker, S. Johnsen, J. McManus, L. Labeyrie, J. Jouzel and G. Bonani, 'Correlations between climate records from North Atlantic Sediments and Greenland ice', *Nature*, 1993, vol. 365, pp. 143–7; G.C. Bond and R. Lotti, 'Iceberg discharges into the North Atlantic on millenial time scales during the Last Glaciation', *Science*, 1995, vol. 267, pp. 1005–10.
[d] W.S. Broecker, 'Massive iceberg discharges as triggers for global climate change', *Nature*, 1994, vol. 372, pp. 421–4.
[e] J. Imbrie, J.D. Hays, D.G. Martinson, A. McIntyre, A.C. Mix, J.J. Morley, N.G. Pisias, W.L. Prell and N.J. Shackleton, 'The orbital theory of Pleistocene climate: support from a revised chronology of the marine ^{18}O record', in A. Berger, J. Imbrie, J. Hays, G. Kukla and B. Saltzman (eds), *Milankovitch and Climate*, Dordrecht, D. Reidel, 1984, pp. 269–306.
[f] D.G. Martinson, N.G. Pisias, J.D. Hays, J. Imbrie, T.C. Moore and N.J. Shackleton, 'Age dating and the orbital theory of the ice ages: development of a high-resolution 0 to 300,000 year chronostratigraphy', *Quaternary Research*, 1987, vol. 27, pp. 1–29.

record, a high degree of precision in dating across different environments is essential to resolve exactly how different climate processes have interacted in the past.

The Last Glacial Maximum: 21,000 years BP

At the maximum of the Last Glacial, during the Late Devensian, the ice-sheet limits in the British Isles were approximately as shown in Figures 4.9 and 4.10. Global sea-level was almost 150 m lower than at the present day, so a wide area of the continental shelf around the British Isles, including most of the North and Irish Seas, was above sea-level.

Beyond the limits of the ice-sheet, periglacial tundra-like conditions prevailed everywhere on land. Periglacial areas are characterised by **permafrost** – permanently frozen ground where only a superficial surface layer may thaw out during the summer season.[22] Three main types of permafrost feature persist after periglaciation: patterned ground, pingos and ice wedges. Patterned ground is produced by segregation of stones into polygons or stripes within the summer-melted zone. Patterned ground develops when average annual air temperature lies between zero and –2°C. Pingos – hills produced by the sub-surface concentrations of ice often driven by hydro-geological conditions – lead on deperiglaciation to circular lakes with ramparts. They form when average annual air temperature lies between –1° and –5°C. Thirdly, when the average annual air temper-ature lies below –6° to –8°C, thermal contraction of land-surface sediments leads to the development of ice wedges. The wedges, which may be arranged in polygonal configuration, become infilled with sand or other material after the ice eventually melts. Patterned ground, pingos and ice wedge casts occur extensively over southern England (see Figure 4.10)[23] beyond the limits of the Devensian ice-sheet. Average annual air temperatures declining to –8°C would be consistent with average July temperatures of approximately 8°C inferred from fossil beetle assem-blages for the time of the Last Glacial Maximum (see Figure 4.8).

Onset of deglaciation: 18,000 to 16,000 years BP

The deep-sea benthic $\delta^{18}O$ record of the Late Quaternary (Figure 4.4) is characterised by a 'saw-tooth' profile. There is a very sharp fall in the propor-tion of ^{18}O in the oceans at the end of each glaciation shortly after they had attained their maximum extent. This is taken to imply the rapid return of ^{16}O to the oceans as a result of rapid melting of the high-latitude ice-sheets. These events are known as 'Terminations'. Termination 1a (18,000 to 16,000 years BP) comprises the initial deglaciation following the Last Glacial Maximum.

Evidence of British Isles climate during Termin-ation 1a is almost completely lacking. In the deep-sea record, however, there is evidence of the seasonal disappearance of the Norwegian/Greenland sea-ice from 20,000 years BP onwards and the submarine Barbados coral reefs record a rapid rise in sea-level from 17,500 to 16,500 years BP as a result of ice-sheet melting. In the Lake District, which is esti-mated to have been covered by *c.* 1,500 m of ice at the Last Glacial Maximum (Figure 4.9), the first evidence of seasonal snow-melt from *valley* glaciers is known from just before 17,500 years BP. This is followed by evidence of snow-bed and fell-field plant communities from 17,000–16,500 years BP.[24] In the south of the British Isles, pollen- and fauna-bearing deposits capable of robust climate interpretation are restricted to the Late and Post Glacial periods from 16,000 years BP onwards (see Chapter 5).

CONCLUSIONS

Although our knowledge of the past climate of the British Isles is diverse and often fascinating, it is also extremely fragmentary. To construct a coherent inter-pretation of climate change in the British Isles over the last million, or even the last 100,000 years, it is essential to make comparisons with the ocean record and terrestrial records outside Britain which reflect continuous global and regional climate changes. Only parts of the global pattern can be detected in the British record. Significant, global

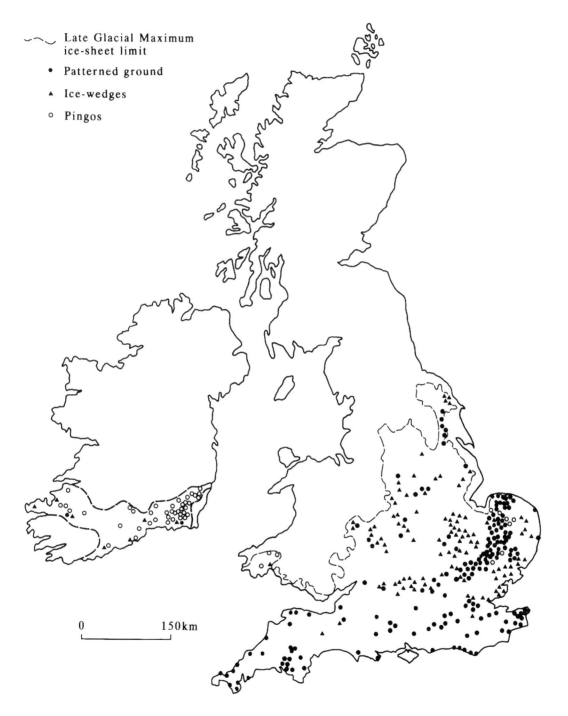

Figure 4.10 Periglacial (permafrost) features beyond the limits of the British Isles ice-sheet at the Last Glacial Maximum. Adapted from Jones and Keen 1993, Figure 9.18.[23]

glacial periods have no certain equivalents in British glacial or periglacial sequences and only a few of the interglacial sequences are known in relation to the considerable number of interglacials indicated by the oceanic record. Is this simply a reflection of the gross inability of terrestrial sedimentation to provide a continuous record of terrestrial climate change, or has the regional sedimentary response to climate change been radically different in successive global climatic cycles?

To answer this question we need to develop and apply further methods for the objective dating of existing sedimentary records and better methods for interpreting the climate signal they contain. We also need to sample new, so far unexplored, sources of sediments in the British Isles.

NOTES

1 S. D'Hondt and M.A. Arthur, 'Late Cretaceous oceans and the cool tropic paradox', *Science*, 1996, vol. 271, pp. 1838–41.
2 B. Molfino, N.G. Kipp and J.J. Morley, 'Comparison of foraminiferal, Coccolithophorid and radiolarian paleotemperature equations: assemblage coherency and estimate concordancy', *Quaternary Research*, 1982, vol. 17, pp. 279–313.
3 M.H. Field, B. Huntley and H. Müller, 'Eemian climate fluctuations observed in a European pollen record', *Nature*, 1994, vol. 371, pp. 779–83.
4 G. Eglinton, S.A. Bradshaw, A. Rosell, M. Sarnthein, U. Pflaumann and R. Tiedemann, 'Molecular record of sea surface temperature changes on 100-year time scales for glacial terminations I, II, and IV', *Nature*, 1992, vol. 356, pp. 423–6.
5 In general the term 'Ice Age' is applied by geologists to extended periods of high-latitude ice sheet development. Each 'Ice Age' exhibits many cycles of ice sheet expansion (called glacials or glaciations) alternating with periods of warming and ice sheet contraction (called interglacials). The term 'Little Ice Age' has, somewhat unfortunately, also been applied to a historic period of cooler climate and mountain glacier expansion, which was an altogether insignificant event compared with the main geological 'Ice Ages'.
6 Cambridge Paleomap Services Ltd, *TimeTrek*, CPS, Cambridge, 1994, version 1.0.2; A.G. Smith, A.M. Hurley and J.C. Briden, *Phanerozoic Palaeocontinental World Maps*, Cambridge, Cambridge University Press,

1981, 102 pp.; A.G. Smith, D.G. Smith and B.M. Funnell, *Atlas of Mesozoic and Cenozoic Coastlines*, Cambridge, Cambridge University Press, 1994, pp. 24–54.
7 A.G. Smith, D.G. Smith and B.M. Funnell, op. cit.; B.U. Haq, J. Hardenbol and P.R. Vail, 'Mesozoic and Cenozoic chronostratigraphy and eustatic cycles', in C.K. Wilgus, B.S. Hastings, H. Posamentier, J.V. Wagoner, C.A. Ross and C.G. St. C. Kendall (eds), *Sea level changes: an integrated approach*, Society of Economic Paleontologists and Mineralogists, Special Publication 42, Tulsa, Okla., 1988, pp. 71–108; B.M. Funnell, 'Global sea-level and the (pen-)insularity of late Cenozoic Britain', in R.C. Preece (ed.), *Island Britain: a Quaternary perspective*, Geological Society Special Publication No. 96, London, 1995, pp. 3–13.
8 B.M. Funnell, op. cit.
9 Ibid.
10 B.M. Funnell, 'Plio-Pleistocene Palaeogeography of the southern North Sea basin (3.75 to 0.55 Ma)', *Quaternary Science Reviews*, 1996, vol. 15, pp. 391–405.
11 A.C. Mix, N.G. Pisias, W. Rugh, J. Wilson, A. Morey and T.K. Hagelberg, 'Benthic foraminifera stable isotope record from Site 849 (0–5 Ma): Local and global climate changes', in N.G. Pisias, L.A. Mayer, T.R. Palmer-Julson and T.H. van Andel (eds), *Proceedings of the Ocean Drilling Program, Scientific Results*, Ocean Drilling Program, College Station, Tex., 1995, vol. 138, pp. 371–412.
12 E.L. Sykes, L.D. Keigwin and W.B. Curry, 'Pliocene paleoceanography: circulation and oceanographic changes associated with the 2.4 Ma glacial event', *Paleoceanography*, 1991, vol. 6, pp. 245–57.
13 D.W. Dansgaard, S.J. Johnsen, H.B. Clausen, D. Dahl-Jensen, N.S. Gunderstrup, C.U. Hammer, C.S. Hvidberg, J.P. Steffensen, A.E. Sveinbjornsdottir, J. Jouzel and G. Bond, 'Evidence for general instability of past climate from a 250 kyr ice-core record', *Nature*, 1993, vol. 364, pp. 218–20.
14 D.Q. Bowen, 'The Pleistocene of North West Europe', *Science Progress*, 1992, vol. 76, pp. 209–23.
15 A.J. Sutcliffe, 'Insularity of the British Isles 250,000–30,000 years ago: the mammalian, including human, evidence', in R.C. Preece (ed.), *Island Britain: a Quaternary perspective*, Geological Society Special Publication No. 96, London, 1995, pp. 127–40.
16 R.G. West, 'Pleistocene forest history in East Anglia', *New Phytologist*, 1980, vol. 85, pp. 571–622.
17 W.H. Zagwijn, 'Vegetation and climate during warmer intervals in the Late Pleistocene of Western and Central Europe', *Quaternary International*, 1989, vols 3–4, pp. 57–67.
18 G.R. Coope, 'Fossil coleopteran assemblages as sensitive indicators of climatic changes during the Devensian

(Last) cold stage', *Philosophical Transactions of the Royal Society of London*, 1977, vol. B280, pp. 313–40.

19 G.S. Boulton, A.S. Jones, K.M. Clayton and M.J. Kenning, 'A British ice-sheet model and patterns of glacial erosion and deposition in Britain', in F.W. Shotton (ed.), *British Quaternary Studies: Recent Advances*, Oxford, Oxford University Press, 1977, p. 234.

20 D. Gordon, P.L. Smart, D.C. Ford, J.N. Andrews, T.C. Atkinson, P.J. Rowe and N.S.J. Christopher, 'Dating of late Pleistocene interglacial and interstadial periods in the United Kingdom from speleothem growth frequency', *Quaternary Research*, 1989, vol. 31, pp. 14–26; A. Baker, P.L. Smart and D.C. Ford, 'North-west European palaeoclimate as indicated by growth frequency variations of secondary calcite deposits', *Palaeogeography, Palaeoclimatology, Palaeoecology*, 1993, vol. 100, pp. 291–301.

21 A. Baker *et al.*, op. cit.; G. Bond, H. Heinrich, W. Broecker, L. Labeyrie, J. McManus, J. Andrews, S. Huon, R. Jantschik, S. Clasen, C. Simet, K. Tedesco, M. Klas, G. Bonani and S. Ivy, 'Evidence for massive discharges of icebergs into the North Atlantic ocean during the last glacial period', *Nature*, 1992, vol. 360, pp. 245–9; G. Bond, W. Broecker, S. Johnsen, J. McManus, L. Labeyrie, J. Jouzel and G. Bonani, 'Correlations between climate records from North Atlantic sediments and Greenland ice', *Nature*, 1993, vol. 365, pp. 143–7; G.C. Bond and R. Lotti, 'Iceberg discharges into the North Atlantic on millennial time scales during the Last Glaciation', *Science*, 1995,

vol. 267, pp. 1005–10.

22 C.K. Ballantyne and C. Harris, *The Periglaciation of Great Britain*, Cambridge, Cambridge University Press, 1994, 330 pp.; R.G.B. Williams, 'The British climate during the Last Glaciation: an interpretation based on periglacial phenomena', in A.E. Wright and F. Moseley (eds), *Ice Ages: Ancient and Modern*, Liverpool, Seel House Press, 1975, pp. 95–117.

23 R.L. Jones and D.H. Keen, *Pleistocene Environments in the British Isles*, London, Chapman and Hall, 1993, 346 pp.

24 W. Pennington, 'The Late Devensian flora and vegetation of Britain', *Philosophical Transactions of the Royal Society of London, Series B*, 1977, vol. 280, pp. 247–71.

GENERAL READING

M. Bell and M.J.C. Walker, *Late Quaternary Environmental Change: Physical and Human Perspectives*, Harlow, Longman Scientific and Technical, 1992, 273 pp.

T.J. Crowley and G.R. North, *Paleoclimatology*, Oxford, Oxford University Press, 1991, 339 pp.

J. Imbrie and K.P. Imbrie, *Ice Ages: Solving the Mystery*, London, Macmillan, 1979, 224 pp.

R.L. Jones and D.H. Keen, *Pleistocene Environments in the British Isles*, London, Chapman and Hall, 1993, 346 pp.

A.E. Wright and F. Moseley (eds), *Ice Ages: Ancient and Modern*, Liverpool, Seel House Press, 1975, 320 pp.

5

RECONSTRUCTING LATE-GLACIAL AND HOLOCENE CLIMATES

Keith Briffa and Tim Atkinson

Hear the voice of the Bard!
Who present, past and future sees;
Whose ears have heard
The Holy Word
That walked among the ancient trees.
William Blake, *Songs of Experience*

INTRODUCTION

This chapter is concerned with the climates of the British Isles from the **Last Glacial Maximum** to the end of the prehistoric period, which ended some 2,000 years ago. All evidence of past climates during our period of interest comes from geological accumulations or sedimentary records and the biological remains that are contained within them. Examples include ice-cores, **speleothems**, marine, river and lacustrine sediments, soils and glacial **moraines**. Biological evidence includes the remains of animals or plants, such as beetle parts, pollen, plant macrofossils and tree megafossils (Table 5.1).

Before attempting to describe what we know of the climates of the British Isles during the last 20,000 years, it is important to note a number of problems that arise when attempting to interpret or synthesise the data that derive from these so-called 'proxy climate' sources. This is pertinent because, while we are entirely dependent on them, each of these sources of information has its particular limitations that frustrate our attempts to present a clear

and full picture of past climates and climate change: limitations in the accuracy with which the evidence can be dated, imprecision in the interpretation of the lags between forcing and response, and even ambiguity in defining the precise nature of the climate variables whose influence is recorded in these remains.

We are used to thinking of climate in terms of the statistical distribution of a specific meteorological variable, representing a measured quantity over some clearly defined period – for example, monthly mean temperature or total annual precipitation. Biological, and even physical, proxy data can rarely be interpreted solely in terms of a single variable such as temperature or precipitation. Realistically, they must be seen as the product of a mixture of thermal and moisture conditions that prevailed over some flexible period – for example during the growing season of a tree or integrated across the accumulation and ablation seasons of a glacier. Often the evidence may show the cumulative influence of conditions that prevailed during years or even decades prior to the formation of the tree-ring or the

Table 5.1 The major sources of palaeoenvironmental and palaeoclimatic data for the Late Glacial and Holocene periods

Palaeoclimate proxy	Primary character of the environment indicated	Aspects of palaeoclimate inferred
Physical evidence: Extent of *former* glaciers		
– Corrie glaciers	Altitude of permanent snowline	Integration of summer ablation (temperature and cloudiness) and winter accumulation (precipitation); directions of snow-bearing winds
– Larger glaciers and ice-sheets	Extent and shape of ice-sheet, flow direction of ice	General indicator of regional climate
– Protalus ramparts	Former snowfields at altitude of former snowline	Confirms evidence of former corrie glaciers
Information within *present* ice-sheets/ice caps		
– Ice-cores	Chemical and isotopic composition of ice and occluded air (greenhouse gas composition of former atmospheres); dust content; accumulation rates	Complex combination of temperature, source region of water vapour; atmospheric circulation patterns; snowfall
– Bore hole temperatures	Profile of ice temperature with depth	General history of mean annual temperature at the surface
Periglacial soil and ground ice structures		
– Involutions	Depth of thaw layer	Degree-days of thaw
– Ice wedge casts		Mean annual temperature
– Open system pingos		Mean annual temperature
Speleothems	Isotopic composition of calcite and included groundwater; luminescence of annual lamination	Cave temperature, isotopic composition of palaeo-precipitation; complex signal related to annual variations in precipitation
Biological evidence: Insect fossils (identified to species level)		
– Beetles	Species range	Monthly mean temperature (with large uncertainty)
– Chironomids	Lake temperature and food supply	Summer temperatures
Mollusc fossils		
– Land snails	Soil type and vegetation	Wetness, dryness
– Aquatic snails and bivalves (fresh or marine)	Water temperature, water chemistry	General warmth/cold

Table 5.1 continued

Palaeoclimate proxy	Primary character of the environment indicated	Aspects of palaeoclimate inferred
Diatoms (in freshwater lake sediments)	Water chemistry, lake productivity	General warmth/cold
Foraminifera (marine sediments)	Water temperature, salinity and food/mineral supply	Seasonal sea-surface and lower ocean temperatures
Vertebrate fossils – Cold blooded – Warm blooded	Species ranges Biotope habitat	Seasonal temperatures Crude indicator of general climate
Plant megafossils (tree remains)	Altitudinal and latitudinal tree lines	Mixture of snowlie, exposure and summer and winter temperatures
Plant macrofossils (identified to species level) – Certain terrestrial plants (some species)	Former plant distributions	summer temperature thresholds (winter for a few species), moisture availability
– Mosses (in peat bogs)	Surface wetness of bog	Water balance in summer
– Aquatic plants (some species)	Character of vegetation	General warmth/cold, summer temperature thresholds
Pollens/spores (identified to genus/family level)	Character of vegetation	General warmth/cold, wetness/dryness
Tree-rings	Tree growth rate, ring density, chemical composition	Variable mixture of temperature and moisture availability

movement of the glacier terminus. Many sources may provide evidence that is fragmentary or discontinuous. Sometimes the 'effect', seen as some change in the proxy data record, may lag behind the 'forcing' climate change by many decades or even centuries. Examples of such delayed responses are changing tree lines or shifts in vegetation zones. Indeed, on time-scales of centuries to millennia, the lack of an equivalent long instrumental-based climate yardstick makes it impossible to compare and calibrate proxy evidence to provide rigorous quantitative estimates of past climates and their uncertainties.

Some data, such as from tree-rings, may be continuous, absolutely dated and of annual or even specific seasonal resolution. Statistical problems associated with assembling long composite chronologies (i.e., those made up of many overlapping series from living and dead trees) can mean, however, that these data may only inform us about climate variability occurring on time-scales up to several decades or centuries. These data provide potentially unique insights into the nature of interannual climate variability and very rapid shifts or abrupt events, but they have limited potential to inform us about the long-term differences in climate, such as occur between one millennium and the next.

Clearly, producing a complete picture of changing climate conditions over many thousands of years is a difficult task, even for an area as small as the British Isles. The diverse evidence of many disciplines must be brought together, interpreted and reconciled. The problem can be likened to the restoration of an old cine film that has been broken into pieces, many of which are lost. Only parts of the surviving individual frames are discernible and we are often not even sure of the strict order in which to place them.

In the following pages we describe selected parts of the 'film', illustrating the changing climate of the British Isles since the last major glaciation, to the extent that it is possible to piece it together at present.

THE END OF THE LAST GREAT ICE AGE: THE LATE-GLACIAL PERIOD

The evidence of beetles

Perhaps our clearest picture of seasonal temperature changes in the British Isles during the transition from glacial to interglacial conditions has been deduced from the presence of assemblages of particular beetle species,[1] identified as fossil remains dated at a number of discrete periods during the last 22,000 radiocarbon years before present (i.e., 22 Ka BP; see Box 5.1).

Like many plants and other animals, particular species of beetle thrive only within certain climate limits. Although they may appear to have complex or diverse geographical distributions, it has been discovered that most European species' ranges, when mapped according to two simple temperature axes (the mean temperature of the warmest month which we will call T_{max}, and the range between the warmest and coldest months in the year, T_{range}), almost invariably represent coherent and clearly definable thermal ranges. Accurate information on the present-day geographical ranges of beetle species can be compared with modern meteorological data to build up a library of two-dimensional climate range 'envelopes'. Wherever two or more beetles are known to co-exist, we may assume that the thermal climate must lie within the area of overlap between their individual climate range envelopes. This is the rationale for the 'mutual climate range' (MCR)

BOX 5.1 RADIOCARBON AND ABSOLUTE TIME-SCALES

Part of the carbon in the atmosphere exists in the form of the radioactive isotope ^{14}C. This is produced when cosmic energy entering the Earth's atmosphere collides with atoms and releases free neutrons, some of which then collide with nitrogen atoms substituting one of the nitrogen protons (which is released as hydrogen) to form ^{14}C. This radioactive carbon is then oxidised to form $^{14}CO_2$ and is rapidly mixed in the atmosphere with the other non-radioactive carbon dioxide and so enters the various carbon reservoirs. In time, the ^{14}C decays back to nitrogen through the release of a beta particle and a neutrino. Over many thousands of years (prior to atomic bomb tests), there has been a general balance between the amount of ^{14}C that decays and the renewed production in the atmosphere.

At any one time, all living things will have the same proportion of ^{14}C in their tissues, either because plants acquire it in photosynthesis or because animals eat plants or other animals that have already assimilated it. For as long as they are alive, the ^{14}C is constantly replenished in their living tissue. As soon as they die, however, this renewal ceases and the ^{14}C then decays at a known exponential rate. By measuring the amount of radioactivity remaining in a sample of old organic matter it is possible to estimate the time that has elapsed since its death. There are many problems and assumptions associated with radiocarbon dating, but they are beyond the scope of our discussion here.

The one major assumption of the technique is that the amount of radiocarbon in the atmosphere has remained constant. Now we now know that there have been short time-scale (annual–decadal) and longer (century–millennial) variations in the

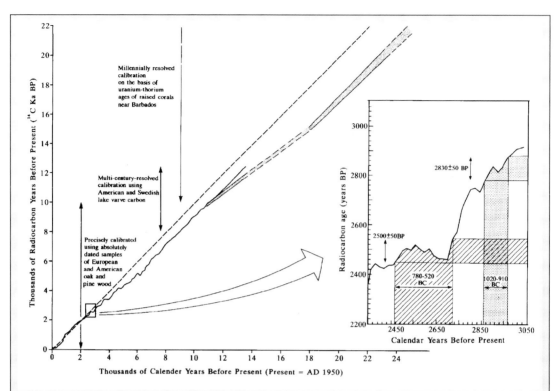

amount of ^{14}C in the atmosphere. Again a detailed discussion is not warranted here other than to say that they relate to changes in solar activity, the Earth's carbon cycle and particularly long-term changes in the Earth's magnetic field. By comparing the amount of radiocarbon in samples of known date, such as in tree-rings and (with less precision) lake varves[a] and corals,[b] the biases between the radiocarbon and actual dates can, however, be calculated and allowance made for these when estimating calendar dates based on the 'calibrated' radiocarbon time-scale.

The figure shows how radiocarbon dates clearly become progressively too young as they get older, and 'plateaux' on the calibration curve can result in an apparent bunching of ^{14}C dates in some periods. For example, an apparent ^{14}C date of 10,000 years BP (the time of the **Younger Dryas/Holocene** boundary – see text) corresponds to a 'real' date of about 11,000 years ago,

while a ^{14}C date of 20,000 years BP implies a date 23,000 years ago. The inset on the figure shows how the small-scale 'wiggles' in the relatively high-precision ^{14}C calibration curve result in variable dating precision. A ^{14}C date of 2,830 ± 50 years BP corresponds to a calendar uncertainty of about 1020–910 BC (2,970–2,860 actual years before present, i.e., 1950), whereas a proximal ^{14}C date with similar fifty-year counting uncertainty, i.e., 2,500 ± 50 years BP, produces a larger calendar range of 780–520 BC (2,730–2,470 actual years before present).

Unless otherwise stated, all the dates in this chapter are radiocarbon dates.

[a] M. Stuiver, A. Long and R.S. Kra (eds), 'Calibration 1993', *Radiocarbon*, 1993, vol. 35, p. 244.
[b] E. Bard, B. Hamelin, G. Fairbanks and A. Zindler, 'Calibration of the ^{14}C timescale over the past 30,000 years using mass spectrometric U-Th ages from Barbados corals', *Nature*, 1990, vol. 345, pp. 404–10.

method of climate reconstruction which has been used to produce Figure 5.1 on p. 90.

In a study[2] of beetle faunas identified from previous analyses of twenty-six mostly late-glacial and Holocene sites in the British Isles, it has been possible to derive MCR estimates of T_{max} and T_{range}. From these, the mean temperature of the coldest month, T_{min}, and the annual mean temperature, T_{annual}, has been deduced for more than fifty discrete time points during the period 15,000 to 8,000 years BP, as well as for another eight time 'snapshots' at about 21,500, 19,500, 18,400, 4,800, 3,000, 2,100 and 400 years BP (Figure 5.1). These data provide detailed quantitative evidence about large, and sometimes very abrupt, temperature changes that occurred in the British Isles as the **Devensian** ice-sheet retreated and the more stable, warmer climates of the Holocene period finally became established. They also provide strong indications that significant shorter time-scale fluctuations in temperature were superimposed on these broader changes.

The beetle data indicate that between 22,000 years BP and 20,000 years BP, mean annual temperatures were in the range 0°C to −15°C, most likely near −12°C. The mean temperature of the warmest month was between 5°C and 14°C, probably near 7°C, and the coldest between −10°C and −35°C, likely near −30°C. The ranges we quote here are those indicated by the full extent of the MCR for the particular beetle assemblage that we know existed at that period. The mean climates were somewhere within these ranges. We can also go some way, however, to estimating more precisely where, within these ranges, the mean climate lay. These 'most likely' values have been derived from regression equations estimating modern observed temperatures from the medians of the temperature ranges indicated by modern beetles living at the same localities. These equations allow us to estimate the real temperatures with a precision of about ±2°C for T_{max} and ±5°C for T_{min}. The full estimated temperature ranges and the 'most probable' values are shown separately for summer and winter in Figure 5.1 and the changing annual cycle of temperatures in the form of estimated **degree days** of frost and thaw is shown in Figure

5.2. For convenience, in the following discussion we will refer to the probable mean temperatures.

There are few data between about 18,000 and 15,000 years BP. Evidence from a site in Gwynedd, North Wales, indicates that very cold conditions (similar to 22,000 years BP) existed at around 14,500 years BP and then again later at 13,000 years BP. Data from two sites in Yorkshire and Cambridgeshire suggest, however, that summers were about 3°C warmer than 22,000 years BP (i.e., a warmest month of about 10°C) and winters up to 15°C warmer (about −15°C) between 20,000 and 18,000 years BP, just at the time when the ice-sheet was probably at its maximum extent.

Between 13,500 and 10,000 years BP, temperatures rose from those characteristic of the last ice age to those generally characteristic of the Holocene: a summer change of roughly 9°C (from about 8.5°C to 17.5°C), a winter change of about 20°C (−20°C to 0°C) and hence a near 50 per cent reduction in continentality (i.e., the summer–winter temperature range declined from about 30°C to 17°C). This glacial/interglacial transition was, however, far from smooth. On the contrary, the beetle data show that it was marked by extreme variability on millennial and even perhaps on century time-scales.

It is not wise to put too much faith in the fine details of the temperature reconstructions shown in Figure 5.1. We do not know what calendar uncertainty might be attached to each radiocarbon date as we have no precise absolute time-scale with which to calibrate the radiocarbon time-scale this far back (cf. Box 5.1). Figure 5.1 also provides an illustration of the T_{annual} change, smoothed to take account of the uncertainty in the individual radiocarbon dates. This can be viewed as an objective, but conservative, representation of the course of temperature variability implied by the raw annual estimates. Many of the individual beetle assemblages are taken from a stratigraphic context, particularly at two sites – one in Glanllynnau in Gwynedd and one from St Bees in Cumbria.[3] This gives a strong indication that at least the *order*, if not the precise *timing*, of events superimposed on the major changes may indeed be real.

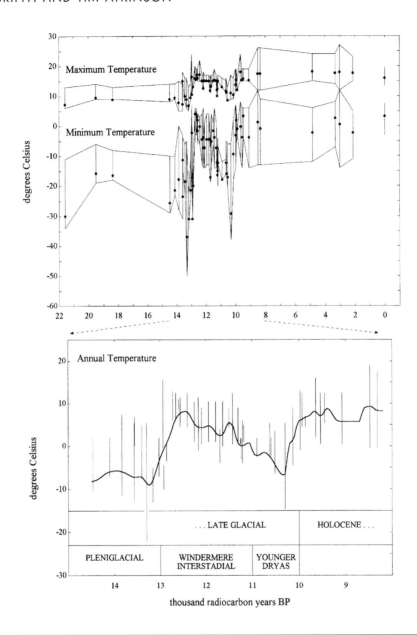

Figure 5.1 Temperature changes in the British Isles during the Late-Glacial and Holocene periods estimated on the basis of beetle remains (see p. 109 Note 1). The vertical bars show the ranges within which the mean temperature of the warmest month, the coldest month and the year lay. Note that these are indicated on a radiocarbon time-scale (see Box 5.1) and do not show the uncertainty associated with each date. The dots indicate the 'most probable' values within the temperature ranges. The bold line in the bottom box shows temporally smoothed estimates which may underestimate the rapidity and magnitude of some of the changes. The values plotted at 0 years BP (i.e., today) are the mean (dot) and maximum range (bar) of the Central England Temperature record over the period 1659–1995 (see Chapter 9).

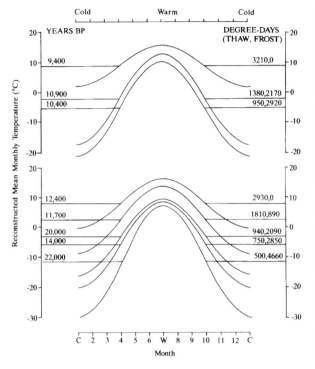

Figure 5.2 An illustration of the estimated annual cycle of monthly mean temperatures in the British Isles reconstructed on the basis of beetle assemblages identified at selected periods during the transition from glacial to Holocene conditions. Also given for comparison are the estimated degree days of thaw and frost (i.e., cumulative degree Celsius temperatures above and below zero respectively). The two sets of curves are plotted on the same scale and are separated only for reasons of clarity.

The smoothed annual curve indicates that temperatures rose abruptly between about 13,000 and 12,000 years BP, but the individual data points indicate a much faster climate amelioration. By 12,700 years BP, seasonal temperatures reached levels near, and in winter even slightly above, maximum values attained later during the Holocene (i.e., winters were at 4°C and summers near 16°C).

Over the following few thousand years (between 12,700 and 10,200 years BP), temperatures oscillated about a generally declining trend. This was associated primarily with an early cooling, followed by

a period of relative stability when temperatures remained above glacial, but below Holocene, levels (summers about 15°C and winters about −5°C). A more abrupt fall occurred at 11,300 years BP, and between 11,300 and 10,300 years BP, temperatures, especially winter temperatures, remained at almost glacial levels. This interval is known as the Younger Dryas (or Loch Lomond Stadial). Finally, after another apparently rapid warming between 10,300 and 10,000 years BP, temperatures attained their general Holocene levels.

The evidence of foraminifera

The gross pattern of temperature change revealed by beetle faunas between 20,000 and 10,000 years BP – from glacial to interglacial, followed by a brief return to a glacial Younger Dryas and a final establishment of a warm Holocene – is in good general agreement with the series of **sea-surface temperature** (SST) changes reconstructed from selected high-resolution North Atlantic sea-bed cores.[4]

Local abundances of different **planktonic foraminifera** in ocean waters are associated with varying water-mass conditions (such as temperature and salinity) so that fossil foraminiferal groups, identified in different core strata, provide information on past ocean temperatures.[5] The resolution and dating control is poor compared to the beetle record, but correlations between cores at many sites across the northern North Atlantic show that the timing and relative magnitude of the main post-glacial warming varied according to location. The first major warming occurred at about 13,000 years BP in the eastern North Atlantic, at 10,000 years BP in the west, and even later at 9,000–7,000 years BP in the north-west, near Labrador. The differences have been explained in terms of the migration of the **polar front** that marks the boundary of cold polar water. The glacial position of the polar front was in an east–west orientation at about 40°N. At around 13,000 years BP temperate waters penetrated the eastern North Atlantic, but not the west, so that the polar front swung northwards, like a gate with its hinge near Cape Cod opening to the north. Later, the front swung south again, occupying a position off the coast of Ireland or south-west England in the Younger Dryas (11,000–10,000 years BP). Finally, in the early Holocene the polar front migrated north, then north-east into the Norwegian Sea as first the Scandinavian and then the **Laurentide** ice-sheet wasted away. The rapidity of the temperature fluctuations shown by the beetles on land support this interpretation of the marine foraminiferal evidence. The magnitude of the winter temperature shifts on land are explicable in terms of extensive formation of sea-ice off the coast of the British Isles during the colder periods, as has also been proposed on the basis of the marine data.

It would be unwise to assume that the detailed succession of British temperature changes through the late-glacial are exactly representative of changes elsewhere, even of other areas in western Europe. Because of their position on the Atlantic seaboard, near the midpoint of the oscillations of the ocean circulation and polar front (see Chapter 2), the British Isles are probably exceptionally sensitive to these changes in regional climate. This point is rein-forced by the results of a comparative MCR study of late-glacial beetle faunas in the British Isles, southern Sweden, western Norway and central Poland.[6] This showed that different patterns of change occurred in each of these areas between 13,000 and 11,000 years BP, possibly associated with their specific proximity to the waning Scandinavian ice-sheet. Interestingly, however, all areas seemed to have experienced a synchronous Younger Dryas between 11,000 and 10,000 years BP.[7]

THE YOUNGER DRYAS IN THE BRITISH ISLES

The cause of the Younger Dryas episode is not fully understood. It coincides with a maximum in summer solar insolation at mid-latitudes (see Box 5.2), when ice-sheet melting should have been at its most rapid. Indeed, well-dated records from submerged tropical coral reefs show that sea-level was rising very rapidly just prior to the Younger Dryas, as the North American and Scandinavian ice-sheets decayed. The rate of ice-sheet decay (as measured by sea-level rise) slowed abruptly at 11,000 years BP and was depressed for around a thousand radiocarbon years, before resuming equally abruptly at around 10,000 years BP and reaching a peak in the early Holocene. This slowing of ice-sheet decay coincided with a marked deterioration in climate in the North Atlantic region, with fainter effects being observed elsewhere in the world. In the British Isles there was a regrowth of small glaciers and ice-sheets. These were the last glaciers to have existed on these islands, the whole country having been ice-free in the Windermere Interstadial (13,000–11,000 years BP – see Figure 5.1). This minor Younger Dryas glacia-tion was much less extensive, however, than that which occurred during the Last Glacial Maximum.

The Younger Dryas glaciers were mostly confined to corries and valley heads in mountain districts, except in the western Highlands where an ice-sheet developed. The moraine deposits left by these glaciers still appear very fresh (Figure 5.3) which has allowed the distribution of Younger Dryas ice to be mapped fairly definitively. Direct dating of the majority of features is poor, however, although in one or two cases the fresh moraines contain disturbed peat which can be radiocarbon-dated to the Windermere Interstadial, thus proving that the glacier that produced them was younger, i.e., of Younger Dryas age. Otherwise, dating these latest glacial features relies upon their fresh appearance and the fact that the oldest sedi-ments found in lakes and tarns within the recon-structed glacier limits are never older than the early Holocene, 9,500–10,000 years BP.

The Younger Dryas also saw a return of **peri-glacial** conditions to the British Isles. Numerous examples of structures formed by ground-ice have been mapped from all over the islands, including soil involutions, striped and polygonally patterned ground, ice wedge casts, frost cracks, and pingos.[8] They appear to have formed in two generations since the retreat of the main British ice-sheet after the Last Glacial Maximum, the first being during the period of cold climate prior to 13,000 years BP and the second during the Younger Dryas (see Figure 5.1).

BOX 5.2 INSOLATION CHANGES THROUGH THE LAST 25,000 YEARS

The regional and seasonal distribution of energy received from the sun at the top of the Earth's atmosphere has not remained constant during geological time. The interplay between the gravitational effects of the Sun, Moon and other planets continuously influence the shape of the path the Earth takes in its orbit around the Sun (eccentricity), change its axial tilt in relation to this orbital plane (obliquity), and cause it to wobble on its rotational axis, so that its orbital position at the times of the equinoxes and solstices also changes (precession of the equinoxes; see also Box 4.2, Chapter 4). These quasi-periodic effects are manifest on relatively long time-scales (on average about 95,800, 41,000 and 21,700 years for eccentricity, obliquity and precession respectively), but they have produced significant changes in the seasonal and regional distribution of global insolation during the last 25,000 years.[a]

In the early Holocene, about 11,000 years ago, perihelion (the time when the Earth is closest to

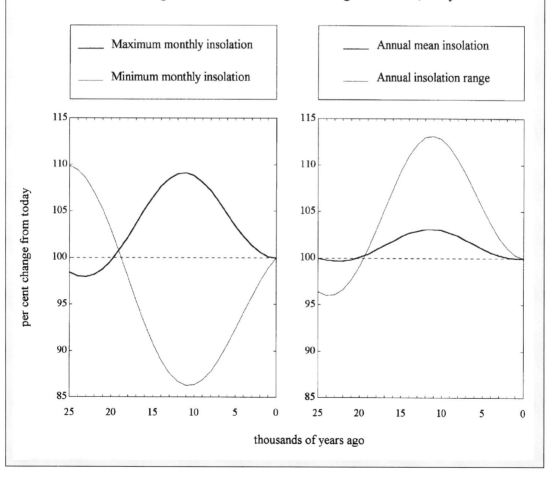

Insolation changes over the British Isles during the last 25,000 years

the sun) occurred in the Northern Hemisphere summer (instead of the Northern Hemisphere winter as it does today) and the Earth's obliquity was greater. The figure, constructed using software provided by André Berger,[b] shows that at the latitudes of the British Isles, maximum summer insolation was then higher by some 9 per cent, while the minimum winter value was lower by about 14 per cent, compared to today. Early Holocene maximum insolation range (between summer and winter) was 13 per cent higher, but the mean insolation over the whole year was only about 2.5 per cent greater than today. Since then, summer insolation has declined steadily and that in winter increased, so there has been a corresponding reduction in the range and in the annual mean.

We would not, however, expect to see a simple or direct correspondence between these orbitally related insolation changes and the evidence for surface temperature changes. The irradiance of the Sun is variable – only slightly so on short time-scales (up to a decade or so), but possibly more so on longer time-scales (centuries to millennia).[c]

The radiation reaching the Earth's surface is also modified by changing stratospheric and tropospheric conditions (e.g., in dust loading, cloudiness, and the concentration and distribution of radiatively active gases). Also, surface climates are the product of atmospheric circulation changes which may be largely random and which are subject to the varying forcing (and damping) effects of the oceans and large ice masses and influenced by complex feedback processes operating on different time-scales.

[a] A. Berger, 'Long term variations of daily insolation and Quaternary climate changes', *Journal of the Atmospheric Sciences*, 1978, vol. 35, pp. 2362–7.
[b] A. Berger, 'A simple algorithm to compute long term variations of daily or monthly insolation', Contribution No. 18, Institute of Astronomy and Geophysics, Université Catholique de Louvain, Louvain-La-Neuve, Belgium.
[c] J. Lean, 'Reconstructions of Past Solar Variability', in P.D. Jones, R.S. Bradley and J. Jouzel (eds), *Climatic Variations and Forcing Mechanisms of the Last 2000 Years*, NATO ASI Series I, 41, Berlin, Springer, 1996, pp. 519–32.

These features give clues as to the mean annual air temperature at the time of their formation. Fossil ice wedge casts indicate former continuous **permafrost**, and active ice wedges do not occur today where the mean annual air temperature is greater than −6°C to −8°C. Their occurrence in glacial sediment *within* the areas occupied by the last ice-sheet at Glacial Maximum implies that temperatures were as low as this after the ice-sheet had retreated, which is in excellent agreement with the mean annual air temperature estimates from beetle evidence shown in Figure 5.1 for the period prior to 13,000 years BP. Almost all British ice wedge casts that can be dated belong to this first late-glacial generation or earlier and the very few that might be of Younger Dryas age are small and imperfectly formed. This suggests that Younger Dryas temperatures may not have been

quite so low as in the full glacial, which agrees both with beetle estimates (Figure 5.1) and with the presence in central England and Wales of remains of a second type of periglacial feature, pingos.

Pingos are hillocks that develop when groundwater freezes to form a lens of ice which may be ten or more metres thick. The ice pushes up the soil above it until it slips off the crest of the growing hillock, exposing the ice beneath. This usually causes the ice-core to melt, leaving a crater-like hollow with a raised rim. These pingo scars occur in groups in a belt across England and Wales. Modern, active examples in Alaska and Spitsbergen occur only in areas where the mean annual air temperature is around −2°C to −5°C. Some East Anglian pingo scars have been radiocarbon dated to the Younger Dryas since older soil formed at around 11,000 years BP

Figure 5.3a Cwm Idwal, Snowdonia. From the vantage point of the Devil's Kitchen Llyn Idwal can be seen, and the terminal and sinuous lateral moraines deposited by a corrie glacier some 11,000 years ago during the Younger Dryas period.

has been incorporated into their structure. Most examples, however, are only partially dated (in the same way as many corrie glaciers are dated) by the fact that the oldest post-pingo sediments in the craters are invariably early Holocene. The presence of pingos in southern Britain in the Younger Dryas implies mean annual air temperatures between $-2°C$ and $-5°C$ which is once again in excellent agreement with beetle estimates (Figure 5.1).

The cold-climate geological features of the Younger Dryas give some information about wind directions and precipitation at the time. From careful mapping of former corrie glaciers it is possible to reconstruct their former equilibrium line altitudes (ELAs). The ELA is the altitude of a glacier surface at which average summer ablation and average winter snow accumulation are in balance. Wherever ELAs have been reconstructed for several former glaciers in the same district they show a marked rise from the south or south-west side to the north-east side of the mountain massif. This occurs despite the fact that ice ablation depends strongly on aspect – south-facing glaciers collect more solar radiation and thus experience more ablation. This pattern points very strongly to orographic **rain-shadow** effects similar to those which occur round the same mountains today. The rise in ELAs from south-west to north-east implies that snow-bearing winds came mainly from the south-west quadrant, as most precipitation does today. An exceptionally careful and detailed study of Younger Dryas glaciers in Skye[9] demonstrated that most snow was deposited by *southerly* winds, but strong westerlies were responsible for blowing fresh snow from one part of a small ice-cap to another. This difference between more southerly winds precipitating snow and westerlies drifting it, hints strongly at a climate in which winter precipitation was brought to the British Isles by Atlantic **depressions**, as it is today.

Regional patterns of precipitation during the Younger Dryas can also be guessed at on the basis of corrie glaciers' ELAs and other snow-line features. A regional map of ELAs over the whole Scottish Highlands (Figure 5.4 inset) shows a rise from 300 m above modern sea-level in the south-west to almost 900 m in the Cairngorms. This pattern mirrors the rain-shadows seen on individual massifs in northern England and Wales. The reality of the pattern has been confirmed by mapping the altitudes of protalus ramparts, which are debris accumulations that occur from permanent snowpatches at altitudes very close to the permanent snow-line. Figure 5.4 shows that the snow-line in St Kilda and the Outer Hebrides lay only a few tens of metres above modern

Figure 5.3b Cwm d'ur Arddu, Snowdonia. A small glacier occupied the hollow to the left, transporting and depositing the areas of debris to the right.

sea-level. Although there may have been a summer temperature (and ablation) gradient from west to east across Scotland, this is unlikely to have been more than one or two degrees at most, and certainly could not account for the differences in snow-line altitude. These must be due to a marked gradient in snow accumulation and thus of precipitation, in winter at least. The islands on the Atlantic seaboard must have experienced very high snowfall compared with eastern Scotland which was much drier, as it is today.

Amounts of precipitation are far harder to estimate than regional gradients. Nevertheless, some constraints may be placed on average annual precipitation by the presence of debris accumulations representing the remains of **rock glaciers** at several locations in the Scottish Highlands. A study of the modern examples in the Alps has concluded that they can form only within a specified range of combinations of mean annual air temperature and average annual precipitation. This means that if the mean annual air temperature can be *independently* estimated

(from beetle evidence) for a fossil example of a rock glacier, then constraints can be placed on precipitation. By this line of argument, a recent study very tentatively contrasts average annual precipitation on the Isle of Jura with that of the Cairngorms during times of rock glacier formation in the Younger Dryas.[10] For Jura, annual precipitation at 350 m above mean sea-level was unlikely to have been greater than 800–1250 mm, which is 50–80 per cent of the present value. By contrast, the Cairngorm precipitation was no more than 375–550 mm at 920 m altitude. This represents a much greater reduction compared to modern levels, at about only 20–30 per cent of present-day precipitation. Although extremely tentative, these values imply a climate that was not only colder but drier than present, with eastern Scotland being starved of precipitation compared to the west. This steeper east–west precipitation gradient compared with the modern pattern may have been due to the effect of the West Highland ice-sheet in 'capturing' snowfall

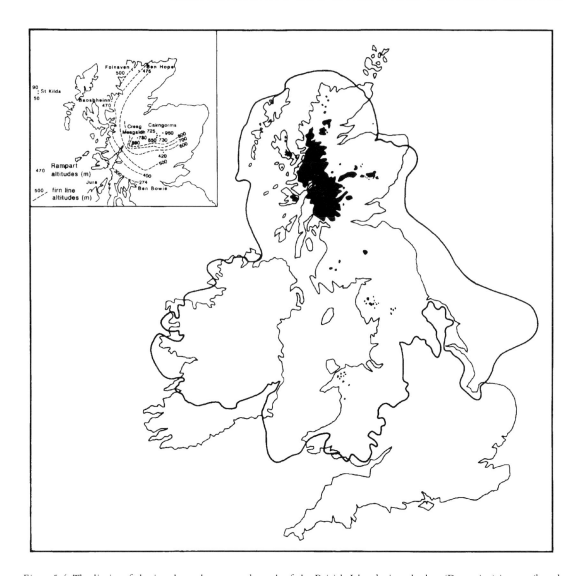

Figure 5.4 The limits of the ice-sheet that covered much of the British Isles during the last (Devensian) ice age (based on the map in D.Q. Bowen, J. Rose, A.M. McCabe and D.G. Sutherland, 'Correlation of Quaternary Glaciations in England, Ireland, Scotland and Wales', *Quaternary Science Reviews*, 1986, vol. 5, pp. 299–340). The dark-shaded regions indicate the maximum Younger Dryas ice extent. The inset shows the altitude of protalus ramparts in Scotland and the inferred equilibrium line altitudes (ELAs) of glaciers at their most advanced positions during the Younger Dryas (after C.K. Ballantyne and M.P. Kirkbride, 'The characteristics and significance of some late glacial protalus ramparts in upland Britain', *Earth Surface Processes and Landforms*, 1986, vol. 11, pp. 659–71).

from **air masses** arriving over Scotland from the south-west.

In summary, the Younger Dryas climate of the British Isles is known in probably more detail than for any other time in the Late Glacial. In England and Wales, mean annual temperatures in the lowlands were around –5°C with the warmest month average about 10°C and the coldest about –20°C. Temperatures at sea-level in Scotland were probably 2–5°C colder than this. The enhanced continentality and much colder winters compared with the present day were most likely due to the presence of continuous sea-ice in the Atlantic west of the islands. The main weather patterns bringing precipitation to the British Isles were Atlantic depressions arriving from the south-west quadrant. Although there is some indication of high precipitation amounts in the extreme western isles of Scotland, general precipitation levels in Scotland were probably less than today's, with a steeper east–west contrast across the Highlands. Very little is known about precipitation patterns in England, though the evidence of former corrie glacier ELAs in Wales suggests that here, too, the main influence was moisture-bearing south-westerly winds.

THE HOLOCENE PERIOD

Early evidence from bogs

There are not enough discrete and well-dated records of beetle assemblages spread throughout the last 10,000 years to enable a detailed British Holocene record to be constructed from beetles. The few data that do exist, viewed together with the foraminiferal data from the North Atlantic,[11] suggest that the magnitude of climate changes were certainly much smaller and probably considerably less rapid than the large, frenetic events of the late glacial period.

The Holocene should not, however, be thought of as a period of unchanging climate. As early as the

1870s, the Norwegian botanist, Axel Blytt, recognised various horizons in peat bogs which he took to be evidence of climate shifts.[12] This stratigraphy was later refined by R. Sernander and cross-referenced to pollen evidence in Swedish bogs.[13] During the first half of the twentieth century, as more pollen evidence from around Europe became available, the vegetation changes that they represented were frequently described in the context of what became known as the Blytt/Sernander scheme of climate subdivision for the late glacial and Holocene (Figure 5.5). This scheme recognises five general climate periods during the Holocene, often interpreted in warmth/wetness terms: the Pre-Boreal (warm and dry); the **Boreal** (warm and dry); the **Atlantic** (warm and wet); the Sub-Boreal (warm and dry) and the Sub-Atlantic (cool and wet). These general climate descriptions, and even their temporal boundaries, are not now considered widely representative of changing bog conditions.

The pollen evidence

Today, largely as a consequence of more refined (and hence more accurate) dating techniques, but perhaps also because 'common wisdom' is now less inclined to expect widespread synchroneity in climate changes, it is recognised that the Blytt/Sernander scheme does not satisfactorily represent the dating of detailed vegetational changes evident in the pollen and plant macrofossil data from many British and European sites. What is known, however, is that climate changes have occurred and are represented in a great many Holocene pollen profiles that have been investigated this century. This multitude of localised site records in the British Isles and throughout Europe has not led, however, to a correspondingly detailed picture of small-scale regional climate variations. To some extent, this is because the dating resolution of many pollen profiles (or parts of profiles) is equivocal. The main reason, however,

Figure 5.5 The 'traditional' view of changing climate and vegetation in northern Europe and the British Isles during the Late Glacial and Holocene Periods (taken from the summary given in Lamb[21]).

DATES		BRITISH ISLES				
		BLYTT-SERNANDER Climatic periods	POLLEN ZONES	Ireland	England and Wales	Forest cover	Archaeology

DATES		BLYTT-SERNANDER Climatic periods	POLLEN ZONES	Ireland	England and Wales	Forest cover	Archaeology
2000 AD — 0 — BC	POST-GLACIAL	SUB-ATLANTIC (Cool/Wet)	VIII	Alder Birch Oak	Alder Oak Elm Birch Beech	Clearing by Man	IRON AGE
2000		SUB-BOREAL (Warm/Dry)	VIIb	Elm decline			BRONZE AGE / NEOLITHIC
4000		ATLANTIC (Warm/Wet)	VIIa	Alder Oak Pine	Oak Elm Linden Alder	FOREST	
6000		BOREAL (Warm/Dry)	VI	Hazel Pine	Pine Hazel		MESOLITHIC
			V	Hazel Birch	Pine		
8000		PRE-BOREAL (Warm/Dry)	IV	Birch	Birch Pine		
	LATE GLACIAL	UPPER DRYAS	III		Tundra Local Birch	GRASS & SEDGE	
10000		ALLERØD	II	Birch	Birch		UPPER PALAEOLITHIC
		LOWER DRYAS	I	Dwarf Willow	Tundra Local Birch	OPEN VEGETATION	
12000		BØLLING					
	FULL GLACIAL	OLDEST DRYAS					
14000		ARCTIC					

is that it is difficult to interpret evidence for changing plant communities at a local scale in terms of a changing climate. The local vegetational changes, even the apparently larger shifts, do not solely reflect local climate forcing.

During the early part of the Holocene (some 11,000 years ago), plant communities in the British Isles, as in much of northern Europe, were still undergoing changes in response to the retreat of the ice-sheet. An ameliorating climate allowed plant taxa to migrate and recolonise areas as they moved north from the positions they had occupied during the glacial. Different taxa migrated at different rates. Some were replaced, not because the climate made them less viable, but because they were less able to compete with others. Hence, vegetation composition in the early Holocene, especially in north-western Europe, may not have been in equilibrium with climate. Even subsequently, local vegetation communities did not stabilise into the same assemblages in response to specific regional climates. Local site ecology (soil types, exposure, drainage, etc.), disturbance history (drainage changes, fire, winds) and to some extent pure serendipity of competitional interaction, produced local plant communities of subtly varying composition. The differences in contemporaneous communities, therefore, make it unwise to overinterpret in terms of climate some of the relatively small temporal differences in plant communities at single locations. Another significant problem, when attempting climatic interpretation of Holocene vegetation changes in Europe, is the confounding influence of humans. Indeed, many pollen records have been studied precisely in order to identify cultural history, such as the timing of forest clearance or adoption of agricultural practices.[14] These activities disguise, or even mimic, the effect of climate change on natural vegetation.

Notwithstanding these problems, important progress is being made in bringing together and synthesising the major features of larger, regional-scale vegetation patterns represented in many of the better-dated European pollen profiles. One recent study was based on pollen profiles from over four hundred sites spread across western Europe,[15] of which thirty are from the British Isles. Maps of pollen percentages were constructed for major plant taxa at four discrete periods: 9,000, 6,000, 3,000 and 500 years BP (i.e., radiocarbon years before present – cf. Box 5.1). These maps were used to infer large-scale vegetation patterns at these times and were interpreted in terms of winter and summer temperature and annual precipitation. This final step was achieved by relating geographical patterns of modern pollen percentages for each taxon to modern climate data. These relationships were then used to make quantitative estimates of each of these variables by finding the combination of values associated with the modern assemblage of plants that corresponded most closely with that reconstructed in the past.

The logic behind constructing such **climate response surfaces** for different subfossil taxa in this way is that the smoothing over such a large spatial scale removes very localised (non-climate) inconsistencies in the patterns and brings out the major features of regional climate forcing. In general, this work provides only very crude climate results – in the case cited above, an indication of the sign of temperature anomalies in January and July and effective moisture (precipitation minus evaporation). At 9,000 years BP, Fennoscandia is reconstructed as warmer than today. The British Isles are not detectably different in January, although it was apparently somewhat cooler in July. All of western Europe, other than the Mediterranean, is estimated to be drier at this time. The only detectable change over the British Isles at 6,000 and 3,000 years BP is a warmer July, although apparently less so during the more recent period.

Recently, an attempt has been made to compile a comprehensive compendium of pollen profiles from many European countries,[16] including some synthesis of the information contained in various so-called reference sites in England,[17] Scotland,[18] Ireland[19] and Wales.[20] Again, there is a potential confounding influence of human impact on the landscape, especially during the late Holocene. Figure 5.6 summarises the (qualitative) information about climate contained in this work, again for generally imprecisely defined periods.

There is some agreement amongst the evidence, suggesting generally warmer conditions in the British Isles consistent with the earlier concept of a Warm/Dry Sub-Boreal period (cf. Figure 5.5) centred on about 4,000 years BP. The Irish evidence for drier conditions (also consistent with the cessation of bog growth in southern England) is in conflict, however, with the evidence of a wet period in Scotland at this time. There is also clear agreement in the English and Scottish evidence for deteriorating (i.e., cooling and/or becoming wetter) conditions at, or after, 3,000 years BP. This is consistent with early ideas of a Sub-Boreal to Sub-Atlantic transition occurring about 2,500 years ago. It has been suggested that this deterioration is the most common feature of many north-west European bog profiles.[21]

The evidence of pine growth

We have described how pollen data generally offer limited potential for accurately representing changes in Holocene climates, since these changes are relatively small in comparison with the climate tolerances of many plant species, and especially when the vegetation may respond only slowly to these changes. It is possible, however, to focus on more selective evidence of changes in vegetation types that *are* potentially more sensitive.[22] One such example is the evidence of past tree-lines, of which Scots pine is a good example. Again, a precise climatic interpretation of this evidence is problematic,[23] but the altitudinal limits and intermittent survival of pine on marginal sites seems to reinforce some of the evidence for climate shifts that we have described above.

Scots Pine (*Pinus sylvestris*) today grows *naturally* only on poor soils in the Scottish Highlands. It is excluded from all other areas basically because of its inability to compete with other trees.[24] During the early to mid-Holocene, however, it was an important and widespread component of the British vegetation. Pine reached its maximum northward extent over the British Isles shortly before 4,000 years BP. Very soon afterwards, it retreated rapidly to its present range and apparently disappeared, except for small isolated

areas where stumps have been found preserved in boggy or alluvial situations. This picture seems to support the notion of a period of widespread relative summer warmth, lasting perhaps until nearly 4,000 years BP, but followed during the next few centuries by a notable climate deterioration. Other evidence of lake sediment chemistry or enhanced peat growth indicates a change to generally wetter and/or colder conditions shortly after 4,000 years BP.[25]

One of the most detailed studies of British tree-lines was carried out in the Cairngorms, Scotland, and involved the collection, radiocarbon dating, and analysis of the deuterium/hydrogen ratios of the cellulose in thirty-eight subfossil pine stumps.[26] This study emphasised how direct interpretation of the altitude of subfossils in terms of former tree-lines is complicated by the need to account for changing conditions (i.e., blanket bog formation) which may or may not be conducive to the preservation of the wood. Even without the preservation uncertainty, it is problematic to interpret the mere presence of pine trees as evidence for specific climate conditions. Nevertheless, this work clearly shows that the elevation of pine subfossils fell dramatically at about 3,500 years BP. Although coincident with population expansion in Scotland, this is interpreted by the authors as unlikely to be attributable to a human cause and most likely to reflect a change to cooler conditions, given the wider evidence of climate change. Later studies of subfossil pines on Rannoch Moor in western Scotland support this.[27] Except for a consistent absence of pine remains at about 5,000–4,800 years BP (suggesting generally colder/wetter conditions), the diversity in pine dates at different sites on Rannoch shows, however, how differences in very localised conditions preclude any simple interpretation of detailed changes on a regional scale. All that this points to is a general survival of pine throughout the Holocene until about 4,000 years BP.

At 6,000 years BP the subfossil pines on the Cairngorms are at relatively low elevation. This would appear not to accord with a general notion of relative warmth at this time. The isotopic ratios of the various wood samples can be interpreted, however, as evidence of relative moisture availability. The

¹⁴C yrs BP	Approx. date BC	N. England (uplands)	S. England (lowlands)	Scotland Cairngorms	Wales	Ireland
0	AD1950					
				D		
1000	AD 1020	Warmer	Warm			
2000	AD10	change to present		D		Cool
	AD/ BC		Cool			Warm
3000	1300- 1200	Cooler		D		
4000	2580- 2480	Warmer	Warm			Drier/ Warmer
5000	3900- 3700			D		
6000	4940- 4860		Warmish			
7000	5060- 4820	Wetter, floods	Coolish			
8000	7060- 6740		Warmish			Drier/ Warmer
9000	8070- 7930	Warmer and drier	Coolish			
10000	9500- 9300	Rapid Warming				

Human Impact/clearance   High treeline

Peat accumulation/bog growth   D= Deterioration

wettest phases were stated as occurring sometime up to 7,300, between 6,200 and 5,800, 4,200 and 3,940, and from 3,300 years BP onwards. The absence of higher-elevation stumps between 6,200 and 5,450 years BP may have been a consequence of water-logged ground. Certainly, blanket bog started to develop in this region at about this time. The pollen evidence discussed above, however, together with the existence of an apparent concentration of pine sub-fossils at Clashgour on Rannoch Moor and evidence of lower lake levels,[28] is at odds with the notion of a wet phase throughout Scotland at this time.

It is also interesting to note that, on the time-scale of the whole Holocene, the overall trend towards lower or retreating tree-lines in Fenno-scandia and the Alps[29] is crudely consistent with reducing summer insolation (cf. Box 5.2). These gradual changes are clearly interrupted between 4,000 and 3,000 years BP, however, by an apparent abrupt and widespread shift towards climates much less favourable for pine growth. Other evidence that is probably relevant, although admittedly somewhat distant to the British Isles, is the well-replicated pattern of northern tree-line change reconstructed in North America from about 6,000 years BP.[30] Based on pollen data from the extreme northern limit of tree growth, this record shows a relatively high-latitude tree-line (i.e., warmth) between 6,000 and 5,000 years BP, and a similarly high position around 4,000 years BP, followed by an initially abrupt and subsequently continuing retreat. The tree-line is at an extreme southern position (i.e., cold and/or wet) at about 2,000 years BP. These data demonstrate a surprising qualitative similarity with the limited British evidence, and also with other European evidence such as the gross changes in glacier fluctuations and lake-level fluctuations in Scandinavia and the Alps.[31]

This evidence is summarised in Figure 5.7. From this, and from Figure 5.6, it is clear that the apparent Holocene cooling that we have said is crudely consistent with the long-term reduction in summer solar radiation is not the whole story. It is also clear that millennial and century time-scale fluctuations, in both temperature and effective moisture conditions, have occurred and that these significant climate changes cannot be attributed to solar insolation alone. Widespread warming or warmth and dryness occurred for a few centuries before or around 4,000 years BP; there was a marked deterioration in warmth (and increasing wetness) between 4,000 and 3,000 years BP; and there is evidence for what might be a short period of very cool/wet conditions during the first few centuries prior to 2,000 years BP. Overall, this picture is fairly consistent with the early concepts of a warm dry Sub-Boreal and cooler, wetter Sub-Atlantic (cf. Figure 5.5) and especially the marked cool/wet transition between them at about 2,500 years BP.

High-frequency climate variability

If we wish to gain insight into environmental or climate change on time-scales less than a century, it is plain that we require evidence that is considerably better resolved than any we have discussed so far. In this final section, we make brief mention of probably the best-known and certainly the best-dated high-frequency palaeoclimate evidence – tree-ring data.[32] Where trees form annual growth rings, time-series of individual tree-ring width measurements can be combined to provide an average record of past growth variations at a site or in a specific region. Constructing such chronologies from samples taken from living trees allows the chronology to be firmly anchored in time (by virtue of the known sampling date). Given sufficiently long ring-width series, greatly extended chronologies can be constructed by matching the unique patterns of year-to-year growth variations in samples representing overlapping

Figure 5.6 A schematic summary of the palynological evidence for Holocene climate changes in England[17], Scotland[18], Wales[20] and Ireland.[19] The evidence is basically qualitative. There is some consensus in the evidence for generally warm/dry conditions around 4,000 years BP and a deterioration in climate between 3,000 and 2,000 years BP. There is only limited evidence for relative warmth around 6,000 years BP.

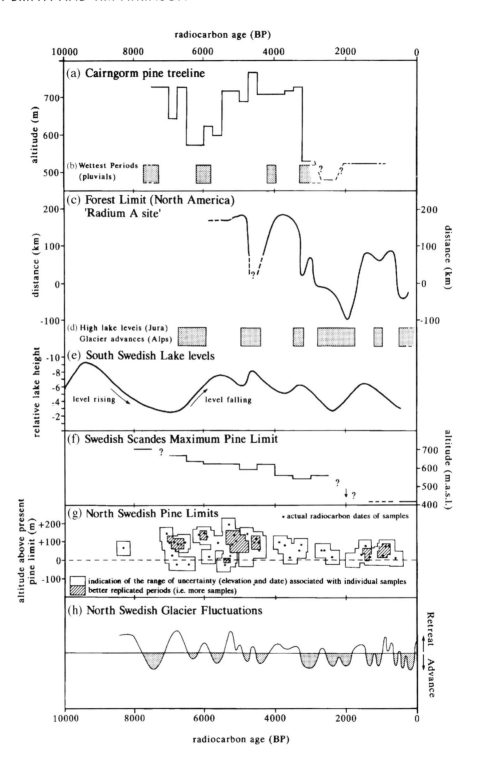

periods, perhaps from surviving dead trees or stumps, combined with measurements of wood samples from historical structures or artefacts and even those from archaeological or geological contexts (e.g., wood preserved in bogs, sediments or in lakes). A great attribute of such chronologies is that they represent a record of changing environmental influences that is both continuous and precisely dated to the year.

Tree-ring chronologies have been constructed for many tree species in a great many areas of the world and some stretch back over thousands of years. It was the painstaking assembly of two such continuous chronologies, made from samples of many generations of oak trees that grew in Northern Ireland and in Germany, that eventually enabled a high-precision European radiocarbon calibration curve to be established (cf. Box 5.1).

The specific nature of the environmental influences that are represented in any tree-ring chronology depends on a great many factors. Principal among these influences are the location and general ecology of the source region. These will dictate the basic climate and the nature of growth response to that climate by the trees. Other factors, however, such as tree-to-tree competition, external interference or damage (e.g., by animals or insects), or site disturbance (e.g., by fire) can all obscure the climate 'signal' contained in tree-ring patterns. By careful selection of sample region, by careful choice of material and by incorporating data from many trees, a robust chronology can be assembled that best represents the underlying pattern of common environmental forcing. The modern part of the chronology can be compared with modern climate data to establish the precise nature of climate influence and quantify how reliable similar inferences for earlier, older parts of the series would be.

There are many statistical and methodological aspects of building chronologies and interpreting the information they contain that are beyond the scope of this discussion,[33] but it is sufficient to say here that very long chronologies, stretching over much of the Holocene, exist in Northern Ireland (oak) and (almost continuously) in northern Sweden and Finland (pine). Each of these contains information on the combined effect of temperature and available moisture limitation on growth.

Figure 5.8 illustrates the variability in a provisional pine chronology being constructed from living trees and subfossil material preserved in Northern Swedish lakes,[34] and also shows a continuous chronology of oak growth mainly recovered from Irish bogs.[35] The annual values are shown to give an impression of the high-frequency variability at both locations, and both curves have also been smoothed and plotted together to highlight the multi-decadal variability. Note that, for the first time in this chapter, here we are dealing with an absolute calendar and the dates refer to actual years before Christ. The absence of very long time-scale variability in Figure 5.8 is a consequence of the chronology construction methods and should not be viewed as evidence for the absence of very long period climate change.[36]

The variability of the curves is partly influenced by changing sample replication. In the pine data sample numbers are low before 5000 BC and around 3000 and 2300 BC. Pine growth in some periods may also be suppressed by wet (and cool) conditions.

Figure 5.7 Selected evidence indicating changing climate conditions during the Holocene in the British Isles, northwest Europe and northern North America. Shown are: (a) the temporal changes in elevation of recovered pine subfossils in the Cairngorms, Scotland[26]; (b) periods of very wet conditions in the Cairngorms implied by deuterium/hydrogen ratios in the subfossil pine wood[26]; (c) changing positions of the northern tree line in North America based on palynological evidence at one representative site 'Radium A'[30]; (d) suggested periods of general glacier advance in the Alps and high lake levels in the Jura, France[31]; (e) representative lake-level changes in southern Sweden (G. Digerfeldt, 'Reconstruction and regional correlation of Holocene lake-level fluctuations in Lake Bysjön, Southern Sweden', *Boreas*, 1988, vol. 17, pp. 165–82); (f) altitudes of recovered pine subfossils in northern Sweden plotted as anomalies with respect to the local elevations today[29]; (g) the changes in the elevation of the highest subfossil pines dated to 500-year periods, in southern Lapland[29]; and (h) schematic summary of the advance and retreat of northern Swedish glaciers.[31] This figure is adapted from the original references cited and plotted on a common radiocarbon time-scale.

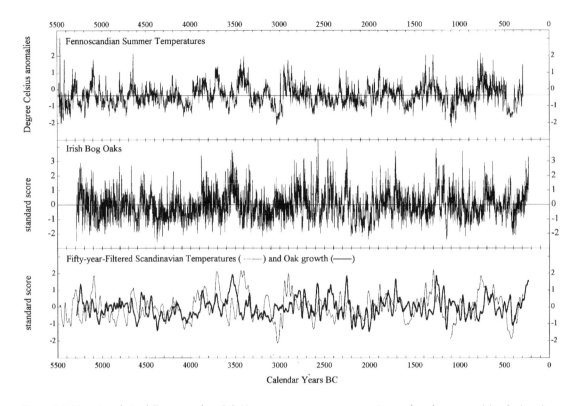

Figure 5.8 Tree-ring-derived Fennoscandian July/August mean temperature estimates based on a provisional pine ring-width chronology being constructed from subfossil remains in northern Swedish lakes,[34] and a mean ring-width chronology comprised of oak samples recovered from Irish bogs[35] – oak data kindly supplied by Professor Mike Baillie of the Palaeoecology Centre, Queen's University, Belfast. The temperatures are plotted as annual degrees Celsius anomalies from a modern base period and the oak chronology as standardised departures from the mean of the whole curve. The small gap shown in the pine-based chronology near 1,100 BC has since been closed, but the chronology is not yet continuous to the present (see text). The dating shown is probably correct to within about fifty years. The lowest box shows the Fennoscandian temperature and Irish oak data smoothed with a filter which emphasises variations on time-scales greater than fifty years.

Essentially, however, these pine data represent summer temperature change in northern Sweden. They show that general warmth prevailed between about 4000 and 3300 BC and that many notable shorter warm periods occurred throughout the record: at 5300, 5100, 4650, 3850, 3700, 3400, 2900, 1300 and at 750 BC.

The Irish oak chronology shown here is a simple average of the ring-width measurements averaged year by year, so again the variability is affected by changing sample size and has some bias associated with the changing age of trees that form the chronology at different times (ring widths of young trees are generally larger than those of old trees). A first-order climatic interpretation of the oak curve would be that high growth reflects warm/dry conditions and low growth a relatively cool/wet climate. These data certainly demonstrate significant climate variability on multidecadal and century time-scales in Ireland and even suggest the presence of a multi-century quasi-climate oscillation through much of the Holocene. Overall, warm and dry conditions appear to have been more frequent over the centuries between 5000 and 4400 BC and between 3900 and 3450 BC. Other periods of notable warm/dry conditions are centred on about 2800, 2450, 1800, 1250 and 700 BC. Longer cool/wet periods are suggested between 4450 and 3900 BC; 3450 and 3100 BC; 2200 and 1900 BC; and between 1150 and 750 BC.

The pine chronology is not yet continuous. No material has been found that bridges the gap in the early centuries BC. This is thought to be only a short gap, though, because the long, earlier 'floating' chronology can be placed quite accurately in time (probably within fifty years either way) on the basis of a number of radiocarbon dates on samples of known *relative* age within the chronology.[37] Material dating to the gap period is proving to be surprisingly elusive, however, not just in northern Sweden but also in Finland.[38] For a long time, the Irish chronology was also discontinuous at this time. Bog oak data do not exist after 229 BC (as shown in Figure 5.8). After that time the chronology is made up from archaeological timbers, and a small gap in the early centuries BC was eventually only bridged using non-Irish material from Carlisle.[39] Also note the apparent lack of subfossil material dated to this period in the Cairngorm and independent Swedish mountain tree-line surveys shown in Figure 5.7. It has been suggested that the demise of the bog oaks in Ireland was associated with increased wetness and that the Irish and Scandinavian gaps are contemporaneous with large 'wash out' phases of wood recovered from riverine gravel deposits in Germany.[40] All of this is consistent with a period of enhanced precipitation. It is a plausible hypothesis that this could have caused a rapid rise in lake and bog levels, leading to severe tree growth disruption and severe distortion of ring-patterns. An inability to identify and measure these samples would prevent the chronology crossing this 'event' in the last few centuries BC.

Swedish samples with just such distorted ring-patterns do in fact exist and appear to date to this period (around 350 BC), but none can as yet be measured with sufficient certainty to be incorporated in the chronology. There is also strong circumstantial

evidence that enhanced volcanic activity occurred in this period and may have played a part in the environmental disruption that clearly occurred at this time, through the effects of volcanic dust and aerosols on climate.[41] Although the modern data are much more detailed than the peat and pollen stratigraphies used by Blytt and Sernander, it is tempting to see in this climatic disruption indicated by tree-rings, the equivalent of the Sub-Boreal/Sub Atlantic transition (see Figure 5.5). Much further development of these Irish and Scandinavian and other very long European chronologies (such as in the Alps, Germany, northern Russia and in Finland) needs to be completed, particularly further work to interpret the detailed environmental signals that each chronology contains. This work is currently underway.

CONCLUSIONS

Within the somewhat arbitrary confines of the period addressed by this chapter (22,000 to 2,000 years BP), British climatic history falls into three clear and distinct epochs. These are the end of the Pleniglacial, the temperate Holocene, and the complex transition between them known as the Late-Glacial. For all three, we rely upon interpreting the climate from the records left in geological accumulations and in the remains of plants and animals. Very little of this evidence reflects climate directly. Rather, climate must be inferred at second remove from what the records tell us of glacial advances and retreats, frozen ground, vegetation communities, tree growth and so forth. Inevitably, non-climatic information is mixed up with the climatic component of the records, so that we can see the problem as being one of how to separate the climatic 'signal' from the non-climatic 'noise'. Paradoxically this is often easier for the Pleniglacial and Late-Glacial than for the Holocene, despite the much greater abundance and detail of the records in the last few thousand years. The paradox is resolved if we recall that the ability of proxy data to detect climatic change is itself limited. For example, the Mutual Climate Range method using beetles simply cannot resolve changes in summer maximum temperature of less than one or two degrees Celsius. Holocene climate change was smaller than this, which is one reason why the beetle record provides no indication of it (Figure 5.1). In summarising the climatic changes which have affected the British Isles in the last 20,000 years, we must keep the question of 'signal-to-noise' ratio constantly in mind.

For the end of the Pleniglacial (20,000–13,000 radiocarbon years BP), direct evidence is rather sparse. Although the ice-sheet that had covered most of the British Isles at the Last Glacial Maximum was in decay, the beetle and periglacial evidence shows that the climate remained extremely cold, with the mean temperature in the warmest months below 10°C, descending to –20°C or even –30°C in the winters. The seas to the west of the islands remained cold with extensive ice covering them in winter. Probably the decay of the ice-sheet came about because the climate was too dry to sustain it as summer insolation (and hence ablation) gradually increased (see Box 5.2). The initial decay may also have involved other causes.

In the Late-Glacial period we have much more abundant evidence, although still sparse by Holocene standards. Fossil beetles demonstrate the large temperature oscillations shown in Figure 5.1. The dramatic warming at about 13,000 years BP left the whole of the British Isles free from ice with a thermal climate initially similar to that of today, but gradually cooling somewhat over the next 1,000 years. Between 11,000 and 12,000 years BP, mean annual air temperature oscillated around 5°C, with warmest months averaging about 15°C and coldest months –5°C. There was then a rather abrupt cooling at the start of the Younger Dryas, with mean annual air temperature dropping to a minimum of around –5°C in the coldest part of the interstadial. Glaciers and a small ice-sheet reappeared on high ground and the climate appears to have been cold, with winter precipitation dominated by westerly storms. Little is known of summer precipitation, but winters seem to have been drier than today and east–west gradients of precipitation across the country much steeper.

For these two periods before the Holocene began, the climatic changes and differences from the present were so large that they can be clearly perceived even with the rather crude proxy-indicators that are available. The 'signal-to-noise' ratio is quite good. This is not because the methods of reconstruction are sensitive (for they are not), but because the climatic 'signal' was so strong. The crudity of the proxy-indicators for these periods means that, so far, we have practically no indication at all of the high-frequency variations that may have occurred from year to year or decade to decade superimposed on the millennial time-scale changes.

In the Holocene period proper we can distinguish two types of proxy-indicators. First are those such as peat bog stratigraphy, preserved pollen and tree-line data, which inform us mainly of millennial changes in climate and at best resolve short episodes to a time-scale of centuries. Second, there are high-resolution records which can resolve interannual changes and give information up to a century time-scale, but for technical reasons cannot on their own tell us of climate changes over millennia.

For the first type of evidence, the Holocene 'signal-to-noise' ratio is poor. There are hints that the very early Holocene climate around 9,000 years BP was slightly cooler and drier than today, whereas the mid-Holocene may have had slightly warmer summers. It is not possible to quantify these changes except to say that they were small, with temperature differences from the present day of certainly no more than about 1°C, which is the resolving power of the methods involved. Despite the fact that summer insolation was falling throughout the Holocene, and winter insolation rising (Box 5.2), there is no clear evidence of a cooling trend in the British Isles climate. That is not to say that such a trend did not occur, merely that 'signal-to-noise' ratio is too poor to detect it properly.

The clearest evidence of long time-scale change in the Holocene comes from peat bog stratigraphy and tree-line data. This tends to suggest that around 4,000 years BP the climate ameliorated for several centuries. It is not clear whether this involved warming or a diminution of precipitation, or both, but in view of the weakness of any obvious temperature signal, drier conditions were almost certainly involved. Tree-lines were higher at this time and there was a widespread slowing or cessation in blanket peat-bog growth, suggesting drier ground conditions on hills in summer. This drier, and possibly warmer, episode may correspond crudely to the early Sub-Boreal of the Blytt and Sernander scheme (Figure 5.5). After about 4,000 years BP there may have been an increase in wetness, as indicated by a country-wide decline in pine pollen and a decline in the tree-line.

All of these changes, although obvious enough in terms of the primary evidence, are very difficult to interpret in clear climatic terms. Most of the vegetation and bog-growth variables appear particularly sensitive to moisture abundance or deficit in the growing season. This suggests that the climatic shifts that occurred may have involved shifts in summer water balance which is itself a composite result of precipitation and evapotranspiration which depends in turn on temperature, windiness and cloudiness. A major task for the future is to resolve the linkages between these factors and the proxy-indicators that are available, so that the climatic signal can be read more clearly.

To some extent the Holocene high-resolution records available from tree-rings in the British Isles confirm the imperfect picture outlined in the preceding paragraphs. Signal-to-noise ratio is only fair (although it is better in Fennoscandian chronologies) and the long chronologies of Figure 5.8 show clear episodes and oscillations between warmer and drier climates and cooler/wetter ones. There are methodological difficulties, however, in correlating the two types of evidence (including the absence of millennial time-scale climate signals in these data). The real importance of the tree-ring record is that it reveals that variation of climate occurs on all time-scales from centuries down to annual. Since even the century-scale oscillations are only barely detectable using other vegetation proxies, the absolute magnitude of these changes is presumably not very great. Tree-ring widths in the British Isles, like peat accumulation, record a rather complex signal involving both temperature and moisture availability, which requires more calibration in future before it can be

unambiguously interpreted. This unravelling of the signals of climate change in the Holocene presents a great scientific challenge for the next decades of palaeoclimatic research.

NOTES

1 T.C. Atkinson, K.R. Briffa and G.R. Coope, 'Seasonal temperatures in Britain during the past 22,000 years reconstructed using beetle remains', *Nature*, 1987, vol. 325, pp. 587–92; T.C. Atkinson, K.R. Briffa, G.R. Coope, M.J. Joachim and D.W. Perry, 'Climatic calibration of coleopteran data', in B.E. Berglund (ed.), *Handbook of Palaeoecology and Palaeohydrology*, London, J. Wiley and Sons, 1985, pp. 851–8; G.R. Coope, 'Fossil beetle assemblages as evidence for sudden and intense climatic changes in the British Isles during the last 45,000 years', in W.H. Berger and L.D. Labeyrie (eds), *Abrupt Climatic Change: Evidence and Implications*, NATO ASI Series C 216, Dordrecht, Reidel, 1987, pp. 147–50.
2 Atkinson *et al.*, 1987, op. cit.
3 G.R. Coope and J.A. Brophy, 'Late glacial environmental changes indicated by a coleopteran succession from North Wales', *Boreas* vol. 1, pp. 97–142; G.R. Coope and M.J. Joachim, 'Late Glacial environmental changes interpreted from fossil coleoptera from St. Bees, Cumbria, N.W. England', in J.J. Lowe, J.M. Gray and J.E. Robinson (eds), *Studies in the Late Glacial of Northwest Europe*, Oxford, Pergamon, 1980, pp. 55–68.
4 W.F. Ruddiman and A. McIntyre, 'The North Atlantic Ocean during the last deglaciation', *Palaeogeography, Palaeoclimatology, Palaeoecology*, 1981, vol. 35, pp. 145–214.
5 J. Imbrie and N.G. Kipp, 'A new micropaleontological method for quantitative paleoclimatology: application to a late Pleistocene Caribbean core', in K.K. Turekian (ed.), *The Late Cenozoic Glacial Ages*, New Haven, Yale University Press, 1971, pp. 171–81.
6 G.R. Coope and G. Lemdahl, 'Regional differences in the late glacial climate of northern Europe based on coleopteran analysis', *Journal of Quaternary Science*, 1995, vol. 10, pp. 391–5.
7 J. Lowe, B. Ammann, H.H. Birks, S. Björck, G.R. Coope, L. Cwynar, J.-L. De Beaulieu, R.J. Mott, D.M. Peteet and M.J.C. Walker, 'Climatic changes in areas adjacent to the North Atlantic during the last glacial–interglacial transition (14–9 ka BP): a contribution to IGGP-253', *Journal of Quaternary Science*, 1994, vol. 9, pp. 185–98.

8 C.K. Ballantyne and C. Harris, *The Periglaciation of Great Britain*, Cambridge, Cambridge University Press, 1994, 330 pp.

9 C.K. Ballantyne, 'The Loch Lomond Readvance on the Isle of Skye, Scotland: glacier reconstruction and palaeoclimatic implications', *Journal of Quaternary Science*, 1989, vol. 4, pp. 95–108.

10 Ballantyne and Harris, 1994. op. cit.

11 W.F. Ruddiman and A.C. Mix, 'The North and Equatorial Atlantic at 9000 and 6000 yr BP', in H.E. Wright, Jr., J.E. Kutzbach, T. Webb, III, W.F. Ruddiman, F.A. Street-Perrott and P.J. Bartlein (eds), *Global Climates Since the Last Glacial Maximum*, Minneapolis, University of Minnesota Press, 1993, pp. 94–135.

12 A. Blytt, *Essay on the Immigration of the Norwegian Flora*, Christiania, 1876.

13 R. Sernander, 'Die schwedischen Torfmoore als Zeugen postglacialer Klimaschwankungen, in *Die Veränderungen des Klimas seit dem maximum der letzten Eiszeit*, Stockholm, Generalstabens Litografiska Anstalt, 1910, pp. 195–246.

14 F.M. Chambers (ed.), *Climate Change and Human Impact on the Landscape: Studies in Palaeoecological and Environmental Archaeology*, London, Chapman and Hall, 1993, 303 pp.

15 B. Huntley and I.C. Prentice, 'Holocene vegetation and climates in Europe', in H.E. Wright, Jr., J.E. Kutzbach, T. Webb, III, W.F. Ruddiman, F.A. Street-Perrott and P.J. Bartlein (eds), *Global Climates Since the Last Glacial Maximum*, Minneapolis, University of Minnesota Press, 1993, pp. 136–68.

16 B.E Berglund, H.J.B. Birks, M. Ralska-Jasiewiczowa and H.E. Wright (eds), *Palaeoecological Events During the Last 15000 Years: Regional Syntheses of Palaeoecological Studies of Lakes and Mires in Europe*, Chichester, John Wiley and Sons, 1996, 764 pp.

17 J. Greig, 'Great Britain – England', in B.E. Berglund, H.J.B. Birks, M. Ralska-Jasiewiczowa and H.E. Wright (eds), *Palaeoecological Events During the Last 15000 Years: Regional Syntheses of Palaeoecological Studies of Lakes and Mires in Europe*, Chichester, John Wiley and Sons, 1996, pp. 15–94.

18 H.J.B. Birks (with contributions by H.H. Birks, P.D. Kerslake, S.M. Peglar and W. Williams), 'Great Britain – Scotland', in B.E. Berglund, H.J.B. Birks, M. Ralska-Jasiewiczowa and H.E. Wright (eds), *Palaeoecological Events During the Last 15000 Years: Regional Syntheses of Palaeoecological Studies of Lakes and Mires in Europe*, Chichester, John Wiley and Sons, 1996, pp. 95–143.

19 F.J.G. Mitchell, R.H.W. Bradshaw, G.E. Hannon, M. O'Connell, J.R. Pilcher and W.A. Watts, 'Ireland', in B.E. Berglund, H.J.B. Birks, M. Ralska-Jasiewiczowa and H.E. Wright (eds), *Palaeoecological Events During the Last 15000 Years: Regional Syntheses of Palaeoecological Studies of Lakes and Mires in Europe*, Chichester, John Wiley and Sons, 1996, pp. 1–13.

20 F.M. Chambers, 'Great Britain – Wales', in B.E. Berglund, H.J.B. Birks, M. Ralska-Jasiewiczowa and H.E. Wright (eds), *Palaeoecological Events During the Last 15000 Years: Regional Syntheses of Palaeoecological Studies of Lakes and Mires in Europe*, Chichester, John Wiley and Sons, 1996, pp. 77–94.

21 H.J.B. Birks and H.H. Birks, *Quaternary Palaeoecology*, London, Edward Arnold, 1981; H.H. Lamb, *Climate: Present, Past and Future. Volume 2: Climatic History and Future*, London, Methuen and Co. Ltd, 1977, 835 pp.

22 H.J.B. Birks, 'Long-term ecological change in the British uplands', in B.M. Usher and D.B.A. Thompson (eds), *Ecological Change in the Uplands*, Oxford, Blackwell, 1988, pp. 37–56; J.J. Lowe, 'Integration of pollen and other proxy data: the Holocene palaeoclimate of the British Isles and adjacent parts of Europe', in B. Frenzel, A. Pons and B. Gläser (eds), *Evaluation of Climate Proxy Data in Relation to the European Holocene*, Paläoklimaforschung, 1991, vol. 6, pp. 37–50.

23 N.V. Pears, 'Interpretation problems in the study of tree-line fluctuations', in J.A. Taylor (ed.), *Research Papers in Forest Meteorology, an Aberystwyth Symposium*, University College Wales, 1972, pp. 31–45.

24 K.D. Bennett, 'The Postglacial history of *Pinus sylvestris* in the British Isles', *Quaternary Science Reviews*, 1984, vol. 3, pp. 133–55.

25 W. Pennington, E.Y. Haworth, A.P. Bonny and J.P. Lishman, 'Lake sediments in northern Scotland', *Philosophical Transactions of the Royal Society of London*, Series B, 1972, vol. 264, pp. 191–294; H.H. Birks, 'Studies in the vegetational history of Scotland. IV. Pine stumps in Scottish blanket peats', *Philosophical Transactions of the Royal Society of London*, Series B, vol. 270, pp. 181–226; – the foregoing as cited by Bennett, 1984, op. cit.

26 A.D. Dubois and D.K. Ferguson, 'The climatic history of pine in the Cairngorms based on radiocarbon dates and stable isotope analysis, with an account of the events leading up to its colonization', *Review of Palaeobotany and Palynology*, 1985, vol. 46, pp. 55–80.

27 M.C. Bridge, B.A. Haggart and J.J. Lowe, 'The history and palaeoclimatic significance of subfossil remains of *Pinus sylvestris* in blanket peats from Scotland', *Journal of Ecology*, 1990, vol. 78, pp. 77–99.

28 S.P. Harrison, L. Saarse and G. Digerfeldt, 'Holocene changes in lake levels as climate proxy data in Europe', in B. Frenzel, A. Pons and B. Gläser (eds), *Evaluation of Climate Proxy Data in Relation to the European Holocene*, Paläoklimaforschung, 1991, vol. 6, pp. 159–69.

29 L. Kullman, 'Orbital forcing and tree-limit history: hypothesis and preliminary interpretation of evidence from Swedish Lapland', *The Holocene*, 1992, vol. 2, pp. 131–7; W. Karlén, 'Lacustrine sediments and tree-limit variations as indicators of Holocene climatic fluctuations in Lapland: northern Sweden', *Geografiska Annaler*, 1976, vol. 58A, pp. 1–34; M. Eronen, 'The retreat of pine forest in Finnish Lapland since the Holocene climatic optimum. A general discussion with radiocarbon evidence from subfossil pines', *Fennia*, 1979, vol. 157, pp. 93–114; V. Markgraf, 'Paleoclimatic evidence derived from timberline fluctuations', in J. Labeyrie (ed.), *Les Méthodes quantitatives d'Etude des Variations du Climat au cours du Pleistocène*, Paris, CNRS, 1974, pp. 67–83.

30 H. Nichols, *Palynological and Paleoclimatic Study of the Late Quaternary Displacement of the Boreal Forest–Tundra Ecotone in Keewatin and Mackenzie, NWT, Canada*, University of Colorado, Institute of Arctic and Alpine Research, Occasional Paper 15.

31 W. Karlén, 'Scandinavian glacial and climatic fluctuations during the Holocene', *Quaternary Science Reviews*, 1988, vol. 7, pp. 199–209; M. Magny, 'Holocene fluctuations of lake levels in the French Jura and sub-Alpine ranges, and their implications for past general circulation patterns', *The Holocene*, vol. 3, pp. 306–13.

32 H.C. Fritts, *Tree Rings and Climate*, New York, Academic Press, 1976, 567 pp., and F.H. Schweingruber, *Tree Rings: Basics and Applications of Dendrochronology*, Dordrecht, Kluwer, 1988, 276 pp.

33 For a general and reasonably up-to-date introduction to the scope and methodology of tree-ring research see E.R. Cook and L.A. Kairiukstis (eds), *Methods of Tree-Ring Analysis: Applications in the Environmental Sciences*, Dordrecht, Kluwer, 1990, 394 pp.

34 Briffa, K.R., 'Mid and Late Holocene climate change: evidence from tree growth in northern Fennoscandia', in B.M. Funnell and R.L.F. Kay (eds), *Palaeoclimate of the last Glacial/Interglacial Cycle*, Swindon, NERC Earth Sciences Directorate, 1994.

35 M.G.L. Baillie, *A Slice Through Time: Dendrochronology and Precision Dating*, London, B.T. Batsford, 1995, 176 pp.

36 K.R. Briffa, P.D. Jones, F.H. Schweingruber, W. Karlén and S.G. Shiyatov, 'Tree-ring variables as proxy-climate indicators: problems with low-frequency signals', in P.D. Jones, R.S. Bradley and J. Jouzel (eds), *Climatic Variations and Forcing Mechanisms of the Last 2000 Years*, NATO ASI Series I, 41, Berlin, Springer-Verlag, 1996, pp. 9–41.

37 Briffa, 1994, op. cit.

38 P. Zetterberg, M. Eronen and K.R. Briffa, 'Evidence on climatic variability and prehistoric human activities between 165 BC and AD 1400 derived from Subfossil Scots Pines (*Pinus sylvestris* L.) found in a lake in Utsjoki, northernmost Finland', *Bulletin of the Geological Society of Finland*, vol. 66.

39 Baillie, 1995, op. cit.

40 Ibid.

41 Ibid.

GENERAL READING

M.G.L. Baillie, *A Slice Through Time: Dendrochronology and Precision Dating*, London, B.T. Batsford, 1995, 176 pp.

C.K. Ballantyne and C. Harris, *The Periglaciation of Great Britain*, Cambridge, Cambridge University Press, 1994, 330 pp.

B.E. Berglund, H.J.B. Birks, M. Ralska-Jasiewiczowa and H.E. Wright (eds), *Palaeoecological Events During the Last 15000 Years: Regional Syntheses of Palaeoecological Studies of Lakes and Mires in Europe*, Chichester, John Wiley and Sons, 1996, 764 pp.

R.S. Bradley, *Quaternary Paleoclimatology: methods of paleoclimatic reconstruction*, Boston, Allen and Unwin, 1985, 472 pp.

N. Roberts, *The Holocene: An Environmental History*, Oxford, Blackwell, 1989, 227 pp.

6

DOCUMENTING THE MEDIEVAL CLIMATE

Astrid Ogilvie and Graham Farmer

History gets thicker as it approaches recent time.

A.J.P. Taylor

INTRODUCTION

The focus of this chapter is on documentary evidence for climate variations in England during the medieval period.[1] England is concentrated upon since there is relatively little written evidence of climate events for this period originating in the more peripheral parts of the British Isles – Scotland, Wales and Ireland. Some evidence from Iceland and, to a lesser extent, Greenland, is also considered.[2] This will serve to place the information from England in a wider geographical and climatic context and provide a basis for the consideration of North Atlantic climate variations during this time. Figure 6.1 shows a map of the North Atlantic region and the location of the relevant countries.

The chapter begins with a discussion of the value and usefulness of documentary data, as well as how they need to be evaluated and analysed. The different kinds of historical sources that contain climate data for the medieval period are then outlined. Finally, the available information regarding climatic conditions in England, Iceland and Greenland during the period from *c.* AD 1200 to 1430 is presented. The last section includes a consideration of the wider perspective of European and North Atlantic climate conditions in medieval times and the question of whether there really was a '**Medieval Warm Period**' as some researchers have suggested.[3]

In order to reconstruct the climate in times before systematic meteorological observations became avail-

able, different kinds of '**proxy climate**' data must be used. These may take the form of, for example: analysis of the annual rings laid down by trees (see Chapter 5); samples of pollen records; marine sediment cores; and deep cores taken from ice-sheets (see Chapter 4). Historical documentary data can also be very useful. Where these are plentiful, and when used appropriately, they can give precise and accurate information on past climatic events. The reconstruction of past climatic regimes using documentary evidence is known as 'historical climatology'.

Rather than going directly to the original sources, many earlier researchers in the field of historical climatology relied on compilations of weather events produced by other authors (i.e., 'secondary sources'). These compilations were produced as early as the sixteenth century and became quite common by the eighteenth century. In the last hundred years or so, some fifteen such compilations have been published, absorbing much of the material contained in earlier ones. These modern compilations list their information in chronological order and contain numerous references extracted from a variety of historical texts and secondary sources. They make accessible, in comparatively few volumes, information that is otherwise dispersed throughout several hundred volumes of source material. As such, they would appear on the surface to be extremely useful. Unfortunately, they invariably reproduce a variety of errors and should only be used with extreme care.[4]

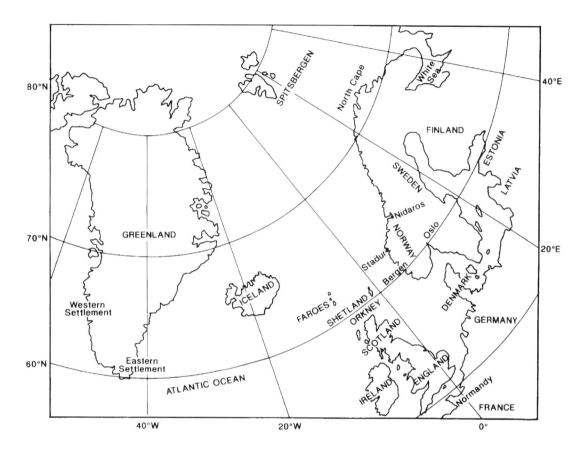

Figure 6.1 The North Atlantic and surrounding countries in medieval times. The English climate data used here are drawn mainly from the southern part of England. The climate data for Iceland come mainly from coastal areas to the north, west and south of Iceland.

One of the acknowledged pioneers in the field of historical climatology is Hubert Lamb, founder of the Climatic Research Unit, and it is the medieval climatic index for England derived by him which has frequently been used by other researchers.[5] This index is based on data of which at least some have been found to be unreliable.[6] Lamb's index is, in turn, based partly on the compilation produced by Charles Britton[7] for the British Isles. This is an unusually careful and accurate example of this type of work. Nevertheless, Britton, no doubt aiming for completeness, has included remarks drawn from weather chronologies written much later than the events they describe and whose authority is uncer-

tain. Both his and Lamb's pioneering efforts must be lauded since they were carried out before the need for careful historical analysis was fully realised. Indeed, it was work initiated by Lamb in the Climatic Research Unit that led to this realisation. In the present analysis, only sources that have been established to be reliable are used. The importance of analysing sources is discussed further below.

THE IMPORTANCE OF SOURCE ANALYSIS

All types of proxy climate data need to go through a process of analysis and evaluation. This is no less true

BOX 6.1: AN EXAMPLE OF SOURCE ANALYSIS

The way in which historians need to unravel the origins of their sources may be described by referring to a specific example. For the year 1279, an Icelandic compiler of weather events, Thoroddsen, notes a severe winter with much ice around Iceland.[a] Thoroddsen's source for this is an earlier compilation by Finnsson, published originally in 1796.[b] Finnsson's source was a manuscript he had found in Copenhagen. This work, now known as *Setbergsannáll*, has been analysed and edited by the Icelandic historian, Jóhannesson.[c] He was able to show that the author of *Setbergsannáll*, Gísli Thorkelsson (1676–1725) actually embellished the sources that were available to him, and even went so far as to invent information when his sources were silent. In short, there is no reliable basis for the reported ice and severe weather in 1279.[d]

[a] Th. Thoroddsen, Árferdi á Íslandi í thúsund ár 1–2, Copenhagen, Hinu íslenzka fraedafjelagi, 1916–17, pp. 32–3. It may be noted that Thoroddsen was a geologist who did a tremendous service to the historical climatology of Iceland by collecting together a vast number of sources of climate information.

[b] H. Finnsson, 'Um mannfaekkun af hallærum á Íslandi', Jón Eythórsson og Jóhannes Nordal sau um útgáfuna, Reykjavík, Almenna Bókafélagid, 1970. (This compilation was first published in 1796.)

[c] J. Jóhannesson, 'Setbergsannáll', in Annales Islandici Posteriorum Sæcularum. Annálar 1400–1800, 1922, vol. 4, pp. 19–21.

[d] This example is explained more fully in A.E.J. Ogilvie, 'The past climate and sea-ice record from Iceland, part 1: data to AD 1780', *Climatic Change*, 1984, vol. 6, pp. 131–52.

of documentary evidence; some historical sources may contain errors, misconceptions and even forgeries,[8] while others provide accurate and reliable information. It is therefore extremely important that all sources to be used for climate reconstruction are carefully analysed in order to ensure their reliability. This procedure is referred to as 'source criticism'. It is a task that usually needs to be undertaken by historians who are, by virtue of their training and familiarity with the subject, better suited to it than climatologists. Source criticism is especially important when using medieval sources; writers from this time did not always value accurate reporting in the same way that modern researchers do.[9] Historical records from later periods are often less problematic; but later compilations, being secondary sources, frequently propagate earlier errors.[10] Clearly, if a documentary source gives incorrect information, any climate reconstruction using it will reproduce this and thus give a false picture of climatic variations. Once the process of source criticism has been completed, and the stage of interpretation of the data been reached, the expertise of climatologists is clearly needed.

A reliable source should ideally be independent, not derivative, and should be contemporary (i.e., written at or near the time of the event). Alternatively, there should be strong evidence regarding its reliability. In order to ensure that these criteria apply, it is necessary to find out certain things about the evidence.[11] First, *who* wrote it? Medieval writers did not always sign their names to their accounts, but if it is known who wrote a particular work, then this can help to place it, both with regard to its date and the location where it was written, and thus help in assessing the reliability of the author. Second, *when* was it written? It is vital to know the date of a particular work. If it was written close in time to the event described it is much more likely to give an accurate description than if it were written many years afterwards (see Box 6.1). Indeed, in order to ensure accuracy, it is almost always essential for an account to be contemporaneous with the events described. Third, *where* was it written? If an account were written close to the events described by an eyewitness observer it is much more likely to be accurate than if it were written far away from it, on the basis of second-hand information or hearsay. Fourth, *why* was it written? A writer may have had a didactic or proselytising mission which was more important to him than conveying precise information.

THE SOURCES AND DATA

Both England and Iceland have a rich medieval literature containing details of many aspects of life including the weather and climate. Of particular interest for our purposes here is information regarding temperature and precipitation. Icelandic sources sometimes also refer to sea-ice reaching the coasts. This ice, originating off the coasts of East Greenland, is an interesting proxy climate indicator in itself.[12] There are no accounts regarding medieval climate that originate specifically in Greenland, but two sources of Norwegian origin give very interesting descriptions of the sea-ice, as well as information on conditions of life amongst the Norse who had settled in Greenland.[13]

The English historical sources, rather more prolific than the Icelandic ones, include medieval annals as well as chronicles and the account rolls kept on manors.[14] The latter, documents of accounts made for the churches who owned these large estates, derive their name from the fact that they took the form of long pieces of parchment rolled up. One of the earliest known diaries specifically of weather events is from England. This journal, kept by William Merle of Lincolnshire, extends from January 1337 to January 1344[15] (see Chapter 7, Figure 7.2).

Except for the beginning of the period, the chronicles' statements are outnumbered by those from the other major type of source – manorial account rolls – and the rolls are major sources for the period after 1300. The longest series of accounts from which data are available is the Winchester episcopal series from which the weather information has been collected and published.[16] This has been supplemented by data from a Hertfordshire manor of Westminster Abbey[17] and a number of details, especially useful in the fifteenth century, from the account rolls of Battle

Abbey relating to manors in Sussex.[18] The account roll evidence is all from south-east or central southern England as far west as Somerset. The great majority of the narrative sources used here also come from southern England, so the two types of sources are comparable. There is still some scope for further work on climatic information from manorial accounts in the Midlands and the North. There are, however, few narrative sources available, outside southern England.

Data first become common in the twelfth century, but occur in usable numbers only from *c.* 1200. They are generally less frequent in the fourteenth century than in the thirteenth, and became scarcer in the early fifteenth century. Presumably this was partly as a result of the disruption to political and economic life caused by the Wars of the Roses in the fifteenth century.

Iceland is particularly renowned for its prose narratives from medieval times – known as the 'family sagas' or 'sagas of Icelanders'. These literary, semi-historical works are ranked among the finest works of literature ever written. Since they are primarily literature rather than history, they are generally not relevant to a survey such as this one. Nevertheless, the intellectual environment that fostered this literary endeavour is a factor that must be considered in evaluating documentary evidence from Iceland. Figure 6.2 shows a fine example of a medieval Icelandic manuscript.

In addition to the family sagas, many fascinating works were also produced which were historical rather than literary in origin. These include other kinds of sagas such as the 'sagas of Icelandic Bishops' and the 'Sturlunga sagas', named after the family who dominated events in thirteenth-century Iceland. Other important sources are the medieval Icelandic annals and works of geographical description.[19]

Iceland was settled, mainly from Norway, in the late ninth century. The first references to weather events which have some claim to reliability commence in the late twelfth century. For the thirteenth century there are a number of very interesting weather descriptions. The fourteenth century has the greatest coverage for the whole period from the

settlement to the latter part of the sixteenth century. It is interesting that this is the reverse of the situation in England. In Iceland, contemporary sources come to an end in 1430, leaving a gap until around 1560. As stated above, there is also a lack of sources in England from the early fourteenth century onwards.[20] As far as Iceland is concerned, the reason for this may be an epidemic (probably the Black Death) which killed a large percentage of the population. In these kinds of circumstance, the clergy as a class often suffered many deaths as they traditionally succoured the dying, and it was the clergy who were also the main class from which writers and scribes were drawn.

STRATEGY AND METHODOLOGY

Even reliable medieval narratives pose a number of problems of interpretation. An example of this is the issue of dating. In both England and Iceland, the Julian, or Old Style, calendar was in use in medieval times. England changed to the New Style, or Gregorian, calendar in September 1752 and Iceland changed somewhat earlier, in November 1700.[21] In order to correct for this, for the period 29 February 1100 to 28 February 1300 seven days should be added to the date. From 29 February 1300 to 28 February 1400, eight days should be added, and from 29 February 1400 to 28 February 1500, nine days.

In Iceland, they also used a dating system that was virtually unique to that country. The year was divided into two halves; summer and winter. Summer began on the Thursday immediately preceding 16 April according to the Old Style calendar. Winter began on the Saturday immediately preceding St Luke's day (18 October, Old Style), or on St Luke's day itself if this were a Saturday. From the beginning of summer, the passing of time was

calculated according to how many 'weeks of summer' had gone by. In addition, the year was divided into twelve months of about thirty days each. The first three months of the year were: *thorri*; *góa*; and *einmánudr*. *Thorri* always began on a Friday between 9 and 16 January (Old Style), and so on. Sometimes reference to the seasons can cause problems when it is unclear just when in a season an event occurred. The precision frequently used in Icelandic dating, however, means that events can usually be dated unambiguously.

In England, too, the terminology used for the different seasons can be problematic. 'Autumn', for example, normally meant the harvest season and was taken to begin on 1 August. Its length could vary from year to year, however, and autumn could be 'late' or 'long this year' (as described, for example, in the *Dunstable Annals*, 1294; *Walter of Coventry*, 1212). 'Winter' was also a flexible term. It could be thought of as beginning about Michaelmas, 29 September (Old Style; as in Matthew Paris, *Chronica Majora*, 1253), and 28 December (Old Style) could be described as 'midwinter' (ibid., 1257). Spring was regarded more as an annual biological event rather than a fixed period. This type of terminological difficulty is actually not so serious, partly because our narratives most frequently refer to named months, and partly because we still have a good enough idea of the dates to allow an interpretation – as long as we can avoid the temptation to be too precise. There is enough overlap between medieval and modern terms to allow the simple classification adopted here.

The adoption of fairly stringent criteria for the acceptance of weather statements in the English chronicles has reduced the available data considerably. Table 6.1 shows the quantity and distribution of our information, and the numbers of unreported seasons are shown in Table 6.2. It should be noted

---

Figure 6.2 This example of an Icelandic manuscript comes from a magnificent vellum codex, *Flateyjarbók*, dating from around 1390. Its name, the 'Flatey book', derives from the fact that it remained for some time on an island named Flatey, in Breidafjord, off the west coast of Iceland. *Flateyjarbók* contains several works, mainly sagas of Norwegian kings. The page shown here concerns the beginning of the Saga of Ólafr Tryggvason. The illuminated letter (P) depicts King Harald Fairhair and his cupbearer.

Table 6.1 Total numbers of seasons with documentary data for England, AD 1200–1429

	Narratives	Account rolls
1200–49	56	22
1250–99	64	76
1300–49	39	94
1350–99	22	85
1400–29	15	35
Total	196	312

Note: Some seasons are described in both types of source.

that, in spite of there being many seasons for which no reliable weather descriptions exist, these are generally seasons with unremarkable weather. The nature of the sources is such that 'normal' or 'near-normal' weather would not be commented on. As a consequence, it is likely that almost all of the severe seasons are included in our dataset.

The different sources used pose different problems of interpretation. In the account rolls, for example, the weather is mentioned when it helps to explain an exceptional agricultural situation. The account rolls therefore tend to record extreme seasons, but mainly those that are agriculturally more sensitive. These data are thus biased towards reporting summer and autumn conditions, but complementary information on extreme winters can be obtained by reference to the narratives. This bias extends to the type of weather most likely to be recorded, dry summers and wet autumns being most frequently noted since they have the most obvious effect on agricultural routine and its incidental expenses.[22] Thus, when all the English records used here are considered for the period 1220–1430, they note ninety-nine dry summers as against thirty-six wet ones, many of the latter coming from the narratives, and seventy-six wet autumns as against only eight dry ones, nearly

Table 6.2 Numbers of unreported seasons by decade for England, AD 1200–1429

	Spring	Summer	Autumn	Winter	Total
1200–09	10	9	10	9	38
1210–19	9	10	8	8	35
1220–29	9	7	6	5	27
1230–39	8	4	7	4	23
1240–49	6	2	3	5	16
1250–59	4	5	3	4	16
1260–69	10	5	7	7	29
1270–79	7	1	8	5	21
1280–89	10	4	6	2	22
1290–99	2	1	4	3	10
Total	75	48	62	52	
1300–09	7	1	4	5	17
1310–19	6	1	5	2	14
1320–29	6	5	6	6	23
1330–39	9	2	5	5	21
1340–49	8	1	5	6	20
1350–59	5	2	4	6	17
1360–69	6	1	4	8	19
1370–79	7	0	3	4	14
1380–89	6	2	2	8	18
1390–99	8	4	7	9	28
Total	68	19	45	59	
1400–09	10	2	7	8	27
1410–19	7	4	8	5	24
1420–29	9	5	5	7	26

all the latter being also noted by chroniclers. If nearly half the summers in this period are described as dry, we may begin to suspect a certain amount of exaggeration in the reporting or interpretation. Long-term trends towards drier or wetter seasons may therefore reflect changes in the balance of source material, particularly account rolls relative to other sources.

A further difficulty arises from the repetition of phrases year after year in accounts from the same manor. Accounts from a well-organised estate can be quite stylised documents, arranged in the same way for decades on end,[23] and when they repeat the same phrase in the same place one suspects that they are being cast in common form.

The best way out of both these difficulties is to accept a season as being extreme only when it is reported by a number of separate sources, or when it is isolated in the record during the periods of less information. An unusually dry summer tends to be recorded by many manors in different areas, often belonging to different lords, and is confirmed by the narratives. Such events are quite clearly different from the 'dry summers' mentioned in cliché or ambiguous form by only one or two manors out of many possible ones. This has led to a small number of modifications being made in the interpretation presented here of the Winchester series, most notably in the spell of dry summers during the 1280s, when only 1285 and 1288 can be accepted as very dry and 1287 need not have been dry at all. Another suspicious run of dry summers occurs in the 1410s, where 1416 looks dubious (the Burghclere accounts simply repeat the previous years' phrases), no evidence is printed for 1417 and in 1418 only one manor mentions dryness.

A further difficulty would arise if the information from the rolls were simply deduced from the supposed effects of the weather on crops – our climatic reconstruction could not then be applied to explain variations in yield, and the weather information itself would be suspect because the weather/crop yield link is not a simple one. The weather information extracted from the accounts, however, takes the form of specific descriptions of weather conditions, even

though proffered to explain agricultural circumstances. The veracity of both data sources – account rolls and chronicles – is further confirmed when their data are compared (see Table 6.2).

PRESENTATION OF THE DATA

Once unreliable data have been discarded, and useful historical sources identified, the next step for climate reconstruction is to extract from these sources any contemporary references to weather. Once these have been collated, they can be used in two principal ways. One is to produce a detailed picture of a particular weather event or season. In cases where the data are numerous enough, a second way of presenting them is to derive time-series of climate indices. These indices may be calculated by assigning a value to each month or season in terms of its warmth/coldness or wetness/dryness. These values may then be summed in order to produce a decadal index (see Figure 6.3). The indices derived for England and Iceland, as well as a general picture of what the climate may have been like in those regions from c. 1200 to 1430, are presented below. Some specific events in England and their sources are presented in Box 6.2. Full details of the Icelandic sources may be found elsewhere.[24]

Weather events and decadal indices for England c. 1200–1439

A data catalogue of climatic events has been compiled for England for the period AD 1200 to 1429. This is too long to present here, but some interesting climate events that occurred during this period may be briefly mentioned. Severe and frosty winters occurred in 1205, 1210, 1234, 1254, 1261, 1271, 1281, 1292, 1306, 1335, 1365 and 1408. In 1205 and in 1282, it was reported that the Thames could be crossed on foot. The river was frozen again in 1310 and 1408. The summers of 1236 and 1238 were unusually hot and dry, whilst the years 1314 to 1318 were very wet, and this led to widespread crop failures.

The data for England are presented here in Tables 6.1 to 6.3 and in Figures 6.3 and 6.4. Table 6.1 shows the total numbers of seasons with documentary data from narrative sources and from the account rolls for the years 1200 to 1439. In Table 6.2, total numbers of unreported seasons are presented by decade (note that lack of reporting is likely to indicate a lack of any extremes in the weather, as pointed out earlier). Table 6.3 gives temperature and precipitation estimates for each month from 1200 to 1439. Decadal indices of summer wetness and winter severity derived from the dataset in Table 6.3 are shown in Figure 6.3, together with indices derived by Lamb.[25] The differences may be explained by Lamb's use of data that are either unreliable or relate to some other part of Europe (or both). Also, although many of the data used here were noted by Britton in his compilation,[26] these data are supplemented by many previously unpublished data and by data drawn from texts unused or unavailable to Britton. In addition, errors in Britton's compilation have been deleted from the present analysis. The annual temperature indices for Iceland and England are compared in Figure 6.4.

Weather events and decadal indices for Iceland c. 1200–1440

It is around 1180 that contemporary climate references first become available in Iceland.[27] From this time, and through to the first years of the thirteenth century, there are several descriptions which suggest that climatic conditions were harsh. Between 1211 and 1232 there are no references to weather. Over the next few years there are sporadic references to severe seasons, including 1233 to 1236. There is virtually no weather information for the 1240s. The 1250s seem to have been variable. The climate of the latter part of the thirteenth century was almost certainly severe. An interesting statement is made

for the year 1287: 'At this time, many severe winters came at once and following them people died of hunger.'[28] The first two decades of the fourteenth century seem to have been relatively mild, although there was some severe weather and sea-ice in the early 1320s. It is possible that the 1330s were mild and also the latter part of the 1350s. The last part of the 1340s seems to have been cold. The 1370s also appear to have been cold on the whole (see Figure 6.4). The sources give the impression of mainly mild weather between c. 1395 to 1440.

Events in Greenland

Greenland was colonised from Iceland in approximately the year 985. Two areas of settlement were established, the so-called Eastern and Western settlements (see Figure 6.1). The Western Settlement ceased to exist around AD 1350, while the Eastern Settlement may have survived into the sixteenth century. It has been suggested that a deterioration in climate played a part in the loss of these Norse settlements. This is possible, although there is no direct evidence for this, and several other factors are likely to have played an equal, or more important, part: for example; conflicts with the native Inuit population; changes in the patterns of European trade; and the structure of Norse Greenlandic society.[29]

SEA-ICE INCIDENCE IN THE NORTH ATLANTIC

The cause of the presence of sea-ice off the coasts of Greenland and Iceland is a complex amalgam of conditions in the Greenland Sea and Polar Basin, the movement of ocean currents, and also local weather conditions. These causes cannot be discussed here, but it should be noted that the incidence of sea-ice

Figure 6.3 Decadal indices of summer wetness (top) and winter severity (bottom) from AD 1220 to 1429 derived from the present analysis[1] (revised index) and from H.H. Lamb.[25] The differences may be explained by Lamb's use of data that are unreliable or relate to some other part of Europe, or both. Shaded decades are wetter (summer) or more severe (winter) than the average.

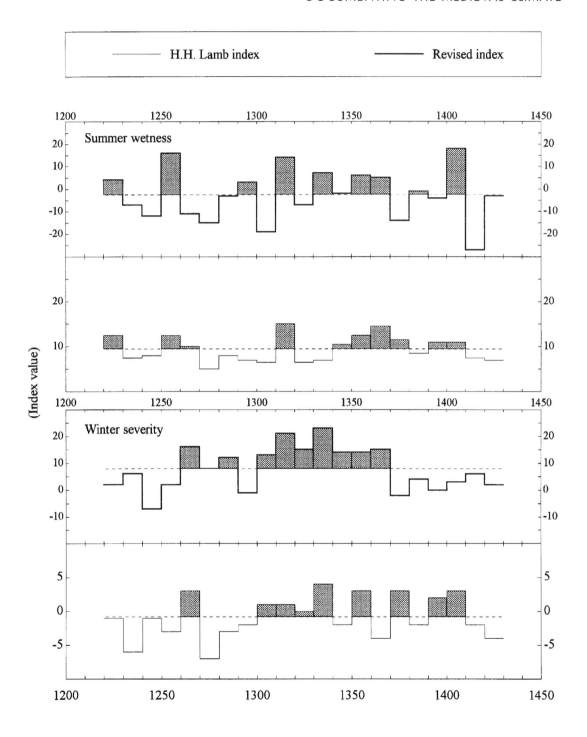

BOX 6.2 DATA CATALOGUE AND SOURCES

The following list some specific events and their sources:

1201   Spring, summer and autumn:
(a) Wet from 20 May to 15 September, crops harmed. Annales de Margan, sire chronica Abbreviate (1066–1232). H.R. Luard (ed.), *Annales Monastici*, Rolls Ser., 1864, i, p. 25.
(b) Storm and flood damage in July. Waverley Annals. Annales Monasterii de Waverleia, AD 1–1291. H.R. Luard (ed.), *Annales Monastici*, Rolls Ser., 1865, ii, p. 253.

1205   Winter:
(a) Frozen from 8 January to 1 April, Thames could be crossed on foot. Ralph of Coggeshall, *Chronicon Anglicanum* (1066–1223). J. Stevenson (ed.), Rolls Ser., 1875, p. 151.
(b) Frozen from 21 January to 29 March, agriculture impossible. Roger of Wendover, *Flores historiarum (from the Creation to 1235)*. H.R. Luard (ed.), Rolls Ser., 1872–83, ii, p. 490.
(c) Severe frost 28 December 1204 to 1 April 1205. Worcester Annals. Annales Prioratus de Wigornia, AD 1–1377. H.R. Luard (ed.), *Annales Monastici*, Rolls Ser., 1869, iv, p. 393.

1210   Winter:
(a) Frost 7 January to 21 February, Severn frozen to 4 miles beyond Gloucester. Worcester Annals. Annales Prioratus de Wigornia, AD 1–1377. H.R. Luard (ed.), *Annales Monastici*, Rolls Ser., 1869, iv, p. 398.
(b) Frost and snow, 8 January to 21 February, ice more than one foot thick. Walter of Coventry, *Memoriale fratris Walteri de Coventria*. W. Stubbs (ed.), Rolls Ser., 1872–3, ii, p. 201.
Spring: cold and windy, Walter of Coventry, *Memoriale fratris Walteri de Coventria*. W. Stubbs (ed.), Rolls Ser., 1872–3, ii, p. 201.

1212   Autumn: severe to 15 September, then heavy rain. Walter of Coventry, *Memoriale fratris Walteri de Coventria*. W. Stubbs (ed.), Rolls Ser., 1872–3, ii, p. 206.

1215   Winter: December mild. *Brut y Tywysogion*. J. Williams ab Ithel (ed.), Rolls Ser., 1860, p. 289.

1220   Winter:
(a) Wet from Christmas (1219) all winter. Worcester Annals. Annales Prioratus de Wigornia, AD 1–1377. H.R. Luard (ed.), *Annales Monastici*, Rolls Ser., 1869, iv, pp. 411–12.
(b) Wet from 16 August to end of December. Worcester Annals. Annales Prioratus de Wigornia, AD 1–1377. H.R. Luard (ed.), *Annales Monastici*, Rolls Ser., 1869, iv, pp. 411–12.

1221   Winter: January cold. Dunstable Annals. Annales prioratus de Dunstaplia, AD 33 to 1297. H.R. Luard (ed.), *Annales Monastici*, Rolls Ser., 1866, iii, p. 53.

1222   Spring: heavy snow, severe frost in April. Waverley Annals. Annales Monasterii de Waverleia, AD 1–1291. H.R. Luard (ed.), *Annales Monastici*, Rolls Ser., 1865, ii, p. 296.
Summer: dry and hot. Ralph of Coggeshall, *Chronicon Anglicanum* (1066–1223). J. Stevenson (ed.), Rolls Ser., 1875, p. 192.
Winter: wet until 9 February 1223. Roger of Wendover, *Flores historiarum (from the Creation to 1235)*. H.R. Luard (ed.), Rolls Ser., 1872–83, ii, p. 74.

1223   Whole year wet. Matthew Paris, *Chronica majora (from the Creation to 1259)*. H.R.

Luard (ed.), Rolls Ser., 1872–83, iii, p. 82.

1223–4 Winter: dry. Waverley Annals. Annales Monasterii de Waverleia, AD 1–1291. H.R. Luard (ed.), *Annales Monastici*, Rolls Ser., 1865, ii, p. 300.

1227 Winter: rivers flood, December 1226, January and February. Worcester Annals. Annales Prioratus de Wigornia, AD 1–1377. H.R. Luard (ed.), *Annales Monastici*, Rolls Ser., 1869, iv, p. 420.

1230 Winter: heavy snow, January. Waverley Annals. Annales Monasterii de Waverleia, AD 1–1291. H.R. Luard (ed.), *Annales Monastici*, Rolls Ser. 1865, ii, p. 421.

1232 Spring–autumn: drought, March to October. *Annales Cambriae*. J. Williams ab Ithel (ed.), Rolls Ser., 1860. p. 79.

1233 Summer: wet. *Flores historiarum (from the Creation to 1326)*. H.R. Luard (ed.), Rolls Ser. 1890. ii, p. 209.

is also highly variable on a variety of time-scales (see Chapter 2).

Evidence for sea-ice variations in the North Atlantic during the medieval period comes from the Icelandic annals and certain sagas, as well as from two interesting sources which were written in Norway. These latter are the *Konungs skuggsjá* (The King's Mirror),[30] composed around 1250, and the *Grænlandslýsing Ívar Bárdarsonar* (the description of Greenland according to Ívar Bárdarson) written probably some time shortly after 1364.[31] The sea-ice sources are discussed below in chronological order.

The *Konungs skuggsjá* contains detailed and accurate descriptions of the Arctic regions. This information was gathered by the author from travellers who had been to these places. Amongst other things, the work contains a very full account concerning sea-ice: 'As soon as one has passed over the deepest part of the ocean, he will encounter such masses of ice in the sea, that I know of no equal to it anywhere else in all the earth . . .'[32] The impression is given from *Konungs skuggsjá* that there was much sea-ice between Iceland and Greenland at the time of writing.

For the year 1261, the Icelandic annals state that there was sea-ice all around Iceland. This is the first recorded reference to such an extreme event. For 1274 and 1275 the annals record the arrival of polar bears and sea-ice.[33] Polar bears often drifted to Iceland on ice floes, so a mention of their presence

can generally be assumed to indicate sea-ice. It is possible that sea-ice only occurred in one of these years and that the other is the result of a copyist's error. In 1306, sea-ice is recorded as having been present all summer in the north.[34] Several of the annals refer to sea-ice in either 1319 or 1320.[35] Sea-ice is said to have lain around the coast all summer and also reached the south coast. The ice also occurred in 1321 and several of the annals refer to polar bears which came on land.[36] In the autumn of 1361, the annal *Gottskálksannáll* states that a polar bear came ashore in Breidafjörd even though there was no ice in the vicinity, which was obviously thought to be strange.[37] In 1374, the annal *Lögmannsannáll*[38] notes that the sea-ice lay off the coasts until 1 September (New Style). One of the Bishops' sagas, *Gudmundar saga biskups Arasonar*, includes a geographical description of Iceland in which glaciers and sea-ice are mentioned. The implication from this Icelandic source is that sea-ice was quite common off the northern coasts at the time the saga was written, *c.* 1350.[39]

Because of its interesting comments on sea-ice the *Grænlandslýsing Ívar Bárdarsonar* has been frequently cited by those interested in the past climate of the North Atlantic region.[40] The work begins by giving detailed sailing directions, from Norway to Iceland, from Iceland to Greenland, from Iceland to Svalbard (Spitsbergen), and from Norway direct to Greenland. As regards the route from Iceland to Greenland, the

Table 6.3 Temperature and precipitation scores for England for each month from AD 1200–1439. (Negative values indicate drier or cooler conditions; positive values indicate warmer or wetter conditions. The value ±2 indicates conditions which are slightly more severe than 'normal', whereas the value ±3 indicates a particularly severe season. Where the values are in parentheses there is a likelihood, rather than definite evidence, of those conditions prevailing.)

Month columns: P = Precipitation, T = Temperature. Seasons: Winter = DJF, Spring = MAM, Summer = JJA, Autumn = SON.

Year	P:D	J	F	M	A	M	J	J	A	S	O	N	T:D	J	F	M	A	M	J	J	A	S	O	N
1200																								
01						2	2	2	2	2														
02																								
03																								
04																								
05													-3	-3		-3								
06																								
07																								
08																								
09																								
1210													-2	-2		-2	-2							
11																								
12										2	2								-2			-2		
13																								
14																								
15																								
16													2											
17																								
18																								
19							(2)	(2)	2	2	2													
1220		2	2	2					2	2	2	2												
21	2						-2	-2					-2											
22							-2	-2		2						-3			2	2				
23	2	2	2	2	2	2	2	2	2	2	2	-2												
24	-2	-2	-2	-2					2	2	2													
25							-3	-3	2	2	2													
26	2	2	2																					
27									2	2	2													
1230		2												-2										
31																								
32				-2	-2	-2	-3	-3	-2	-2	-2													
33								2	3	3	3													
34	-2	-2							2				-2	-2										
35							-2	-2																
36		3	3						-2	-3	-3	-2					2		2	2	2			
37					2																			
38	2	2	2	2	2	2	-3	-3	3	3	3								3	3				
39	2	2	2	2																				
1240		-2	-2	-2	2	2	2	2	2	2	2	2												
41	2				-2	-2	-2	-2	-2	-2	-2					2	2		2	2		2	2	
42						-2	-2	-2	-2	-2	-2		-2				2		2	2		2	2	
43																								

Table 6.3 continued

Year	Precipitation: Winter			Spring			Summer			Autumn			Temperature: Winter			Spring			Summer			Autumn		
	D	J	F	M	A	M	J	J	A	S	O	N	D	J	F	M	A	M	J	J	A	S	O	N
44									−2	−2	−2													
45							(−2)	(−2)	2	2	2													
46							2	2	2	2	2													
47				2	2	2	−3	−3	(−2)	(−2)	(−2)	(−2)				−2	−2	−2						2
48	(−2)	(−2)	(−2)	(−2)			−3	−3					2	2	2	2								3
49							(−2)						3	3	2	3	−2							
1250																								
51																								
52			−2	−2	−2	−2	−2	−2	−2	2	2					2	2	2	2	2				
53					−2	−2	−2	2	2															
54							(−2)	3	(2)	(2)	(2)	(2)	−2	−2		−2	−2	−2	−2	−2	−2	−2	−2	−2
55			3	3	−3	−2	2																	
56									2	3	3	3												
57	3	3	2	2	2	2	2	2	2	2	2	2												2
58	2	2	2				2	2	2	2	2		2	2	−2	(−2)	(−2)	(−2)	(−2)					
59									−2	2	2													
1260		2	2				−2	−2													−2	−2	−2	
61														−2	−2									
62																								
63							−3	−3																
64																								
65																								
66							−2	−2																
67																								
68							−2	−2	2	2	2		(−2)	(−2)	(−2)									
69			2				−2	−2	2	2	2		−2	−2	−2									
1270					2	2							−2	−2	−2									−2
71							(−2)	(−2)					−2	−2	−2	−2								(−2)
72							−3	−3	(−2)	−2	−3	−3	(−2)	(−2)	(−2)									
73				3			3	3																
74					(−2)	(−2)	−3	−3								−3								
75						(−2)	2	2		2	2	2												
76	2	2	2	2	−2	−2	−2	−2																
77			3	3	−2		−3	−3											2					
78							−3	−3																3
79							−2	−2					3	3	3	3								
1280												2												
81		2	2												−3									
82													−3											
83							3	3	2	2														
84	(−2)	(−2)	(−2)				−3	−3	(2)	(2)	2	2										2	2	
85	2	2	2	2			−3	−3					2	2	2	2								
86				−2			−2	−2					−2	−2	−2									−2
87							−2	−2					−2	−2	−2	−2								
88							−2	−2		(2)			(2)	(2)	(2)				2	2	(2)			−2
89							−3	2	2	2	2	2	−2	−2	−2	−2								2
1290	2	2	2	−2			−2	2	2	2	2		2	2	2									
91				−2			−3	−3	2	2	2													
92									2	2	2	2	−2	−2	−3	−2								

Table 6.3 continued

Year	Precipitation: Winter D J F	Spring M A M	Summer J J A	Autumn S O N	Temperature: Winter D J F	Spring M A M	Summer J J A	Autumn S O N
93		2	-2 2 2		(2)(2)(2)			
94		2 2	2 2 2	2 2 2	-2 -2			
95	2 2 2	2						
96		-2 -2	2 2	2 2				
97		-3	-3 -3	-2 -2				
98		2	-3 -3					
99	2	2					2 2	
1300			-3 -3					-2
01		(2)(2)(2)	(2)	(2)(2)	-2			
02		2	-3 -3					2 2
03		-2	-2 -2 2	2 2 2				
04	2 2 2	2	-3 -3 -3	-3				
05	-2	-2 -3	-3 -3 -2	-2 -2	-2 -2	-2	2	2 2
06			-3 -3		-2 -3 -2	-2 -2		
07		(-2)(-2)	(-2)(-2)	3 3				
08	2	-3	-3 2	2 2				
09			-3 -3 (2)	(2)-2 -2				
1310		-2	-2 -2	-2 -2	-3			
11		-2	-2		(-2)			
12		-2 -2	-2					
13	(2)(2)(2)		-2 -2 2	2 2 2				
14			2 2	2 2 3	-2 -2 -2	-2		
15	2 2		3 3 3	3 3 3	-2 -2			
16	3	2 2	2 2 2	2 2				
17	2 2 2				-2 -2 -2			
18		2	-3 -3				-2	
19			(2) 3	3 3				
1320			2	2 2 2				
21		2 2	-2 -2					
22			(2)	(2)(2)	-3	-3 -3		
23		(2)(2)		-2 -2				
24			(2)	(2)(2)				
25			-3 -3	-2				-2
26	-2 -2 -2	-2	-3 -3		-2 -2 -2	-2		
27	-2		-2 -2					-2
28					-2 -2 -2	-2		
29	(2)(2)(2)		-3 -3					-2
1330		2	2 2 2	2 2 2	-2 -2 -2	-2		
31		-2 -2 -2	-2					
32			-2 -2					
33	(-2)(-2)(-2)		-3 -3					
34								-2
35			3	3 -2 -2	-3 -3 -3	-2		
36			-3 -3					
37			(-2) 2 3	3 3 2				
38	2 2		(-2)(-2) 3	3 2 2				
39	2 2 2		(2)	(2) 2	-2 -3 -3	-2		
1340		-2	-2 -2		-3 -3	-3	2 2	
41								

Table 6.3 continued

Year	Precipitation: Winter D J F	Spring M A M	Summer J J A	Autumn S O N	Temperature: Winter D J F	Spring M A M	Summer J J A	Autumn S O N
42		(-2)	-2 -2 2	2 2	-2	-2 -2		
43			-3 -3					
44			-3 -3					
45	(2)(2)(2)		-2 -2 2	2 3 3				
46	2 2 2	2 2	-2 -2	-2 -2				
47		(2)(2)	(-2)(-2) 2	2 2				
48			2 2	2 2 2				
49	2 2		2 2 2	2 2	-2 -2 -2			
1350		(2)(2)	-2 -2 2	2 2				(-2)
51		(2)(2)		-2 -2	(-2)(-2)(-2)	(-2)		
52		-2 -2	-3 -3					
53	-2 -2 -2	-2 -3 -3						
54					-2 -3 -3	-2		
55		-2	-2 2 3	3 3				
56		-2 -3 -3	-3 2 2	2 2				
57	(2)(2)	(2) -2	-2 -2					
58			-2 -2					
59			(-2)(-2) 2	2 2				
1360			-3 -3					
61			-3 -3					
62			-3 -3 2	2 2 2				
63		2 2	2 2		-2 -2 -2	-2		
64	2	2			-3 -3 -3	-3		
65			2	2 2				
66		-2	-2 2 2	2 2(2)				
67	(2) 2 2	2	2 2(2)	(2)(2)				
68		(-2)(-2)	(-2)(-2) 2	2 2				
69				(2)				
1370	(2)(2)(2)	(2)(2)(2)	-2 -2 2	2 2				
71	-2	-2 -2	-2 -2 2	2 2	-2	-2	-2 -2	
72			-2 -2 3	3 2			2 2	
73		2 2	-2 -2 3	3 3 2				2
74	2 2 2		(-2)(-2)		2 2 2			
75			-3 -3(-2)(-2)(-2)				3 3 (2)	(2)(2)
76		(2)(2)(2)	(-2)(-2)(-2)(-2)(-2)(-2)				(2)	(2)(2)
77	(-2)(-2)(-2)	(-2)(-2)(-2)	(-2)(-2)(3)	(3)(3)				
78			-2 -2					
79			2					
1380		(2)(2)	(2) 2	2 3 3				
81	3	(2)(2)						
82	-2	-2 -2	3 3 3	3				
83	2		-3 -3 -3	-3				
84		-2 -2 -2	-3 -3 -2	2 2				
85	2 2 2		-2 -2		2	2 2	2 2 2	2
86			(-2)(-2) 2	2				
87			2	2 2			2 2	-2
88		-2 -2	-2 -2 2	2 2	-2 -2 -2	-2		
89			2	2 2				
1390	-2	-2 -2	(-3)(-3)				2 2 2	2

Table 6.3 continued

	Precipitation:				Temperature:			
Year	Winter D J F	Spring M A M	Summer J J A	Autumn S O N	Winter D J F	Spring M A M	Summer J J A	Autumn S O N
91			(−3)(−3) 2					
92								
93							2 2	
94			−3 −3					
95		2 2 2				2 2 2		
96		2 2 2	(2)(2) 2	2 2				
97			−3 −3					
98								
99			2	2 2				
1400			2 2	2 2				
01			2 2 3	3 3				
02			2 2 2	2 2				
03			(2)(2)					
04			2 2 2	2 2				2
05					2 2 2			
06								
07			−3 −3					
08					−3 −3 −3	−3		
09			(2)					
1410	−2 −2	−2	−3 −3					
11								
12			−2 −2					
13	−3 −3	−3	−3 −3					−2
14			−3 −3		2 −2 −2	−2 −2		
15	2 2 2	2 2	−3 −3 −3	−3 −3				
16	2	2 2	−3 −3					
17			−3 −3					
18			−3 −3					
19			−3 −3 2	2 2		−2 −2		
1420			−3 −3					
21			3	3 3				
22		3 3	3 −2 −3	−3 −3				
23			−3 −3 2	2 2 2				
24	2 2 2		−3 −3					
25								
26								
27				−2				
28	−2 −2 −2	2 2 2	3 3 3	2 2 2				
29					−2	−2		
1430								
31	2							
32			2 2					
33								
34								
35								
36								
37								
38								
39								

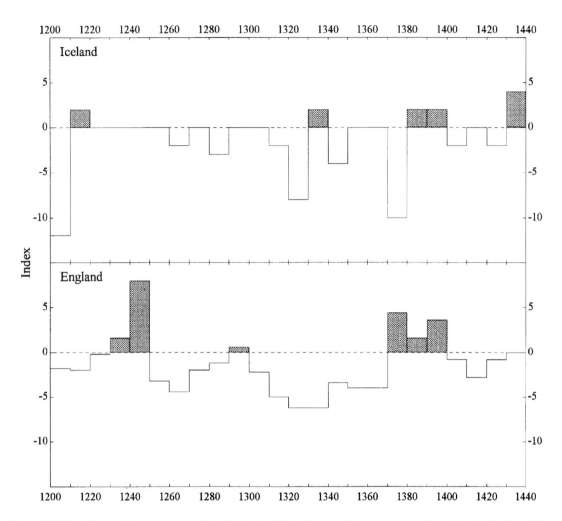

Figure 6.4 Decadal annual temperature indices for Iceland (data from northern, western and southern Iceland) and for England[1] (data from southern England) for the period AD 1200 to 1439. The Icelandic data are indicators of annual temperatures obtained by subtracting the number of cold seasons from mild seasons. For the English data, the values shown are the sum of the seasonal indices divided by four. Shaded decades are warmer than the average. Note that this index is closely related to the inverse of the winter severity index shown in Figure 6.3.

statement is then given: 'This is our old sailing route, but now ice is come out of the deeps of the north-east . . . [and] no one sails this old route without putting their life in danger . . .'[41] This comment on sea-ice has often been taken as a strong indication for an increase in sea-ice *c.* 1350. The earliest time that the work could have been written, however, is *c.* 1364 and, furthermore, it is very likely that the account regarding sea-ice is, in fact, a later interpolation. If this is correct then we cannot be sure what period it refers to, or indeed if it refers to a long-term or short-term (transient) change.[42] As noted above, there are better sources which do, in fact, suggest that there was high sea-ice incidence *c.* 1350.

CONCLUSIONS

The climate of the North Atlantic region is driven by local winds and ocean currents as well as by the general large-scale atmospheric and oceanic circulations (see Chapter 2). On longer time-scales of centuries and more, a similarity in climatic regimes would be expected in areas centred in this region. On shorter time-scales local effects would be more marked. The existence of an interesting 'see-saw' in winter temperatures between Greenland and northern Europe may also be noted. This tendency for winter temperatures to be low over northern Europe when they are high over Greenland and the Canadian Arctic has been documented.[43] Out of interest, an attempt to compare data on medieval winter temperatures for England and Iceland was made here. Unfortunately the data available were not sufficient to do a full analysis, but some years could be noted as corresponding to this pattern (for example, in 1254) while other winters were similar in both countries in terms of the contemporary descriptions of mildness or severity. An example of the latter is the year 1290, when a very cold winter is described in both England and Iceland.

The years from around the ninth to the fourteenth centuries, the 'medieval period' in European perspective, have been characterised as having been as warm as, or warmer than, today and the period has come to be known as the Medieval Warm Period (also the Little Climatic Optimum or Medieval Warm Epoch). Following this, many researchers have discerned a period (e.g., the sixteenth to eighteenth centuries; see Chapter 9) much colder than today which has come to be known as the **Little Ice Age**. Foremost among these researchers is Hubert Lamb who argued that, amongst other regions, the British Isles, Iceland and southern Greenland were favoured by a prolonged warm phase from around the tenth to the thirteenth centuries which facilitated the colonisation of the latter two places by Norse peoples.[44] While aspects of this picture may be correct, as further research is done the details appear more complex.[45] Interestingly, the century from 1260 to 1360, based on our improved historical dataset,

appears to have been one of relatively cold winters in England (see Figure 6.3). Lamb's suggestion of a 'Medieval Warm Period' is not supported by the documentary data, neither his nor ours. This is not necessarily a contradiction because his basis for a 'Medieval Warm Period' is largely biological and phenological. Nevertheless, if such a 'Medieval Warm Period' did exist, it was clearly less well-defined and climatologically more complex than has popularly been believed.

Apart from the pioneering work by Lamb mentioned above, work on medieval climate indices for Europe has been undertaken by several other researchers, notably by Pierre Alexandre,[46] whose work centred on Belgium and neighbouring regions. In another recent study, Zongwei Yan and colleagues[47] compare reliable historical datasets from continental Europe for the medieval period with, for example, tree-ring data for Fennoscandia[48] and glacier oscillations in the Swiss Alps.[49] In summary, they observe that the whole of western Europe seems to have experienced a drying and warming trend around AD 1200. They also suggest that, during the entire period from AD 1200 to 1426, temperatures remained at a more or less stable level while precipitation increased. They also argue strongly for caution in using the traditional term the 'Medieval Warm Period', suggesting, as we do here, that this conceptual scenario is being challenged as more detailed data become available.

Overall, the English data analysed here show a long time-scale cooling c. 1240 to c. 1340, warming c. 1510, and thereafter cooling. The data from Iceland suggest a mainly variable climate during the early to mid-twelfth century, but a distinctly colder climatic regime during the latter part of the twelfth and early thirteenth century. This picture is not dissimilar to that for England. The fourteenth century in Iceland appears to have been very variable, but the 1320s and 1370s, and possibly the 1340s and 1360s, were almost certainly relatively cold (see Figure 6.4). In terms of decade-to-decade changes in the fourteenth century, there appears to be an out-of-phase relationship between Iceland and England, whereas the period between 1395 to 1430 seems to have been relatively mild in Iceland, similar to the situation in England.

As stated above, it has been suggested by some researchers that a warm phase in medieval times was followed by a climatic phase that was generally so cold that it has become traditionally known as the 'Little Ice Age'. As more research is done into the climatic history of the past thousand years, it becomes clearer that, as for the hypothesised 'Medieval Warm Period', the situation was more complex than has been previously thought and that this term also should be used advisedly.

NOTES

1   This analysis of documentary climate data for England in the medieval period is based on work undertaken in the Climatic Research Unit by a number of researchers, among them T.M.L. Wigley, G. Farmer, R. Mortimer, M.J. Ingram, D.J. Stern and A.E.J. Ogilvie. Two specific publications/reports on this work may be mentioned. These are: G. Farmer and T.M.L. Wigley, 'The reconstruction of European climate on decadal and shorter time scales', 1984, Final Report to the Commission of the European Communities under Contract No. CL-029–81-UK(H), unpublished report, Climatic Research Unit, Norwich, UK; T.M.L. Wigley, G. Farmer and A.E.J. Ogilvie, 'Climatic reconstruction using historical sources', in A. Ghazi and R. Fantechi (eds), *Current Issues in Climatic Research*, Proceedings of the EC Climatology Programme Symposium, Sophia Antipolis, France, 2–5 October 1984, Commission of the European Communities, Dordrecht, Reidel, 1986.

2   For a fuller discussion see A.E.J. Ogilvie, 'Climatic changes in Iceland AD *ca.* 865 to 1598', in G.F. Bigelow (presenter), *The Norse of the North Atlantic, Acta Archaeologica*, 1991, vol. 61, pp. 233–51; A.E.J. Ogilvie, 'Historical accounts of weather events, sea-ice and related matters in Iceland and Greenland,' AD *ca.* 1250 to 1430', in B. Frenzel (ed.), 'Documentary climatic evidence for 1750–1850 and the 14th century', *Palaeoclimatic Research/Paläoklimaforschung 23, Special Issue 15*, 1997 (in press), Mainz, The European Science Foundation and the Academy of Sciences and Literature.

3   See, for example, H.H. Lamb, 'The early medieval warm epoch and its sequel', *Palaeogeography, Palaeoclimatology, Palaeoecology*, 1965, vol. 1, pp. 13–37; H.H. Lamb, *Climate: Present, Past and Future. Vol. 2:*

Climate History and the Future, London, Methuen, 1977. However, see also Ogilvie, op. cit., 1991, pp. 247–9, and M.K. Hughes and H. Diaz, 'Was there a 'Medieval Warm Period', and, if so, where and when?', in M.K. Hughes and H.F. Diaz (eds), *The Medieval Warm Period*, Reprinted from *Climatic Change*, 1994, vol. 26, pp. 109–42, Dordrecht, Kluwer Academic Publishers.

4   For a critique of the main compilations see W.T. Bell and A.E.J. Ogilvie, 'Weather compilations as a source of data for the reconstruction of European climate during the medieval period', *Climatic Change,* 1978, vol. 1, pp. 331–48.

5   Lamb, op. cit., 1977, Appendix 5, Table 4.

6   Bell and Ogilvie, op. cit.

7   C.E. Britton, 'A meteorological chronology to AD 1450', Meteorological Office Geophysical Memoirs, No. 70, 1937, London, His Majesty's Stationery Office.

8   T.F. Tout, 'Medieval Forgers and Forgeries', in *The Collected Papers of Thomas Frederick Tout*, vol. 3, Historical Series no. LXVI, Publication of the University of Manchester, no. CCXXI, Manchester, Manchester University Press, 1934, pp. 117–43.

9   Ibid.

10   See, e.g., A.E.J. Ogilvie, 'Documentary evidence for changes in the climate of Iceland, AD 1500 to 1800', in R.S. Bradley and P.D. Jones (eds), *Climate since AD 1500*, London, Routledge, 1992, pp. 92–117.

11   The importance of source analysis with regard to historical climatology has been discussed by e.g., Th. Vilmundarson, 'Evaluation of historical sources on sea ice near Iceland', in T. Karlsson (ed.), *Sea Ice*, Proceedings of an International Conference, Reykjavík, National Research Council, 1972, pp. 159–69; W.T. Bell and A.E.J. Ogilvie, op. cit., 1978; M.J. Ingram, D.J. Underhill and G. Farmer, 'The use of documentary sources for the study of past climates', in T.M.L. Wigley, M.J. Ingram and G. Farmer (eds), *Climate and History*, Cambridge, Cambridge University Press, 1981, pp. 180–213; Ogilvie, op. cit., 1991.

12   P. Bergthórsson, 'An estimate of drift ice and temperature in 1,000 years', *Jökull*, 1969, vol. 19, pp. 94–101. See also A.E.J. Ogilvie, 'The past climate and sea-ice record from Iceland, part 1: data to AD 1780', *Climatic Change* 1984, vol. 6, pp. 131–52 (esp. pp. 133, 146).

13   These sources, *Konungs Skuggsjá* (The King's Mirror) and *Grænlandslýsing Ívar Bárdarsonar* (Ívar Bárdarson's Description of Greenland) are discussed more fully in Ogilvie, 1996, op. cit.

14   The origins of, and differences between, some of these sources are discussed in Bell and Ogilvie, 1978, op. cit.

15   William Merle's journal is discussed in Chapter 7 of this book. See also R. Mortimer, 'William Merle's

weather diary and the reliability of historical evidence for medieval climate,' *Climate Monitor*, 1981, vol. 10(2), pp. 42–5.

16 J.Z. Titow, 'Evidence of weather in the account rolls of the Bishopric of Winchester 1209–1350', *Economic History Review*, 2nd ser., 1960, vol. 12, pp. 360–407; J.Z. Titow, 'Le climat à travers les rôles de compt-abilité de l'évêché de Winchester (1350–1450)', *Annales ESG*, 1970, vol. 25, pp. 312–50.

17 D.V. Stern, 'A Hertfordshire manor of Westminster Abbey: an examination of demesne profits, corn yields and weather evidence', Unpublished Ph.D. thesis, King's College, London, 1978.

18 P.F. Brandon, 'Late-medieval weather in Sussex and its agricultural significance', *Institute of Geographers, Transactions No. 54*, 1971, pp. 1–17.

19 These different sources are all fully discussed in Ogilvie, 1991, op. cit.

20 This situation improved in England after the end of the fifteenth century when political and social order was restored after the accession of Henry Tudor in 1485.

21 A detailed discussion on dating may be found in Lamb, 1977, op. cit., p. 49. A discussion specifically on Icelandic dating may be found in A.E.J. Ogilvie, 'Climate and society in Iceland from the medieval period to the late eighteenth century', Unpublished Ph.D. thesis, School of Environmental Sciences, University of East Anglia, Norwich, 1982, p. 35.

22 Titow, 1960, op. cit., p. 361.

23 J.Z. Titow, *Winchester Yields*, Cambridge, Cambridge University Press, 1972, pp. 5–7.

24 Ogilvie, 1991, op. cit.; Ogilvie, 1996, op. cit.

25 Lamb, 1977, op. cit.

26 Britton, 1937, op. cit.

27 All these data and their sources are discussed more fully in Ogilvie, 1991, op. cit.

28 Translated from G. Storm, *Islandske Annaler Indtil 1578*. Reprinted by Norsk Historisk Kjeldeskrift Institutt, Oslo, 1977. Originally published in 1888, Udgivne for det norske historiske kildeskriftfond, Christiania, p. 260.

29 The loss of the Norse Greenland colony is discussed in detail in P.C. Buckland, T. Amorosi, L.K. Barlow, A.J. Dugmore, P.A. Mayewski, T.H. McGovern, A.E.J. Ogilvie, J.P. Sadler and P. Skidmore, 'Bioarchae-ological and climatological evidence for the fate of Norse farmers in medieval Greenland', *Antiquity*, 1996, vol. 70, pp. 88–96. See also L.K. Barlow, J.P. Sadler, A.E.J. Ogilvie, P.C. Buckland, T. Amorosi, J.H. Ingimundarson, P. Skidmore, A.J. Dugmore, and T.H. McGovern. 'Interdisciplinary investigations of the end of the Norse Western Settlement in Greenland' (unpublished).

30 L.M. Larson *The King's Mirror (Speculum Regale – Konungs Skuggsjá)*. New York, Scandinavian Mono-graphs Vol. 3, The American–Scandinavian Foun-dation, 1917. For a discussion of this work see Ogilvie, 1996, op. cit.

31 F. Jónsson *Det gamle Grønlands beskrivelse af Ívar Bárdarson (Ivar Bårdsson)*, Udgiven efter Hånds-krifterne, Copenhagen, 1930. For a discussion of this work see Ogilvie, 1996, op. cit.

32 Larson, 1917, op.cit., p.138.

33 Storm, op. cit., p. 332.

34 Ibid. pp. 53, 148, 201, 340.

35 Ibid. pp. 152, 204, 267, 345.

36 Ibid. pp. 152, 205, 267, 345.

37 Ibid. pp. 358–9.

38 Ibid. p. 280.

39 Translated by Ogilvie from the text as published by G. Vigfússon and J. Sigurdsson *et al.*, *Biskupa Sögur – 2*. Hinu Íslenzka Bókmentalfélagi, Copenhagen, 1878, p.5.

40 See, for example, Lamb, 1977, op. cit., p.6.

41 Jónsson, 1930, op. cit., pp. 17–18.

42 For a more detailed discussion of this see Ogilvie, 1996, op. cit.

43 H. Van Loon and J.C. Rogers, 'The see-saw in winter temperatures between Greenland and Northern Europe. Part 1: General Description', *Monthly Weather Review*, 1978, vol. 106, pp. 296–310; J.C. Rogers and H. Van Loon, 'The see-saw in winter temperatures between Greenland and Northern Europe. Part 2: Some oceanic and atmospheric effects in middle and high latitudes', *Monthly Weather Review*, 1979, vol. 107, pp. 509–19.

44 Lamb, 1965, op. cit.; Lamb 1977, op. cit.; H.H. Lamb, *Climate: History and the Modern World*, London, Methuen, 1982.

45 See, for example, Ogilvie, 1991, op. cit.; Ogilvie 1996, op. cit.; Hughes and Diaz, 1994, op. cit.

46 P. Alexandre, *Le Climat au Moyen Age en Belgique et dans les régions voisines (Rhénanie, Nord de la France)*, 1976, Liège, Publication No. 50 du Centre Belge d'Histoire Rurale.

47 Z. Yan, P. Alexandre and G. Demarée, 'Some seasonal climatic scenarios in continental western Europe based on a dataset of medieval narrative sources, AD 708–1426', Brussels, *Institut Royal Météorologique de Belgique, Publication scientifique et technique No. 003*, 1996 (in press).

48 K.R. Briffa, P.D. Jones, T.S. Bartholin, D. Eckstein, F.H. Schweingrüber, W. Karlén, P. Zetterberg and M. Eronen, 'Fennoscandian summers from AD 500: temperature changes on short and long timescales', *Climate Dynamics*, 1992, vol. 7, pp. 111–19.

49 J.M. Grove and R. Switsur, 'Glacial geological evidence for the Medieval Warm Period', in M.K.

Hughes and H. Diaz (eds), *The Medieval Warm Period*, Reprinted from *Climatic Change*, 1994, vol. 26, pp. 143–70.

GENERAL READING

W.T. Bell and A.E.J. Ogilvie, 'Weather compilations as a source of data for the reconstruction of European climate during the medieval period', *Climatic Change*, 1978, vol. 1, pp. 331–48.

R.S. Bradley and P.D. Jones (eds), *Climate since AD 1500*, London, Routledge, 1992.

M.K. Hughes and H.F. Diaz (eds), 'The Medieval Warm Period', *Climatic Change* (Special Issue), 1994, vol. 26, Dordrecht, Kluwer Academic Publishers.

T.M.L. Wigley, M.J. Ingram and G. Farmer (eds), *Climate and History*, Cambridge, Cambridge University Press, 1981.

Part 3

MONITORING THE PRESENT

We, which now behold these present days,
Have eyes to wonder, but lack tongues to praise.
William Shakespeare, Sonnet 106

7

OBSERVING AND MEASURING THE WEATHER

A Brief History

John Kington

Good order is the foundation of all things.
Edmund Burke

INTRODUCTION

Weather varies enormously over the Earth and is of the greatest geographical significance for human endeavour. Travellers and writers have from the earliest times described the almost endless varieties of weather to be found from place to place, and also from time to time at the same place. Simple observations of the weather, encaptured in traditional weather lore and poetry, are as old as literature itself and the Mesopotamian, Egyptian, Greek and Hebrew civilisations all contain numerous references to their meteorological knowledge. As an example of how this rather subjective knowledge can be used in the scientific process, Chapter 6 examined in some detail how literary sources from the Middle Ages can be used to reconstruct climatic variations over tens and hundreds of years. Following the early meteorological investigations by the Greeks – notably Aristotle and Hippocrates – and the written records of the medieval scholars and travellers, the rise of western science in the wake of the period of geographical exploration in the fifteenth and sixteenth centuries led to a new thirst and rigour for meteorological observation (see Figure 7.1). Observation duly led to

explanation and the establishment of the first principles of scientific weather forecasting.

This chapter follows the history of weather observation and measurement, with particular emphasis on the British Isles, from the Renaissance through to the present century. During the nineteenth century, many of the key figures responsible for setting up standardised weather observing networks came from these islands. Nevertheless, this innovation only came about due to the wider European legacy of new meteorological recording instruments and international co-operation between learned societies and scholars. This brief historical review ends in the era of modern observing systems and satellites, but considers along the way the continuing role of the amateur observer. The ways in which these evolving weather observing systems contributed to the development of weather forecasts is discussed in more detail in Chapter 14.

THE PRE-INSTRUMENTAL PERIOD

During the Middle Ages texts from Greek and Muslim sources became accessible in western Europe

Tycho Brahe in seiner Sternwarte „Uranienburg" auf der Insel Hveen
Nach einem Kupferstich zu Tycho Brahes „Astronomiae instauratae Mechanica" vom Jahre 1602

Figure 7.1 One of the 'earliest climatologists', Tycho Brahe (1546–1601), in his observatory on the island of Hveen (now Ven, Denmark) (taken from the 1602 edition of his *Astronomiae Instauratae Mechanica*).

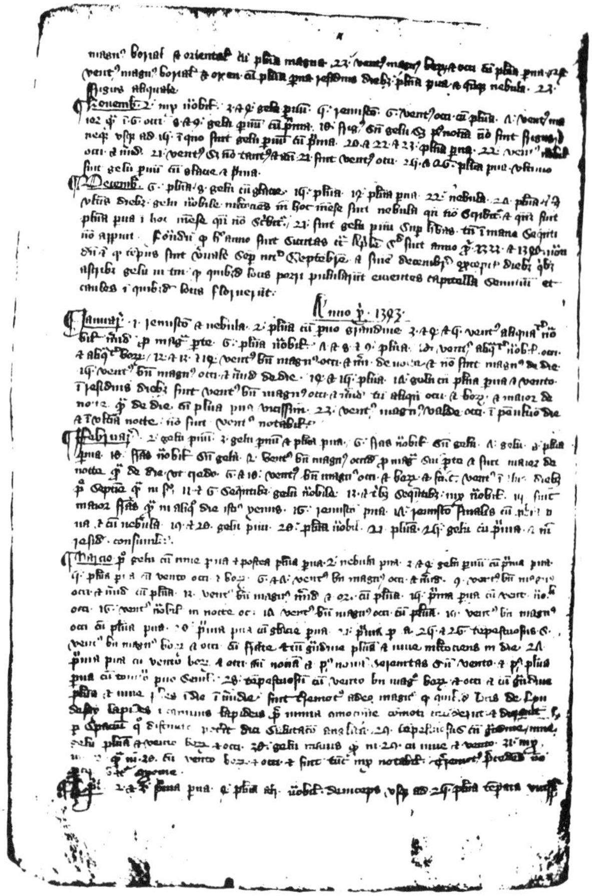

Figure 7.2 A page of William Merle's weather journal (1337–44); the monthly accounts, recorded in Latin, contain details about individual days.[1]

through Latin translations. Amongst these volumes were manuscripts claiming to predict human destiny, as well as natural events such as the weather, from the motions of the stars, planets, sun and moon. These texts were enthusiastically received in learned circles as being novel and promising for meteorology. Up to that time it had only been possible to forecast the weather for short periods ahead using empirical rules based on 'signs' related to the appearance of the sky. The new astrological hypotheses held out hope that it might be possible to make longer-range weather predictions according to the motions of

heavenly bodies. This study was termed astro-meteorology and, with its mystical origins, came to enjoy the patronage of many authorities, including some notable figures in the history of science such as Johannes of Toledo and Justus Stöffler.

Despite this false hope of centuries past, a bonus for current climatic research has emerged from these early attempts to predict the weather by astrological methods: daily meteorological observations began to be made on an increasing scale in Europe during the Middle Ages.

The manner of making the entries, sometimes in

the margins of astronomical tables and almanacs, appears to suggest that, at first, there was not a great concern about keeping a detailed record of the weather in its own right. Rather, an attempt was made to establish possible links between certain astronomical phenomena (such as eclipses, planetary conjunctions and phases of the moon) and particular types of weather – as well as compiling meteorological statistics to assess the success or failure of previous astrometeorological predictions.

Over the period from the thirteenth century to the early 1600s, a gradual change can be detected in the manner of recording, with the astrological entries becoming less frequent whilst the meteorological observations become more continuous and orderly. A good example of the latter type of record is the journal of William Merle, rector of Driby, Lincolnshire and Fellow, perhaps, of Merton College, Oxford.[1] Merle has the distinction of being the author of the earliest known systematic register of the weather. His journal, *Temperies aeris Oxoniae pro septennio*, extends over a seven-year period from January 1337 to January 1344 (see Figure 7.2). Merle's example was more or less followed during the following three centuries with the result that even before the basic observing instruments were invented in the seventeenth century, there were many learned people who were laying the foundations of meteorology, as an exact science, by making systematic reports of eye-observable weather elements such as the state of the sky, wind direction and precipitation.

THE BEGINNING OF INSTRUMENTAL OBSERVING

The invention of the thermometer by Galileo in 1597 and the barometer by his pupil, Torricelli, in 1643 made it possible to begin instrumental meteorological observing in the early seventeenth century. Great interest was shown in these new instruments, as it was hoped that their response to changing atmospheric conditions would provide the means of predicting the weather using the scientific method

of **induction** advocated by Francis Bacon in the early 1600s.

It was soon realised that the value of such exact and quantifiable data would be greatly enhanced if readings at various places could be made simultaneously, and in 1653 the first attempt to establish a network of meteorological observing stations was made in Italy under the patronage of the Grand Duke of Tuscany, Ferdinand II, founder of the Accademia del Cimento. Standardised instruments were dispatched from Florence to about a dozen stations, mostly situated in northern Italy, and a uniform procedure for making the observations was devised. Although the network ceased to function after the Academy was disbanded in 1667, it did set the pattern for later attempts. In fact, the idea was taken up again in the 1660s by the newly founded Royal Society of London when Robert Hooke proposed a scheme to compile a history of the weather (see Figure 7.3).

The aim of establishing a network of meteorological stations was further pursued by the Royal Society in 1723 when its Secretary, James Jurin, issued an invitation to the scientific community at large to form a world-wide system of weather observations. Jurin's interest in this undertaking was both meteorological and medical; earlier he had studied medicine at Leyden under Hermann Boerhaave, who was interested in the association between weather and public health. Like many other early attempts, however, the momentum was lost after a few years and it was not until another half-century or so had passed that the time was ripe for such efforts to succeed.

In the 1770s, following the lead of meteorologically inclined physicians earlier in the century, medical authorities in France decided to make a systematic study of weather and public health. As a result, the Société Royale de Médecine was established in 1778, under the patronage of Louis XVI, to maintain a regular correspondence on meteorological and medical matters with doctors throughout the French kingdom. Felix Vicq d'Azyr, physician to Marie-Antoinette, was appointed Secretary-General and, together with the meteorologist-cleric, Louis

ROYAL SOCIETY. 179

A
SCHEME
At one View reprefenting to the Eye the Obfervations of the Weather for a Month.

Dayes of the Month and place of the Sun. Remarkable houre.	Age and fign of the Moon at Noon.	The Quarters of the Wind and its ftrength.	The Degrees of Heat and Cold.	The Degrees of Drinefs and Moyfture.	The Degrees or Preffure.	The Faces or vifible appearances of the Sky.	The Notableft Effects.	General Deductions to be made after the fide is fitted with Obfervations: As,
14 II 12.46	27 ♂ 9. 46. Perigeu.	W 2. 3. 3½ W.S.W.1	9 12 16 10 7	5 8 9	29½ 29⅜ 29⅔	Clear blew but yellowish in the N. E. Clowded toward the S. Checker'd blew.	A great dew. Thunder, far to the South. A very great Tide.	From the laft Q of the Moon to the Change the Weather was very temperate, but cold for the feafon; the Wind pretty conftant between N.& W.
15 II 13.40	18 ♂ 24.51.	N. W. 3 N. 2	28½ 8 7	29½ 9	29⅒	A clear Sky all day, but a little Check'd at 4. P. M. at Sun-fet red and hazy.	Not by much fo big a Tide as yefterday. Thunder in the North.	A little before the laft great Wind, and till the Wind rofe at its higheft, the Quick-filver continu'd defcending til it came very low; after wch it began to re-afcend, &c.
16 14 37	N.Moon. at 7. 25' A. M. II 10. 8.	S.	10	10 28½		Overcaft and very lowr-ing.	No dew upon the ground, but very much upon Marble-ftones, &c.	
&c.	&c.	&c.	&c.	&c.	&c.	&c.	&c.	

Z 2   D I.

Figure 7.3 An example from the Royal Society scheme for making weather observations, as suggested by Robert Hooke in the 1660s.[3]

Cotte, became actively involved in establishing a network of weather stations which, by the mid-1780s, comprised over seventy stations and had been extended beyond France to include correspondents in other parts of Europe as well as in America and Asia (see Figure 7.4). This was the kind of organisation envisaged by Antoine Lavoisier when he suggested that in order to prepare a forecast a meteorologist would need to have simultaneous daily observations of the principal weather elements.

Unfortunately, the means of transmitting the observations to a central forecasting office where they could be processed rapidly enough to keep pace with the weather were lacking at that time, and meteorologists had to wait several more decades for advances in both communications and weather studies before Lavoisier's idea could be realised. Sadly, the Société Royale de Médecine was suppressed in 1793 by a French Revolutionary decree. However, the original manuscripts of daily observations made in the 1780s have been preserved in the archives of the Académie de Médecine in Paris, where, after nearly two centuries of oblivion, they were brought to light in 1965 by Jean Meyer and Emmanuel Le Roy Ladurie at the Sorbonne.[2]

The network of the Société Royale de Médecine was not the only such system established in the 1780s for the advancement of weather observing. Mannheim, capital of the Rhineland Palatinate, had developed into an influential centre for the arts and sciences during the Enlightenment. In 1780 the Prince-Elector, Karl Theodor, founded the Societas Meteorologica Palatina and appointed his Court Chaplain, Johann Hemmer, as its director. Besides investigating medico-meteorological relationships, scientists in the eighteenth century were also concerned that progress towards finding a satisfactory method of predicting the weather had been disappointingly slow since the appearance of the barometer and thermometer in the early 1600s. Consequently, members of the Societas Meteorologica Palatina hoped that the analysis of standardised weather data by statistical methods would show that variations of atmospheric behaviour were subject to a regularity comparable to that expressed in the laws of nature discovered earlier by Johann Kepler and Isaac Newton. Each station of the Mannheim weather network was supplied, gratis, with a set of instruments together with detailed instructions, written in Latin, on observational procedure. Registers were dispatched annually to Mannheim for publication in the *Ephemerides* of the Society, which were issued to all participating observers (see Figure 7.5).

From a nucleus of about a dozen stations, mostly located in central Europe, the network rapidly

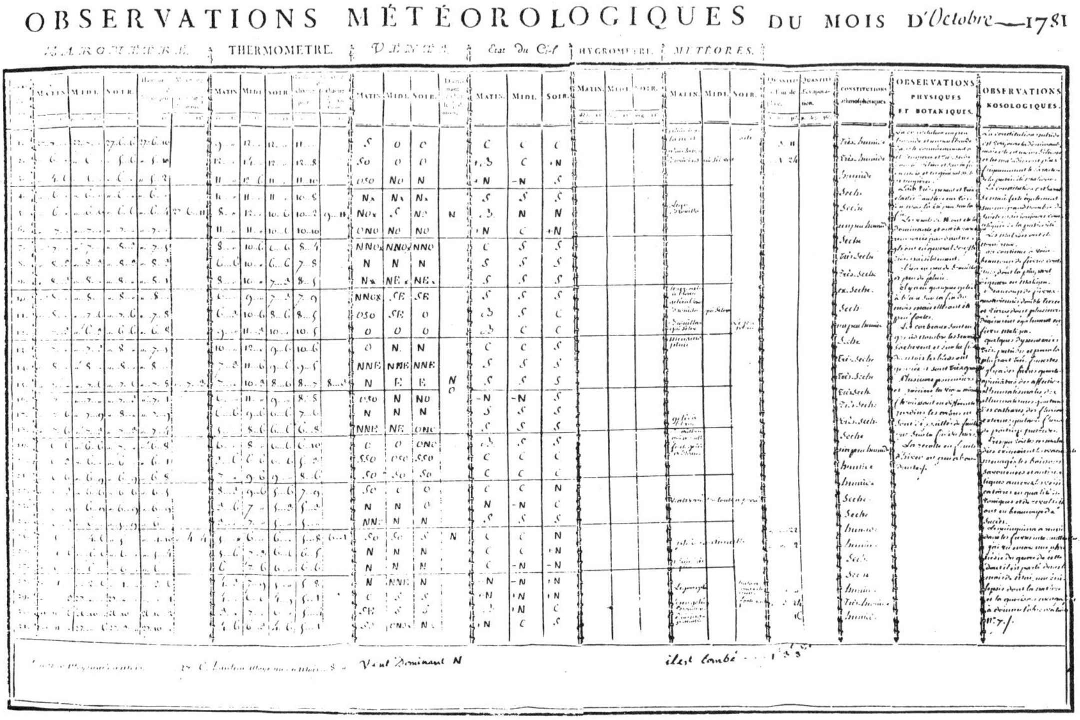

OBSERVATIONS MÉTÉOROLOGIQUES DU MOIS D'*Octobre*—1781

Figure 7.4 An extract from a meteorological register of the Société Royale de Médecine including thrice-daily observations of pressure (Paris inches), temperature (degrees Réaumur), wind, state of the sky, and significant weather; recorded at Dijon in October 1781.[3]

expanded, so that by the mid-1780s it included over fifty observatories extending from Russia across Europe to eastern North America. The observers, mostly physicists, astronomers, and clerics, were associated with the various scientific academies, learned societies, and observatories which had been established in many European cities during the Enlightenment. Although Hemmer died in 1790 the activities of the Society continued for a further five years. Unfortunately, with the publication of the twelfth volume of the *Ephemerides*, containing data for 1792, the series was brought to a close. The Society was facing increasing financial difficulties, and the final blow which led to its disbandment was the fall of Mannheim to French Revolutionary forces in 1795.[3]

Besides the weather stations organised by the two European scientific societies discussed above, a large number of private observers were also recording daily meteorological observations during the eighteenth century. In the British Isles, these more individual efforts were made mostly by physicians, clerics and country landowners. Although working in isolation, they sometimes corresponded with one another about their mutual interests in meteorology and natural philosophy, the Royal Society providing a centre for more formal discussion and exchange of ideas. One of the best examples of these individual meteorological observers was the Rutland squire, Thomas Barker, who, beginning in 1736, maintained an instrumental weather register at Lyndon Hall, near Oakham, for over sixty years (see Figure 7.6).[4]

OBSERVATIONES BUDENSES

Autore Bruna.

Horae obſervationis ordinariae 7 mat. 2 pom. 9 veſp.

Januarius.

Dies.	Barom.	Therm. intern.	Therm. extern.	Hygr.	Declin.	Ventus.	Pluvia.	Evap.	Danub	Luna.	Coeli fac.	Meteora.
	dig. lin. dec.	gr. dec.	gr. dec.	gr. dec.	gr. min.	direct. vires.	lin. 64tae	lin. dec	ped. dig.			
1	27, 1, 5	—0, 9	—4, 8	20, 4	15, 54	WNW 1					=- cin.	
	2, 0	—0, 3	—1, 0	20, 8	54	WNW 1				♑	==	
	1, 3	—0, 8	—1, 9	16, 4	54	WNW 2					==	
2	26, 11, 0	—0, 5	—0, 3	14, 2	15, 54	WNW 2					==	‡‡ h. 7 mat.
	10, 9	—2, 0	—4, 1	19, 0	51	WNW 4	0, 51			♒	==	
	27, 1, 0	—3, 3	—7, 7	21, 2	51	NW 3					=- cin.	
3	27, 3, 6	—4, 2	—7, 4	24, 0	15, 51	WNW 2					==	
	3, 8	—3, 0	—4, 5	25, 5	54	NNW 1				♒	==	
	3, 6	—3, 0	—5, 6	23, 2	54	NNW 2					==	
4	27, 3, 8	—4, 2	—8, 0	22, 4	15, 51	NW 3					==	‡‡ nocte praec. & ante meridiem per intervalla.
	5, 0	—4, 1	—7, 3	21, 8	51	NW 3	1, 49			♓	==	
	6, 4	—4, 3	—7, 0	25, 3	48	NNW 2					==	
5	27, 7, 3	—4, 9	—7, 5	25, 2	15, 48	WgN 2					==	
	8, 0	—4, 4	—6, 6	25, 7	51	NW 2				♓	=- a.	
	8, 6	—5, 3	—9, 0	25, 6	51	WNW 3					⊙	
6	27, 9, 0	—5, 4	—10, 5	26, 5	15, 54	WNW 1					⊙	
	9, 1	—4, 6	— 7, 5	25, 7	54	NW 1				♈	⊙	
	9, 1	—5, 0	—11, 6	23, 3	51	NW 1					⊙	
7	27, 8, 4	—6, 0	—12, 5	22, 1	15, 51	WNW 1				☽ h. 1 m. 58 pom.	⊙	
	7, 0	—5, 3	—10, 3	23, 2	54	SW 2					=- a.	
	6, 2	—6, 5	—11, 9	22, 9	51	SO 2				♈	==	
8	27, 4, 5	—7, 0	—10, 9	22, 0	15, 51	SgO 1					==	‡‡ h. 7 mat. & h. 2 pom.
	4, 2	—6, 5	— 8, 2	20, 9	48	SO 1	1, 12			♉	==	
	4, 3	—6, 4	— 7, 7	18, 2	48	SO				☾	==	
9	27, 4, 3	—6, 3	—6, 9	17, 9	15, 48	SO 1					==	‡‡ nocte praec.
	4, 3	—5, 9	—4, 7	19, 3	45	SO	0, 38			♉	==	
	4, 3	—5, 9	—4, 9	17, 4	45	SO 1					==	
10	27, 3, 6	—5, 6	—3, 0	14, 6	15, 45	OgN 1					==	∴ n. mane.
	3, 4	—5, 0	—1, 2	13, 7	48	OgN 1				♉	==	
	3, 6	—4, 7	—1, 2	13, 2	48	OgN					==	∴ veſperi.
11	27, 3, 4	—4, 3	—0, 7	12, 5	15, 54	OgS 1					==	
	3, 0	—3, 4	1, 5	12, 3	51	OgN 1				♊	=- a.	
	3, 0	—2, 4	1, 9	13, 1	51	WgS 2					==	
12	27, 2, 7	—1, 0	2, 3	16, 0	15, 48	SgW 1					==	‡‡ h. 7 mat.
	2, 8	—0, 9	2, 8	14, 9	48	SW 1	0, 43			♊	==	
	2, 9	—0, 8	2, 0	13, 4	45	SSW 1					==	
13	27, 4, 4	—0, 4	2, 2	13, 9	15, 45	SW 1					=- c.	
	5, 0	1, 2	3, 8	13, 3	48	SSW 1				♋	-θ- a.	
	5, 2	1, 0	2, 5	11, 4	48	SSW 1						∴ veſperi.
14	27, 5, 0	0, 8	1, 7	19, 5	15, 51	NW 1				☺ h. 1 m. 51 pom.		∴ mane.
	4, 8	1, 4	2, 9	9, 3	51	NW 1					==	
	4, 8	1, 6	1, 4	10, 0	54	NW 1				♋	-θ- a.	
15	27, 5, 0	1, 0	0, 0	10, 2	15, 51	NW 1						∴ mane & poſt meridiem.
	4, 3	1, 8	0, 6	9, 6	51	NW				♌		
	3, 3	2, 0	2, 2	9, 5	48	SgW 2					==	

Figure 7.6 Two pages from the weather journal of Thomas Barker including twice-daily observations of pressure (English inches), temperature (degrees Fahrenheit), cloud movement, wind, state of the sky, and significant weather; recorded at Lyndon Hall, Rutland in June and July 1783.[4]

By the end of the Enlightenment meteorology had reached an impasse; the application of purely statistical methods to the analysis of standardised weather data had not resulted in the hoped-for breakthrough in understanding atmospheric behaviour. The key that eventually opened up the way to further progress was the mapping of observations over a large area so as to give a synoptic view of the general weather situation. This concept was first presented by Heinrich Brandes in 1820. As a systematic meteorological network did not then exist in Europe, Brandes made use of the extensive data collected in the 1780s. From his analysis, he clearly discerned, for the first time, the relation between pressure and wind flow – one of the fundamental principles of meteorology, later expressed as Buys Ballot's law, from which pressure systems, such as **depressions** and **anticyclones**, came to be recognised.

Figure 7.5 An extract from the *Ephemerides* of the Societas Meteorologica Palatina illustrating thrice-daily observations of pressure (Paris inches), temperature (degrees Réaumur), humidity, wind, state of the sky, and significant weather; recorded at Budapest in January 1786.[3]

Figure 7.7 The Durham University Observatory, designed by Anthony Salvin in 1840 and opened in 1843, was one of a number of observatories which were established at this time in the British Isles following the introduction of regular daily observations at Greenwich under James Glaisher.

THE ADVENT OF THE MODERN INSTRUMENTAL PERIOD

In 1832, Samuel Morse conceived the idea of the electric telegraph and by about 1840 he had made it possible for the system to be used as a workable apparatus for rapid communications. This, together with its almost immediate application by meteorologists, revolutionised meteorology as dramatically as the invention of the thermometer and barometer some 200 years earlier. For instance, in 1842 Karl Kreil suggested that meteorological observations should be collected by telegraphy as a basis of forecasting. This plan was put into operation at almost the same time in Britain and America. In 1848, James Glaisher used observations collected by telegraphy from a network of stations in the British

Figure 7.8 Rear-Admiral Robert FitzRoy (1805–65). From a posthumous painting (based on a photograph) by Francis Lane in 1882. It hangs in the Admiral's Residence at the Royal Naval College, Greenwich.

Isles (see Figure 7.7) to produce newspaper weather reports; three years later **synoptic maps** were printed at the Great Exhibition in London.

In 1854, Robert FitzRoy (see Figure 7.8) was appointed Head of the Meteorological Department of the Board of Trade – the first official meteorological service in the British Isles. By 1857, a series of daily weather maps was being prepared based on simultaneous observations made at a number of stations within the area of western Europe. During the spring of 1859 it was decided to increase the number of observations and to enlarge the area to

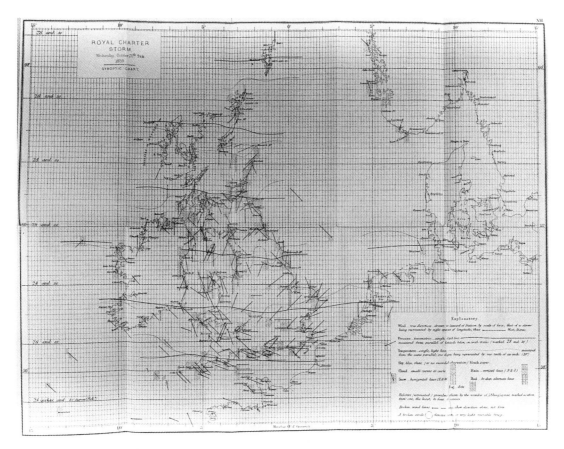

Figure 7.9 An example of the synoptic charts published by FitzRoy following the 'Royal Charter Storm' of 25–26 October 1859, illustrating early methods of mapping weather observations.[5]

be analysed on a trial basis. In addition to the established observatories, ships' captains, lighthouse keepers and private individuals were invited to assist in the scheme by sending their daily registers at monthly or quarterly intervals to FitzRoy in London. The experiment was planned to last for about a year, ending in October 1860. In America, Matthew-Fontaine Maury offered to co-operate and arranged for observations made along the Atlantic and Gulf coasts of the United States to be collected.

It was quite remarkable that not long after these plans had been made two severe storms in close succession passed directly over the British Isles on 25–26 October and 1 November 1859.[5] The former

became known as the 'Royal Charter Storm', after the steam clipper, on a voyage from Melbourne to Liverpool, was driven ashore on the coast of Anglesey with 400 people drowned. Between 21 October and 2 November 1859 over two hundred vessels were wrecked in waters around the British Isles in severe gales with a loss of 800 lives.

Following in the wake of these disasters, FitzRoy published a special series of synoptic charts using the data collected from his earlier appeal. These maps, covering the eastern North Atlantic–European sector from 21 October to 2 November 1859, well illustrate the early methods of synoptic weather representation (see Figure 7.9).

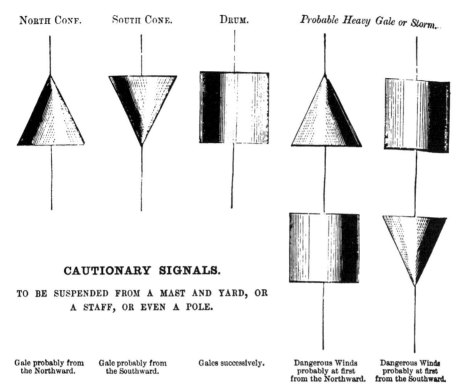

NORTH CONE.   SOUTH CONE.   DRUM.   *Probable Heavy Gale or Storm.*

CAUTIONARY SIGNALS.

TO BE SUSPENDED FROM A MAST AND YARD, OR
A STAFF, OR EVEN A POLE.

Gale probably from   Gale probably from   Gales successively.   Dangerous Winds   Dangerous Winds
the Northward.   the Southward.   probably at first   probably at first
                                                                                    from the Northward.   from the Southward.

NIGHT SIGNALS.
(instead of the above)
LIGHTS IN TRIANGLE OR SQUARE.

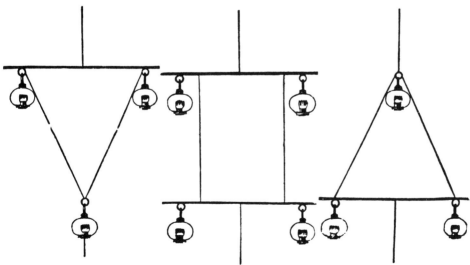

Four lanterns and two yards, each not less than four feet long, will be sufficient — as only one signal will be used at night.
These signals may be made with any lanterns, showing either white, or any colour, but *alike*.
Red is most eligible. Lamps are preferable to candles. The halyards should be good rope, and protected from chafing.
The lanterns should hang *at least* three feet apart.

The importance of using telegraphic reports for transmitting warnings of the approach of storms was indicated to the Board of Trade by the British Association for the Advancement of Science in 1859. In the following year, FitzRoy arranged to receive daily weather reports by telegraphy from fifteen British stations and to exchange the data via Paris for reports from continental stations. From 5 September 1860 official meteorological summaries began to be published in *The Times*, followed shortly by several other newspapers. These have been continued to the present day. In February 1861, storm warnings began to be transmitted by telegraphy to coastal stations, whereby a system of 'cautionary signals' using cones and drums by day and lanterns by night were hoisted from a mast and yard (see Figure 7.10).

The same year, 1861, Francis Galton, co-founder of the British Meteorological Society, invited meteorologists in Britain and the European mainland to co-operate in a scheme for making simultaneous observations in order to obtain a series of daily weather maps over the area between 42°N to 62°N and from the westernmost limits of the British Isles to 21°E during the month of December 1861. Participating correspondents were requested to send their reports by post to Galton.

Galton was concerned that the large number of observations being made in the area – over three hundred fully instrumental reports, thrice daily, besides those of lighthouse keepers and independent observers – were not being combined and utilised to their full potential (see Figure 7.11). He pointed out the difficulties involved in having to apply to several meteorological institutes both in the British Isles and on the Continent for access to the original data, the problem of reducing instrumental readings to a uniform set of units and, finally, the task of plotting the material on a series of maps.

Although acknowledging the efforts made by Fitz-Roy at the Meteorological Department in collecting daily weather reports by telegraphy, he believed that they were insufficiently numerous, extended or frequent enough to allow a proper synoptic analysis to be made. In order to show his fellow meteorologists that his proposal was feasible, Galton decided to carry out a trial run of synoptic mapping over the European region, so far as meteorological stations extended, for a limited period of one month. In the event, he received a great deal of co-operation and published full results in his *Meteorographica* in 1863. This was another milestone in the history of observing, but the way ahead was still not without setbacks.

When the supervision of the Meteorological Department was transferred from the Board of Trade to the Royal Society in 1867, the issue of storm-warnings was discontinued. Daily forecasts, also initiated by FitzRoy, had already been stopped after his tragic death by suicide in 1865. Fearing the kind of criticism which had been levelled at FitzRoy for using the concept that depressions develop on the boundary between tropical and polar **air masses**, the new Meteorological Office attempted to continue the service by simply using empirical forecasting rules. These methods were not very successful however, and forecasts based on synoptic charts were re-started in 1869 and became a regular feature of the 'Daily Weather Report' from March 1872. Furthermore, following representations by the Board of Trade, the storm-warning service was also resumed in February 1874, although the use of the drum was discarded and no distinction was now made between gale- and storm-force winds.

Autographic instruments

In 1868, autographic instruments were first installed, on the suggestion of the Meteorological Committee, for recording pressure, temperature, wind and humidity at seven observatories in the British Isles – namely, Kew, Stonyhurst, Falmouth, Aberdeen, Glasgow, Armagh and Valentia. The resulting autographic

Figure 7.10 The system of 'cautionary signals' introduced by FitzRoy for use at coastal stations to warn shipping of expected gales and storms; cones and drums were hoisted by day and lanterns at night.[5]

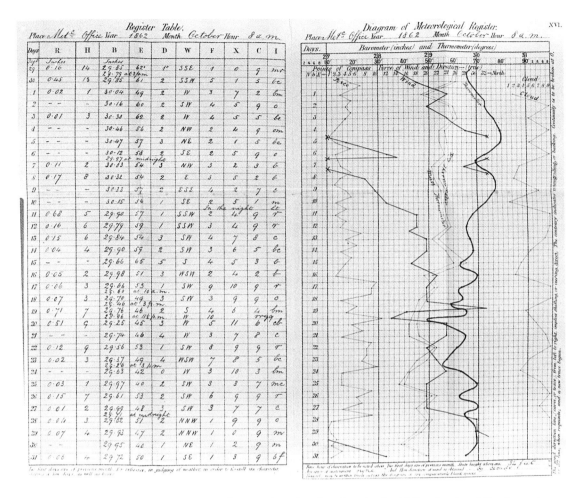

Figure 7.11 An extract from a detailed type of meteorological register kept in the 1860s, showing daily observations of rain (R), hours of rain (H), pressure (B), temperature (E), wet-bulb depression (D), wind direction (W), wind force (F), extreme wind force (X), cloud (C) and weather (I). The observed weather elements were plotted in graphical form on the right-hand page of the record.[5]

records, or **meteorograms**, showed, on occasions, sudden simultaneous changes of weather elements. Unfortunately, the potentially important synoptic details shown on these meteorograms, clearly indicating on occasions the presence of fronts and air masses, were mostly disregarded by meteorologists at the time. The publication of these records was discontinued for reasons of economy in 1880. If this valuable observational material had been thoroughly investigated in the late nineteenth century, perhaps more

realistic models of mid-latitude depressions might have been developed a generation earlier than actually occurred.

Early upper-air observations

The first upper-air soundings, using balloons and kites, were made during the 1890s by Leon Teisserenc de Bort and Richard Assmann in Europe and Abbott Lawrence Rotch in the United States. In the British

Isles, the first ascents with **meteorographs** constructed by William Henry Dines, were made in 1907. The instrumental data of pressure, temperature and humidity from these soundings resulted, in due course, in a better understanding of the vertical structure of weather systems and to the discovery of the **stratosphere**.

Synoptic observations in the First World War

During the First World War neutral Norway was cut off from the exchange of meteorological data by the combatants and synoptic weather forecasting became virtually impossible. However, this was the spur for Vilhelm Bjerknes to establish a very dense network of observing stations in Norway, together with 'indirect aerological reports' based on observations of clouds and other **hydrometeors** helping to make up for the lack of upper-air soundings.

The detailed analysis of synoptic maps that began to be constructed at Bergen from 1918 revealed the fine structure of weather systems that had been mostly overlooked by meteorologists in the late nineteenth century. As already mentioned, traces from the autographic instruments, installed at seven observatories in the British Isles from 1868 to 1880, could have provided evidence of this detail a generation earlier but, unfortunately, the material was never thoroughly investigated.

The development of upper-air station networks

The increasing number of aircraft and pilot balloon soundings made in the 1920s, allowed earlier speculations, using 'indirect aerology' about the vertical structure of weather systems, to be compared with actual observations. The rapid expansion of upper-air station networks in the early 1930s led to a more dynamic approach in weather forecasting, taking into account both surface and upper-level circulation patterns.

With the introduction of the **radiosonde** as a practicable instrument in the late 1930s, networks of upper-air stations, measuring pressure, temperature, humidity and wind velocity, were established

over extensive areas during and soon after the Second World War. This major advance in observing made it possible to construct upper-level charts for the entire Northern Hemisphere, which resulted in the discovery of **long waves** in the upper westerlies and associated jet streams.

TODAY'S WORLD OF OBSERVATIONS

The launching of SPUTNIK 1 in 1957 by the former Soviet Union made the idea of obtaining a global view of weather from space a practical proposition, and in 1960 the first fully equipped meteorological satellite, TIROS 1, was launched by the United States. Acting in response to President John F. Kennedy's 1961 proposal for an international weather prediction programme, the United Nations requested the World Meteorological Organisation (WMO) and the International Council of Scientific Unions to develop a global meteorological system. The resolution of the United Nations led to the formation of two major programmes: a global meteorological system undertaken by the WMO and a global research programme jointly undertaken by the WMO and the International Council of Scientific Unions.

In 1963, a new global meteorological system in which actual and potential developments in satellites, computers, automatic weather stations (see Figure 7.12) and telecommunication techniques would be incorporated, was endorsed by the Fourth Congress of the WMO at Geneva under the name of World Weather Watch, and implementation of the programme was begun in 1968. World Weather Watch is a global meteorological system which comprises the facilities and operations of the national weather services of all the WMO member-states. In the basic World Weather Watch observation system, simultaneous surface-based observations are made at standard times every day at over 9,000 land stations, by about 7,000 ships at sea and around 850 upper-air stations. Space-based observations are made continuously by several **geostationary** and polar-orbiting satellites.

Three world meteorological centres, located at Moscow and Washington, DC, in the Northern

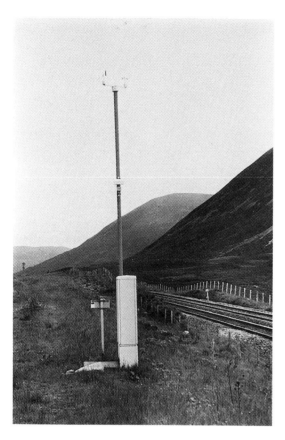

Figure 7.12 An example of an Automatic Weather Station. Automatic Weather Stations are now reliable and accurate enough to be used to supplement the surface observing network.

Hemisphere, and at Melbourne in the Southern Hemisphere, collect, process and disseminate these observations and prepare analyses and forecasts on a global basis. Regional meteorological centres, such as Bracknell in the British Isles, New Delhi and Tokyo, prepare predictions for more limited areas, while national meteorological centres are responsible for weather services to their own countries. Two other current WMO world-wide observing systems include the Global Atmosphere Watch for monitoring the chemical and physical constituents and properties of the atmosphere (including dispersion, transport, chemical transformation and deposition of atmos-

pheric pollutants), and the newly established Global Climate Observing System to meet the long-term observational requirements for an enhanced description of the Earth's climate.[6] How these developments in weather-observing systems have contributed to the enhancement of weather forecasting is described in Chapter 14.

THE ROLE OF THE AMATEUR OBSERVER

For much of its history, meteorology has owed its development to the activities of amateur enthusiasts – in fact, our knowledge of weather and climatic conditions in historical times is mostly drawn from records kept by amateur observers. However, since the establishment of national weather services in the mid-nineteenth century, meteorology has become an increasingly complex science with professional meteorologists now having a vast array of instrumental and technical resources at their command.

Nevertheless, there is still room for the amateur in meteorology today. Indeed, apart from the fact that amateur observations may become accepted into the network of official meteorological stations, watching the weather can be a rewarding activity in its own right by providing an absorbing interest in the various processes at work in the atmosphere.

Contact with other meteorologists, both amateur and professional, can be made through membership of meteorological societies such as the Royal Meteorological Society and the Climatological Observers Link. The latter group, which was set up in 1970, comprises over three hundred meteorologists, both amateur and professional, situated at stations mostly in the British Isles, but with a few in continental Europe and even as far afield as Australia. Information on weather elements, including pressure, temperature, wind, rainfall and sunshine, observed in the Climatological Observers Link network of stations, is presented in a monthly bulletin. This publication also contains synoptic reviews of the weather over the British Isles, together with brief descriptions of conditions in North America, global summaries and occasional articles by members.[7]

Figure 7.13 An experimental sketch of cloud formations by John Constable, 1823. Tate Gallery, London.

One of the chief attractions of meteorology for the amateur is that it can be pursued at different levels of commitment, with or without instruments, according to the means and inclination of the individual; the all-important common factor being a lively curiosity about the weather. A basic task of any observer is to make a careful survey of the sky. With good visibility and an unobstructed skyline it is sometimes possible to observe clouds above the horizon at a distance of 240 km (150 miles). As a result, the effects of large-scale weather features, such as **warm** and **cold fronts**, can often be observed from a single station by the appearance of the sky; the various cloud-forms providing visual guides as to processes at work in the atmosphere.

It was not until the beginning of the nineteenth century that a generally accepted method of observing the clouds was devised. This was the classification introduced by Luke Howard in 1802–3.[8] He identified four main types of clouds according to their appearance, giving them the Latin names, *cirrus* (streak cloud), *cumulus* (heap cloud), *stratus* (sheet cloud) and *nimbus* (rain cloud). The present international classification of clouds has been developed from the original system of Howard. It is also noteworthy that Howard's classification influenced the English landscape artist, John Constable, who sketched a large number of cloud studies in the early 1820s (see Figure 7.13). Later in the century, Clement Ley, another diligent amateur watcher of the skies, advocated the importance of clouds as weather 'signs', returning, in effect, full circle to the methods used by medieval observers – as mentioned at the beginning of this chapter.

Another age-old weather-related activity which has been pursued by many amateur observers is

phenology. This study involves observing and recording annually recurring events in nature such as the leafing and flowering of plants, migration of birds and behaviour of insects. Being manifestations of prevailing climatic conditions, phenological events generally occur in a seasonal order that is substantially constant. However, the individual dates can vary widely from year to year according to the vicissitudes of the prevailing weather.

Phenological records kept over a number of years can provide valuable data from which the lateness or earliness of a series of seasons can be compared and analysed. In 1875 a network of phenological stations was established by the Royal Meteorological Society and reports were published annually for over seventy years. Individual records have extended over even longer periods, the most outstanding example being that set up by Robert Marsham at Stratton Strawless, Norfolk, in 1736. This series was maintained by Marsham himself for sixty years and continued, with only a few gaps, by his descendants into the present century.[9]

By providing natural indicators of atmospheric conditions, phenological observations can be applied to studies relating agriculture to weather and climate; a long-period record, comprising data from a group of mature plants and trees growing in one locality, can provide a valuable contribution to research of this kind.

Finally, although amateur observers may be less well-endowed with instrumental and technical facilities than their professional colleagues, they may, on occasion, be favoured with recording rare or outstanding atmospheric phenomena which may have escaped the open network of official stations and the broader viewpoint of weather satellites.[10]

NOTES

1 E.N. Lawrence, 'The earliest known journal of the weather', *Weather*, 1972, vol. 27, pp. 494–501.

2 J.-P. Desaive, J.-P. Goubert, E. Le Roy Ladurie, J. Meyer, O. Muller and J.-P. Peter, *Médecins, climat et épidémies à la fin du XVIII siècle*, Paris, Mouton and École Pratique des Hautes Études, Sorbonne, 1972.

3 J. Kington, *The Weather of the 1780s over Europe*, Cambridge, Cambridge University Press, 1988.

4 J. Kington (ed.), *The Weather Journals of a Rutland Squire: Thomas Barker of Lyndon Hall*, Oakham, Rutland Record Society, 1988.

5 R. FitzRoy, *The Weather Book: a manual of practical meteorology*, London, Longman, Green, Longman, Roberts, and Green, 1863.

6 Anon., 'Notes and news', *Meteorological Applications*, 1994, vol. 1, pp. 93–5.

7 R. Brugge, 'The Climatological Observers Link', *Weather*, 1994, vol. 49, pp. 35–6.

8 See J. Kington, 'A century of cloud classification', *Weather*, 1969, vol. 24, pp. 84–9, and J.E. Thornes, 'Luke Howard's influence on art and literature in the early nineteenth century', *Weather*, 1984, vol. 39, pp. 252–5.

9 J. Kington, 'An application of phenological data to historical climatology', *Weather*, 1974, vol. 29, pp. 320–8.

10 J. Kington, 'Forecasting', in R. Hardy, P. Wright, J. Gribbin and J. Kington, *The Weather Book*, London, Michael Joseph, 1982.

GENERAL READING

E.G. Bilham, *The Climate of the British Isles*, London, Macmillan, 1938.

W.E. Knowles Middleton, *Invention of the Meteorological Instruments*, Baltimore, Johns Hopkins University Press, 1969.

Meteorological Office, *Meteorological glossary* (Met. O. 985), London, Her Majesty's Stationery Office, 1991.

Meteorological Office, *Observer's handbook* (Met. O. 933), London, Her Majesty's Stationery Office, 1982.

8

CLASSIFYING THE WINDS AND WEATHER

P. Mick Kelly, Phil Jones and Keith Briffa

The atmosphere knows no boundaries and the winds carry no passports.
Sir Crispin Tickell

INTRODUCTION

What caused the bitterly cold conditions of the winter of 1962–3? Why were the summers of 1976 and 1995 so hot and dry? What is the cause of varying levels of pollution over Scotland? The immediate answers to these questions can be found in some disturbance of the atmospheric circulation – the wind patterns, **depressions** and **anticyclones** that shape the weather of the British Isles (see Chapter 2). The importance of circulation of the atmosphere in determining the weather was recognised early on in the history of weather forecasting. So much so that, in the second half of the nineteenth century, the need to acquire data from beyond national frontiers to anticipate the approach of weather systems led to the sharing of data world-wide (see Chapters 7 and 14). This system of data exchange formed the foundations of the success of present-day weather forecasting and has enabled the regular monitoring of climate change.

The early weather forecasters relied not on high-speed computers, but on rules and procedures based on the likely progression of the weather from one day to the next. The process was more sophisticated than the shepherd's prognosis based on a 'red sky at night', but was derived from a similar process of observation, tested over the years. As understanding of the nature of the atmospheric circulation improved, so did the sophistication of the forecasting techniques (see Chapter 14). An active **warm front**, embedded in the cyclonic motion of a mid-latitude depression, would bring rain, a rise in temperature and a shift in wind direction. High pressure over the country meant dry, stable conditions, warm in summer but cold in winter. Any observation of regular, systematic behaviour, or a systematic relationship, could provide the basis of a weather forecasting rule. Similar relationships can provide a means of understanding the immediate cause of episodes of extreme climate and of longer-term climatic fluctuations over the British Isles.

Consider the weather maps shown on television each day (Figure 8.1). Many look similar. A depression might be approaching or crossing the British Isles moving towards the east, signalling changeable conditions. An anticyclone may squat over the region, almost stationary and bring a period of settled weather. Strong northerly flow might dominate the area, drawing cold, Arctic air south over the region. Circulation 'types' such as these recur frequently and can be used to characterise, and hence classify, not only the weather of a particular day but also the average climate of a longer period, a season, a year or a decade.

Classification systems can be applied to a range of different weather and climate variables and many systems have been developed. The intention, though, is always the same. It is to provide a concise summary of the main characteristics of the weather or atmospheric circulation over a period of time.

Figure 8.1 Penny Tranter, a BBC weather presenter and graduate of the School of Environmental Sciences at the University of East Anglia, in front of a weather forecast chart. Photograph courtesy of the BBC (© BBC Weather Centre).

WEATHER TYPES

The history of weather classification goes back to the dawn of humanity, when the first forecasting rules were developed to assist the planting and harvesting of crops. These rules were generally based on an understanding of the progression of the seasons and of key episodes – periods of characteristic weather – at different stages of the cycle. In modern times, it was the understanding gained during the nineteenth century of the relationship between air flow, pressure systems – such as depressions and anticyclones – and the resulting weather that led to the development of the first classification systems. In the 1880s, for example, Ralph Abercromby[1] studied the effect of the dominant air flow, from the north, south, east or west, on the weather of the British Isles, and found that each wind direction led to characteristic temperature and rainfall or snowfall conditions. Ernest Gold,[2] some forty years later, developed a more sophisticated approach, taking account of the overall character of the main weather system – depression or anticyclone – affecting the region.

Understanding of the characteristics and mechanisms of the large-scale atmospheric circulation grew during the early twentieth century as meteorological databanks extended both in space and time. With the advent of regular observation through the depth of the atmosphere, the importance of the flow of air many kilometres above the ground in steering the **synoptic systems** which determine the weather at the surface was recognised. This finding led to the development of classification systems which also took account of the movement, or steering, of depressions and anticyclones over a period of days. The technique was pioneered in the 1930s by the German meteorologist Franz Baur, thus providing the basis for a classification system characterising large-scale circulation patterns over Europe and the surrounding region, the **Grosswetterlagen**,[3] which is still in regular use today. At about the same time, the Scandinavian meteorologist Tor Bergeron was undertaking pioneering studies of the role of **air masses**. The air mass concept provided another means of classifying weather types, again based on characteristics of the large-scale atmospheric circulation but this time emphasising the origin of the air affecting the region under classification.

Before the advent of computers, producing a weather-type classification was a laborious process.

The work had to be undertaken manually with no assistance from the electronic aids available today. It was a matter of painstaking study of many daily weather maps, during which the regular occurrence of certain weather types would be noted, a classification system developed and each day assigned to a particular category. Now, the process can be largely automated, with statistical pattern recognition techniques identifying recurring types and rules based on these results used to classify each day. As a result, the number of classification systems has increased dramatically. Roger Barry and Allen Perry[4] identify seven fundamental approaches in their comprehensive discussion of the development of weather classification systems. Four are based on atmospheric circulation patterns and the remainder deal directly with the weather elements of temperature, rainfall, and so on.

This chapter focuses on **circulation types** as these provide the most effective basis for defining and explaining climate variability. Conceptually, the approach used in typing weather elements is similar. The classification system may be based on typical geographical patterns of temperature, rainfall or any other variable. One pattern may be characterised by colder than normal conditions in the north of the region while the south remains warm. Some schemes contain types based on multiple weather variables. A type may be defined by cold and wet in the west, dry in the east and warm in the south. In other cases, the types are defined by a complex of conditions at a single point, as in the pioneering example of this approach undertaken by Eugene Federov for Moscow.[5]

As far as circulation typing systems are concerned, the four variants listed by Barry and Perry can be further split into two categories: static and **kinematic** (or mobile). Methods falling into the first category view the weather map as a snapshot, frozen in time. Each daily map is categorised either in terms of the dominant circulation feature at the time (a depression over the region, for example) or else in terms of the overall pattern of the circulation on the day (such as high pressure to the south-west and low pressure over Scandinavia). The kinematic approach

considers the weather on a particular day as the current manifestation of a longer-term process shaping, for example, the steering of the synoptic features that determine the weather on a particular day. To categorise a particular day, it may therefore be necessary to consider what has happened on previous days and what happens next. This is the case with the Grosswetterlagen where a type has to persist for at least two days.

THE LAMB CLASSIFICATION

The definitive classification of daily weather types for the British Isles was developed by Hubert Lamb during the late 1940s and 1950s, based on earlier work by Rodney Levick.[6] The final version of the catalogue was published in 1972[7] and it continues to be updated at regular intervals. A listing of the catalogue from 1972 to 1995 is provided in Appendix B. The classification is based on surface synoptic charts depicting the state of the atmospheric circulation close to the ground and, when they became available, charts at the **500 hPa level** depicting the flow at height in the atmosphere. In the Lamb Classification of Daily Weather Types, the synoptic pattern is characterised by a series of basic, or pure, circulation types which may be combined to form other, hybrid types. The catalogue is subjective – that is, it is based on the judgement of an experienced analyst – although an automated procedure has been developed which provides a close match to the original scheme (see Box 8.1). Each day over the period from 1861 to the present has been classified. The classification was undertaken after shuffling the years at random to avoid any bias on the part of the analyst. John Kington has applied the scheme to information for the 1780s, and a number of other shorter periods of interest prior to 1861 have been classified as well, resulting in a valuable record of historic changes in the atmospheric circulation in the vicinity of the British Isles.[8]

The Lamb Classification contains eight directional types – north (denoted N), north-east (NE), east (E), south-east (SE), south (S), south-west (SW), west (W) and north-west (NW) – defined by the general air flow and the motion of the synoptic systems embedded in that flow. The motion of the synoptic systems is determined by the circulation flow at height in the atmosphere, rather than at the surface, so the definition of types in terms of steering means that this classification system is attempting to characterise the overall atmospheric circulation rather than surface features alone. Equally, the scheme is not necessarily based solely on conditions on a particular day, but on features of the circulation that may be consistent over a series of days. There are three non-directional types. The anticyclonic (A) and cyclonic (C) types occur when high or low pressure, respectively, dominates the British Isles (see Figure 8.2). In the case of the cyclonic type, the central isobar of the depression should extend over a large section of the British Isles. The final 'non-directional' type is termed unclassifiable (U); because of the complexity of the atmospheric circulation, even the most comprehensive classification scheme will fail to type some days!

The directional and non-directional types may be combined to categorise more complex circulation patterns. For example, the cyclonic westerly (CW) type corresponds to a situation where the dominant steering of synoptic features is from west to east over the British Isles with depressions passing over the north of the area. Another hybrid, anticyclonic westerly (AW), for example, indicates that high pressure dominates the southern part of the area with, most likely, depression tracks diverted well to the north. If the British Isles were only affected by westerly flow, with no depressions or anticyclones in the area, the type would be westerly.

The seven pure circulation types, considered by Lamb to be the most fundamental, are the cyclonic and anticyclonic non-directional types and the westerly, north-westerly, northerly, easterly and southerly directional types. Typical synoptic charts for six of these types are shown in Figure 8.3. The average frequencies of all twenty-seven types over the period 1861 to 1990 are given in Table 8.1. Definitions for the two non-directional types and the westerly type have already been given. The north-westerly type is generally associated with a displacement north or north-east of the

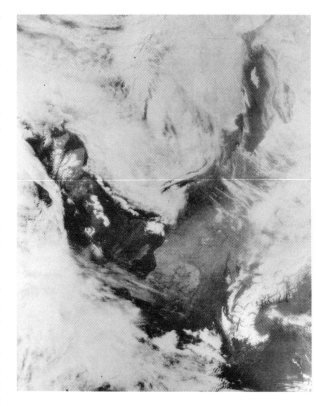

Figure 8.2 Visual satellite image of the British Isles on 28 February 1977 at 0924 GMT. This day is classified as 'A' in the Lamb Catalogue (see Appendix B). The anticyclone is located over western Europe, with a ridge of high pressure extending over the British Isles. The mountains of Britain are snow covered. Courtesy of the University of Dundee.

Azores anticyclone with depressions passing from the region of Iceland towards the North Sea (see Figure 8.4). In the case of the northerly type, high pressure can extend from Greenland as far south as the Azores to the west of the British Isles and, to the east, depressions may travel south from the Norwegian Sea. The easterly type is typified by an anticyclone over Scandinavia extending over the Norwegian Sea towards Iceland. This represents a reversal of the normal westerly flow and any cyclonic activity is displaced south over south-west Europe and the western Atlantic. Finally, the southerly weather type is accompanied by high

BOX 8.1 AUTOMATING THE LAMB CLASSIFICATION

Arthur Jenkinson and Peter Collison of the UK Met. Office have developed an automated procedure for identifying the Lamb weather types from gridpoint mean-sea-level pressure.[a] The method has been extended by Phil Jones and colleagues who have compared the results with the original subjective approach.[b] Measures of the character of the atmospheric circulation over the British Isles are calculated from the pressure data resulting in estimates of the dominant direction and speed of the flow and the **vorticity** (the spin of the air, related to whether a cyclone or anticyclone is present). Particular values of these parameters can then be associated with specific Lamb types. For example, in the case of the westerly type, the flow is, self-evidently, from the west and the vorticity is low. If the vorticity is strongly positive, a cyclone is present. For vorticity values that are moderately positive the air flow direction is also considered, resulting in a hybrid cyclonic-directional type. For negative vorticity values, the same rules are applied to the basic and hybrid anticyclonic types.

Thorough comparison has shown that the agreement between this automated procedure and the original, subjective classification is generally good although some discrepancies occur, particularly prior to the 1960s. For example, the decline in the westerlies is less marked in the automated classification. This failure to agree on the most frequent weather type is disturbing. It may reflect the changing amount of information available to the subjective analyst over time and shifts in convention in drawing the surface charts. The most likely explanation, though, is that it reflects a shortcoming of the automated procedure. The automated procedure underplays the kinematic, or steering, element of the Lamb Classification, being based on only one chart a day with no account taken of adjacent days' data and making use of surface charts alone. The use of surface charts introduces a bias in that days on which the steering of synoptic systems, determined by the flow at height in the atmosphere, is, say, westerly will tend towards south-westerly flow at the surface.

[a] A.F. Jenkinson and B.P. Collison, 'An initial climatology of gales over the North Sea', *Synoptic Climatology Branch Memorandum No. 62*, Bracknell, Meteorological Office, 1977.
[b] P.D. Jones, M. Hulme and K.R. Briffa, 'A comparison of Lamb circulation types with an objective classification scheme', *International Journal of Climatology*, 1993, vol. 13, pp. 655–63.

pressure over northern and central Europe with depressions stagnating to the west of the British Isles or moving north or north-east in that area.

RELATIONSHIPS WITH TEMPERATURE, PRECIPITATION AND OTHER WEATHER VARIABLES

How can the Lamb Classification assist in answering the questions raised in the introduction to this chapter? Why were the winter of 1962–3 and the summers of 1976 and 1995 extreme? The Lamb types for these periods are shown in Table 8.2. The winter of 1962–3 was the most severe since 1740 (see Figure 8.5 and Chapter 9) and the cause of the prolonged period of cold conditions with some snowy intervals is clear from the record of daily weather types. The season was dominated by a mix of anticyclonic and easterly conditions, with sufficient cyclonic days to ensure a supply of snow (though precipitation levels overall were not high for the season). What is immediately striking about the record for the summer of 1976 is the dominance of anticyclonic types, particularly during the month of August. In fact, anticyclonic conditions

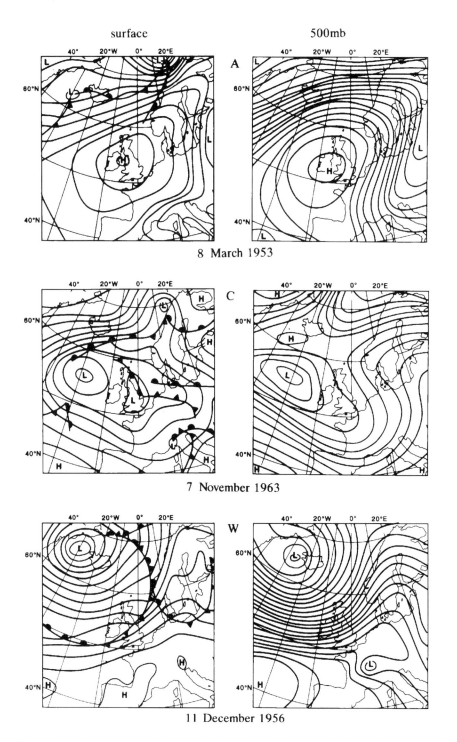

Figure 8.3 Typical synoptic patterns associated with six of the most important Lamb weather types: anticyclonic (A); cyclonic (C); westerly (W); northerly (N); easterly (E); and southerly (S). The left-hand chart shows the surface pattern

surface   500mb

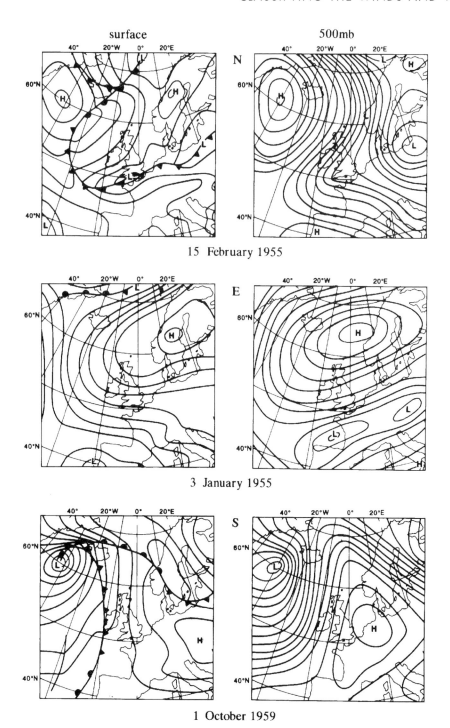

15 February 1955

3 January 1955

1 October 1959

and the right-hand chart the upper air (500 hPa) pattern.

Table 8.1 Average annual and seasonal frequencies of the Lamb weather types over the period from 1861 to 1990

Type	Annual		Winter		Spring		Summer		Autumn	
	Days	(%)	Days	(%)	Days	(%)	Days	(%)	Days	(%)
U	14.3	(3.9)	3.4	(3.7)	3.9	(4.2)	3.5	(3.8)	3.5	(3.9)
A	66.3	(18.2)	14.4	(16.1)	17.3	(18.8)	17.3	(18.8)	17.2	(18.9)
ANE	5.0	(1.4)	0.8	(0.8)	1.7	(1.8)	1.4	(1.5)	1.2	(1.3)
AE	8.9	(2.4)	1.7	(1.9)	3.2	(3.5)	2.1	(2.3)	1.8	(2.0)
ASE	3.4	(0.9)	0.8	(0.9)	0.9	(1.0)	0.7	(0.7)	1.0	(1.1)
AS	4.0	(1.1)	1.2	(1.3)	0.8	(0.9)	0.8	(0.9)	1.1	(1.2)
ASW	3.1	(0.9)	1.0	(1.1)	0.6	(0.6)	0.7	(0.7)	0.8	(0.9)
AW	17.0	(4.7)	4.2	(4.7)	3.4	(3.7)	5.1	(5.6)	4.2	(4.6)
ANW	5.4	(1.5)	1.1	(1.2)	1.3	(1.4)	1.8	(2.0)	1.1	(1.2)
AN	7.4	(2.0)	1.4	(1.6)	2.4	(2.6)	2.1	(2.3)	1.5	(1.7)
NE	3.4	(0.9)	0.6	(0.7)	1.3	(1.4)	0.8	(0.9)	0.6	(0.7)
E	12.8	(3.5)	3.0	(3.3)	5.4	(5.9)	1.8	(2.0)	2.6	(2.9)
SE	6.3	(1.7)	2.1	(2.3)	1.9	(2.1)	0.7	(0.7)	1.6	(1.8)
S	15.4	(4.2)	4.9	(5.5)	4.0	(4.3)	2.3	(2.5)	4.3	(4.7)
SW	10.3	(2.8)	3.6	(4.0)	1.9	(2.1)	1.9	(2.1)	2.8	(3.0)
W	67.4	(18.4)	20.7	(23.0)	12.3	(13.3)	16.4	(17.9)	17.9	(19.7)
NW	13.8	(3.8)	3.4	(3.7)	3.1	(3.4)	4.2	(4.6)	3.1	(3.4)
N	17.0	(4.7)	3.3	(3.6)	5.2	(5.6)	4.4	(4.8)	4.2	(4.6)
C	47.5	(13.0)	9.6	(10.7)	12.0	(13.1)	14.8	(16.1)	11.0	(12.1)
CNE	1.4	(0.4)	0.3	(0.3)	0.5	(0.5)	0.2	(0.4)	0.3	(0.3)
CE	4.0	(1.1)	0.9	(1.0)	1.3	(1.4)	0.8	(0.9)	1.0	(1.1)
CSE	1.7	(0.5)	0.4	(0.4)	0.6	(0.7)	0.2	(0.2)	0.4	(0.5)
CS	4.6	(1.2)	1.2	(1.3)	1.3	(1.4)	0.9	(1.0)	1.2	(1.3)
CSW	2.4	(0.7)	0.7	(0.7)	0.4	(0.5)	0.6	(0.6)	0.7	(0.8)
CW	14.4	(3.9)	3.8	(4.3)	2.6	(2.9)	4.1	(4.5)	3.8	(4.2)
CNW	3.2	(0.9)	0.7	(0.7)	0.9	(1.0)	1.0	(1.1)	0.6	(0.7)
CN	4.8	(1.3)	1.0	(1.1)	1.5	(1.7)	1.2	(1.3)	1.1	(1.2)

had been prevalent earlier in the year, as the record for spring testifies. The drought of the summer months represented the culmination of a period of deficient rainfall that began the previous year. A similar pattern is seen in the spring and summer of 1995 with anticyclonic days frequent throughout the two seasons.

Numerous studies have shown that there is a correlation between the Lamb types and the various weather elements. Figure 8.6, for example, shows an early study of the relationship between the different Lamb types and rainfall at Southampton.[9] Most of the rainfall is associated with the higher frequency westerly and cyclonic types, although there is a seasonal cycle in the relative importance of the different types. This type of analysis has been extended[10] to derive the contribution of each of the Lamb daily weather types to precipitation (rain plus snow) amounts at a series of stations over the British Isles. A cyclonic day results in an average precipitation amount of 4.2 mm over the network, while an anticyclonic day results in a mere 0.8 mm. The geographical patterns of daily precipitation amounts associated with the three major Lamb types – westerly, anticyclonic and cyclonic – are shown in Figure 8.7. As John Sweeney and Greg O'Hare note,[11] the pattern associated with the westerly type shows a clear gradient from west to east, reflecting the

Figure 8.4 Visible satellite image of the British Isles on 17 September 1978 at 0935 GMT. This day is classified as 'ANW' in the Lamb Catalogue (see Appendix B). The rippled appearance of the cloud over Scotland and the Midlands is caused by wave clouds which appear when the air flow, in this case a light north-westerly, is disturbed by passing over high ground. Courtesy of the University of Dundee.

Figure 8.5 The frozen River Cam at Cambridge during the winter of 1962–3. This winter was the coldest in the British Isles since 1740.

motion of the cyclones responsible for the precipitation, and from south to north associated with the pressure gradient driving the westerlies. **Rain-shadow** effects can be seen downstream of the Wicklow Mountains and the Pennines, and precipitation levels are high over the western Highlands in Scotland as a result of the orographic influence. Not surprisingly, anticyclonic days are characterised by low precipitation amounts throughout the country with only the extreme north-west likely to be affected by frontal activity. Depressions may take any number of tracks over the region on days classified as cyclonic, so on these days precipitation is enhanced throughout the entire region although a maximum occurs over the northern Irish Sea.

The Lamb Classification contains a mass of information and, for the sake of convenience, indices are often used to portray the main variations evident in the data. Roy Murray and Rod Lewis[12] have, for example, defined the P, S, C and M indices which measure the degree of Progression (mobile, westerly patterns), Southerliness, Cyclonicity and Meridionality, respectively, of the synoptic situation over a month, season or year. A simple index of westerly day frequency has also frequently been used to monitor the changing strength of the North Atlantic **zonal flow**.

Table 8.2 Lamb Classification for winter 1962–3 and spring–summer 1976 and 1995. (See Appendix B for a full listing of the Lamb Catalogue from 1972 to 1995.)

	Winter 1962–3			Spring–Summer 1976						Spring–Summer 1995					
Day	Dec	Jan	Feb	Mar	Apr	May	Jun	Jul	Aug	Mar	Apr	May	Jun	Jul	Aug
1	AS	E	NE	A	C	AW	C	ASE	NW	W	W	AS	NW	A	AE
2	S	E	AE	A	CN	W	AN	AE	NW	CW	W	A	U	N	A
3	S	E	N	AS	W	CW	A	E	ANW	C	U	A	C	N	ANE
4	A	C	U	S	W	U	A	AE	ANW	W	A	A	C	AN	ANE
5	AS	E	S	SE	NW	E	AW	A	A	C	W	A	ANW	W	A
6	AS	E	CS	SE	NW	A	AW	E	A	W	W	A	NW	W	AN
7	SW	E	CSE	E	N	A	A	E	A	C	ANW	U	N	AW	A
8	CW	AE	S	E	A	A	S	E	A	C	A	CN	N	A	A
9	W	AE	SE	A	A	A	W	C	A	SW	A	N	N	ASE	A
10	NW	AE	E	AS	AW	NW	AW	A	A	S	AW	ANE	N	E	ASE
11	CW	AE	E	W	U	CW	ASW	AS	A	AW	AW	NE	N	SE	S
12	N	AE	A	C	A	C	AW	S	A	A	A	N	ANE	S	SW
13	N	ANE	A	CE	U	W	AW	CS	A	W	A	N	NE	S	W
14	NW	AN	S	U	CN	ASW	AW	SW	A	CNW	NW	N	N	C	AW
15	CNW	NW	SE	C	A	SW	A	SW	AE	CNW	NW	A	N	C	A
16	N	U	E	C	A	SW	A	C	A	W	NW	CNE	W	C	A
17	CW	A	AE	A	A	SW	A	AW	A	CW	C	CNE	W	CSW	A
18	NW	AE	AE	U	A	C	W	U	A	W	N	N	W	U	A
19	W	E	E	ASE	AE	C	CW	W	A	NW	N	CNW	SW	ASW	AE
20	W	E	N	S	AE	C	U	NW	A	A	N	A	AW	SW	A
21	ANW	SE	NW	S	AE	W	A	NW	A	A	C	AS	A	W	A
22	A	A	A	U	E	ASW	AS	ANW	E	A	NE	S	A	A	A
23	ASE	A	A	A	ANE	A	ASW	ANW	SE	C	S	A	AW	NW
24	A	A	A	A	NE	S	A	NW	E	W	E	C	A	A	W
25	A	A	A	W	ANE	C	A	NW	A	NW	E	CS	A	ASE	CW
26	N	AN	AS	W	NE	U	A	A	ANE	N	NE	S	AE	CSE	C
27	CNE	A	AS	W	ANE	A	A	A	ANE	N	A	S	AE	C	NW
28	NE	A	SE	AW	A	SE	A	AN	AE	C	A	S	A	ASW	N
29	E	AN		W	A	C	A	ANW	E	A	AE	C	A	S	NW
30	CE	CNE		W	A	C	A	N	C	ASW	A	C	A	U	AN
31	CE	NE		CW		U		N	NW	W		NW		E	A

Table 8.3 shows the degree of correlation between three selected indices and the Central England Temperature and England and Wales Precipitation records (discussed in Chapters 9 and 10, respectively). The westerly days index, the W index, is a count of the days when westerly flow occurred. The cyclonicity index, Murray and Lewis's C index, is a measure of the balance between cyclonic and anticyclonic days and the southerly index, the S index, is a measure of the balance between southerly and northerly flow. As far as precipitation is concerned, the C index, not surprisingly, shows the strongest correlations, consistent through all four seasons. There is also a significant, but much weaker, correlation between the W index and precipitation in winter (cf. Figure 8.7 and the related discussion above) and between the S index and precipitation in winter. The southerly Lamb type produces the highest precipitation of any of the types as warm tropical air, moving polewards, cools over the waters of the North Atlantic ocean. Uplift over

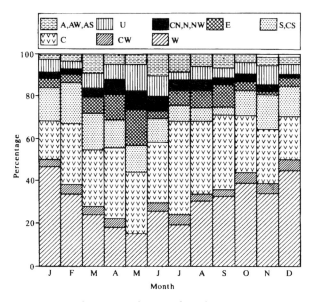

Figure 8.6 The contribution of Lamb types to precipitation at Southampton, 1921–50 percentage contribution to the monthly mean (from R.G. Barry 1967[9]).

Table 8.3 Correlations between seasonal indices based on the Lamb-type frequencies and Central England Temperature (CET), and England and Wales Precipitation (EWP), over the period from 1861 to 1990. (For convenience, only correlations significant at the 5 per cent level are presented.)

	CET	EWP
Winter		
W index	+0.72	+0.25
S index	+0.46	+0.27
C index		+0.84
Spring		
W index		
S index	+0.45	
C index	−0.33	+0.78
Summer		
W index		
S index	+0.28	
C index	−0.60	+0.84
Autumn		
W index		
S index	+0.66	
C index		+0.75

Notes: The indices are calculated by scoring each day according to type as follows: For the W index used here, westerly days score +2 and south-westerly and north-westerly days score +1. This index is a measure of how strong the westerlies are. It is a slightly different formulation from the westerly circulation index used in Figure 9.7. For the S index, southerly, cyclonic southerly and anticyclonic southerly days score +2; pure and hybrid south-westerly and south-easterly days score +1; northerly, cyclonic northerly and anticyclonic northerly days score −2; and pure and hybrid north-westerly and north-easterly days score −1. This index is a balance between southerly and northerly flow. For the C index, pure cyclonic days score +2, hybrid cyclonic days score +1, pure anticyclonic days score −2 and hybrid anticyclonic days score −1. This index is a balance between cyclonic and anticyclonic conditions.

the mountains of Ireland, Wales and the Scottish Highlands then results in heavy precipitation.[13]

In the case of temperature, the correlations with the indices are strongly dependent on season. For example, there is a strong positive correlation between westerliness, the W index, and Central England Temperature in winter. Any enhancement of the maritime influence in this season will result in amelioration of the winter conditions. In spring, the dominant influence on temperature is the degree of southerliness and, to a lesser extent, anticyclonicity. During summer, the relative influence of these two factors switches, with anticyclonicity becoming the more important. The balance between southerly and northerly days exerts an influence on seasonal temperatures throughout the year, continuing through autumn. While strong relationships do exist between certain Lamb weather types and temperature and precipitation, it should be noted that these relationships may change with time.[14] This behaviour could, for example, be the result of climate variation in other areas, altering air-mass characteristics.

The Lamb Classification has been used to identify more subtle characteristics of the weather and circulation. Trevor Davies and his colleagues[15] studied changing acid pollution levels in Scottish precipitation using information in the classification to define the origin of the air mass from which the snow or

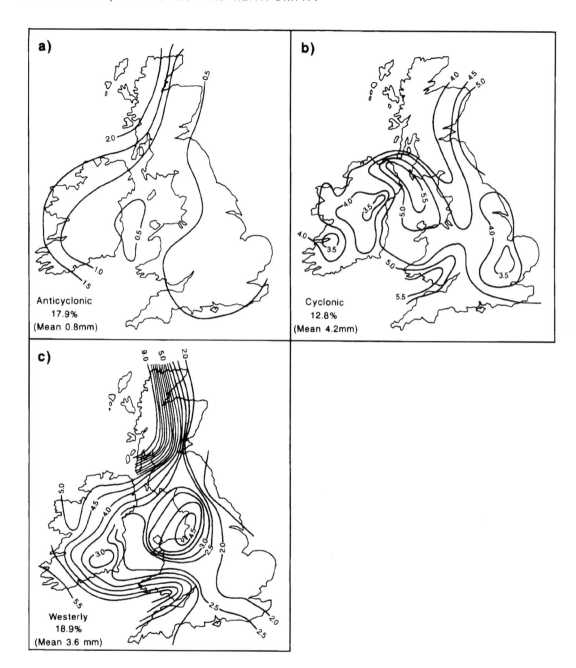

Figure 8.7 Mean daily precipitation (mm) associated with (a) anticyclonic, (b) cyclonic and (c) westerly types (from J.C. Sweeney and G.P. O'Hare 1992[10]).

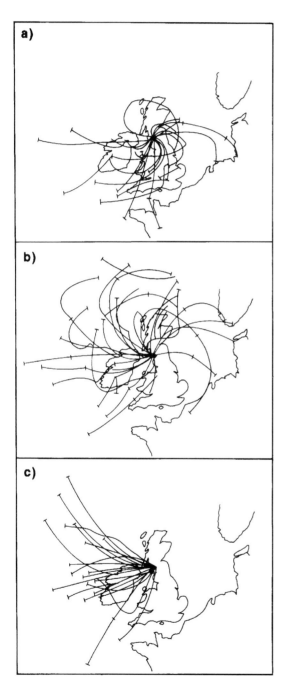

a)

b)

c)

rain originated. Cyclonic types indicated an air mass trajectory which passed over pollution sources to the south over industrial areas of England (particularly during the first half of the year), while westerly days suggested that the air came from relatively unpolluted areas over the neighbouring Atlantic Ocean (Figure 8.8). Precipitation acidity was high during the former events, but low during the latter. This kind of analysis has also identified eastern Europe as a major pollution source for parts of the British Isles.

These examples illustrate the value of the Lamb Classification in providing information regarding the mechanisms that shape extreme or unusual events. The classification scheme can also provide valuable indicators of the course of the seasonal cycle and of long-term fluctuations in the climate of the British Isles.

THE ANNUAL CYCLE IN THE CIRCULATION OVER THE BRITISH ISLES

As can be seen in the annual average frequencies of the major Lamb types presented in Table 8.1, three types dominate with a cumulative frequency of 181 days, nearly half the year. The westerly type is the most common with an annual frequency of 67.4 days over the period from 1861 to 1990. The anticyclonic type follows closely with an average frequency of 66.3 days a year; in third place is the cyclonic type with an average frequency of 47.5 days a year. Given their high frequency of occurrence, these three types alone can be used to define the main features of the annual cycle in the atmospheric circulation over the British Isles.

Figure 8.8 Back trajectories (start-time 1200 GMT) associated with days classified as (a) the cyclonic type during the months of January–May 1980; (b) the cyclonic type during the months of June–August 1980; and (c) the westerly type during all months 1980. The trajectories associated with the cyclonic type in the period June–August have a more maritime track than during earlier months (from T.D. Davies *et al.* 1990[15]).

Figure 8.9 shows the frequencies of the westerly, anticyclonic and cyclonic types, as percentages, for each day, again calculated over the period from 1861 to 1990. For reference, the corresponding annual cycles in the daily Central England Temperature and England and Wales Precipitation records are shown in Figure 8.10. As can be seen, the westerly type is most common during the winter months. The frequency of westerly days drops quite abruptly during the early weeks of February and remains low during late winter and spring, reaching a minimum during May. The effects of the transition to lower westerly frequencies during February can be seen in the precipitation record which also shows a marked fall at this time of year (cf. Figure 8.7, which illustrates the strong link between the westerly type and precipitation). A pronounced shift in the atmospheric circulation over the Atlantic–European sector often occurs in mid-June, marking an abrupt shift to higher frequencies of westerly days.

Marked phases of the annual cycle when particular weather conditions tend to recur, or when transitions between periods of consistent weather occur, are known as 'singularities'[16] and provided meteorologists with useful forecasting rules in the pioneering days of the science. Singularities tend to occur as the large-scale atmospheric circulation adjusts, in a well-defined and often abrupt manner, to the changing pattern of seasonal heating and cooling. The mid-June singularity marks the onset of the 'European monsoon' as the upper westerlies are displaced south over the region with the movement of the centre of the **circumpolar vortex** towards the Atlantic sector. The return of the westerly winds is accompanied by greater precipitation in years when this singularity is distinct. From late June onwards, the frequency of westerly days remains relatively constant until winter conditions are established in early December.

The annual cycle in the average frequency of the anticyclonic type is not as pronounced. In this case, frequencies are lowest from late October to the end of January when a transition to high frequencies occurs. There are some intriguing signs of singularities in the seasonal record of this type. For example,

frequencies are above the prevailing seasonal level for a prolonged period from the end of May through to the onset of the European monsoon in mid-June. These are the driest weeks of the year in Ireland and Scotland (particularly in the north and west), testament to the prevalence of anticyclonic conditions. A similar period of enhanced anticyclonic frequencies occurs during the second and third week of September, a notably dry time in eastern and central England. As far as the cyclonic type is concerned, frequencies are highest from late April to the end of August with the maximum in the annual cycle reached in the latter month before a rapid drop to lower levels occurs in September. There is a rather intriguing tendency for cyclonic conditions to be prevalent towards the end of the first week in May, but this is probably a random quirk of the statistics and not related to the May Day holiday weekend!

The phases of the annual cycle defined by the frequency of westerly, anticyclonic and cyclonic types correspond well to the **natural seasons** for the British Isles, defined by Hubert Lamb[17] on the basis of spells of consistent weather. These natural seasons are more in tune with the rhythm of the weather than the conventional three-month seasons which rather arbitrarily divide the year into equal portions.

The natural season of autumn, running from September to November, can be characterised by frequent anticyclonic conditions, a moderate level of westerly days and a low frequency of cyclonic days. Early winter (or forewinter) extends from late November through to early February and is a time of frequent westerly days and a relatively low frequency of both anticyclonic and cyclonic conditions. Late winter, or early spring, sees a drop in westerly frequency and a rise in the occurrence of anticyclonic conditions, although cyclonic days remain low in frequency. This season extends from February through to the end of March. Spring, or early summer, is marked by low frequencies of westerly days, a relatively high level of anticyclonic conditions and rising frequencies of cyclonic days. The natural season of spring begins in early April and lasts until early June. The onset of the European

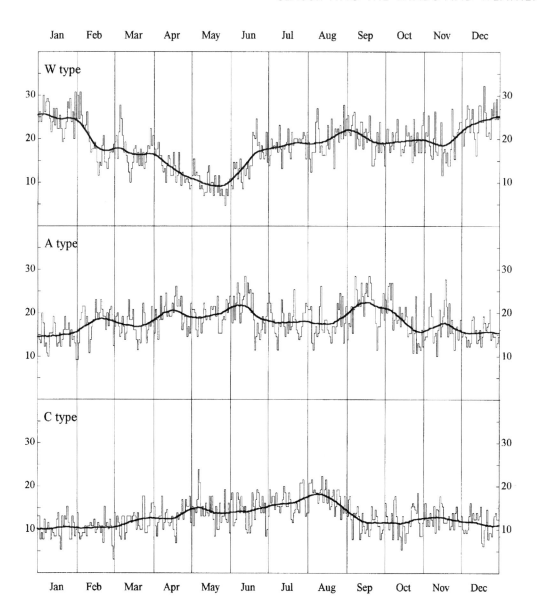

Figure 8.9 Daily average frequency, as a percentage of the years in which the type occurred on the specified day, for the westerly, anticyclonic and cyclonic Lamb types over the period from 1861 to 1990. The daily values are plotted as departures from the underlying seasonal trend, which is defined by the smooth, filtered curves.

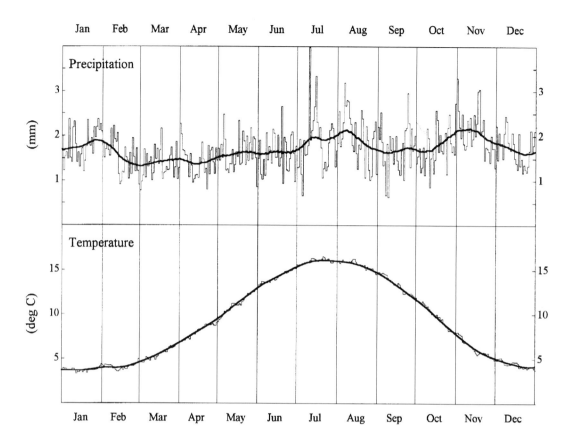

Figure 8.10 Daily average Central England Temperature over the period 1861 to 1990 and England and Wales Precipitation over the period 1931 to 1980. The daily values are plotted as departures from the underlying seasonal trend, which is defined by the smooth, filtered curves.

monsoon during June marks the beginning of high summer, with westerly days moderately frequent, anticyclonic conditions relatively infrequent and cyclonic frequencies reaching a peak in early August. The final two months of the high summer, from mid-July onwards, are characterised by higher rainfall than the first month (Figure 8.10). It is, perhaps, not coincidental that the onset of this period occurs close to St Swithun's Day – 15 July – when the occurrence of rainfall is popularly supposed to presage forty days of continuous rain, 'In this month is St Swithun's Day / On which if that it rain they say / Full forty days after it will / More or less some

rain distil'. Could this tale reflect an ancient appreciation of the nature of rainfall seasonality?

Changes in the timing of these natural seasons, and in their character, can radically affect the character of a particular year and, if persistent, may result in longer-term climate change.

THE PAST RECORD OF CIRCULATION CHANGES

The present-day record of the Lamb Classification extends back to 1 January 1861, but daily weather types have also been assigned to earlier periods in the course of the reconstruction of synoptic charts undertaken by John Kington and by Hubert Lamb. The classification for October 1781, for example, shows that this month was dominated by anti-cyclonic conditions.[18] This was one of the driest Octobers in the England and Wales Precipitation record (see Chapter 10). Gilbert White, writing at Selbourne in Hampshire, reported that 'the upland villages are in great distress for want of water' on the 25th and that 'men sow their wheat in absolute dust' on the 26th. The drought can clearly be attributed to the dominance of high pressure through the month. In fact, the decade of the 1780s, coinciding with a surge in the early industrialisation of British cities, was notable for many months of extreme climate (see Chapter 9). The Lamb Classification for the period suggests it was a period when the westerly winds were low in frequency with **blocking** anticyclones often bringing extreme weather of one kind or another to the region.[19] The Lamb Classification for historic periods has been used as a basis for discussion of the influence of weather on events such as the Irish Potato Famine in the 1840s and the Spanish Armada of 1588, broadening understanding of the circumstances of the times.[20]

The three indices based on the Lamb Classification introduced earlier are now used to define the major trends affecting the circulation over the British Isles in the period since 1861 – the start of the modern record. The westerly days record is shown in Figure 8.11. After reaching a peak early in the twentieth century, the frequency of westerly days fell during the 1950s and remained low until the late 1980s when a slight rise in frequency occurred. These trends reflect larger-scale changes in the strength of the zonal flow across the North Atlantic and western European sectors.

There is no overall trend in the C Index, the measure of the balance between cyclonicity and anticyclonicity, although peak cyclonicity has occurred at intervals of approximately fifty years. Note the very cyclonic year of 1872, which was by far the wettest year in the England and Wales Precipitation series over the period since 1766 when the record

begins (see Chapter 10). This index is strongly correlated with the precipitation data, as has been noted earlier, and decade-to-decade fluctuations in the C Index and the England and Wales Precipitation record are in good agreement. The final index – the S index – reflects the character of the **meridional flow** over the British Isles, which is the balance between southerly and northerly conditions. It is clear from the record that there has been a marked rise in southerly days, and/or decline in northerly days in recent decades as evidenced by the increase in the S Index after 1970. Some further assessment of trends in the inter-relationships between Lamb types is provided in Box 8.2.

TOWARDS MORE DETAILED CLASSIFICATION

The Lamb Classification has proved its value in many applications, but it does have certain limitations. Arguably, the main weakness lies in its broad-scale nature. It represents the circulation over a very wide area and this means that it can be difficult to relate the information in the catalogue to local conditions which, on any given day, may vary considerably across the British Isles. The classification is most relevant to studies of the character of the weather towards the centre of the British Isles.

In an attempt to remedy this limitation, Julian Mayes[21] has developed a set of regional classifications for four areas of the British Isles: Ireland, Scotland, south-west England and south-east England. Using a typing scheme analogous to that developed by Lamb, although based on surface charts alone, the four circulation type catalogues should provide the opportunity for a more detailed assessment of conditions in the represented areas. Analysis of indices based on the regional catalogues suggests that, on this regional scale, the major changes in airflow frequency over the period since the 1950s have been an increase in westerliness over Scotland from the middle of the 1970s onwards and a particularly marked period of anticyclonicity towards the end of the 1980s over south-east England.

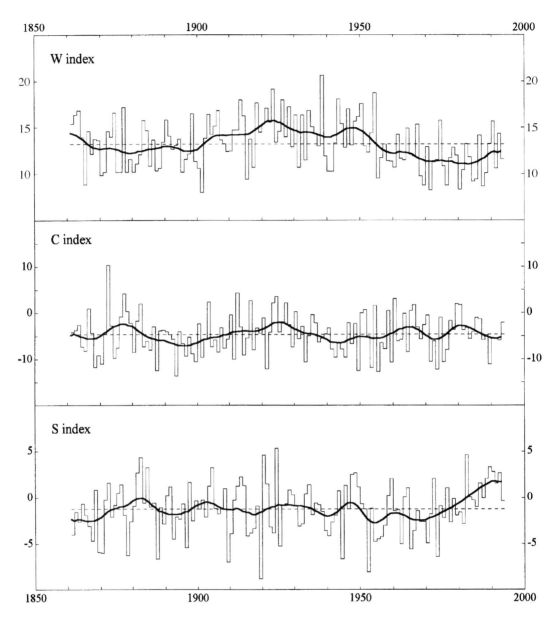

Figure 8.11 Annual averages of the monthly values for the W index, the C index, and the S index derived from the Lamb Catalogue. See Table 8.3 for a note on these indices. The smoothed curve shows the result of filtering the records to reveal long-term variations. The dashed lines are the long period averages.

BOX 8.2 RELATIONSHIPS BETWEEN THE LAMB TYPE FREQUENCIES

Are there systematic relationships between the frequencies of the different Lamb types? Is a westerly winter one which is necessarily low in anticyclonic days? Are southerly summers likely to be more or less anticyclonic? There are statistical means of identifying compensatory relationships such as these in an objective fashion. **Principal Component Analysis**, for example, has been applied to the data in the Lamb Classification to see whether or not there are systematic relationships between the weather-type frequencies which might throw light on the mechanisms underlying the variations from year to year and decade to decade.[a] This method of statistical analysis is a powerful means of isolating relationships within a complex dataset.

When applied to the Lamb Classification, the technique reveals that the dominant relationship is between the frequency of westerly and anticyclonic types. The rise and fall of the westerlies over the twentieth century has been accompanied by an inverse trend in the frequency of anticyclonic blocking conditions. The second most important relationship suggests that when the combined frequency of westerly and anticyclonic conditions is low then there is a rise in the frequency of cyclonic days (and vice versa). This situation seems to have been characteristic of recent decades. The third most important relationship is an index of the balance between cyclonic and/or anticyclonic conditions and the north-westerly weather type. When the former two types are low in frequency, a rise in the north-westerly weather type frequency occurs. There is also a relationship which is similar to that underlying the S index, the relative frequency of the southerly and the combined northerly and north-westerly weather types.

[a] P.D. Jones and P.M. Kelly, 'Principal component analysis of the Lamb Classification of daily weather types: Part 1, annual frequencies', *Journal of Climatology*, 1982, vol. 2, pp. 147–57; K.R. Briffa, P.D. Jones and P.M. Kelly, 'Principal component analysis of the Lamb Classification of daily weather types: Part 2, seasonal frequencies and update to 1987', *International Journal of Climatology*, 1990, vol. 10, pp. 549–63.

NOTES

1 R. Abercromby, 'On certain types of British weather', *Quarterly Journal of the Royal Meteorological Society*, 1883, vol. 9, pp. 1–25: see also R. Abercromby, *Weather: a Popular Exposition of the Nature of Weather Changes from Day to Day*, London, Kegan Paul, 1887.

2 E. Gold, 'Aids to forecasting: types of pressure distribution, with notes and tables for the fourteen years 1905–1918', *Geophysical Memoirs (London)*, 1920, vol. 2(16), pp. 149–74.

3 F. Baur, *Musterbeispiele europäischer Grosswetterlagen*, Wiesbaden, Dieterich, 1947.

4 R.G. Barry and A.H. Perry, *Synoptic Climatology: Methods and Applications*, London, Methuen, 1973. See also B. Yarnal, *Synoptic Climatology in Environmental Analysis*, London, Belhaven Press, 1993.

5 E.E. Federov, 'Climate as a totality of weather', *Monthly Weather Review*, 1927, vol. 55, pp. 410–13; P.E. Lydolph, 'Federov's complex method in climatology', *Annals of the Association of American Geographers*, 1959, vol. 49, pp. 120–44.

6 R.B.M. Levick, 'Fifty years of British weather', *Weather*, 1950, vol. 5, pp. 245–7.

7 H.H. Lamb, 'British Isles weather types and a register of the daily sequence of circulation patterns', *Geophysical Memoirs (London)*, 1972, vol. 16 (116), 85 pp.

8 H. H. Lamb, 'British Isles daily wind and weather patterns 1588, 1781–86, 1991 and shorter early sequences (in 1532, 1570, and other years, notably 1688, 1689, 1694, 1697, 1703, 1717, 1783–84, 1791, 1795, 1822, 1825, 1829, 1845, 1846, 1849, 1850, 1854–5)', *Climate Monitor*, 1994, vol. 20, pp. 47–71.

9 R. G. Barry, 'The prospect for synoptic climatology: a case study', in R.W. Steel and R. Lawton (eds),

Liverpool Essays in Geography, London, Longmans, 1967, pp. 85–106.

10 J.C. Sweeney and G.P. O'Hare, 'Geographical variations in precipitation yields and circulation types in Britain and Ireland', *Transactions of the Institute of British Geographers N.S.*, 1992, vol. 17, pp. 448–63.

11 Ibid.

12 R. Murray and R.P.W. Lewis, 'Some aspects of the synoptic climatology of the British Isles as measured by simple indices', *Meteorological Magazine*, 1966, vol. 98, pp. 201–19. See also R. Murray and P.R. Benwell, 'PSCM indices in synoptic climatology and long range forecasting', *Meteorological Magazine*, 1970, vol. 99, pp. 232–45.

13 Murray and Benwell, op. cit.

14 A.H. Perry and R.G. Barry, 'Recent temperature changes due to changes in the frequency and average temperature of weather types over the British Isles', *Meteorological Magazine*, 1973, vol. 102, pp. 73–82.

15 T.D. Davies, P.M. Kelly, P.B. Brimblecombe, G. Farmer and R.J. Barthelmie, 'Acidity of Scottish rainfall influenced by climatic change', *Nature*, 1986, vol. 322, pp. 359–61; T.D. Davies, G. Farmer and R.J. Barthelmie, 'Use of simple daily atmospheric circulation types for the interpretation of precipitation composition at a site (Eskdalemuir) in Scotland, 1978–1984', *Atmospheric Environment*, 1990, vol. 24A(1), pp. 63–72.

16 H.H. Lamb, *Climate, Present, Past and Future. Volume 1: Fundamentals and Climate Now*, London, Methuen, 1972. There are minor differences in the start and end of the seasons identified by Lamb and evident in the current analysis, but this is presumably the result of the extended analysis period used here.

17 H.H. Lamb, 'Types and spells of weather around the year in the British Isles', *Quarterly Journal of the Royal Meteorological Society*, 1950, vol. 76, pp. 393–438.

18 J.A. Kington, 'An examination of monthly and seasonal extremes using historical weather maps from 1781: October 1781', *Weather*, 1976, vol. 31, pp. 151–8.

19 P.D. Jones and P.M. Kelly, 'Principal component analysis of the Lamb Classification of daily weather types: Part 1, annual frequencies', *Journal of Climatology*, 1982, vol. 2, pp. 147–57.

20 H.H. Lamb, *Weather, Climate and Human Affairs*, London and New York, Routledge, 1988; H.H. Lamb, *Historic Storms of the North Sea, British Isles and Northwest Europe*, Cambridge, Cambridge University Press, 1991.

21 J.C. Mayes, 'Regional airflow patterns in the British Isles', *International Journal of Climatology*, 1991, vol. 11, pp. 473–91.

GENERAL READING

R.G. Barry and R.J. Chorley, *Atmosphere, Weather and Climate*, 6th edn, London and New York, Routledge 1995.

R.G. Barry and A.H. Perry, *Synoptic Climatology: Methods and Applications*, London, Methuen, 1973.

B. Yarnal, *Synoptic Climatology in Environmental Analysis*, London, Belhaven Press, 1993.

THE CHANGING TEMPERATURE OF 'CENTRAL ENGLAND'

Phil Jones and Mike Hulme

Or wallow naked in December snow
By thinking on fantastic summer's heat?
William Shakespeare, *Richard II*

INTRODUCTION

Temperature is the fundamental measure of the state of the climate system. Past periods, especially in mid-latitudes, are generally referred to as cold or warm epochs (e.g., **Ice Age**, **Hypsithermal**, **Medieval Warm Epoch** and **Little Ice Age**). Changes in temperature, when sustained over a period of time, therefore tell us something about changes in the functioning of the climate system. Chapter 7 looked at the history of meteorological instrumentation and observations, indicating the key figures and developments over the last 400 years. While the geographical variations in temperature over the British Isles have been described in Chapter 3, this chapter considers what the historical measurements reveal about the variation of temperature since the seventeenth century. Trends in British Isles temperatures will also be compared with those for other land areas of the Northern Hemisphere, and the causes of colder and warmer years discussed in relation to larger-scale changes in the atmospheric circulation (cf. Chapter 2). How this most important period of our climatic history relates to the **Holocene** and to the medieval period has been addressed, respectively, in Chapters 5 and 6.

The British Isles have the longest **homogeneous** temperature series for any region in the world. This

has led many to imply a much larger geographical sphere of inference from these long records than is probably justified. This issue is addressed later in the chapter. The homogeneous record, referred to as the 'Central England' Temperature (CET) record, was developed by the late Professor Gordon Manley[1] (see Figure 9.1). The record extends on a monthly basis back to 1659, the year following the death of Oliver Cromwell, and on a daily basis back to 1772.[2] These two series represent, respectively, the monthly and daily mean temperature of the English Midlands; that is, the average of the daily maximum and minimum temperature. The 'Central England' concept of Manley was of a three-station average with a northern station representing the Lancashire Plain and two others at the western and eastern extremities of the south Midlands (Figure 9.2). Over the last few years, the daily and monthly series have been updated by the United Kingdom Met. Office by using temperatures from Rothamsted, Malvern and an average of Ringway (Manchester) and Squires Gate (Blackpool).

Changes of stations in a long record like this are inevitable due to closures and sometimes to missing data. Recently, it has also been recognised that even though the sites and their data were carefully scrutinised, temperatures may be artificially raised

HISTORICAL VARIABILITY OF TEMPERATURE

The seasonal and annual values for the entire period of the Central England Temperature record, 1659 to 1995, are plotted in Figure 9.3. The seasons used are the standard three-month definitions common to climatology: winter is December to February, spring is March to May, etc. The data prior to 1721 are considered less reliable by both Manley and others. In common with most temperature series for non-tropical regions in the Northern Hemisphere, variability from year to year is greatest in winter and least in summer.

Temperatures have risen since the late seventeenth century (Figure 9.4), with the two most recent decades being about 0.8°C warmer than those at the beginning of the record. The warming has not been continuous, however, and is clearly only poorly approximated by a straight line. Despite this, we have calculated trends for each season for three periods starting in 1659, 1753 and 1901, all extending to 1995 (Table 9.1). The first and last of these periods represent the entire record and the twentieth century respectively, while the second represents the period since the change to the Gregorian calendar. David Thomson[4] has clearly shown Manley's inability to correct the weekly data available to him for the eleven-day change in the calendar which occurred in 1752. Before that time, there would have been a slight difference in the timing of the annual cycle of temperature compared to now. Annual temperatures have risen by nearly 0.7°C since the start of the record

Figure 9.1 Professor Gordon Manley, MA, D.Sc. (1902–1980), creator of the 'Central England' Temperature record.

by up to 0.5°C at some sites because of increasing urbanisation. Corrections of between −0.1°C and −0.2°C have been applied to the measured temperatures since 1980 to allow for this effect.[3] The daily series has been derived using some of the same stations as Manley, and the monthly average of the daily series equates to Manley's original monthly averages. The Central England Temperature record is both the most studied climate record in the world and the envy of climatologists in other regions. Seasonal and annual values of the record are listed in Appendix D.

Table 9.1 Temperature change in degrees Celsius explained by a straight-line trend fitted to the Central England Temperature record for three different periods.

Period	1659–1995	1753–1995	1901–1995
Winter (DJF)	1.12*	1.10*	−0.11
Spring (MAM)	0.72*	0.39	0.42
Summer (JJA)	0.14	0.02	0.63*
Autumn (SON)	0.75*	0.96*	0.93*
Annual	0.68*	0.62*	0.47*

* Warming is statistically significant at the 95 per cent level.

Figure 9.2 The Radcliffe Observatory at Oxford was used for systematic meteorological observations from the early nineteenth century and was one of the stations used by Gordon Manley in compiling his Central England Temperature record. It continues to observe to the present day.

and by nearly 0.5°C this century. In the earlier centuries the warming was greatest in winter, but in the twentieth century autumn has seen the greater warming.

The warmest calendar year of the entire record occurred in 1990, with a mean temperature of just over 10.6°C. This was 1.1°C warmer than the average of the 1961 to 1990 period. The second warmest year occurred in 1949 with a value which also rounds to 10.6°C. The next four warmest years were 1995, 1989, 1959 and 1921, each with a mean temperature of 10.5°C. The warm years of 1989 and 1990 had substantial impacts on the environment of the

British Isles[5] and raised awareness in the region about the prospects and consequences of future climate change. This awareness was further stimulated when the warmest twelve-month period of the entire Central England record occurred between November 1994 and October 1995 with a mean temperature of 11.1°C. The warmest and coldest individual months of the Central England Temperature record since 1721 are listed in Table 9.2.

The two coldest years in the record occurred in 1740, with an average temperature of only 6.8°C, and in 1879 with 7.4°C. The absolute range in annual temperature between the warmest and coldest

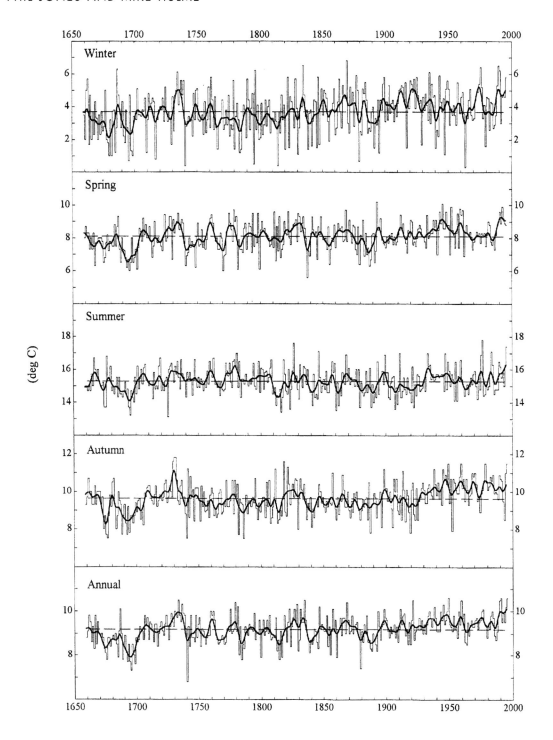

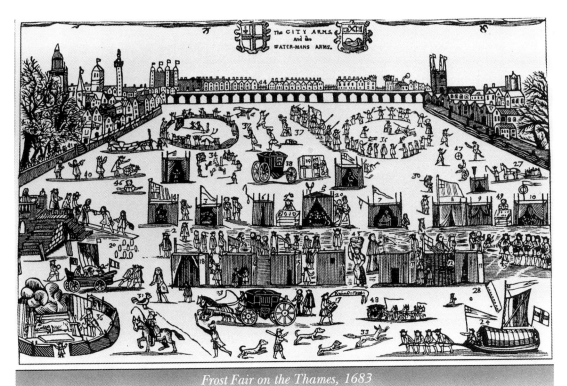

Figure 9.4 A frost fair on the River Thames at London during the winter of 1683–4. This winter recorded a mean air temperature of –1.2°C in the Central England Temperature record, the coldest in the 340-year series. Courtesy of the UK Met. Office.

years is therefore just 3.8°C, much less than the year-to-year temperature ranges for other regions at equivalent latitudes. The most anomalous year of the entire record was 1740, the birth year of James Boswell, the Scottish essayist. The year was over 2.4°C colder than either 1739 or 1741. No other year has been so different from the previous or subsequent years. All months in 1740, except September, were below the 1961 to 1990 average, with May and October being the coldest such months in the entire record. The year must have caused great distress at the time, particularly given the very warm period

from 1727 to 1738, the first ten years of the reign of George II and the warmest ten-year period in the Central England record prior to the twentieth century.

Although most temperature series from other sites in the British Isles are significantly correlated with Central England Temperature, the strength of the relationship weakens, particularly in the north of the British Isles (Figure 9.5). Three long station-series are chosen to represent the extremities of the Isles: Valentia in south-west Ireland, Stornoway on the Outer Hebrides and Lerwick in the Shetlands.

Figure 9.3 Seasonal and annual temperatures for 'Central England', 1659 to 1995. On this and all subsequent time series plots in this chapter, the smooth bold line is a filter designed to highlight decade-average changes. The horizontal dashed lines are the long-period average.

Table 9.2 Monthly and annual temperature extremes since 1721 (warmest and coldest) in the Central England Temperature record, expressed as degrees Celsius and in standard deviation unit (σ) departures from the 1961–90 average. (The years before 1721 in the CET are generally considered less reliable than those after 1721.)

	1961-90 (°C)	1961-90 σ (°C)	Year of warmest month	°C	(σ)	Year of coldest month	°C	(σ)
J	3.8	2.1	1916	7.5	(1.8)	1795	−3.1	(−3.3)
F	3.8	2.0	1779	7.9	(2.0)	1947	−1.9	(−2.8)
M	5.7	1.4	1957	9.2	(2.5)	1785	1.2	(−3.2)
A	7.9	1.0	1865	10.6	(2.8)	1837	4.7	(−3.4)
M	11.2	1.0	1833	15.1	(4.1)	1740	8.6	(−2.7)
J	14.2	1.1	1846	18.2	(3.5)	1909, 1916	11.8	(−2.1)
J	16.1	1.2	1983	19.5	(2.8)	1816	13.4	(−2.1)
A	15.8	1.2	1995	19.2	(2.9)	1912	12.9	(−2.5)
S	13.6	0.9	1729	16.6	(3.3)	1807	10.5	(−3.4)
O	10.6	1.2	1969	13.1	(2.1)	1740	5.3	(−4.6)
N	6.6	1.1	1994	10.1	(3.2)	1782	2.3	(−3.9)
D	4.7	1.8	1934, 1974	8.1	(2.0)	1890	−0.8	(−3.1)
Yr	9.5	0.5	1990	10.6	(2.2)	1740	6.8	(−5.1)

Valentia has been an official observatory (magnetic and weather) since 1869, Stornoway a lighthouse (since 1873) and since the late 1940s an airport site, while the Lerwick record began in 1905 and was also a magnetic observatory site for many years. To extend the Lerwick record back to the 1820s, we have appended the long record from the Orkney Islands which extends from 1827 to 1906. An adjustment was made to the Orkney record to bring it to the level of the Lerwick site. Annually, the Orkneys site was 0.6°C warmer than Lerwick, based on comparisons with Stornoway, Aberdeen and Thorshavn (Faroes), and so its data were reduced by this amount. The sites at Stornoway may also need a slight correction for the airport move in the 1940s, and another move in 1930, but these adjustments have not been applied here.

The annual, winter and summer time-series from the early nineteenth century to the present are shown in Figures 9.6a–c for these three peripheral stations, together with the Central England Temperature series. There is clearly some strange behaviour in the 'Lerwick' record during summers before 1870. These exceptionally warm summers are likely the result of the thermometer (on the Orkneys), which was prob-

ably located on a north wall, being affected by direct solar radiation in early morning and late evening during the long summer days. A similar influence may have partly affected spring and autumn temperatures also. For high latitude countries such as Scandinavia, Iceland and Greenland, this is a serious problem for the construction of homogeneous time-series before louvred screens of the Stevenson type were introduced in the late nineteenth century.

Variability in temperature from year to year is slightly less for the three peripheral British Isles sites compared to the Central England Temperature series, particularly during winter. This is because of their coastal locations compared to the inland domain of the Central England Temperature – coastal temperatures vary less from year to year than inland temperatures because of the moderating influence of the oceans (see Chapters 2 and 3). Agreement between the temperature series for the three peripheral stations tends to arise because of similar year-to-year extremes rather than because of coincident trends over time. This observation is confirmed by the correlations between the four annual series over the 1869 to 1995 period (Table 9.3). Correlations have been calculated for both annual and decade-average

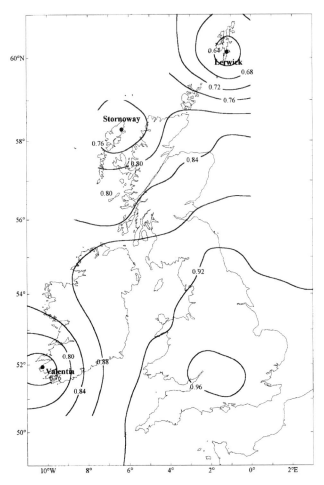

Figure 9.5 Correlation between 20 individual sites over the British Isles and the Central England Temperature record. The correlations are based on annual temperature data for the 1931 to 1990 period. The locations of Valentia, Stornoway and Lerwick, station series used elsewhere in the chapter, are shown.

temperatures and the results reflect greater similarities between the four series on the year-to-year time-scale than from decade to decade. For example, annual temperatures at Stornoway correlate with the Central England Temperature at 0.65, but when the temperatures are averaged by decades the correlation falls to 0.39.

Both Scottish sites tend to show cooling temperatures from about 1940, whereas the Central England Temperature shows a slight rise since this time. This reflects a pattern of cooler temperatures over the last

fifty years over the northern North Atlantic which has affected eastern Canada, Greenland, Iceland and Scandinavia and, to a lesser degree, northern Scotland. Over the same period, warming has been evident over western North America, most of Asia, and central and eastern Europe.[6] The Scottish sites have therefore been more affected by the cooling region than have either the Central England Temperature or Valentia. Since the mid-1980s, however, the cooling area in the North Atlantic has diminished in size and much warmer temperatures

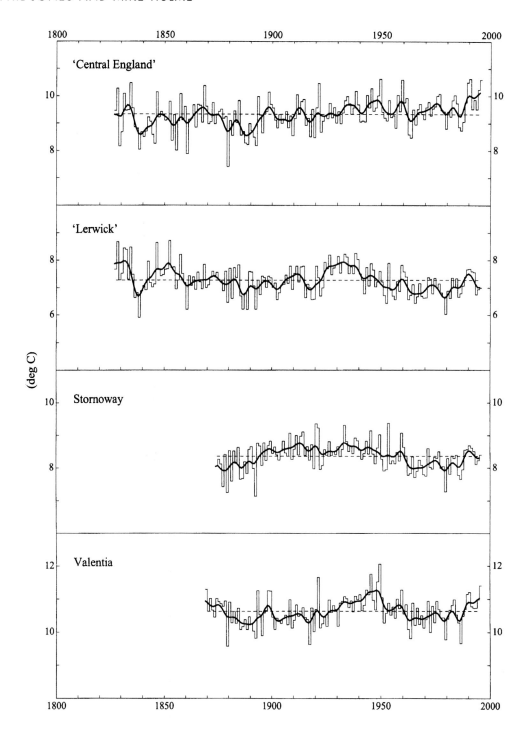

Figure 9.6*a* Annual time-series for Central England Temperature and for the three peripheral stations of Valentia, Stornoway and 'Lerwick'. 'Lerwick' has been extended by merging it with the record for the Orkneys (see p. 178).

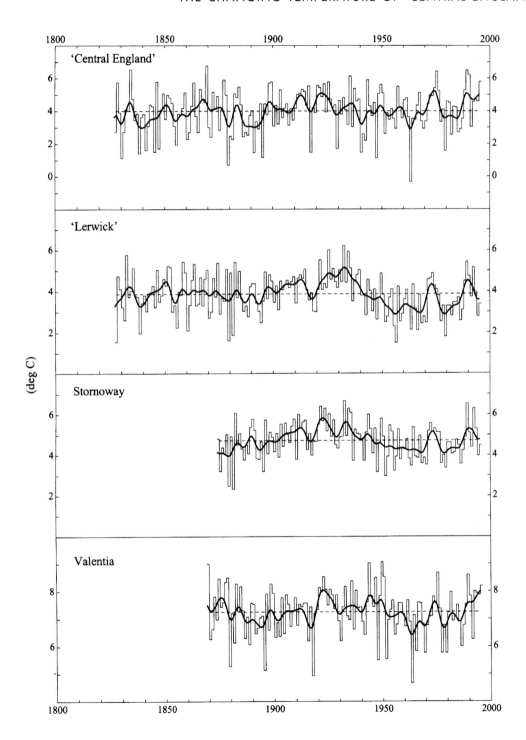

Figure 9.6b Winter time-series for Central England Temperature and for the three peripheral stations of Valentia, Stornoway and 'Lerwick'. 'Lerwick' has been extended by merging it with the record for the Orkneys (see p. 178).

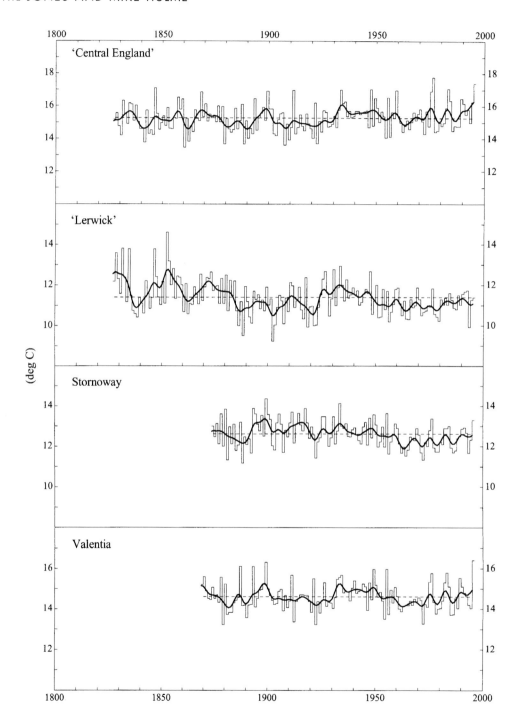

Figure 9.6c Summer time-series for Central England Temperature and for the three peripheral stations of Valentia, Stornoway and 'Lerwick'. 'Lerwick' has been extended by merging it with the record for the Orkneys (see p. 178).

Table 9.3 Correlations between annual Central England Temperature (CET) and the three peripheral stations, Valentia (V), Stornoway (S) and Lerwick (L). (Correlations are given for both annual and decade-average data. All series extend from 1869 to 1995, except for Stornoway which begins in 1874. Compare with Figure 9.5.)

Time-scale	Annual				Decade-average		
	V	S	L		V	S	L
CET	0.78	0.65	0.54	CET	0.71	0.39	0.33
V		0.63	0.51	V		0.47	0.55
S			0.78	S			0.68

have been recorded over Scandinavia and northern Scotland. Conditions remain anomalously cool over Iceland and Greenland.

Correlations between Central England Temperature and the two Scottish sites are markedly lower than between Central England Temperature and Valentia. The distances between all three sites and Central England Temperature are comparable (about 800 km), yet correlations for Valentia are in the range 0.71 to 0.78 but for the Scottish sites are only 0.33 to 0.65. This is to be expected given the general west-to-east progression of the atmospheric circulation across the British Isles (Chapter 2). The reduction with distance in the temperature correlations will therefore be greater in the meridional (north–south) than in the zonal (west–east) direction (cf. Figure 9.5).

RELATIONSHIPS BETWEEN CET AND ATMOSPHERIC CIRCULATION

Similarities between the year-to-year variations in the seasonal time-series for the three selected stations (Valentia, Stornoway and Lerwick) and for the Central England Temperature clearly suggest that, often, in any one year the whole of the British Isles is being influenced by changes in the atmospheric circulation. Chapter 8 discussed weather patterns and the Lamb weather types for the British Isles. Changes in weather-type frequency clearly influence the average annual and seasonal temperatures. Cooler years contain fewer southerly weather types and greater num-

bers of northerly days. Winters will tend to be cooler and summers warmer with more anticyclonic days and less cyclonic days. Circulation and temperature relationships will not always be so apparent by considering only annual averages, and seasonal and/or monthly series should be studied to help define the underlying relationships.

Another major circulation feature which has been shown to be of importance to the temperature of the British Isles is the **North Atlantic Oscillation** (NAO). This is the pressure difference between Ponta Delgada on the Azores and Stykkisholmur in Iceland.[7] It is a measure of wind strength across the eastern North Atlantic, with larger pressure differences implying greater average wind speeds. The NAO series is available from 1865, the start of the Azores record. Over this period, the series is strongly correlated with westerly days over the British Isles, the correlation ranging from 0.75 in winter to 0.43 in summer.

The seasonal correlations between the various atmospheric circulation statistics and Central England Temperature are shown in Table 9.4, and the annual, winter and summer time-series of Central England Temperature, NAO and the number of westerly days are plotted in Figure 9.7. Most correlations with Central England Temperature are as expected. In all seasons, northerly air flow brings cooler temperatures, southerly airflow warmer ones. Anticyclonic weather brings warmer summers, while in winter easterly weather brings cool conditions and higher numbers of westerly days milder temperatures. Stronger flow from the Atlantic as measured

Table 9.4 Seasonal correlations between the Central England Temperature and totals of the seven basic Lamb weather types (1861–1995) and the North Atlantic Oscillation (1865–1995). (Note: these circulation indices are different to those used in Table 8.3.)

	A	C	N	E	S	W	NW	NAO
Winter	−0.26	−0.06	−0.57	−0.63	0.28	0.71	0.06	0.67
Spring	0.22	−0.35	−0.43	−0.18	0.34	0.31	0.01	0.45
Summer	0.62	−0.44	−0.35	0.19	0.19	−0.30	−0.10	0.14
Autumn	−0.05	−0.03	−0.64	−0.12	0.46	0.16	−0.08	0.38
Annual	0.16	−0.22	−0.44	−0.25	0.25	0.24	−0.07	0.46

by the NAO brings warmer weather, although the influence is negligible in summer.

For the Central England Temperature to warm in the short term, there ought to be a change in the mix of **circulation types** over the British Isles with, for example, southerly flow dominating with respect to northerly. In the longer term, however, a sustained warming over several decades would gradually increase **sea-surface temperatures** around the British Isles, bringing gradually less cold air from adverse directions such as the north. Changes in circulation types would then not be necessary to create a warmer climate for the region.

RELATIONSHIPS BETWEEN CET AND NORTHERN HEMISPHERE TEMPERATURES

The available record of average Northern Hemisphere temperature for land areas commences in the mid-nineteenth century.[8] Extension further back in time is not really possible because too few areas of the hemisphere have long enough temperature series. Seasonal and annual values of this record are listed in Appendix D. Despite the clear and proven limitation of hemispheric temperature series, which are based on data from only a few locations, attempts are regularly made to indicate that long temperature records from individual sites, and from the British Isles in particular, have wider significance. It has been argued that the British Isles, because of its location in the path of hemispheric westerlies, are indicative of larger-scale conditions. We can test this assertion.

The annual, winter and summer time-series of the Central England Temperature and Northern Hemisphere average temperature over land areas are shown in Figure 9.7. Over the period 1851 to 1995, the seasonal correlations between the two series are all between 0.39 and 0.43. For all seasons except winter, correlations tend to be much higher when the temperatures are averaged by decades, rising to 0.76 for autumn and to 0.75 for the year as a whole. This suggests that Central England Temperature trends can be used, with caution, as an indicator of hemispheric temperature trends further back in time.

Two notes of caution, however, need sounding. First, the coherence between Central England Temperature and Northern Hemisphere land temperature is greater the longer the time-scale considered. For shorter periods there are occasions when the two records are out of phase. Winters during the years of the Second World War in the early 1940s, for example, were relatively mild when averaged over the land areas of the Northern Hemisphere, yet were cold in the Central England Temperature series. Second, the relationship does not imply that temperatures through all regions of the hemisphere are so correlated. The previous section remarked on the cooling over the last fifty years in the Greenland/ Iceland sector of the North Atlantic. Decade-average temperatures for this region (which is far larger than the British Isles) actually show *inverse* correlations with hemispheric temperatures when calculated over the twentieth century. There is, therefore, no easy alternative to monitoring temperatures in as many

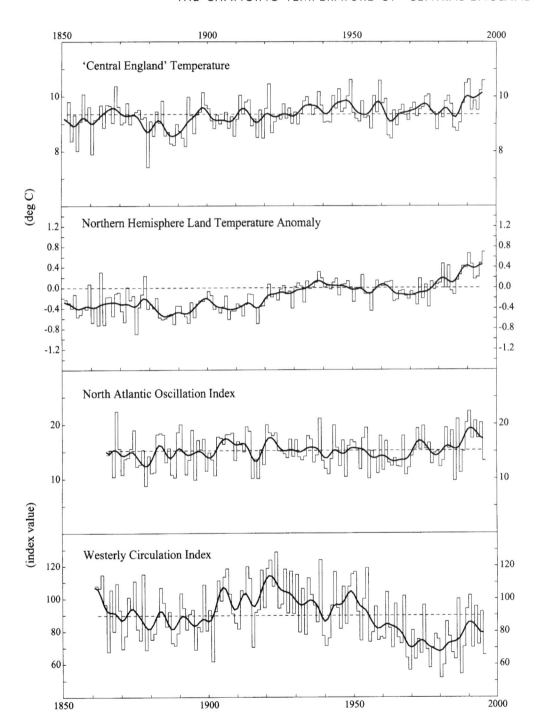

Figure 9.7a Annual time-series for Central England Temperature, the Northern Hemisphere temperature anomaly using land stations, the North Atlantic Oscillation index and the number of westerly days over the British Isles according to the Lamb Catalogue. See Appendix D for listings of these data.

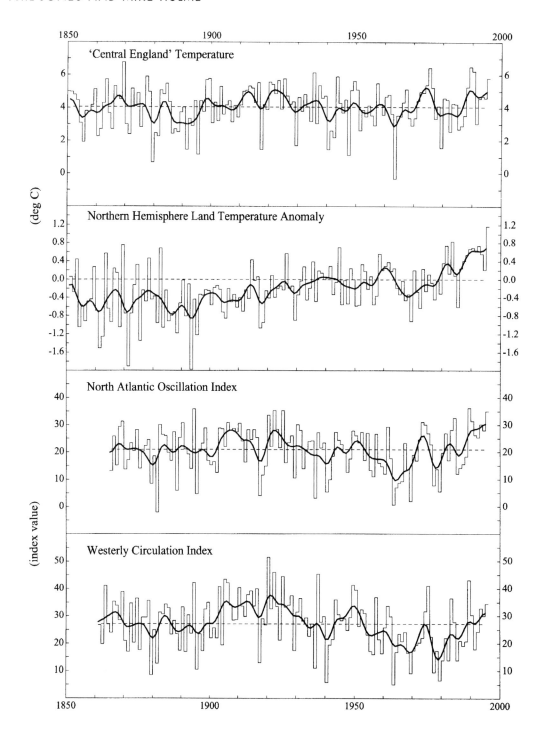

Figure 9.7b Winter time-series for Central England Temperature, the Northern Hemisphere temperature anomaly using land stations, the North Atlantic Oscillation index and the number of westerly days over the British Isles according to the Lamb Catalogue. See Appendix D for listings of these data.

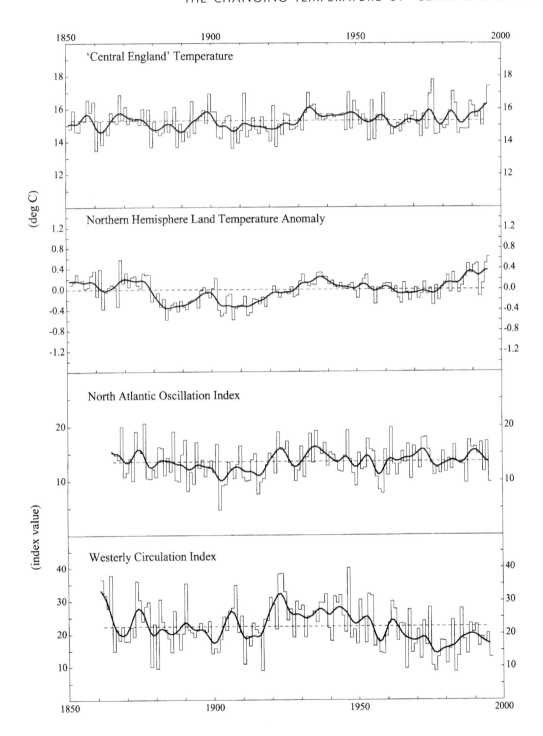

Figure 9.7c Summer time-series for Central England Temperature, the Northern Hemisphere temperature anomaly using land stations, the North Atlantic Oscillation index and the number of westerly days over the British Isles according to the Lamb Catalogue. See Appendix D for listings of these data.

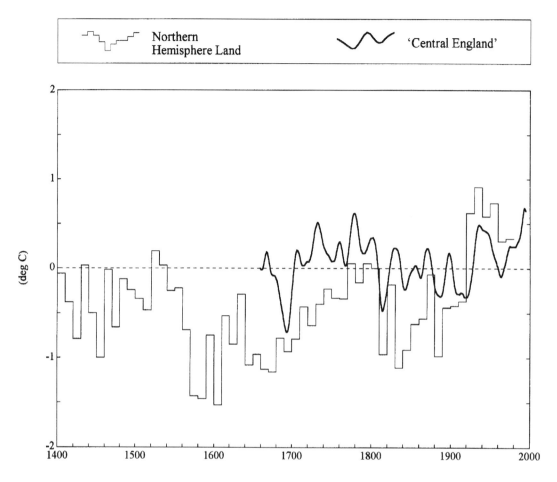

Figure 9.8 Decadal values of 'summer-average' temperature anomalies for the Northern Hemisphere land area using up to sixteen proxy climate series for the period 1400 to 1969 compared to the decade-average Central England Temperature anomalies for summer, 1660 to 1989. The anomalies are calculated from the 1861 to 1960 averages.

locations as possible to establish a faithful record of hemispheric and global temperatures.

The relationship between proxy temperature series for individual sites and larger-scale temperature averages for the hemisphere has been addressed by a number of authors. There is clearly no reason why any region should necessarily be more indicative of hemispheric conditions than any other, except that the larger the region is, the more likely its temperature trend should follow the hemispheric temperature trend. Similarly, in the temporal domain,

variations in temperature over a small region are more likely to conform with hemispheric variations the longer the time-scale considered. The Central England Temperature may seem to coincide fairly closely with hemispheric temperature trends during spring, summer and especially autumn on a decade-by-decade basis since 1850, but it does not necessarily follow that such a situation prevailed over the earlier period from 1650 to 1850.

Our most likely chance of estimating robust hemispheric temperature series prior to the mid-nineteenth

Table 9.5 Notably very hot (>23.5°C) and very cold (<−8.0°C) days in the daily Central England Temperature record

Very hot days		Very cold days	
25.2	29 July 1948	−11.9	20 January 1838
24.9	1 August 1995	−10.8	25 December 1796
24.7	4 July 1976	−9.3	4 January 1867
24.6	3 July 1976	−9.2	8 January 1841
24.6	2 August 1995	−8.9	25 January 1795
24.5	13 July 1808	−8.9	19 January 1823
24.4	14 July 1808	−8.8	9 February 1816
24.4	3 August 1990	−8.5	15 January 1820
24.0	5 July 1852	−8.5	12 December 1981
24.0	13 July 1923	−8.4	27 December 1798
23.9	24 July 1818	−8.4	23 January 1963
23.9	5 July 1976	−8.3	8 February 1895
23.9	2 August 1990	−8.2	28 December 1798
23.7	15 July 1825	−8.2	15 January 1838
23.7	12 July 1923	−8.2	24 January 1963
23.7	28 July 1948	−8.1	21 January 1881
23.7	8 August 1975	−8.0	15 January 1929

century is to use the Central England Temperature as one of a number of temperature indicators around the hemisphere. An instrumental-based series such as the Central England Temperature will be more accurate than one based on tree-rings, ice-cores or corals, but no more important, given its small representative area. Non-instrumental or proxy records of the past are generally specific to the 'summer' or to the growing season.[9] Ray Bradley and Phil Jones constructed an estimate of 'summer' hemispheric land temperatures back to 1400 on a decade-average basis using up to sixteen different proxy records.[10] This is compared with the Central England Temperature (Figure 9.8). The decadal correlation between the Central England Temperature and the average of the other fifteen proxies located elsewhere in the Northern Hemisphere is 0.43 over the thirty-one decades from 1660 to 1969. This is somewhat lower than would be expected from the earlier correlations between the Central England Temperature and the instrumental Northern Hemisphere temperature record from 1851. This again emphasises the caution that should be applied when interpreting the Central England

Temperature record as an indicator of larger, hemispheric-scale temperature fluctuations.

DAILY TEMPERATURE VARIABILITY

In this final section of the chapter we analyse the daily Central England Temperature record[11] to consider day-to-day temperature variability and extreme daily temperatures. The daily record runs from 1772, the year Captain James Cook left on his second voyage of exploration, to the present. Daily Central England Temperature values for nine selected years (1816, 1868, 1879, 1921, 1949, 1963, 1989, 1990 and 1995) are shown in Figure 9.9. Each sequence of daily values is shown compared to the average for the 1961 to 1990 period. The nine selected years include the four warmest years in the entire record (1990, 1949, 1995 and 1989), the coldest year in the daily record (1879, the year of the Tay Bridge disaster), the coldest summer in the daily record (1816 – often called 'the year without a summer'[12]) and the third coldest winter of the entire record (1962–3). The recent

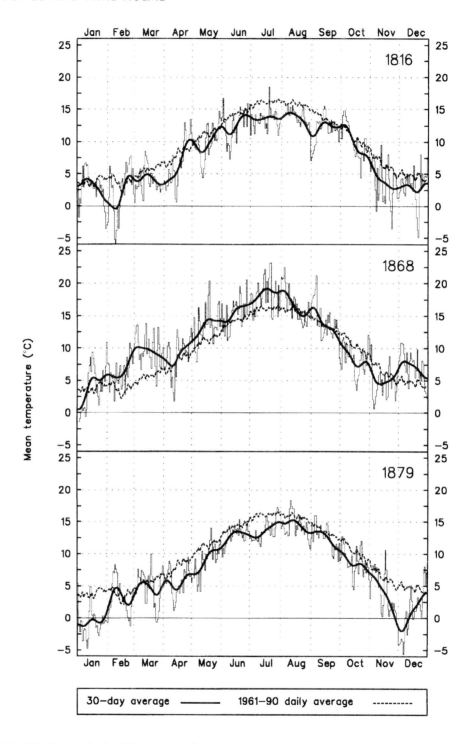

Figure 9.9 Daily Central England Temperatures for nine selected years – 1816, 1868, 1879, 1921, 1949, 1963, 1989, 1990 and 1995. Each sequence is shown compared to the average daily temperature for the 1961 to 1990 period (dashed line). The smooth bold line shows temperature variations on the monthly timescale. See Appendix C for equivalent plots for *all* years from 1961 to 1995.

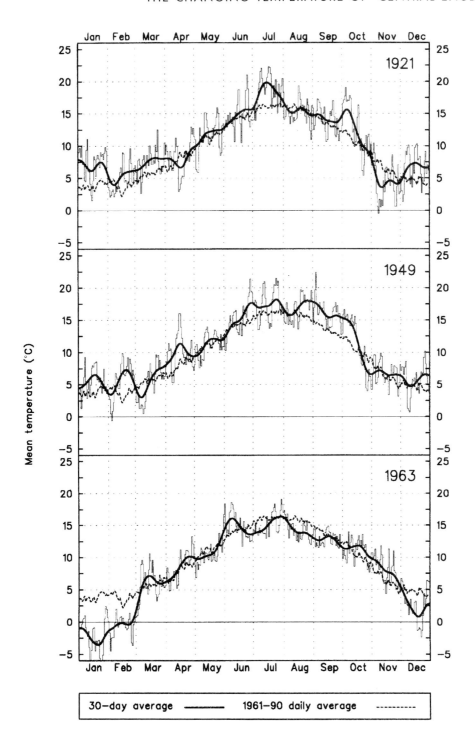

Figure 9.9 Daily Central England Temperatures for nine selected years (cont.)

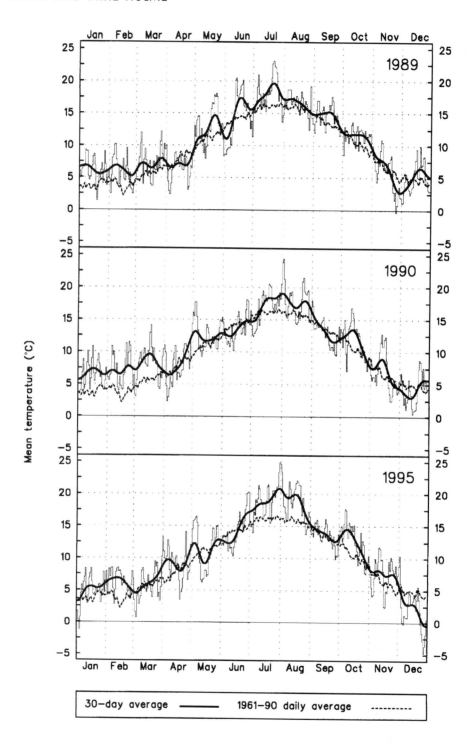

Figure 9.9 Daily Central England Temperatures for nine selected years (cont.)

Figure 9.10 The River Thames frozen at Marlow in January 1987. The winter of 1986–7 was not extremely cold, but a short spell of very cold weather in the middle of January caused the river to freeze. The mean daily CET fell to −7.7°C on 12 January and remained below zero for a week. This photograph was taken on the 18 January.

warm years all show warmer than 1961 to 1990 average temperatures throughout most of the year, with only three or four short excursions below the average. The short cold spell in late December 1995 prevented this year from breaking the record warmest annual temperature for the Central England Temperature established in 1990. An average December would have given 1995 an annual temperature of 10.7°C, which would have been 0.1°C warmer than 1990. Cold years such as 1879 and 1816 were generally below the average temperature for all but a few short periods. All the years show periods when the weather was anomalously cool or warm for long periods. Superimposed on these persistent anomalies are fluctuations typically of five to ten days' duration associated with the passage of fronts.

The two hottest days of the entire Central England Temperature series have been 29 July 1948 (25.2°C) and 1 August 1995 (24.9°C). Other notably hot days (>23.5°C) are listed in Table 9.5. These days have all occurred in July or August. The coldest day in the entire series occurred on 20 January 1838 (−11.9°C). Other notably cold days (<−8.0°C) are also listed in Table 9.5. Most of these very cold days have fallen in January. The warmest sequence in the entire record occurred from 26 June to 7 July 1976, with twelve consecutive days recording a mean temperature above 22°C. The coldest sequence occurred from 14 January to 26 January 1881, the month that postal orders were first issued in Britain, where ten of the twelve days had mean temperatures below −5°C.

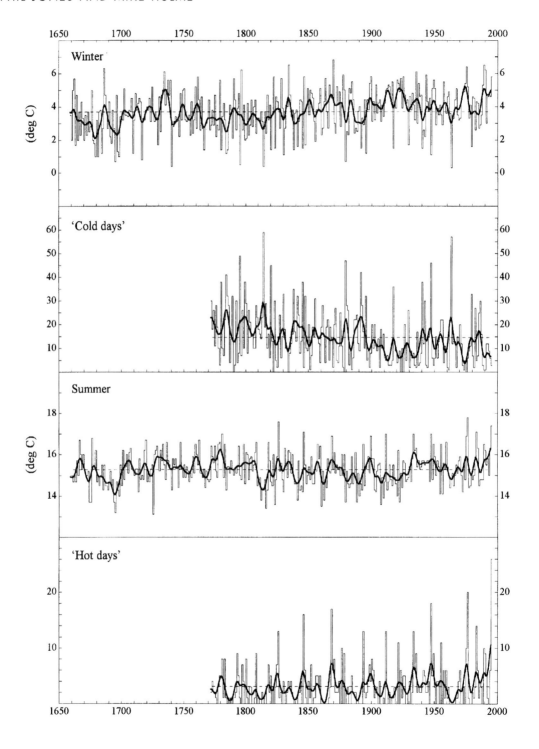

Notably cold and warm days or spells of one or two weeks' duration need not be characteristic of the season as a whole (Figure 9.10) and may not be reflected therefore in the average seasonal temperatures shown in Figure 9.3. The number of 'cold' days each winter with mean temperature below 0°C and the number of 'hot' days each summer with mean temperature above 20°C are plotted in Figure 9.11. Since 1772, the number of cold days correlates with winter temperatures to –0.89, but the number of days warmer than 20°C correlates with summer temperatures to only 0.73. Clearly, a hot summer month (e.g., July 1983) need not always be associated with extreme high temperatures, and a few extreme high temperatures in summer (e.g., July 1948) need not lead to a very warm summer month. The summers with the greatest number of hot days were 1976 which recorded twenty such days and 1995 with twenty-six hot days. Plots of daily Central England Temperature for each year from 1961 to 1995 are shown in Appendix C.

CONCLUSIONS

The British Isles have the longest instrumental record of temperature for any region in the world: the Central England Temperature series which extends back to 1659. Temperatures over the British Isles have risen by about 0.7°C since 1700 and by about 0.5°C since the turn of the present century. The warming this century has been larger in the southern, compared to the northern, half of the islands. The warmth has occurred in all four seasons, but is most pronounced in the twentieth century in summer and autumn. The variability in temperature from year to year can be related to changes in the atmospheric circulation both at the local scale, as measured by weather types over the British Isles, and on a larger scale through the strength of the zonal

circulation over the North Atlantic, as measured by the North Atlantic Oscillation.

The Central England Temperature series is related to average land temperatures over the Northern Hemisphere, but probably no more than would be expected from any other region of comparable size in the hemisphere. The relationship appears to strengthen when temperatures are averaged over decades or more, but considerable caution should be exercised in extrapolating trends from the long Central England Temperature record to the hemisphere as a whole.

Three of the four warmest years in the Central England Temperature record have occurred since the late 1980s (1990, 1995 and 1989). Recent years have also seen many record-breaking warm months and warm days. Global and hemispheric temperatures are also at record levels and scientific opinion, based on the similarities between observed and predicted patterns of climate change,[13] suggests a discernible influence on global temperatures from human activities (see Chapter 15). It seems likely that the Central England Temperature series will record its first year with mean temperature above 11°C within the next few years.

NOTES

1 G. Manley, 'Central England Temperatures: monthly means 1659 to 1973', *Quarterly Journal of the Royal Meteorological Society*, 1974, vol. 100, pp. 389–405.
2 D.E. Parker, T.P. Legg and C.K. Folland, 'A new daily Central England temperature series', *International Journal of Climatology*, 1992, vol. 12, pp. 317–42.
3 Ibid.
4 D.J. Thomson, 'The seasons, global temperature, and precession', *Science*, 1995, vol. 268, pp. 59–68.
5 M.G.R. Cannell and C.E.R. Pitcairn (eds), *Impacts of the Mild Winters and Hot Summers in the United Kingdom in 1988–1990*, London, HMSO, Dept of the Environment, 1993, 154 pp.

Figure 9.11 The number of days with mean temperature below 0°C each winter ('cold' days) and the number of days with mean temperature equal to or in excess of 20°C ('hot' days) each summer. All data are from the daily Central England Temperature record. Also shown are the winter and summer values of the monthly Central England Temperature record.

6 D.E. Parker, P.D. Jones, C.K. Folland and A. Bevan, 'Interdecadal changes of surface temperature since the late nineteenth-century', *Journal of Geophysical Research*, 1994, vol. 99, pp. 14373–99.

7 H. van Loon and J.C. Rogers, 'The see-saw in winter temperatures between Greenland and Northern Europe, Part 1: General Description', *Monthly Weather Review*, 1978, vol. 106, pp. 296–310.

8 P.D. Jones and K.R. Briffa, 'Global surface air temperature variations over the twentieth century: Part 1, spatial, temporal and seasonal details', *The Holocene*, 1992, vol. 2, pp. 163–79.

9 K.R. Briffa and P.D. Jones, 'Global surface air temperature variations over the twentieth century: Part 2, Implications for large-scale high-frequency paleoclimatic studies', *The Holocene*, 1993, vol. 3, pp. 82–93.

10 R.S. Bradley and P.D. Jones, 'Little Ice Age summer temperature variations: their nature and relevance to recent global warming trends', *The Holocene*, 1993, vol. 3, pp. 367–76.

11 Parker, Legg and Folland, op. cit.

12 C.R. Herrington (ed.), *The year without a summer: world climate in 1816*, Ottawa, Canadian Museum of Nature, 1992, 576 pp.

13 B.D. Santer, K.E. Taylor, T.M.L. Wigley, J.E. Penner, P.D. Jones and U. Cubasch, 'Towards the detection and attribution of an anthropogenic effect on climate', *Climate Dynamics*, 1995, vol. 12, pp. 77–100; J. Houghton, L.G. Meira Filho, B.A. Callander, N. Harris, A. Kattenberg and K. Maskell (eds), *Climate Change 1995: the Science of Climate Change*, Cambridge, Cambridge University Press, 1996, 572 pp.

GENERAL READING

R.S. Bradley and P.D. Jones (eds) *Climate Since AD 1500*, 2nd edn (paperback), London and New York, Routledge, 1995, 706 pp.

M.G.R. Cannell and C.E.R. Pitcairn (eds), *Impacts of the Mild Winters and Hot Summers in the UK in 1988–1990*, UK Department of the Environment, HMSO, London, 1993, 154 pp.

M.J. Tooley and G.M. Sheail (eds), *The Climatic Scene*, London, George Allen and Unwin, 1983, 306 pp.

10

PRECIPITATION VARIABILITY AND DROUGHT

Phil Jones, Declan Conway and Keith Briffa

If you took away everything in the world that had to be invented,
there'd be nothing left except a lot of people getting rained on.
Tom Stoppard, *Enter a Free Man*

INTRODUCTION

Although temperature is the primary measure of the state of the climate system, precipitation (rainfall and snowfall) and its variability is vital to ecosystems and human activity. This is particularly true for agriculture and for water supply systems. If too little rain falls over extended periods, agricultural yields decline, produce increases in price and water rationing may be needed in certain regions. Too much rainfall may also reduce crop yields and lead to damaging floods that affect lives and property (Figure 10.1). This chapter uses the extensive precipitation data for the British Isles to examine changes in rainfall and snowfall variability on different geographical and historical scales. We also examine the spatial distribution and severity of major drought periods in the region of the last 150 years.

For precipitation, as with temperature (see Chapter 9), the British Isles have some of the longest records in the world. The first precipitation measurements were made by Richard Towneley in Burnley, Lancashire, in 1677.[1] Since that time, at least 15,000 rain gauges have been in operation in the British Isles for periods ranging from a few months to 200 years or more. At present there are about 8,000 operational gauges, making the British Isles probably the most densely gauged region in the world.

The impetus behind such a comprehensive monitoring of precipitation largely rests with George James Symons (see Figure 10.2) who, in the mid-1860s, founded the British Rainfall Organisation and established the publication *British Rainfall*.[2] In this publication, precipitation observations from all over the British Isles were listed in a timely fashion. The volume has been published annually since 1865, with publication now being the responsibility of the United Kingdom Met. Office. Apart from listing precipitation totals from gauges all over Britain and Ireland, it also included articles on severe rainfall events, droughts, snowfall and rain-gauge design. In the early years – from the 1860s to 1880s – observers competed to operate the gauge that caught the most precipitation in one year in the British Isles. For a number of years therefore, the Lake District (which had the wettest location with an annual average of about 4,400 mm), the Ben Nevis region and Snowdonia were better gauged than at any time since. Through the continuing efforts of Symons, Hugh Robert Mill and Carle Salter, the British Rainfall Organisation remained, for many years, separate from the UK Met. Office. Eventually, however, in

WOFULL NEWS FROM WALES
GREAT FLOOD IN MONMOUTHSHIRE, 1607

Figure 10.1 A contemporary woodcut illustration from 1607 showing the great floods which afflicted North Devon and Monmouthshire in January of that year. The result was great damage to property and livestock and serious loss of life.

July 1919, the responsibility for the archives passed to the UK Met. Office, where it remains today.

THE GEOGRAPHICAL VARIABILITY OF PRECIPITATION

The development of a 1961 to 1990 climatology for the British Isles was discussed in Chapter 3. For precipitation, there are about 8,000 gauges with enough data to calculate thirty-year averages. This is almost an order of magnitude greater than is available for other meteorological variables. The geographical distributions of average seasonal and annual precipitation totals and rain-day counts were

also described in Chapter 3. In this section we comment on the geographical patterns of the year-to-year variability of precipitation.

The standard deviations of precipitation totals are shown in Figure 10.3 for the four standard seasons and for the year as a whole. The variability of seasonal precipitation is greatest in western Scotland and least in southern Britain and much of Ireland. There is a strong annual cycle in variability in the higher precipitation regions of the west, with the winter season recording standard deviation values twice those of summer. In the drier regions of eastern England precipitation variability is roughly the same value (about 60 mm) in all seasons. The map for the annual standard deviation shows a similar pattern to

Figure 10.2 George James Symons, FRS (1838–1900) founder in 1863 of the British Rainfall Organisation.

the winter season, with the highest variability over western Scotland. The only other region with variability in excess of 140 mm is the south-west coastal fringe of Ireland. Only the eastern third of England has an annual precipitation variability less than 100 mm, and there is a marked gradient over England of increasing variability from south-east to north-west. The greater variability of precipitation in the north and west of England means that surface water resource schemes (i.e., reservoirs) are more prone to short summer droughts than elsewhere in

the region. This was witnessed in 1995 in the Yorkshire Water region. Over Ireland, in contrast, there is practically no gradient in variability except in the south-west.

THE HISTORICAL VARIABILITY OF PRECIPITATION

Although precipitation varies over short distances more than does temperature, attempts have also been made to derive regional average precipitation series. The earliest serious attempt was the development of the England and Wales Precipitation series in 1931.[3] The series extended back on a monthly basis to 1727, the year Sir Isaac Newton died. The choice of domain seems to have been influenced by the publication in the 1920s of a wheat price series for the same region and the desire to know how much variation in this price index might be explained by precipitation. The choice is somewhat odd in that there were no gauges in Wales until the late eighteenth century and very few in upland areas of western and northern England until about 1840.

Despite most climatologists knowing about the lack of long-term homogeneity in this 'official' UK Met. Office England and Wales Precipitation series, little was done about the problem until the early 1980s. Only in 1996 has the non-homogeneity of the official UK Met. Office England and Wales Precipitation series been acknowledged.[4] A new England and Wales Precipitation (EWP) series, extending back to 1766, had earlier been developed by the Climatic Research Unit,[5] using seven stations in each of five regions. The regions were defined from a **principal component analysis** of over sixty long **homogeneous** precipitation series covering the British Isles and extending from 1861 to 1970. These series had originally been developed by Dick Tabony.[6]

In addition, three regional series were developed for Scotland and one for Northern Ireland (NI). An All Ireland (AI) series was also constructed. The nine regional and four country series are listed in Table 10.1 and their construction discussed in Box 10.1.

Winter

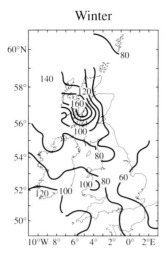

Spring

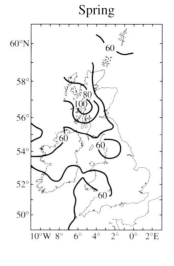

Summer

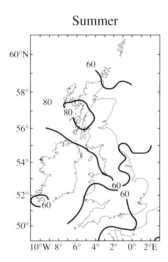

Autumn

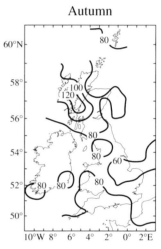

Annual

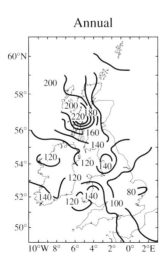

Figure 10.3 Standard deviation (in mm) of seasonal and annual precipitation totals based on the 1961 to 1990 period.

Table 10.1 The precipitation regions of the British Isles analysed in this chapter, together with the length of record

Abbreviation	Period	Region
SEE	1931–1995	South-east England
SWE	1931–1995	South-west England and South Wales
NWE	1931–1995	North-west England and North Wales
NEE	1931–1995	North-east England and south-east Scotland
CE	1931–1995	Central and eastern England
SS	1931–1995	South-west and South Scotland
NS	1931–1995	North-west and North Scotland
ES	1931–1995	East Scotland
NI	1931–1995	Northern Ireland
AI	1841–1995	All Ireland
EWP	1766–1995	England and Wales
S	1931–1995	Scotland
SMTH	1757–1992	Scotland[9]

Most of the analyses in this chapter use the long homogeneous single-site series developed by Tabony[7] and the regional and country series developed by the Climatic Research Unit.[8] The nine regional series are also available as daily precipitation series from 1931 to 1983.

Recent work by Keith Smith[9] has updated earlier work by Arthur Jenkinson in developing a Scottish (SMTH) average precipitation series back to 1757. The earliest years are based on only one gauge in either southern Scotland or Carlisle and we begin the series for plotting purposes in 1800. It would be feasible to extend the EWP series back to 1700 using gauge data in the south-east of England, especially the London region, as was originally done by the UK Met. Office.[10] At present, however, we have produced a series only from the year when at least one gauge is available in all five England and Wales regions.

The boundaries of the coherent precipitation regions and the locations of the seven sites in each region are shown in Figure 10.4. The choice of seven sites to represent each of the regions of Britain is principally related to data availability in the period before 1900. It can also be shown, however, that the addition of extra stations would not greatly improve the accuracy of the regional series (see Box 10.2).

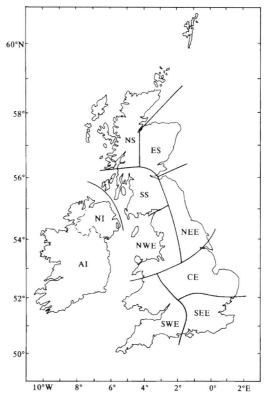

Figure 10.4 Regions of coherent precipitation variability for Britain and Ireland (after Wigley *et al.*[5]).

BOX 10.1: CONSTRUCTION OF THE REGIONAL PRECIPITATION SERIES

Lists of the stations used in each of the regions have been published elsewhere. This chapter updates the series to the end of 1995. Since the series were last updated, several stations have closed and these have been replaced with nearby alternatives. For each alternative station, a correction factor must be calculated from the ratio of the annual average catch between the new and the closed gauge to correct the new gauge to the annual precipitation at the site used earlier. This is necessary since the nine regional series representing the United Kingdom are simple averages of the seven gauges located within them. The larger-scale country series for England and Wales and Scotland are weighted averages of the regions thus:

$$EW = 0.275 \, SEE + 0.288 \, SWW + 0.265 \, CEE + 0.158 \, NWW + 0.128 \, NEE$$
$$\text{and} \quad S = 0.379 \, SS + 0.270 \, NS + 0.529 \, ES.$$

The weights were determined by regression analysis between the regional averages and the values published in *British Rainfall*. The All Ireland series has been developed during this study based on a simple average of fourteen gauges (eleven in the Republic of Ireland and three in Northern Ireland).

Seasonal time-series of precipitation are shown for England and Wales (EWP, 1767 to 1995) in Figure 10.5, for Scotland (SMTH, 1758 to 1995) in Figure 10.6, and for All Ireland (AI, 1841 to 1995) in Figure 10.7. Figure 10.8 shows the annual series for England and Wales, Scotland and All Ireland, while Figure 10.9 shows the regional series of annual precipitation since 1931.

For England and Wales (Figure 10.5), seasonal precipitation totals show considerable variability from one year to the next and occasionally from decade to decade. There is evidence of a shift to slightly higher winter precipitation during the 1860s, resulting in wetter winters over the last 130 years of the series compared to the first 100 years. Summer precipitation shows evidence of a slight decline in totals over the last 100 years. The 1995 summer was the driest on record, yielding 7 mm less than the summer of 1976 (66.9 mm cf. 73.9 mm). The seasonal and annual totals for this series are listed in Appendix D.

For Scotland (Figure 10.6), variability in spring precipitation is clearly lower than in the other three seasons. Such a feature is also present in England and Wales, but is less clear than in Scotland. Similar shifts and trends to those in England and Wales are evident in Scottish winter and summer precipitation totals. The wettest consecutive winters on record in Scotland occurred in 1988–9 and 1989–90. Several recent seasons have seen record breaking totals. For all four seasons, at least two of the six wettest seasons recorded between 1869 and 1995 have occurred since 1980. The All Ireland series (Figure 10.7) also shows some features evident in the England and Wales and Scotland series. Variability from year to year is lowest in spring and there is some tendency towards drier summers and wetter winters.

In earlier work,[11] it was found that three of the five driest summers in England and Wales, north-west England and north-east England occurred between 1975 and 1985, while four such summers occurred in south-west England. For wet extremes, three of the five wettest winters occurred between 1975 and 1986 in north-east England and central-east England, and three of the five wettest springs in England and Wales, north-west England, central-east England and south-east England. More recent data have shown that these clusters of extreme seasons have not continued into the 1987 to 1995 period.

The precipitation time-series show significant correlations between neighbouring regions. When

BOX 10.2: THE ACCURACY OF REGIONAL PRECIPITATION SERIES

It has been shown that when averaging time-series together, the average inter-series correlation ($\bar{r}$) determines the accuracy of the average.[a] If R_m is the correlation between the m-site average and the true average (using an infinite number of rain-gauges), it can be shown that,

$$R^2_m \approx \frac{m\bar{r}}{1 + (m - 1)\bar{r}}$$

As an example, the $\bar{r}$ value for daily data in the SE region is 0.624, so that R^2_m (the variance of the true average explained with $m = 7$) is 0.92. Seven sites therefore explain 92 per cent of the daily precipitation variability over the region. Increasing the number of gauges to eight increases R^2_m by only 1 per cent. The time and expense of obtaining an additional gauge is therefore not warranted. Table 10.2 gives $\bar{r}$ and R^2_m values for each of the nine regions of the United Kingdom using daily and monthly precipitation values.

Table 10.2 Average inter-series correlations ($\bar{r}$) and the variances explained by the regional seven-gauge networks (R^2_m) for daily and monthly precipitation totals

Region	Daily		Monthly	
	$\bar{r}$	R^2_m	$\bar{r}$	R^2_m
NWE	0.53	0.89	0.75	0.95
NEE	0.49	0.87	0.69	0.94
CE	0.52	0.88	0.74	0.95
SWE	0.55	0.90	0.77	0.96
SEE	0.62	0.92	0.81	0.97
SS	0.67	0.93	0.85	0.98
NS	0.47	0.86	0.70	0.94
ES	0.52	0.88	0.63	0.92
NI	0.63	0.92	0.73	0.95

[a] T.M.L. Wigley, K.R. Briffa and P.D. Jones, 'On the average value of correlated time series, with applications in dendroclimatology and hydrometeorology', *Journal of Climate and Applied Meteorology*, 1984, vol. 23, pp. 201–13.

the regions are further apart, however, correlations are much lower. We can quantify this geographical variability by examining the correlations one region at a time (Figure 10.10). These annual correlations are lower than the seasonal correlations, particularly in autumn, although the differences in correlation between the seasons are small. As would be expected, correlations between the summer series show the greatest variability over the British Isles with both high and low correlations. Western regions correlate more strongly with other western regions and similarly for eastern regions. Lower precipitation correlations in the east–west direction rather than in the north–south direction is in contrast to temperature which showed higher correlations in the east–west direction (see Chapter 9). The stronger north–south precipitation correlations relate to the frontal nature of the majority of British Isles precipitation. There is generally only a weak relationship between precip-itation in the Scottish regions and most parts of England and Wales. North-west England correlates the highest with all other regions and north Scotland the least.

DAILY PRECIPITATION VARIABILITY

Frequencies of wet days

A dry day is not normally defined simply as a day with zero precipitation. Instead a low cut-off value is used, such as 0.1, 0.2 or 1.0 mm. It is particularly important to have a non-zero threshold in evaluating spells of dry weather, since a single day with less than 1 mm of precipitation in the middle of a twenty-day dry spell would have no practical significance. A very low threshold such as 0.1 or 0.2 mm is prone to observer bias. In the daily time-series produced from the regional series (the average

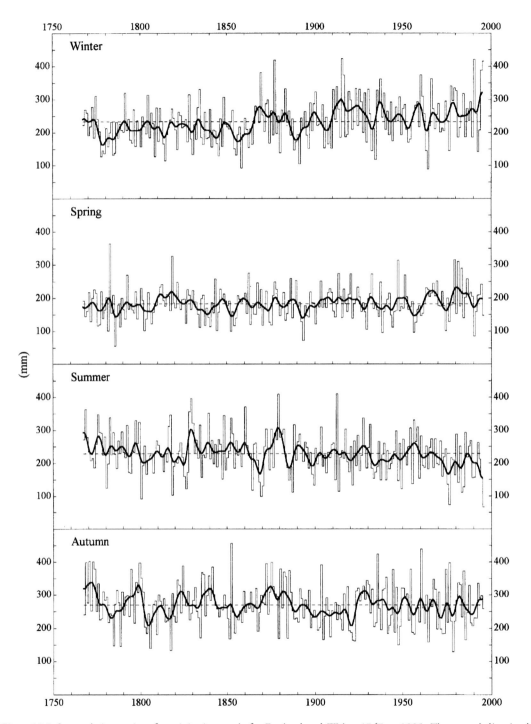

Figure 10.5 Seasonal time-series of precipitation totals for England and Wales, 1767 to 1995. The smooth line in this and subsequent plots in this chapter is a filter which suppresses variations on time-scales of a decade or less.

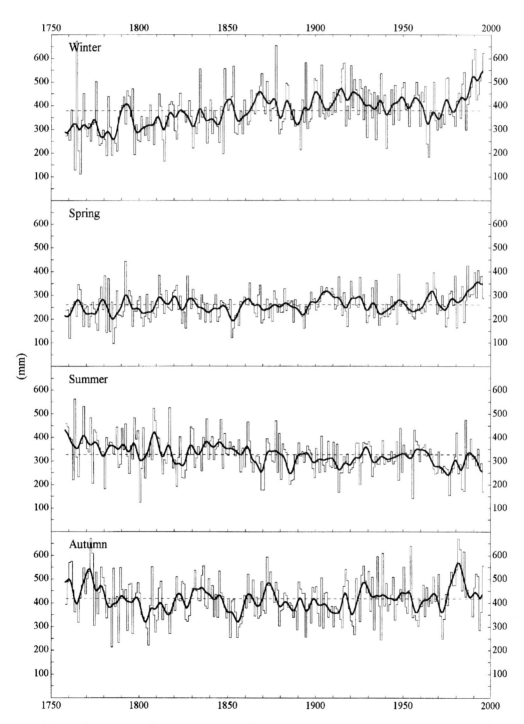

Figure 10.6 Seasonal time-series of precipitation totals for the Scotland (SMTH) series produced by Smith,[9] 1757 to 1992. The values for 1993 to 1995 are taken from the Scotland series (S) produced by the authors and re-scaled to give the same mean value as SMTH.

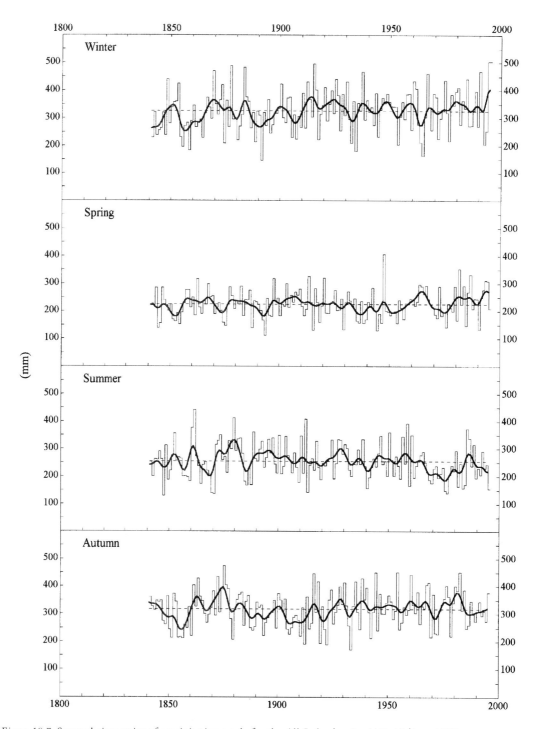

Figure 10.7 Seasonal time-series of precipitation totals for the All Ireland series (AI), 1841 to 1995.

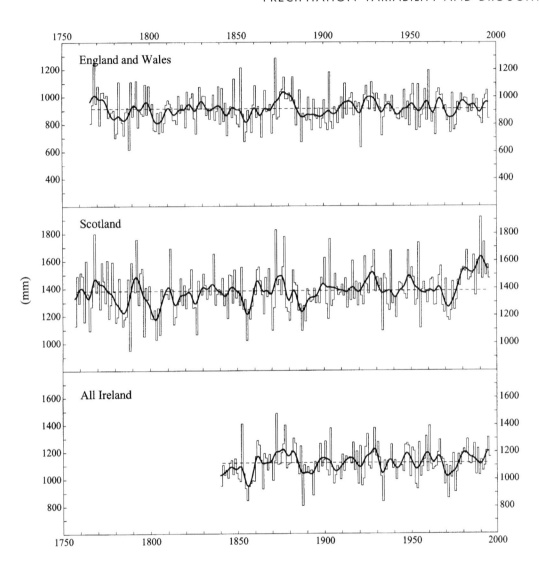

Figure 10.8 Annual precipitation series for England and Wales, Scotland (SMTH) and for All Ireland.

of seven evenly spaced gauges) for Britain and Northern Ireland we therefore use a 1 mm threshold for wet days. The number of wet days is least in the south-east and central parts of England and is most (nearly twice as great) in northern and western Scotland. The time-series for wet days (the inverse being dry days) are characterised by considerable variability from year to year, some variability from decade to decade, but no longer-term trends (Figure 10.11). The similarities between the regions imply that wet days are wetter in the wetter regions and that wetter seasons have both more wet days and more precipitation per wet day. Calculations of one- and two-day extreme precipitation totals from these data also do not reveal any longer-term trends.

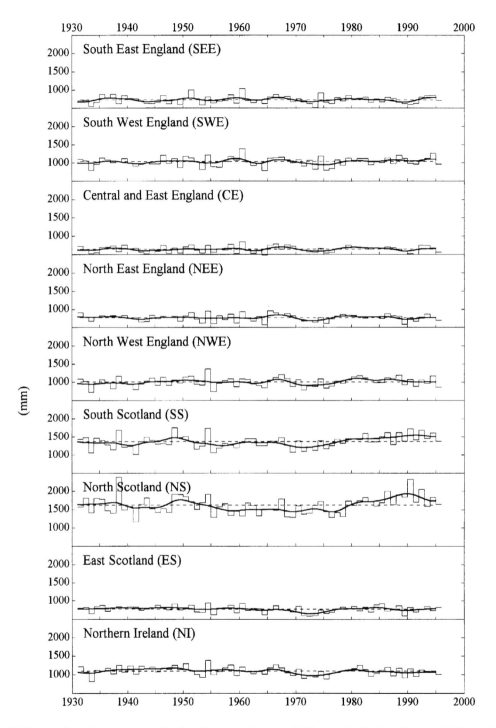

Figure 10.9 Annual precipitation series for the nine regional series of Britain and Northern Ireland, 1931 to 1995.

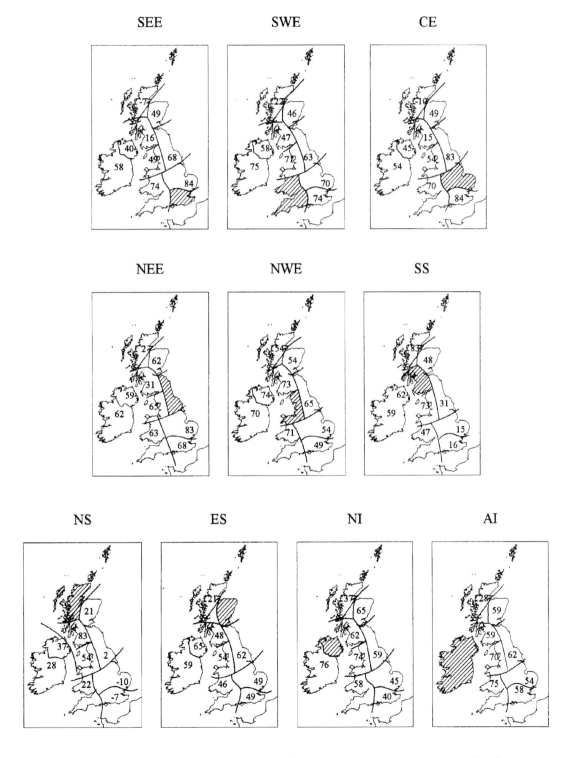

Figure 10.10 Correlations on an annual basis between each of the regional precipitation time-series. The reference series for each correlation map is shaded. The correlation coefficients are multiplied by 100.

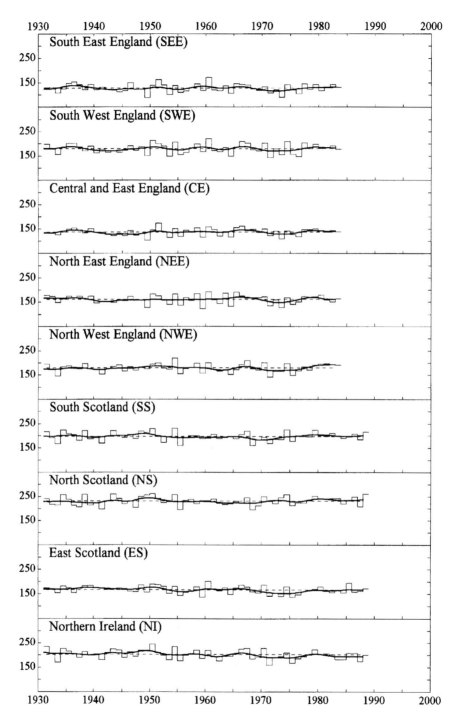

Figure 10.11 Annual counts of wet days (> 1 mm) for each of the nine regions of Britain and Northern Ireland, 1931 to 1983 or 1988 .

Dry and wet spells

For each of the nine regional series, the number of spells of duration 1, 2, 3 and so on, days, have been counted over the 1931 to 1983 period, using 1 mm as the dividing value between wet and dry days. Figure 10.12 shows some examples of the average annual number of spells of different durations. Regional differences in the lengths of dry spells are less pronounced than in wet spells, which tend to be markedly longer in the wetter regions of the British Isles (see Box 10.3). Although longer drier spells may be expected to occur during spring, the length of record used here (only fifty-three years and hence few very long spells) does not allow the seasonality in long spell lengths to be considered.

SNOWFALL

Referred to as solid precipitation in many colder countries, snowfall is difficult to measure. It is also difficult to intercompare different events. Standard UK Met. Office rain-gauges are less efficient at measuring snowfall than rainfall since even light winds reduce the number of flakes that enter the gauge and they often freeze and/or are buried by the snow. When excessive drifting occurs the only reliable means of estimating the water equivalent precipitation are often *in situ* depth and density measurements.

Assessing whether there have been long-term changes in snowfall occurrence and amounts is difficult because few long series exist. The annual *Snow Survey of Great Britain* publications in the mid-1980s listed both the number of days with snow lying and with snow falling for ten sites in the United Kingdom back to 1946–7. Longer records exist for a few sites.

The longest record of winter snowiness for the country was published by Mervyn Jackson,[12] based on earlier work by Leo Bonacina.[13] In the above publication, each winter between 1875–6 and 1974–5 was classified by one person – Bonacina – according to intentionally arbitrary definitions: 'very snowy', 'snowy', 'average' or 'little'. Similarly, months with notable falls in part of the country were classified as 'snowy'. Only one extensive fall over a large part of the country, that may only last a few days, is therefore sufficient for a month to be called snowy. Here we have updated this series to the present, extending it by twenty years. It is based on Jackson up to 1978–9 and thereafter on our own assessments from the annual *Snow Survey*, *Weather Log* and from Richard Wild and colleagues.[14]

BOX 10.3 THE NATURE OF DRY AND WET SPELLS

The slope of the curves in Figure 10.12 is related to the probability that a dry (wet) day will occur given that n previous days have been dry (wet). This probability is referred to as the conditional transition probability, $R(n)$. If $R(n)$ is constant it means that the slope of the curve will be a straight line, which would be the case if precipitation on a given day depended only on the day before in **Markovian** fashion. If the curve is concave (as in all the examples in Figure 10.12) this means that $R(n)$ must increase with increasing n. The concave curvature of the plots indicates that the probability of a wet or dry spell continuing increases as the length of the spell increases, at least for the first five to ten days. If the curve were convex the reverse would hold. The likely reason for this is the persistence of certain types of weather situation over the British Isles (such as **blocking**) once such situations have become established. Dry sequences may therefore be considered as coming from two different populations: one related to persistent circulation modes such as blocking, and the other to the more usual **progressive** modes where short dry and wet spells are interspersed.

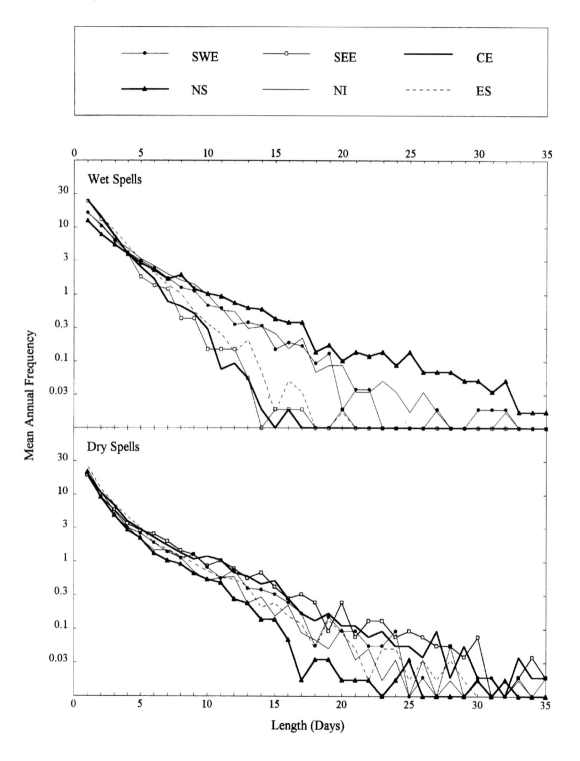

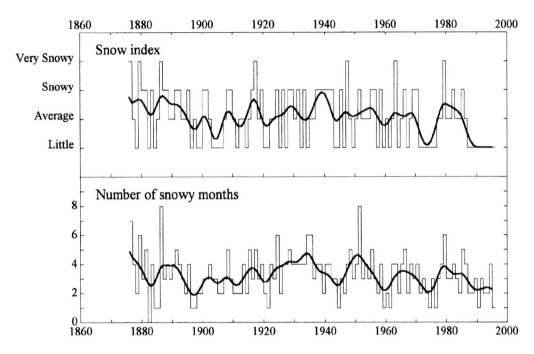

Figure 10.13 Snowy winter classification for 1875–6 to 1994–5: (top) overall winter rating on the four-category scale, and (bottom) the number of the months classified as snowy per winter.

Figure 10.13 shows the two series, the count of the number of snowy months each winter and the four-category classification of each winter as a whole. The correlation between the two series is 0.66 and their correlation with the 'extended' – November to April – winter Central England Temperature is –0.55 (number of snowy months) and –0.69 (the four-category classification). Correlations are slightly higher with the 'extended' winter definition as opposed to the conventional December to February winter definition. The snowiest and most widespread winters since 1875–6 were 1875–6, 1878–9, 1885–6, 1916–17, 1946–7, 1962–3 and 1978–9. The most severe of these was 1946–7 which Jackson[15] considers the most severe since 1813–14 (see Figure 10.14). A discussion of other notable snow events of the early nineteenth century and earlier is given by Gordon Manley.[16]

Very snowy winters do not necessarily last all winter. Both 1946–7 and 1962–3 had only four months classified as snowy. Contrast this with the 'snowy' winter of 1950–1 in which all eight 'winter' months were rated as snowy! This winter had the longest number of days of snow lying of any winter since 1945–6 in upland Britain. The most remarkable feature of Figure 10.13 is the lack of snowy winters since the mid-1980s. The nine winters from 1986–7 to 1994–5 have all been classified as having 'little' snow, although this sequence was broken by the 1995–6 winter which we provisionally classify as 'snowy'. This is easily the longest sequence of winters with 'little' snow in the series, the previous

Figure 10.12 Mean annual frequency of wet and dry spells of different lengths for six sample regions of Britain and Northern Ireland, 1931–95.

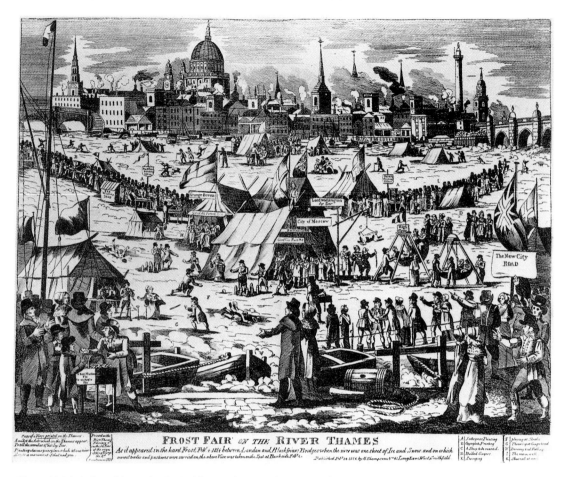

Figure 10.14 A frost fair on the River Thames during February 1814, in the middle of one of the snowiest winters in the last 200 years. The caption reads, 'Between London and Blackfriars Bridges when the river was one sheet of ice and snow and on which several trades and pastimes were carried on.' Since this time, changes in the number and positioning of bridges, in the hydrology and in the thermal characteristics of the river, mean that the Thames can no longer completely freeze over.

longest being five winters in the early 1970s. Snowy months also were infrequent during the last ten years with an average only two months per winter being rated as snowy.

DROUGHTS

Numerous definitions have been proposed to classify droughts, both in the British Isles and abroad. The most important of these 'deficits of rainfall' are those that impact water resources, and hence riverflow and crop production. Water resource problems, brought about by fluctuations in the climate, are much more regular and pronounced in England compared to Wales, Scotland or Ireland. This is because of the greater population in England (and hence greater demand for water) and because riverflow tends to be lower due to lower precipitation and higher evapotranspiration. Within England and Wales there are two different types of water resource schemes, surface reservoirs in the northern and western upland regions

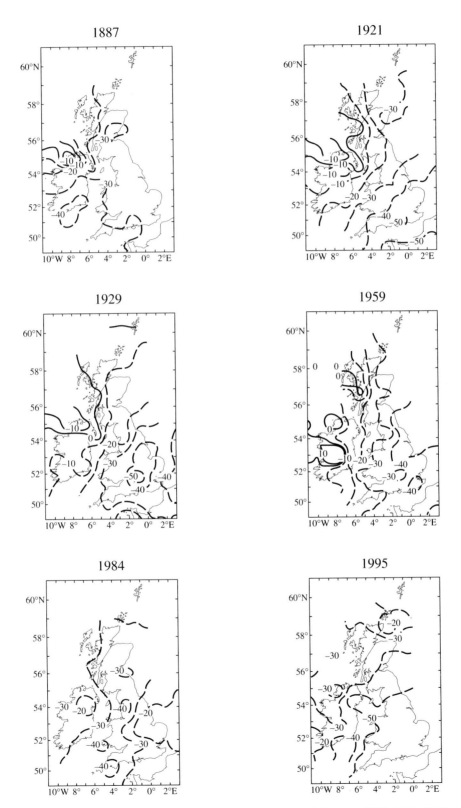

Figure 10.15 Six major droughts in the British Isles of eight to ten months' duration: 1887, 1921, 1929, 1959, 1984 and 1995. Precipitation deficits are shown as percentage deviations from the long-term average.

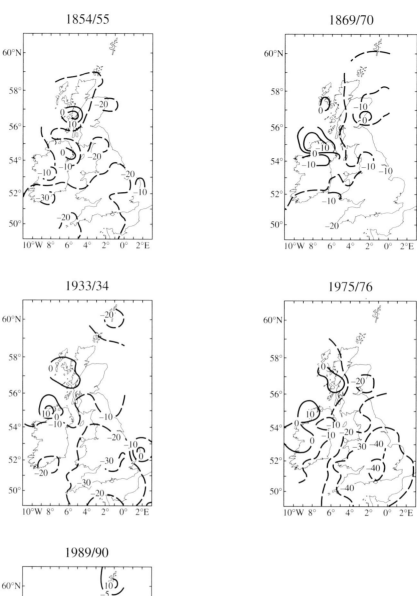

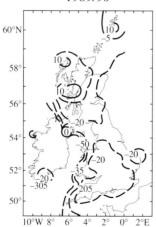

Figure 10.16 Five major droughts in the British Isles of fifteen to eighteen months' duration: 1854–5, 1869–70, 1933–4, 1975–6 and 1989–90. Precipitation deficits are shown as percentage deviations from the long-term average.

and groundwater reserves in central and eastern England. The responses of these two types of systems give two characteristic drought types in the region – short droughts of up to nine months ending in the autumn (affecting the upland surface reservoirs as occurred in 1995) and fifteen- to eighteen-month droughts (generally two summers with an intervening dry winter with reduced groundwater recharge) which have their greatest impact in the south-eastern region of the British Isles.

The most extreme of these two types of drought to have affected the British Isles since 1850 are plotted in Figures 10.15 and 10.16. These maps show the patterns of six shorter-duration droughts during 1887, 1921, 1929, 1959, 1984 and 1995, and the patterns of five longer-duration droughts during 1854–5, 1869–70, 1933–4, 1975–6 and 1989–90.[17] The drought maps are expressed as per cent deviations in precipitation from the 1961 to 1990 average. All the shorter-duration droughts have some region of the southern British Isles with a 40 per cent or more precipitation deficit – 1887 (southeast Ireland), 1921 (southern and eastern England), 1929 (central and eastern England), 1959 (north-eastern and eastern England and eastern Scotland), 1984 (south-western England and Wales and north-western England) and 1995 (central and northern England). The longer-duration droughts (Figure 10.16) tend to have some regions with deficits of over 20 per cent – 1854–5 (southern Britain and Ireland), 1869–70 (south-western England), 1933–4 (southern Britain and Ireland), 1975–6 (England and Wales and southern Scotland) and 1989–90 (southern England and Wales).

The most severe droughts in the England and Wales Precipitation series were in 1921 (short duration) and 1975–6 (long duration; Figure 10.17). The 1921 event was the most severe drought over western Europe this century.[18] **Return periods** have been estimated for recent droughts in reports which discuss their effects in much more detail for the regions of England and Wales (also looking at river-flow and groundwater levels).[19] The effects of the short, but extreme, April to August drought in 1995 are only just being assessed.[20]

Figure 10.17 A reservoir near Church Stretton, Shropshire, at the end of the 1976 summer drought. This was the culmination of eighteen months of very dry weather in England and Wales, the driest such sequence since records commenced.

It is apparent that the climate of England and Wales is such that a 1-in-50 year 'water-resources' drought may occur somewhere in England and Wales every six to eight years. Such droughts seem to be occurring more frequently, especially over the last twenty years. The various time-series shown in Figures 10.5 to 10.9 indicate trends to less precipitation, particularly in summer, highlighting the need in a changing climate to continually reassess return period estimates. Increases in the frequency of extreme events and in precipitation variability have occurred during this period which has put tremendous pressure on water resource systems. Over the same time, demand for water has also increased, so that supply problems occur more frequently, and to satisfy this demand additional resources must continually be created.

CONCLUSIONS

The British Isles is probably the most densely gauged region for precipitation measurements in the world. Long homogeneous regional series for England and Wales, Scotland and the whole of Ireland indicate less year-to-year variability in spring precipitation compared to the other seasons. In all three regions there are tendencies for winters to have got wetter and summers drier, particularly in the last twenty years. The last ten winters, except for the winter of 1995–6, have seen very little snow.

The decrease in summer precipitation in recent years has led to more frequent droughts. When combined with the increasing demand for water in most regions of the British Isles it is not surprising that water resources and hence water availability have been badly affected.

NOTES

1 J.M. Craddock, 'Annual rainfall in England since 1725', *Quarterly Journal of the Royal Meteorological Society*, 1976, vol. 102, pp. 823–40.
2 G.J. Symons, 'On the rainfall in the British Isles', *Report of the 35th meeting of the British Association for the Advancement of Science*, 1865, pp. 192–242.
3 F.J. Nicholas and J. Glasspoole, 'General monthly rainfall over England and Wales, 1727 to 1931', *British Rainfall*, 1931, pp. 299–306.
4 M.R. Woodley, 'A review of two national rainfall series', *International Journal of Climatology*, 1996, vol. 16, pp. 677–87.
5 T.M.L. Wigley, J.M. Lough and P.D. Jones, 'Spatial patterns of precipitation in England and Wales and a revised, homogeneous England and Wales precipitation series', *Journal of Climatology*, 1984, vol. 4, pp. 1–25; T.M.L. Wigley and P.D. Jones, 'England and

Wales precipitation: a discussion of recent changes in variability and an update to 1985', *Journal of Climatology*, 1987, vol. 7, pp. 231–46; J.M. Gregory, P.D. Jones and T.M.L. Wigley, 'Precipitation in Britain: an analysis of area-average data updated to 1989', *International Journal of Climatology*, 1991, vol. 11, pp. 331–45; P.D. Jones and D. Conway, 'Precipitation in the British Isles: an analysis of area-average data updated to 1995', *International Journal of Climatology*, 1997, vol. 17, in press.
6 R.C. Tabony, 'A set of homogeneous European Rainfall series', *Met. O. 13 Branch Memorandum No. 104*, Meteorological Office, Bracknell, 1980; R.C. Tabony, 'A principal component and spectral analysis of European rainfall', *Journal of Climatology*, 1981, vol. 1, pp. 283–94.
7 Tabony, 1980, 1981, op. cit.
8 Wigley *et al.*, op. cit.; Wigley and Jones, op. cit.; Gregory *et al.*, op. cit.; Jones and Conway, op cit.
9 K. Smith, 'Precipitation over Scotland, 1757–1992: some aspects of temporal variability', *International Journal of Climatology*, 1995, vol. 15, pp. 543–56. See references in this paper to the work of Arthur Jenkinson.
10 Nicholas and Glasspoole, op. cit.
11 Wigley and Jones, op. cit.; Gregory *et al.*, op. cit.
12 M.C. Jackson, 'A classification of the snowiness of 100 winters – a tribute to the late L.C.W. Bonacina', *Weather*, 1976, vol. 32, pp. 91–7.
13 L.C.W. Bonacina, 'Chief events of snowfall in the British Isles during the decade 1956–65', *Weather*, 1966, vol. 21, pp. 42–6 (see records for earlier decades in *British Rainfall* in 1927, 1936, 1948 and 1955).
14 R. Wild, G. O'Hare and R. Wilby, 'A historical record of blizzards/major snow events in the UK and Ireland, 1880–1989', *Weather*, 1996, vol. 51, pp. 82–91.
15 Jackson, op. cit.
16 G. Manley, 'Snowfall in Britain over the past 300 years', *Weather*, 1969, vol. 24, pp. 428–37.
17 G.J. Symons, 'Historic droughts', *British Rainfall*, 1887, pp. 23–35; C.E. Wright and P.D. Jones, 'Long period weather records, droughts and water resources', in *Optimal Allocation of Water Resources, IAHS Publ. No. 135*, 1982, pp. 89–99; J.C. Doornkamp, K.J. Gregory and A.S. Burns, 'Atlas of the drought in Britain, 1975–76', *Institute of British Geographers*, 1980, 86 pp.; T.J. Marsh and M.L. Lees, 'The 1984 drought', *Hydrological Data UK Series*, Institute of Hydrology, 1985; T.J. Marsh, R.A. Monkhouse, N.W. Arnell, M.L. Lees and N.S. Reynard, 'The 1988–92 drought', *Hydrological Data UK Series*, Wallingford, Institute of Hydrology, 1994, 80 pp.
18 K.R. Briffa, P.D. Jones and M. Hulme, 'Summer moisture variability across Europe, 1892–1991: An analysis

based on the Palmer Drought Severity Index',
International Journal of Climatology, 1994, vol. 14, pp.
475–506.
19 Marsh and Lees, op. cit.; Marsh *et al.*, op. cit.
20 T.J. Marsh and P.S. Turton, 'The 1995 drought – a
water resources perspective', *Weather*, 1996, vol. 51,
pp. 46–53.

GENERAL READING

J.C. Doornkamp, K.J. Gregory and A.S. Burns, 'Atlas of
the drought in Britain, 1975–76', *Institute of British
Geographers*, 1980, 86 pp.

T.J. Marsh, R.A. Monkhouse, N.W. Arnell, M.L. Lees and
N.S. Reynard, *The 1988–92 Drought*, Wallingford,
Natural Environment Research Council, 1994, 76 pp.

G. Sumner, *Precipitation: Process and Analysis*, Chichester,
John Wiley and Sons, 1988, 455 pp.

11

WIND

Resource and Hazard

Jean Palutikof, Tom Holt and Andrew Skellern

Ill blows the wind that profits nobody.
William Shakespeare, *Henry VI, Part III*

INTRODUCTION

The wind climate of the British Isles has important economic consequences for the region. On the one hand, wind is a resource to be exploited. For example, of the energy-hungry nations, the United Kingdom has one of the richest wind resources and, under government legislation to broaden the base of non-polluting energy production, premium prices are currently paid for wind-generated electricity. As a result, installed capacity for wind turbines in England and Wales rose from about 2.5 megawatts (MW)[1] at the end of 1991 to 190 MW at the end of 1995.[2] To exploit the wind resource successfully, information on the geographical and historical variability of the average wind (commonly specified as the hourly mean wind speed and direction) is required. This topic is considered in the first part of this chapter.

On the other hand, wind is a hazard. Large wind storms cause extensive damage to commercial and domestic property and trees (Figure 11.1). The frequency and severity of large wind storms is of great concern to insurance companies and the forestry industry amongst others. It is estimated that the cost to the insurance industry of the October 1987 storm was around £1.2 billion,[3] and that around fifteen million trees were uprooted.[4] Adequate planning for

such disasters requires knowledge of the historical record of extreme wind events (often expressed in terms of the maximum three-second gust in a one-hour period). This aspect of wind is considered in the second part of this chapter.

WIND CLIMATOLOGY OF THE BRITISH ISLES

The location of the British Isles, on the west coast of a large continental landmass between 50°N and 60°N, leads to a climate dominated by the **polar front** (see Chapter 2). The instability of this front causes **depressions** to form, tracking across the North Atlantic and, at the longitudes of the British Isles, following a preferred route which passes between Iceland and Scotland. As these depressions move across the Atlantic, they follow a life cycle which, by the time they reach the longitudes of the British Isles, means that they are usually in a phase of maturity or decay, manifest as an **occlusion**.

It is this 'typical' climatology which is the primary constraint on the average wind field of the British Isles. At well-exposed sites, both the prevailing wind and the highest wind speeds generally lie in the south-west quadrant of the compass (Figure 11.2). Speeds tend to be highest in the north-west of the British Isles (closest to the depression tracks),

Figure 11.1 Tree damage to a house and car at Addlestone, Surrey, following the October 1987 storm. The frequency and severity of large wind storms is of great concern to insurance companies and the forestry industry, amongst others.

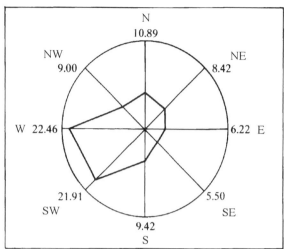

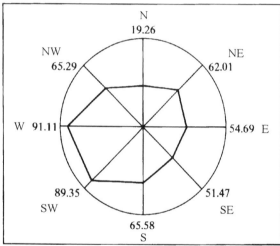

Figure 11.2 Percentage of wind observations by direction (top), and mean wind speed by direction (tenths of a metre per second; bottom) for High Bradfield in the Pennines, a typical exposed upland site. The site location is shown in Figure 11.9.

decreasing towards the south and east. The secondary constraint is the geography of the country. First, larger **drag coefficients** over land lead to the highest wind speeds being observed at the coast, decreasing inland. Second, acceleration effects over orography lead to wind speeds increasing with altitude. These effects are clearly demonstrated in Figure 11.3[5] which shows average wind speeds over the British Isles.[6] In this map, broad wind speed bands are shown and within each band the local geography dictates the actual wind speed range. See Appendix B for a more detailed map of average wind speeds.

Although this broad pattern is typical, many permutations of weather and geography exist which affect the wind regime on a day-by-day basis. An annual cycle of higher wind speeds in winter and at the equinoxes, and lower speeds in summer, reflects

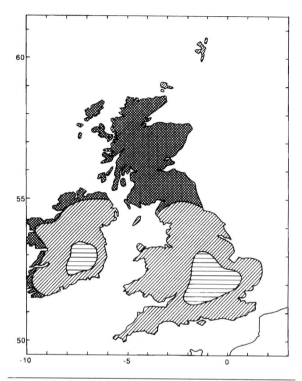

	Sheltered terrain	Open plain	Sea coast	Open sea	Hills & ridges
▓	>6.0	>7.5	>8.5	>9.0	>11.5
▨	5.0-6.0	6.5-7.5	7.0-8.5	8.0-9.0	10.0-11.5
▭	4.5-5.0	5.5-6.5	6.0-7.0	7.0-8.0	8.5-10.0

Figure 11.3 Mean annual wind speed (ms⁻¹) over the British Isles at 50 m above the ground (modified from Troen and Petersen[5]).

Figure 11.4 One of the wind turbines at Llidiart y Waun, Powys. The typical hub height for a large wind turbine is about 50 m.

the seasonally varying strength of the large-scale atmospheric circulation. Depressions may suddenly be rejuvenated on the eastern side of the Atlantic due, for example, to a local **sea-surface temperature** maximum. Such explosive deepening,[7] which may be difficult to forecast, led to the storms of 15–16 October 1987 and 25 January 1990.[8] The path of depressions also varies, and the intense damage experienced in the October 1987 storm was due in part to the fact that the depression centre crossed the British Isles much to the south of the usual track. Conversely, **blocking** situations, when

the progress of depressions across the Atlantic is impeded by the presence of a large stationary **anticyclone** over continental Europe, can lead to low wind speeds over the British Isles.

Local circulations of thermal origin may develop in sheltered places when the large-scale circulation is weak. A diurnal cycle is therefore common, with higher wind speeds in the afternoon in response to increased local turbulence due to heating from the ground below. Sea-breezes close to the coast also lead to afternoon maxima,[9] and **katabatic winds** on long mountain slopes[10] may lead to night-time wind maxima during periods of otherwise calm conditions.

WIND AS A RESOURCE: THE ANALYSIS OF AVERAGE WIND SPEEDS

Geographical variability

The World Meteorological Organisation standard measuring height for wind speeds is 10 m above the ground. The broad features of the variability of average wind speeds over the United Kingdom at 50 m height was shown in Figure 11.3. This height was selected to be a typical hub height for a large wind turbine (Figure 11.4). At the 10 m height speeds will be lower than this – average annual wind speeds above open level terrain range between about 6.5 ms^{-1} on the exposed coast of north-west Scotland (e.g., 6.9 ms^{-1} at Benbecula) to below 4 ms^{-1} at inland sites in south-east England (e.g., 3.8 ms^{-1} at Kew Gardens).

The number of **anemometers** in the United Kingdom as recorded in the *Monthly Weather Report* for 1991,[11] and their distribution by altitude band are shown in Table 11.1. It is immediately clear that the number at any significant altitude (above 100 m, say) is very small. This is in part because many UK Met. Office anemometer sites are at airports (civil and military) which tend to be located in low-lying, flat terrain. There has been little recent change in the size of the anemometer network, which in 1978 stood at 142 sites.[12] We have already noted the relationship between wind speed and altitude, and the biased distribution of anemometer sites causes difficulties in trying to monitor the true geographical variability of wind speeds in the British Isles. Despite this difficulty, there is a genuine need for information on wind speeds in high-altitude areas, not only because these have high wind-energy potential but also to assess the safety of structures (radio masts, etc.) which require installation on high ground for good signal reception.

Table 11.1 Geographical distribution of UK Met. Office anemometers (AMSL = above mean sea level)

	All	100–200 m AMSL	>200 m AMSL
England	83	13	7
Wales	7	3	1
Scotland	44	5	6
Northern Ireland	11	2	0
Channel Islands	2	1	0
All UK	147	24	14

This lack of high-altitude information regarding the wind climatology is a problem common to many countries, and researchers have sought to exploit the known relationship between height change and wind speed to develop numerical models which predict wind speeds from an underlying terrain elevation map. These models vary greatly in their sophistication and consequently in the size and power of the computing facilities they require. The simplest are the mass-consistent class of models.[13] The basic underlying theory is that as air is forced up and over a hill, the height of the top of the **boundary-layer** rises, but to a smaller extent than the relative height of the hill. This 'squeezes' the layer of air between the ground and the boundary-layer top and, to maintain mass consistency, the speed of movement must increase (see Figure 11.5).[14] More complex non-linear models simulate turbulent flow over hills by solving the mass-continuity equations, the momentum conservation equations (Navier–Stokes) and the energy conservation equations. Various simplifications and assumptions are required to arrive at a unique solution, and the computing requirements (and sophistication) of the model are determined in part by the number of these assumptions.[15]

A wind-speed prediction model is of limited value if it produces inaccurate estimates and/or estimates with wide confidence limits. Table 11.2 shows some results from a mass-consistent model (COMPLEX) applied at two sites, separated by 5 km, in the northern Pennines.[16] This table demonstrates two features typical of model results. First, the model

FLOW OF AIR OVER :

Shallow topography

Steep topography

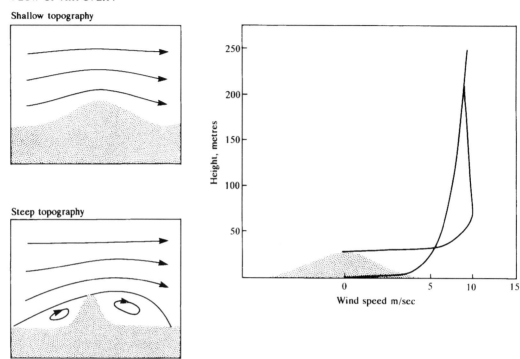

Figure 11.5 Representation of the movement of air over shallow and steep topography (left). The effect of air movement over shallow topography on the vertical wind speed profile is shown (right).[14] The deformation of the wind-speed profile over steep topography is less predictable.

estimates are within 1 ms⁻¹ of observations in all seasons except summer. Predictions are generally less accurate in the slacker circulation systems of the summer months. Second, the predictions are all under-estimates. Great Dun Fell is a hilltop site, and models commonly under-predict hilltop wind speeds and over-predict valley-bottom wind speeds.[17]

Lying in complexity somewhere between the

Table 11.2 Wind-speed predictions (ms⁻¹) at Great Dun Fell in the northern Pennines using the mass-consistent model COMPLEX initialised with wind data from Moor House

	Winter	Spring	Summer	Autumn
Observed	12.8	8.9	8.9	11.6
Predicted	11.9	8.2	7.6	11.1

simple mass-consistent models and the fully non-linear turbulent flow models, the Wind Atlas Analysis and Application Programme (WAˢP) is a wind flow model suitable for use on a Personal Computer. A simple dynamic model, widely used by wind engineers,[18] WAˢP can simulate the effects of roughness changes, obstacles to the flow, and changes in atmospheric stability. In a study to validate the performance of this model at a number of sites in the United Kingdom, it was found that only half the estimates were within 10 per cent of the true value. Particular problems identified were the assumption of a **Weibull frequency distribution** of wind speeds, and the need to specify **surface roughness length** accurately because of the sensitivity of the model result to this parameter.[19] These problems are typical of this class of models.

In the absence of particularly impressive results from simple models, attempts have been made to use statistical techniques for wind-speed prediction. Unlike modelling approaches, these empirical techniques require some data from the site for which a prediction is sought. Wind farm developers will, typically, measure wind speeds at their proposed site for periods no longer than a year. They then extrapolate from this short time-series to a ten-year period by developing regression equations (commonly one for each compass sector) between measurements from a nearby long-term UK Met. Office anemometer record and their own measurements. This is known as the measure–correlate–predict method (MCP).[20] The success of this method is determined in part by the strength of the correlation between the data from the UK Met. Office site and the proposed wind turbine site.

The relationship between the separating distance and the strength of the correlation between monthly mean wind speeds at pairs of UK Met. Office sites is shown in Figure 11.6. For each pair there are twelve correlations, one for each calendar month. It is clear that the strength of the correlation deteriorates rapidly with distance, particularly where the intervening terrain is complex. Pairs of sites separated by more than 100 km of complex terrain are unlikely to share more than 50 per cent of their variability in monthly wind speeds in common.

There are, therefore, real difficulties in trying to predict wind speeds at specific sites, whether by numerical or statistical models, and there is a substantial research effort directed towards improvement of prediction tools, particularly numerical models.[21] It is essential that the long-term recording network of anemometers remains at least at its current density, and that effort is directed towards improvement of this network in upland areas of complex terrain.

Historical variability

Whether one is interested in wind as a resource, or as a hazard, it is important to understand how it varies over the long term. Wind farm developers, for example, require information on the profitability of

a scheme over its total lifespan, which is of the order of twenty years. They will commonly have a prediction of the average ten-year wind speed, based on a short record of on-site observations and the MCP method. Although this method is widely accepted throughout the industry, and ignoring the issue of its accuracy, we are still left with the question as to how representative the ten-year period is within the long-term record.

Long time-series of observations are a possible source of information on long-term wind speed variability. The UK Met. Office archives, however, contain very few long records of monthly mean wind-speed.[22] The principal reason is that, until recently with the widespread adoption of automated logging, mean wind speeds had to be calculated from the **anemograph trace**. This was a time-consuming and skilled operation, with associated high labour costs. In the absence of an anemograph, readings from an indicating dial anemometer or a **Beaufort scale** estimation of the effect of the wind on nearby objects (Figure 11.7) were the only source of information. Figure 11.8 shows, for the month of March, the long records for seven sites which could be extracted.

Even where long wind-speed series exist, questions arise with respect to their homogeneity, or reliability, over time.[23] One of the sites plotted in Figure 11.8 is Eskdalemuir, in the Southern Uplands of Scotland, with a record extending from 1911 to 1985. This record shows a clear downward trend in wind speeds from around 1920 to the mid-1960s. Whereas the average March wind speed during 1916 to 1925 is 5.6 ms^{-1}, the average for 1956 to 1965 is only 3.6 ms^{-1}. Is this trend real, or an artefact of the record? *The Observatories' Year Book* for 1922 has this to say about the site of the Eskdalemuir instrument, which stood at that time 7 m above the roof of the main observatory building:

> Apart from the surrounding hills the exposure of the vane-head is tolerably free in all directions save to the west where at a distance of some 130 ft. (40 m) is a rather large building of which the height is somewhat greater than that of the Main Building.[24]

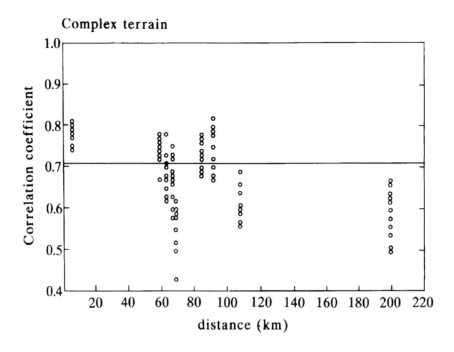

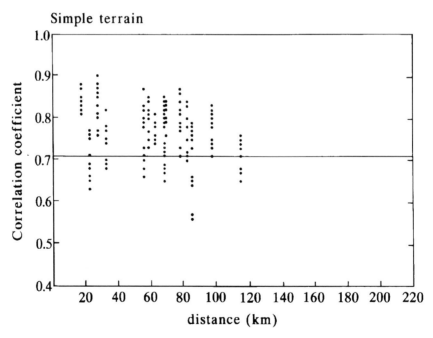

Figure 11.6 Correlations between monthly mean wind speeds at neighbouring UK Met. Office anemometer sites. Terrain classification refers to the nature of the terrain separating the site-pairs. Correlations decay with distance more rapidly over complex terrain than over simple terrain.

Figure 11.7 'The Beaufort Scale, revised 1906'. One of a collection of Late Victorian and Edwardian humorous glass slides of meteorological interest held by the Royal Meteorological Society.

Yet, by 1955, a report of a Met. Office site inspector stated that the exposure of the anemometer was 'extraordinarily bad' with 'trees to all sides'. The downward trend in wind speeds between 1920 and the mid-1960s can be at least partially explained by the growth of surrounding trees. The site was moved in 1968, at which time an electrical cup generator (ECG) anemometer was installed, with an anemograph. A step-like jump can be seen in Figure 11.8 at this point. This is unlikely to be entirely due to the site change: mean wind speeds derived from ECG anemometers are consistently higher (by between

0.5 and 1.0 ms^{-1}) than those from pressure-tube anemometers.[25]

Eskdalemuir is the exception rather than the rule in the amount of information that is available. For most long records there is insufficient information to judge whether the observed trends are climatic in origin, or whether they arise from such factors as a change in exposure or instrument type.

In an attempt to overcome this problem, an investigation has been made of the possibility of reconstructing surface wind speeds from the **geostrophic wind**, calculated in turn from a dataset of gridded daily mean sea-level pressure. The reasoning behind this study was that the pressure data should be more **homogeneous**, and that any discontinuities in the record would be easier to trace. The pressure dataset extends back to 1881 with few missing data.[26] The main findings of this study are summarised below.

The monthly geostrophic wind speed was computed for thirty-eight anemometer sites in the United Kingdom for which surface wind speed data were available. The correlation coefficient between the anemometer wind speed and the geostrophic wind speed was then calculated. Sixteen sites were found to have correlation coefficients of 0.9 or above, and a further eight to have coefficients between 0.85 and 0.9. The sixteen highly correlated stations are chiefly at exposed locations, either on the west coast (e.g., Benbecula and St Mawgan), or on inland hilltops (e.g., Great Dun Fell and High Bradfield). By contrast, many of the fourteen stations with correlation coefficients below 0.85 are at sheltered locations, such as inland sites in the southern half of Britain (e.g., Elmdon and Kew), or along the east coast (e.g., Gorleston and Lossiemouth).

The twenty-four highly correlated stations (shown in Figure 11.9) were used for the reconstruction from the mean sea-level pressure data. For each station twelve regression equations were developed, one for each month, using the monthly geostrophic wind speed as the predictor and the monthly anemometer wind speed as the dependent variable. Typically, there were only ten to fifteen years of reliable surface wind speed data available for construction of the regression equations, but correlation coefficients of

228

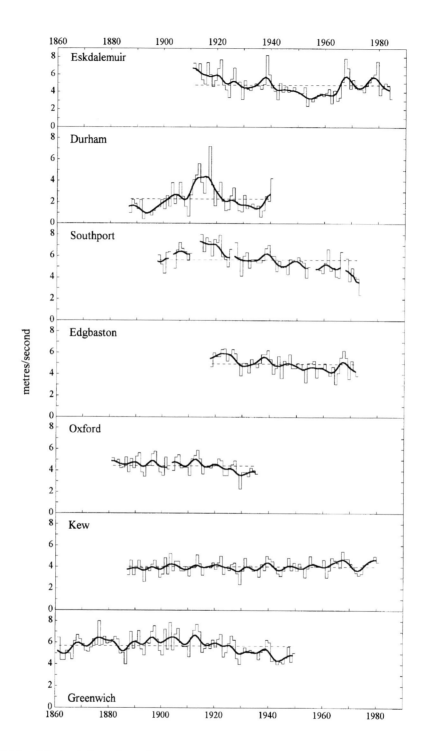

Figure 11.8 Mean March wind speeds (ms⁻¹) at seven stations in the United Kingdom with long instrumental records. The dashed lines show the long-term average and the smooth bold lines result from applying a filter which emphasises variations on time-scales longer than ten years.

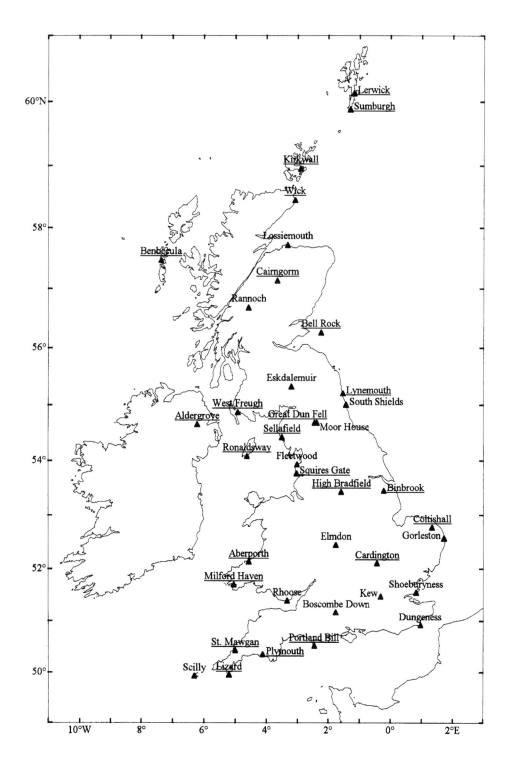

Figure 11.9 Locations of sites in the UK with long-term wind data. The twenty-four sites with high correlations between near-surface and geostrophic monthly mean wind speeds are underlined.

230

BOX 11.1 APPLICATION OF THE RECONSTRUCTED SURFACE WIND SPEEDS

The study described has demonstrated that long-term variability in wind speeds does occur. The question still remains as to whether these variations have any economic significance, for example, to wind farm developers. Would variations in wind speeds over the lifespan of a wind turbine invalidate economic projections based on a ten-year mean wind speed calculated by the MCP method?

(a)

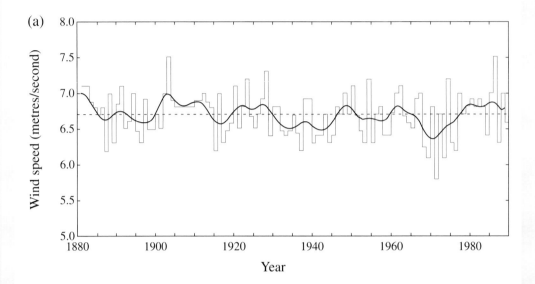

(b)

Relative Price per Kilowatt Hour at Aberporth

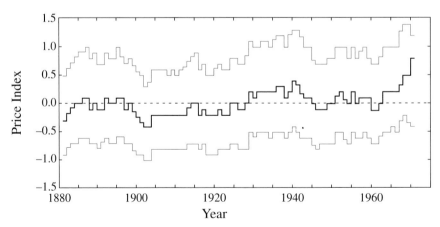

Reconstructed annual mean wind speed for Aberporth, 1881 to 1989, using the mid-range model parameters (ms^{-1}; top); and relative price per kilowatt-hour required to give a 20 per cent annual return on capital invested for overlapping twenty-year periods (relative price index, per cent from average, bold line; bottom). The upper and lower price curves show the 95 per cent confidence bands based on uncertainties in the reconstructed wind field (not shown in the top graph). Location of Aberporth is shown in Figure 11.9, p. 229.

A simple economic model was constructed to investigate the implications of the long-term variability in wind speeds as shown, for example, in Figure 11.10. The basic parameters of the model are:

turbine type:	300 kW rated power
initial investment:	£300,000
annual running cost:	£9,000
turbine life expectancy:	20 years
required annual return on capital:	20%

In fact, these parameters are of little significance – what is important is the relative differences in time, not the absolute values.

First, the annual power production from a typical wind turbine is calculated from the reconstructed wind speeds. The model then takes these values and calculates the price (in pence per kWh) required over the twenty-year life of the turbine to give the required annual return on capital. The calculation is made for overlapping twenty-year periods from 1881 onwards. The last period is 1970 to 1989, giving a total of ninety lifespans for which required price can be calculated. The figure shows some results from Aberporth, on the west coast of Wales. At this site, for the mid-range reconstructed wind speed, the model gives a maximum required price for any twenty-year period of 7 per cent above the average and a minimum required price of 6 per cent below the average. This range doubles if the upper and lower 95 per cent confidence reconstructed wind speeds (not shown in the figure) are used in the price calculation. These figures may be interpreted as the maximum possible error – e.g., a wind turbine project planned on the basis of one required price might, in the worst possible scenario of falling wind speeds, fall short of estimated capital return by this amount.

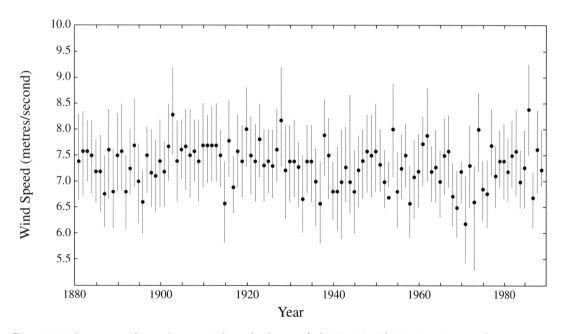

Figure 11.10 Reconstructed annual mean wind speeds (dots; ms⁻¹) for High Bradfield in the Pennines (for location see Figure 11.9), shown with the upper and lower 95 per cent confidence limits (vertical lines).

0.85 and above are statistically significant at the 1 per cent level or better for this range of cases. Historic values of the monthly site geostrophic wind speed back to 1881 were then substituted in the twelve equations to reconstruct the historic anemometer wind speeds. Prior to this stage, inhomogeneities in the pressure dataset were removed.[27] Reconstructed annual mean wind speeds for a typical upland site, High Bradfield, are shown in Figure 11.10. An application in which these reconstructed wind speeds may be used is illustrated in Box 11.1.

Figure 11.11 shows a comparison between observed and reconstructed March wind speeds for Eskdalemuir, discussed earlier. The persistent downward trend between 1911 and 1960 noted in the observations is not present in the reconstruction. In the absence of other evidence, this would suggest that the downward trend is an artefact of the record, and not climatological.

WIND AS A HAZARD: STORMS AND HIGH WIND SPEEDS

It is a current perception in industries susceptible to damage from extreme wind speeds that the frequency

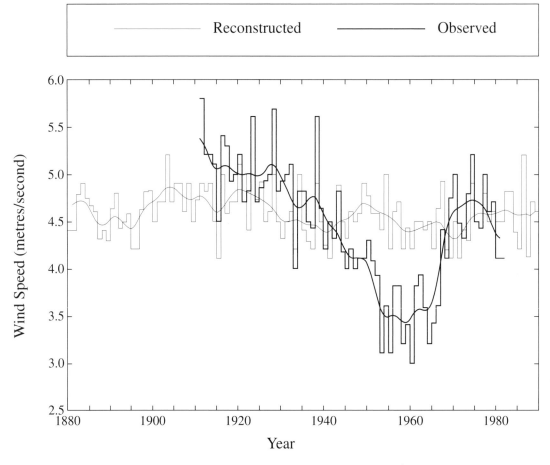

Figure 11.11 Comparison of observed and reconstructed annual mean wind speeds (ms⁻¹) for Eskdalemuir. The smooth lines result from applying a filter which emphasises variations on time-scales longer than ten years.

and/or intensity of wind storms is increasing.[28] It is very difficult to prove or disprove such a hypothesis in an objective and rigorous manner. This is obviously a matter of some concern, and a number of attempts have been made to compile time-series of storm frequency for analysis.

One of the longest of these time-series is the catalogue of severe storms produced by Hubert Lamb for north-west Europe which extends back to the mid-1500s.[29] He classified storms according to a Storm Severity Index (SSI), calculated as the product of the cube of the maximum surface wind speed, the greatest area affected, and the duration of occurrence. The most severe storms affecting the British Isles over the last 500 years include the storms of December 1792 (SSI 10,000–20,000), February 1825 (about 12,000), November 1694 (about 10,000) and December 1703 (9,000).[30] For the twentieth century, the most severe event is the October 1987 storm (8,000), with storms in 1953, 1976 and 1981 all having an SSI value of 6,000. (Note that the final year of the catalogue is 1989, although some information on the 1990 storms is appended.)

The catalogue is based on historical sources in the early period and, more recently, on **synoptic maps**. It is not really suitable for analysis with respect to changes in wind storm occurrence, since the series cannot be regarded as homogeneous, nor was it ever intended for this purpose. It is described by the author as a 'historical investigation of great storms'. In the early years, and apart from the most severe events, it must be largely a matter of chance whether or not a particular storm is recorded (Figure 11.12). The change in sources, and improvements in the observations and analysis techniques underlying the synoptic charts, will lead to discontinuities in the record.

Changes in the methods of compiling synoptic charts may also have affected the study of Holger Schinke, who made a daily count of the number of cyclones with a central pressure less than or equal to 900 hPa in the North Atlantic–European sector, based on United States and German weather maps covering the period 1930 to 1991.[31] This count shows an increase in the number of cyclones, but this increase is not gradual. Three periods are identified.

Figure 11.12 A woodcut illustrating the storms which ravaged England in the autumn and winter of 1612–13. The sub-heading on the title page of the pamphlet containing the woodcut reads, 'By terrible storms and tempests, to the losse of lives and goods of many thousands of men, women and children. The like by Sea and Land, hath not beene seene, nor heard of in this age of the World.'

First, cyclone frequency was relatively constant until the mid-1930s, after which there was a rapid increase up to 1945. The second period, 1945–70, is a period of fairly constant frequencies. Finally, from 1970 onwards, cyclone frequencies increase sharply in the early 1970s and then level off. It has been suggested that this trend is due more to inhomogeneities in the charts, arising from improvements in observations and analysis, than it is to a real increase in cyclone frequency.[32]

The Gale Index

Time-dependent trends in storminess over the United Kingdom can also be investigated using a

Number of Gales

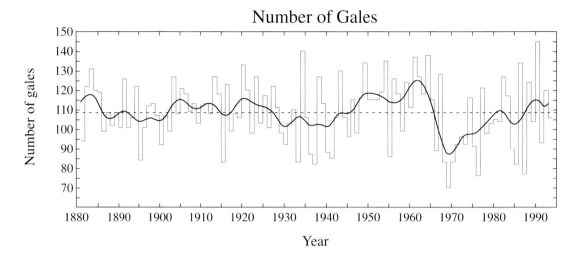

Maximum Wind Speed

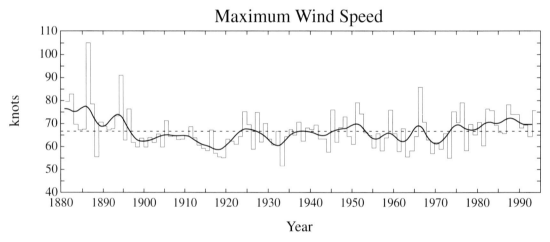

Severe Gales

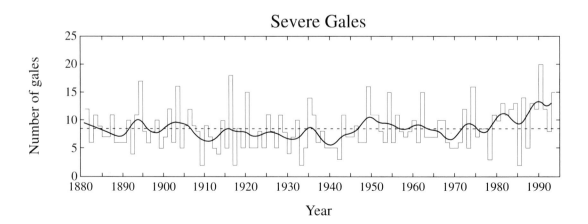

Table 11.3 Frequency distribution of wind speeds over northern England, calculated from the Gale Index

Wind speed (knots)	Frequency (%) 1881–1993	1970–1993
30–39	67.12	64.90
40–49	24.83	25.25
50–59	6.38	7.22
60–69	1.41	2.17
70–79	0.23	0.47
80–89	0.02	
90–99	0.01	
100–109	0.01	

Table 11.4 Exceedence frequencies for wind speeds over northern England, calculated from the Gale Index

Wind speed (knots)	Exceedence frequency (%) 1881–1993	1970–1993
30	100.00	100.00
40	32.88	35.11
50	8.05	9.85
60	1.67	2.64
70	0.26	0.47
80	0.03	0.00
90	0.02	
100	0.01	
110	0.00	

Gale Index. This is based on measures of westerly and southerly flow and **shear vorticity** in the geostrophic wind, which are in turn calculated from sea-level pressure data. Arthur Jenkinson and Peter Collison[33] provided guidelines for the calculation of this index, based on a daily dataset of sea-level pressure gridded on a 5° latitude by 10° longitude grid, extending back to 1881.[34] Inevitably, with such a coarse grid, there must be question marks surrounding the ability of the Gale Index to capture intense cyclonic systems. Such intense systems, however, are a feature more common to tropical than to mid-latitudes. Jenkinson and Collison successfully compare the performance of their method against the great European storm of 1976.

The Gale Index approach has been used by a number of authors to calculate long time-series of gale occurrence.[35] A recent update is presented in Figure 11.13 for the northern British Isles (centred on 55°N 5°W) from 1881 to 1993.[36] This shows the number of gales per year (defined as events with wind speeds in excess of 15.5 ms^{-1}),[37] the maximum wind speed per year (a measure of storm intensity), and the number of severe gales per year (events with wind speeds over 25.8 ms^{-1}), respectively.

The graph of storm frequency shows a sharp decline in the decade of the 1960s, followed by a gradual recovery to levels more typical of the early decades of the century. The largest number of gales in a single year is found in 1990. The time-series of storm intensity has largest values at the beginning and end of the record, i.e., at the end of the nineteenth and of the twentieth centuries. The rising trend from 1970 to 1990 takes maximum wind speeds to levels above those common for most of the century, but there is some indication of a return to frequencies more typical of the twentieth century in the last few years of the record. The final graph of Figure 11.13, of severe gales, shows a clear upward trend since around 1970. This is in agreement with the finding by James Hurrell[38] that the **North Atlantic Oscillation** has shown an upward trend since around that date (see Chapter 9).

Tables 11.3 and 11.4 confirm the information plotted in Figure 11.13. Over recent years there has been a reduction in the number of storms with wind speeds in the range 30–39 knots (15.5–20.0 ms^{-1}),

Figure 11.13 Gale Index for the northern British Isles.[36] Upper graph: the number of gales per year. Middle graph: the maximum wind speed each year (in knots). Lower graph: the number of severe gales per year. Dashed lines are the long-term average, and the smooth curves result from applying a filter which emphasises variations on time-scales longer than ten years.

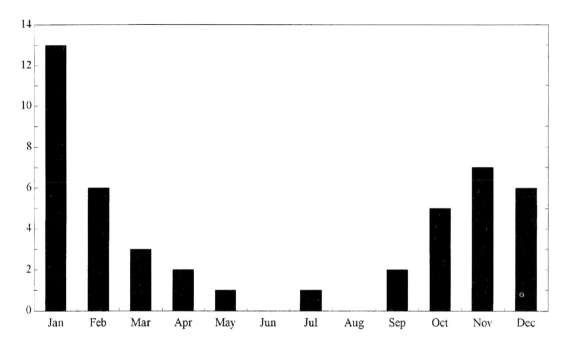

Figure 11.14 The average number of severe storms in each month between 1920 and 1990, taken from the Storm Catalogue.[39]

and an increase in the frequency of storms with higher wind speeds up to 70 knots (40.7 ms⁻¹). The data for the complete record show, however, that storms of much greater severity have occurred before 1970 over the northern British Isles. Table 11.4 presents this information in a slightly different form, as the exceedence frequency, or percentage of gales where wind speeds were greater than the threshold shown in the first column. This shows that the number of storms with wind speeds greater than 60 knots (30.9 ms⁻¹) and less than 80 knots (41.2 ms⁻¹) has approximately doubled over the last twenty years.

A Storm Catalogue for the United Kingdom

A catalogue of storms has been compiled, extending back to 1920, specifically for the United Kingdom and intended primarily for use by the insurance industry.[39] This list of forty-seven severe storms, shown in Table 11.5, was obtained by inspection of the literature. January is the worst month for storms, followed by November and February (Figure 11.14). No storms were recorded in June or August.

Synoptic charts were consulted in order to assign an index of severity. This was an absolute measure, based on maximum recorded wind speed, duration, etc., and the rankings are shown in the first column of Table 11.6. It is a characteristic of a damaging storm that it is unusual within the context of the background wind climate of the affected region. Absolute measures are of limited use, since a storm that would be perceived as relatively harmless in the north of Scotland would have much more damaging consequences if it tracked across southern England. The index rating for each storm was therefore weighted by reference to the expected one-in-fifty-year gust (i.e., the gust speed which is likely to be exceeded once in fifty years, on average) for the area affected. The rankings of storm severity according to this weighted index are shown in the second column of Table 11.6. We find, for example, that the October 1987 storm ranks only twenty-seventh in the

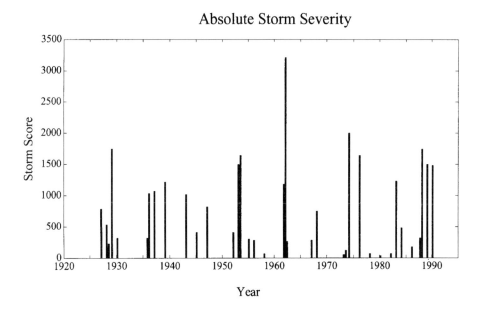

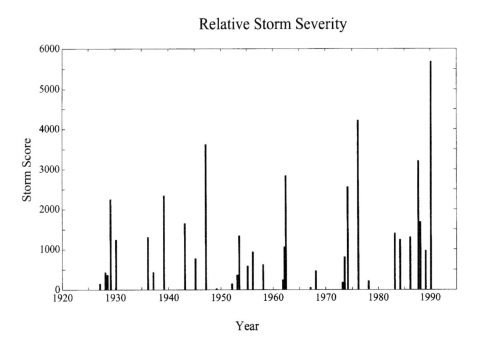

Figure 11.15 Time-series of individual storm scores, taken from the Storm Catalogue.[39] Upper graph shows storms according to their absolute severity; lower graph shows storms according to their relative severity (see text for explanation).

Table 11.5 Storm Catalogue for the period from 1920 to 1990

Year	Month	Date	Principal source[a]
1920	January	26–27	Lamb
1927	January	28	Lamb
1927	October	28–29	Hammond, Lamb
1928	January	6–7	Lamb
1928	November	16–17	Lamb
1928	November	23–25	Lamb
1929	December	5–7	Hammond, Lamb
1935	September	16–17	Hammond
1935	October	18–19	Hammond
1936	October	26–27	Lamb
1937	January	16–22	Lamb
1938	October	4	Hammond
1938	November	23–24	Hammond, Lamb
1943	April	7	Hammond
1945	January	18	Hammond
1947	March	16	Hammond
1949	February	9–10	Lamb
1951	December	30	Lamb
1952	December	17	Lamb
1953	Jan./Feb.	31–1	Hammond, Lamb
1954	November	26–30	Hammond
1954	December	21–23	Lamb
1956	July	29	Hammond, Lamb
1957	November	4	Hammond
1961	September	16–17	Hammond, Lamb
1962	January	11	Hammond
1962	February	16–17	Hammond, Lamb
1962	May	16	Hammond
1967	March	6	Lamb
1968	January	14–15	Hammond, Lamb
1972	November	12–13	Lamb
1973	April	2–3	Lamb
1974	January	12	Hammond, Lamb
1974	January	27–28	Hammond, Lamb
1976	January	2	Hammond, Lamb
1978	January	11–12	Hammond, Lamb
1979	December	4–5	Lamb
1981	November	23–25	Lamb
1983	February	1	Lamb
1984	January	13	Hammond
1986	March	24	Hammond
1987	October	16	Hammond, Lamb
1988	February	9–10	Lamb
1989	February	13	Lamb
1989	December	16–17	Lamb
1990	January	25	Hammond, Lamb
1990	February	26	Hammond, Lamb

[a] H.H. Lamb, *Historic Storms of the North Sea, British Isles and Northwest Europe*, Cambridge, Cambridge University Press, 1991, 204 pp. J.M. Hammond, 'The strong winds experienced during the late winter of 1989/90 over the United Kingdom: historical perspectives', *Meteorological Magazine*, 1990, vol. 119, pp. 211–19. J.M. Hammond, 'storm in a teacup or winds of change?', *Weather*, 1990, vol. 45, pp. 443–8.

Table 11.6 Ranking of storms in the Storm Catalogue, 1920 to 1990, by absolute and by relative severity

Ranking by absolute severity			Ranking by relative severity		
16–17	February	1962	25	January	1990
12	January	1974	2	January	1976
23–25	November	1928	16	March	1947
9–10	February	1988	16	October	1987
31	January–1 February	1953	16–17	February	1962
2	January	1976	12	January	1974
17	December	1952	23–24	November	1938
13	February	1989	23–25	November	1928
25	January	1990	17	December	1952
11	January	1962	26	February	1990
1	February	1983	9–10	February	1988
23–24	November	1938	7	April	1943
16–17	September	1961	1	February	1983
26–27	October	1936	31	January–1 February	1953
18–19	October	1935	16–17	September	1935
7	April	1943	24	March	1986
26	February	1990	13	January	1984
16	March	1947	5–7	December	1929
28	January	1927	11	January	1962
14–15	January	1968	16–17	November	1928
28–29	October	1927	13	February	1989
13	January	1984	29	July	1956
18	January	1945	18–19	October	1935
30	December	1951	2–3	April	1973
16–17	September	1935	18	January	1945
5-7	December	1929	4	November	1957
16	October	1987	26–30	November	1954
26–30	November	1954	14–15	January	1968
29	July	1956	26–27	October	1936
6	March	1967	28–29	October	1927
16	May	1962	6–7	January	1928
21–23	December	1954	16–17	September	1961
6-7	January	1928	11–12	January	1978
24	March	1986	12–13	November	1972
16–17	November	1928	30	December	1951
4	October	1938	28	January	1927
27–28	January	1974	16–17	December	1989
2-3	April	1973	4	October	1938
26–27	January	1920	21–23	December	1954
4	November	1957	6	March	1967
23–25	November	1981	26–27	January	1920
11–12	January	1978	9–10	February	1949
12–13	November	1972	16	May	1962
16–17	December	1989	27–28	January	1974
4–5	December	1979	4–5	December	1979
9–10	February	1949	23–25	November	1981
16–22	January	1937	16–22	January	1937

absolute severity list, but moves up to fourth place in the relative severity ranking.

The time-series of the scores of individual storms according to the absolute and relative severity indexes are shown in Figure 11.15. The absolute severity time-series is dominated by the February 1962 event.[40] This storm was active over northern England and Scotland, areas of high one-in-fifty-year return-period wind speeds. It is much less important in the relative severity time-series (rank order fifth in the second column of Table 11.6), in which the 'Burns' Day' storm of 25 January 1990 is the most important. From both graphs, it is clear that the late 1980s saw many severe storms, with a frequency not observed since the late 1920s. The relative severity index also suggests an increase in the severity of individual storms in recent years.

It is important to recall that, although all possible steps were taken to ensure that the Storm Catalogue is homogeneous, there are still large potential (and unmeasurable) sources of error. For example, more information, both media and meteorological, is now available in comparison to the early decades of the series, making it difficult to specify the characteristics of the early storms with the same accuracy as later events. Furthermore, the compilation of the Catalogue was carried out in the context of the requirements of the insurance industry, so that the indexes of severity are biased towards damage caused, rather than towards meteorological severity. It is of interest, however, that the broad findings of this study, which was carried out by literature review, and with the needs of the insurance industry primarily in mind, are in broad agreement with the Gale Index analysis described earlier, which was based on meteorological data.

CONCLUSIONS

The wind climate of the British Isles is characterised by high wind speeds, particularly at west coast and exposed upland locations. This provides both the United Kingdom and the Republic of Ireland with an abundant non-polluting and renewable energy resource. The total accessible resource for the United Kingdom is estimated by the Department of Trade and Industry to be about 340 tera-watt hours (TWh) per year onshore, and about 380 TWh per year offshore. Under a scenario of 'heightened environmental concern', wind energy could contribute between 20 and 30 TWh per year of electricity by 2025. This would be around 8 per cent of the total electricity needs of the United Kingdom.[41]

Successful exploitation of the wind resource requires information on the geographical and historical variability of wind speeds, as summarised in this chapter. This information, however, is insufficient in itself. Successful development also requires sensitive planning to minimise the impact on the local environment. Areas of high wind resource in the British Isles often coincide with areas of environmental importance, and conflicts are frequently reported between wind farm developers and interest groups of local residents, nature conservancy organisations and recreational countryside users.

Inevitably, a high average wind speed environment implies a relatively high frequency of gale-force winds. The preferred track for depressions in the eastern Atlantic passes to the north of the British Isles, and buildings in northern Scotland are constructed to withstand much higher wind loadings than those in southern England. Problems arise when intense depressions deviate from their normal track, as in the October 1987 storm, with devastating consequences. It is a perception amongst interested parties, such as the insurance industry, that wind storms over the British Isles have increased in severity and frequency in recent years. There is evidence to support this perception (see the earlier discussion of the Gale Index). Current levels of scientific knowledge do not allow us to establish whether such a trend is likely to continue in the future. If it is a response to natural climate variability, then we may expect levels of storm occurrence to return to 'normal' levels in the near future. If it is a response to some external forcing, such as the enhanced **greenhouse effect**, then the future is more uncertain. Current thinking on greenhouse warming and storminess is dealt with in Chapter 15.

NOTES

1 'Windfarms in UK – update', *Wind Directions*, 1994, vol. 13(4), pp. 12–13.

2 J. Massey, 'Heading for a busy year of development', *Windpower Monthly*, 1996, vol. 12(2), pp. 33–4.

3 Society of Fellows Study Group, *The Impact of Changing Weather Patterns on Property Insurance*, London, Chartered Insurance Institute, 1994, 87 pp.

4 C.P. Quine, 'Damage to trees and woodlands in the storm of 15–16 October 1987', *Weather*, 1988, vol. 43, pp. 114–18.

5 This map is modified from I. Troen and E.L. Petersen, *European Wind Atlas*, Risø National Laboratory, Roskilde, 1989, 656 pp.

6 A comprehensive description of the wind climatology of the UK is given by N.J. Cook, *The Designer's Guide to Wind Loading of Building Structures. Part 1: Background, Damage Survey, Wind Data and Structural Classification*, Garston and London, Building Research Establishment and Butterworths, 1985, 371 pp.

7 J. Namias, 'Factors relating to the explosive North Atlantic cyclone of December 1986', *Weather*, 1987, vol. 42, pp. 322–5.

8 J.P. Palutikof, 'Recent storms over the UK: the greenhouse effect or natural variability?', in J. Rew and C. Sturge (eds), *Chatset Directory of Lloyd's of London*, London, Chatset, 1990, pp. 70–3.

9 J.E. Simpson, *Sea Breeze and Local Wind*, Cambridge, Cambridge University Press, 1994, 234 pp.

10 T.D. Davies, J.P. Palutikof, X. Guo, L. Berkovsky and J.A. Halliday, 'Development and testing of a two-dimensional downslope wind model', *Boundary-Layer Meteorology*, 1995, vol. 73, pp. 279–97.

11 UK Meteorological Office, *Monthly Weather Report*. Summary for 1991, 1992, vol. 108, pp. 289–341.

12 X. Guo, 'An assessment of mesoscale wind modelling techniques in complex terrain', Ph.D. thesis, School of Environmental Sciences, University of East Anglia, Norwich, 1989, 301 pp.

13 X. Guo and J.P. Palutikof, 'A study of two mass-consistent models: problems and possible solutions', *Boundary-Layer Meteorology*, 1990, vol. 53, pp. 303–32.

14 T.R. Hiester and W.T. Pennell, *The Siting Handbook for Large Wind Energy Systems*, New York, WindBooks, 1981.

15 One of the more sophisticated is BLASIUS, described by N. Wood and P. Mason, 'The pressure force induced by neutral, turbulent flow over hills', *Quarterly Journal of the Royal Meteorological Society*, 1993, vol. 119, pp. 1233–67. A simpler model, widely used by wind farm developers, is WASP, described by Troen and Petersen, op. cit. A full literature survey is given by D. Hollis

and J. Ashcroft, 'Techniques for estimating the local wind climatology', *ETSU W/11/00400/REP*, Harwell, Energy Technology Support Unit, 1995, 61 pp.

16 Guo and Palutikof, op. cit.

17 See also results from two simple dynamic models, MS3DJH and CONFORM, p. 146 in P. Hannah, 'Application of wind modelling techniques in complex terrain', Ph.D. thesis, School of Environmental Sciences, University of East Anglia, Norwich, 1993, 365 pp.

18 Troen and Petersen, op. cit.

19 Hollis and Ashcroft, op. cit.

20 W.E. Bardsley and B.F.J. Manly, 'Regression-based estimation of long-term mean and variance of wind speed at potential aerogenerator sites', *Journal of Climate and Applied Meteorology*, 1983, vol. 22, pp. 323–7. See also Hollis and Ashcroft, op. cit.

21 Numerical models of air flow are, in addition, widely used in air pollution studies. See P. Zannetti, *Air Pollution Modelling*, New York, Van Nostrand Reinhold, 1990, 444 pp. In addition, models exist for thermal circulations, such as downslope winds (see Davies *et al.*, op. cit.) and sea-breeze circulations (see Simpson, op. cit.).

22 J.P. Palutikof, T.D. Davies and P.M. Kelly, 'An analysis of seven long-term wind-speed records for the British Isles with particular reference to the implications for wind power production', in A.D. Garrad (ed.), *Wind Energy Conversion 1985*, London, Mechanical Engineering Publications, 1985, pp. 235–40.

23 J.P. Palutikof, T.D. Davies and P.M. Kelly, 'A data bank of wind speed records for the British Isles and offshore waters', in P Musgrove (ed.), *Wind Energy Conversion 1984*, Cambridge, Cambridge University Press, 1985, pp. 414–25.

24 UK Met. Office, *The Observatories' Year Book 1922*, London, HMSO, 1925, 212 pp.

25 S.G. Smith, 'Comparison of wind speeds recorded by pressure-tube and Meteorological Office electrical cup generator anemographs', *Meteorological Magazine*, 1981, vol. 110, pp. 288–301.

26 See P.D. Jones, T.M.L. Wigley and K.R. Briffa, 'Monthly mean pressure reconstructions for Europe (back to 1780) and North America (to 1958)', *Technical Report TR037*, DOE/ER/60397-H1, US Dept. of Energy, Carbon Dioxide Research Division, Washington, DC, 1987, 99 pp., and P.D. Jones, 'The early twentieth century Arctic high – fact or fiction?', *Climate Dynamics*, 1987, vol. 1, pp. 63–75.

27 This work is fully described by J.P. Palutikof, X. Guo and J.A. Halliday, 'Climate variability and the UK wind resource', *J. Wind Engineering and Industrial Aerodynamics*, 1992, vol. 39, pp. 243–9, and by J.P. Palutikof, J.A. Halliday, X. Guo, R.J. Barthelmie and T.J. Hitch, 'The impact of climate variability on

the UK wind resource', *ETSU WN 6029*, Harwell, Energy Technology Support Unit, 1993, 148 pp. (plus appendices).

28 Society of Fellows Study Group, op. cit.

29 H.H. Lamb, *Historic Storms of the North Sea, British Isles and Northwest Europe*, Cambridge, Cambridge University Press, 1991, 204 pp.

30 The SSI values refer to the severity of the storm in its entirety, and not with respect to the British Isles alone. Many of these storms were particularly severe over the North Sea region, causing damage in the Netherlands, Denmark, northern Germany and southern Scandinavia.

31 H. Schinke, 'On the occurrence of deep cyclones over Europe and the North Atlantic in the period 1930–1991', *Beiträge zur Physik der Atmosphäre*, 1993, vol. 66, pp. 223–37.

32 H. Schmidt and H. von Storch, 'German Bight storms analysed', *Nature*, 1993, vol. 365, p. 791.

33 A.F. Jenkinson and B.P. Collison, 'An initial climatology of gales over the North Sea', *Synoptic Climatology Branch Memorandum No. 62*, Bracknell, UK Met. Office, 1977, 18 pp.

34 Jones *et al.*, op. cit., Jones, op. cit.

35 S.G. Smith, 'An index of windiness for the United Kingdom', *Meteorological Magazine*, 1982, vol. 111, pp. 232–47; J.M. Hammond, 'The strong winds experienced during the late winter of 1989/90 over the United Kingdom: historical perspectives', *Meteorological Magazine*, 1990, vol. 119, pp. 211–19; M. Hulme and P.D. Jones, 'Temperatures and windiness over the UK during the winters of 1988/89 and 1989/90 compared to previous years', *Weather*, 1991, vol. 46, pp. 126–35.

36 This is taken from T. Holt and P.M. Kelly, *Western European Gales, 1881–1993: a Statistical Assessment*, Report to Harvey Bowring and Others, Norwich, Climatic Research Unit, University of East Anglia, 1995, 31 pp.

37 The Gale Index is calibrated in knots, and the conver-

sion into ms⁻¹ leads to untidy numbers. One knot = 0.515 ms⁻¹.

38 J.W. Hurrell, 'Decadal trends in the North Atlantic Oscillation: regional temperatures and precipitation', *Science*, 1995, vol. 269, pp. 676–9.

39 J.P. Palutikof and A.R. Skellern, *Storm Severity over Britain*, Report to Commercial Union, Norwich, Climatic Research Unit, University of East Anglia, 1991, 102 pp.

40 C.J.M. Aanensen and J.S. Sawyer, 'The gale of February 16th 1962 of the West Riding of Yorkshire', *Nature*, 1963, vol. 197, pp. 654–6.

41 The statistics in this paragraph are taken from Department of Trade and Industry, *An Assessment of Renewable Energy for the UK*, HMSO, London, 1994, as reported in 'Government publishes renewables strategy', *Wind Directions*, 1993, vol. 13, no. 4, pp. 15–16.

GENERAL READING

N.J. Cook, *The Designer's Guide to Wind Loading of Building Structures. Part 1: Background, Damage Survey, Wind Data and Structural Classification*, Garston and London, Building Research Establishment and Butterworths, 1985, 371 pp.

M.P. Coutts and J. Grace (eds), *Climate and Trees*, Cambridge, Cambridge University Press, 1995, 485 pp.

Review – the magazine of new and renewable energy, published by the Department of Trade and Industry, Editorial Office: Room 803, Bridge Place, 88–89 Eccleston Square, London SW1V 1PT.

Royal Meteorological Society, Weather – special issue devoted to articles on the storm of 15–16 October 1987, *Weather*, 1988, vol. 43, pp. 66–142.

Windpower Monthly, an independent windpower news journal published by Torgny Møller, Vrinners Hoved, 8420 Knebel, Denmark.

12

THE AIR THAT WE BREATHE
Smogs, Smoke and Health

Peter Brimblecombe and Graham Bentham

Herein is not only a great vanity, but a great contempt of God's good gifts,
That the sweetness of man's breath, being a good gift of God,
Should be wilfully corrupted by this stinking smoke.
James I of England and VI of Scotland

INTRODUCTION

The air of cities has been polluted for thousands of years.[1] Early concerns about air quality in urban areas stemmed from the use of wood and coal as fuels, both industrial and domestic. During the years of the Industrial Revolution smoke pollution was considered a necessary part of progress. Air quality in cities tended to be worse in winter months when cold stagnating anticyclonic conditions trapped air and resulted in a build-up of smoke and associated pollutants, especially sulphur dioxide. Combined with fog, this resulted in the infamous 'pea-souper' smogs of London, where visibility was frequently drastically reduced. It was not until the 1950s, however, when a particularly severe smog resulted in 4,000 excess deaths in London, that an important Act was passed by Parliament to force the clean-up of urban air.

Although these 'pea-souper' smogs are now a thing of the past, air quality in cities is still generally poor (Figure 12.1). This is because vehicle exhaust emissions, due to the large amounts of traffic in urban areas, lead to nitrogen oxides smogs in winter and the build-up of low-level ozone in summer. The effect of these pollutants on human health has not been categorically identified, although air pollution from traffic has often been blamed for the major increase in the incidence of asthma that has occurred in recent decades. The introduction of catalytic converters into the vehicle fleet may go some way to remedying these problems.

EARLY HISTORY OF AIR POLLUTION

Although early cities burnt large quantities of wood, contemporary complaints were often related to odour not smoke. Nevertheless there were adverse reactions to smoke, most especially with the shift from wood to coal as a fuel. In medieval Britain, coal became important as a fuel following the depletion of conveniently usable wood supplies.[2] The unfamiliar smell of coal smoke led to early fears about its health risk through the belief that disease was carried in malodorous air (**miasmas**). The use of coal by thirteenth-century London made its citizens aware of problems of air pollution at an early date. Initially,

Figure 12.1 Fog in London, 2.30 p.m. on 30 November 1982. Calm anticyclonic conditions caused air pollution levels to rise to unusually high concentrations for several hours during the day. Although the sources of modern pollution are largely different from forty or a hundred years ago, the quality of city air is often still poor.

coal was used by lime-burners and blacksmiths and it was not until the sixteenth century that the widespread construction of chimneys allowed its use as a domestic fuel. A transition to coal was virtually complete in London by the early seventeenth century, but it was delayed until the nineteenth century in some other cities of the British Isles.

Industrialisation

The Industrial Revolution, and in particular the development of the steam engine, led to the possibility of smoke pollution on a much larger scale than before. The steam engine forced air pollution to be taken more seriously. As early as 1800 the Commissioners of Police in Manchester appointed a nuisance committee that looked at the methods available for abating smoke from steam engines. Although the steam engine provoked protest it also caused the early enthusiasts of smoke abatement to

focus on point sources. By the mid-nineteenth century smoke was generally regarded as an undesirable aspect of urban life, although perhaps a 'necessary evil'.

The belief that smoke should be eliminated became embodied in laws that arose from the sanitary reform of the nineteenth century. The early attempts to abate air pollution lacked real power or any clear mechanism for enforcement, but the Public Health Acts of 1872 and 1875 tried to set up administrative mechanisms for the control of nuisance. Even where the administrative procedures were well defined and enthusiastically followed, the lack of appropriate smoke control technology seemed to prevent both the administrators and industrialists from achieving a substantial improvement in air quality.[3]

Laws of the nineteenth century suffered not only from the weaknesses noted above but also lacked a scientific basis, and monitoring was rarely seen as a

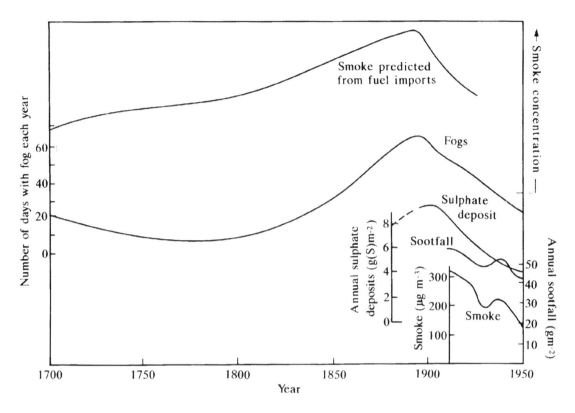

Figure 12.2 Air pollution in London since 1700, comparing predicted smoke concentrations (arbitrary scale) with fog days and later sulphate, soot and smoke measurements.

significant part of improving the urban atmosphere. The most widespread observations made last century were simple descriptions of smoke from chimneys and the reports of the inspectors would suggest that such observations were not particularly useful when trying to gauge the improvements. The few quantitative measurements that we have come from enthusiasts and amateurs, which means that they are sporadic and often inaccurate.[4]

THE TWENTIETH CENTURY

The early twentieth century saw no rapid improvement in perceived conditions, but there was a growing recognition of the need for monitoring. The fledgling United Kingdom Met. Office was involved in the London Fog Inquiry in the opening years of this century, and later in the formation of the earliest national air pollution monitoring network. This network was established during the First World War using **deposit gauges** that were to be used for more than fifty years. Control passed, however, to the Department of Scientific and Industrial Research in the 1930s and various government groups became involved in the design of a range of air pollution monitoring equipment: the **jet dust counter**, **lead candle** (to measure sulphur dioxide deposit) and the **bubbler**. Instruments such as these form the basis of the longest systematic record of air pollution in the British Isles, which can be linked with other surrogate information such as fog frequency and concentrations modelled from fuel use (Figure 12.2).

The longest quantitative records available are those for the deposit of soot and some soluble materials which begin early this century in urban areas. By the 1930s there were limited measurements of sulphur dioxide and soot in the atmosphere. Such observations were much more widespread in the years that followed the London smog of 1952. Early scientific work was only just beginning to influence policy when the Second World War intervened and pressures for cleaner air were delayed. The London smog of December 1952, with its 4,000 excess deaths, prompted a clear response and the passage of the Clean Air Act of 1956 through Parliament. This Act necessitated an air pollution monitoring network which was dedicated to policing its regulations.

The Clean Air Act is often proclaimed as a success, but it may have simply reinforced broader changes in fuel used which were already lowering the concentration of smoke and sulphur dioxide in British cities. Perhaps the most notable changes were the reduction in domestic coal use (as it was replaced by gas and electricity) and the growing importance of liquid fuels used in the private automobile fleet. Other influences at work in this period would have included the shifts of industry into suburban and rural locations, development of large rural power stations with tall stacks, industrial decline, and more recently a shift from coal to gas for electricity generation. Reductions in traditional pollutants, namely sulphur dioxide and smoke, masked an increase in the new polluting components of urban air, nitrogen oxides and ozone.[5] The warm dry summer of 1976, for example, saw massive tongues of ozone-rich air across the British Isles, which heralded a new regional air pollution problem. Other issues, most notably a preoccupation with wrangles over **acid rain**, focused monitoring attention on the rural precipitation networks, which meant only a limited growth in the number of urban monitoring sites.

The inability of the United Kingdom monitoring network of the 1980s to examine a broad range of relevant contemporary air pollutants was increasingly embarrassing and ultimately the Government White Paper *This Common Inheritance*[6] sparked off a new interest in the urban atmosphere. It led, in the 1990s, to the creation of a centralised real-time monitoring network, the Enhanced Urban Network, and its subsequent development into the Automatic Urban Network. There are also important non-urban elements to new networks (see Table 12.1 and Figure 12.3), most notably the rural ozone monitoring sites.

Long-term records of atmospheric phenomena are vitally important. In meteorology such records were initially established through the interest of individual observers (see Chapter 7), since visible weather processes have always attracted large numbers of enthusiasts. Records of air pollution, by contrast, have been a victim of changing priorities within monitoring agencies. Thus, interpreting long-term change in air quality can be more of a problem than with classical weather observation. There are hopes that the centrally controlled Automatic Urban Network will have some degree of stability and continuity.

Increasingly, air quality in the British Isles will be regulated under European Union directives. The Council Directive on *Ambient Air Quality Assessment and Management*[7] will promote a set of daughter directives to define the type of monitoring or modelling required for different air pollutants, the provision of public information and the setting of alert thresholds and limit values. Although there is a special focus on urban centres with a population greater than 250,000, the directive hopes to consider the effects of air pollution beyond simply human health. In addition to the air pollutants currently covered in directives the new regulations will cover benzene, polycyclic aromatic hydrocarbons, carbon monoxide, cadmium, arsenic, nickel and mercury.

THE POLLUTANTS

Sulphur dioxide

Sulphur dioxide (SO_2) concentrations in urban air have long been driven by the use of coal in the British Isles. Thus the high concentrations experienced in coal-burning cities of Britain late last century and through the first half of this century are no longer found (see Figure 12.2). The 1956 Clean

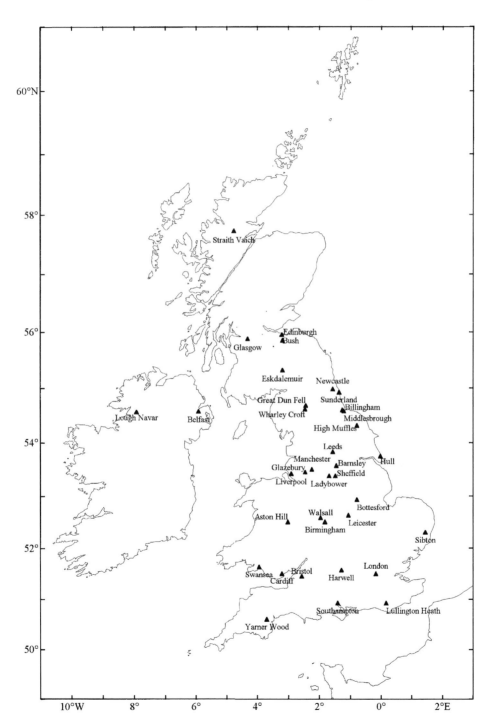

Figure 12.3 Location of sites in the Automatic Networks which monitor contemporary air pollutants. Table 12.1 lists the pollutants measured.

Table 12.1 Sites in the Automatic Networks which monitor contemporary air pollutants. see Figure 12.3 for a map of locations. The ● indicates sites at which the following pollutants are measured: O_3 – ozone; NO_x – nitrogen oxides; CO – carbon monoxide; SO_2 – sulphur dioxide; PM_{10} – small particles with a diameter less than 10 μm; HC – hydrocarbons

No.	Site name	O_3	NO_x	CO	SO_2	PM_{10}	HC
1	Straith Vaich	●	●		●		
2	Edinburgh Centre	●	●	●	●	●	
3	Edinburgh Med S						●
4	Glasgow		●	●			
5	Bush	●					
6	Eskdalemuir	●					
7	Belfast Queens Uni						●
8	Newcastle Centre	●	●	●	●	●	
9	Sunderland				●		
10	Great Dun Fell	●					
11	Belfast Centre	●	●	●	●	●	
12	Belfast East				●		
13	Wharley Croft	●					
14	Billingham		●				
15	Lough Navar	●					
16	Middlesbrough						●
17	High Muffles	●					
18	Leeds Centre	●	●	●	●	●	
19	Hull Centre	●	●	●	●	●	
20	Barnsley				●		
21	Manchester		●	●			
22	Glazebury	●					
23	Liverpool Centre	●	●	●		●	
24	Sheffield	●	●	●			
25	Ladybower	●	●		●		
26	Bottesford	●					
27	Leicester Centre	●	●	●	●	●	
28	Walsall		●				
29	Aston Hill	●					
30	Birmingham Wardend	●	●	●	●	●	●
31	Birmingham Centre	●	●	●	●	●	
32	Sibton	●					
33	Harwell	●					
34	London UCL						●
35	London Bloomsbury	●	●	●	●	●	
36	London Bridge Place	●	●	●	●		
37	London Cromwell Road		●	●	●		
38	West London		●	●			
39	Cardiff East						●
40	Cardiff Centre	●	●	●	●	●	
41	London Bexley	●	●	●	●	●	
42	London Eltham						
43	Bristol Centre	●	●	●	●		●
44	Southampton Centre	●	●	●	●		
45	Lullington Heath	●	●		●		
46	Yarner Wood	●	●				●
47	Swansea	●	●	●	●	●	
48	Bristol East						●
49	Leeds						●

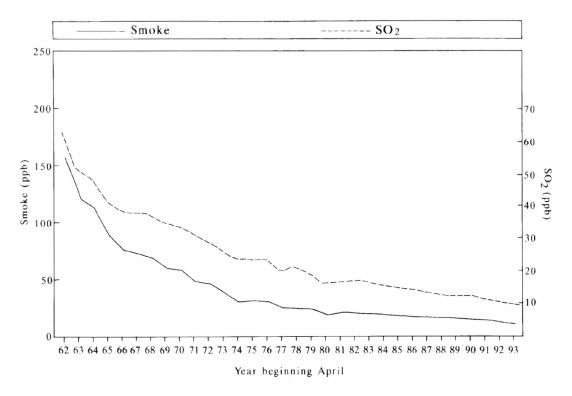

Figure 12.4 Annual average concentrations of smoke and sulphur dioxide for the United Kingdom from 1962 to 1993.[8]

Air Act aimed to control only smoke, but it is possible to argue that pressures to reduce smoke emissions also help lower the emission of sulphur dioxide. In particular, the movement away from the localised use of coal in both the domestic and industrial sectors can reduce sulphur dioxide emissions simultaneously with smoke.

The high sulphur content in fuel and diesel oils has provoked some concern, but regulations have pre-empted any substantial increase in sulphur emissions from this source and legislative pressure now requires oil to have a low sulphur content. Special, low-sulphur diesel is now more widely available, so it appears likely that sulphur concentrations in the urban atmosphere over British cities will probably remain reasonably low. Sulphur dioxide levels are typically between 5 and 15 parts per billion (ppb), but cities can still experience short periods of high

sulphur dioxide concentrations from industrial plumes. Extended pollution episodes, however, are more likely to be found where large amounts of coal are burnt. Belfast is the most notable example of a city with elevated winter sulphur dioxide concentrations, largely because of the non-availability of natural gas as a domestic fuel.

Smoke and particles

Smoke emissions, the central concern of the 1956 Clean Air Act, have declined considerably. It is still monitored along with sulphur dioxide as part of the Basic Urban Network of more than 150 sites.[8] Improvements in smoke and sulphur dioxide concentrations have been marked (see Figure 12.4) and 1993–4 was significant in that for the first time no site in the United Kingdom exceeded any of the

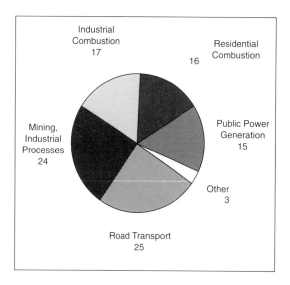

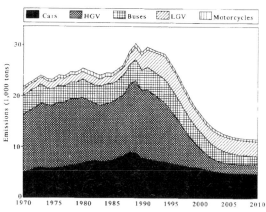

Figure 12.5b Estimated United Kingdom road transport emissions of PM_{10}.[13]

Figure 12.5a The relative importance of PM_{10} sources in the 1990s for the United Kingdom.[13]

limit values under the European Directive. Ten years previously twelve sites had exceeded the limit values.

Sulphur dioxide and smoke were traditionally considered together because of a synergistic relationship with health effects. With the decline in these two pollutants it has become important to decouple their effects. Recent studies, most significantly those of Douglas Dockery,[9] have stimulated an interest in the health effects of fine particles such as PM_{10} or $PM_{2.5}$.[10] Unfortunately PM_{10} has been monitored for only a short time in the British Isles so it is not possible to assess long-term trends. The annual urban background concentrations are currently between 20 and 30 micrograms per cubic metre (μgm^{-3}). There are few exceedences of the 100 μgm^{-3} daily average, but there are widespread exceedences of the daily average level of 50 μgm^{-3}.[11] The Expert Panel on Air Quality for the United Kingdom recently recommended that 50 μgm^{-3} be set as a standard.[12]

Traffic represents a very important source of PM_{10} in urban areas (86 per cent in London), but for the country as a whole mining and industrial processes are also important, as shown in Figure 12.5.[13] There

is likely to be some reduction in emissions from heavy goods vehicles as these sources become more regulated.

Nitrogen oxides

Nitrogen oxides are produced by combustion processes and are released predominantly as nitric oxide (NO), but oxidation in the atmosphere converts this to nitrogen dioxide (NO_2). The two pollutants can be treated together and are called nitrogen oxides for convenience (NO_x).

Traffic is responsible for most of the nitrogen oxides in urban air. There has been no distinct trend in concentrations over the past ten years (see Figure 12.6), but road transport emissions increased sharply through the 1980s to reach a peak in the early 1990s. The comparative stability of concentrations at the monitoring sites may be because these were initially placed at places of high pollution and high traffic congestion where little extra traffic growth was in evidence. Declines in emissions are expected over the next few years as catalytic converters, required of new cars as of 1993, become incorporated into the vehicle fleet.

The development of passive samplers, such as the diffusion tube, has enabled studies of average nitrogen dioxide concentrations to be made over wide

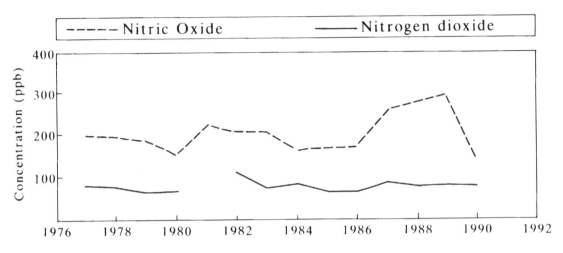

Figure 12.6 Trends in nitric oxide and nitrogen dioxide at London Victoria from 1976 to 1990.[18]

geographical areas (see Plate 8) and for these concentrations to be modelled[14] using various surrogates such as population. Perhaps the most important air quality change in recent years has been the recognition of winter-time nitrogen oxide smogs. These are quite different from the sulphur-laden winter smogs of earlier decades, although the stagnant atmospheric conditions under which they originate are similar. Emissions from urban traffic can raise concentrations of nitric oxide to many hundreds of parts per billion, levels which may increase urban mortality.[15] If high nitric oxide concentrations are coupled with low temperatures the oxidation of nitric oxide, normally an unimportant reaction, can become a significant source of nitrogen dioxide.

Part of the reason for this is because the **rate constant** for this reaction increases as temperature falls, as in winter. At low concentrations of nitric oxide – below about 50 ppb – the production of nitrogen dioxide is limited by the amount of ozone present, but when the nitric oxide concentration reaches several hundred ppb the second order reaction becomes important (see Box 12.1). The square dependence of nitrogen dioxide production on nitric oxide concentration means that in winter smogs it might well be worth issuing alerts and attempting to reduce traffic and other relevant urban sources.

Carbon monoxide

More than 90 per cent of the emissions of carbon monoxide (CO) come from the transport sector. This means that the concentrations are at their highest in urban areas with high traffic flow (see Plate 8) and this traffic also controls the diurnal variation in carbon monoxide concentration. In particular, at roadside sites, carbon monoxide shows high daytime concentrations with peaks during the morning and evening rush hours.

The World Health Organisation guidelines for short time periods (i.e., the one hour guideline of 25 parts per million (ppm)) are not exceeded at sites in the British Isles, but measurements at the side of Cromwell Road, London, have certainly approached this limit under conditions of winter-time stagnation. A more stringent eight-hour guideline of 10 ppm was exceeded at least once a year at 45 per cent of monitored urban sites and at 71 per cent of road-side sites between 1980 and 1991–2. In 1992 and 1993, however, Belfast was the only site to record exceedences. The fragmentary nature of the carbon monoxide record makes concentration trends difficult to establish, but the record of emissions is shown in Figure 12.7.

BOX 12.1 RATES OF REACTION

Consider the reaction

$$A + B \rightarrow C$$

The rate of this reaction may be proportional both to the concentration of A raised to some power x and to the concentration of B raised to some power y:

$$\text{Rate} \propto [A]^x[B]^y$$

which implies that

$$\text{Rate} = k[A]^x[B]^y$$

where k is the rate constant for this reaction at a fixed temperature. The square brackets indicate concentration. This reaction is of order x with respect to A and of order y with respect to B. The overall reaction is said to be of order $x + y$. The order of reaction is usually determined experimentally.

The reaction:

$$2NO + O_2 \rightarrow 2NO_2$$

is a third order reaction. The rate of the reaction is given by:

$$\text{Rate} = k_3[NO]^2[O_2]$$

and k_3 indicates the rate constant for a third order reaction. The reaction is second order with respect to NO and first order with respect to O_2, giving, overall, a third order reaction.

There are two main reactions which produce NO_2 in the atmosphere:

$$2NO + O_2 \rightarrow 2NO_2$$
$$NO + O_3 \rightarrow NO_2 + O_2$$

The first reaction usually occurs very slowly and so it is that latter reaction with ozone which usually predominates.[a] If the concentration of NO is low then it is the concentration of O_3 which limits the rate of production of NO_2. If, on the other hand, concentrations of NO are high then the first reaction becomes important. As the rate of this reaction is proportional to the square of the concentration of NO, then a small increase in NO will result in a large increase in the rate of NO_2 production.

[a] J.S. Bower, G.F.J. Broughton, J.R. Stedman and M.L. Williams, 'A winter NO_2 smog episode in the UK', *Atmospheric Environment*, 1994, vol. 28, pp. 461–75.

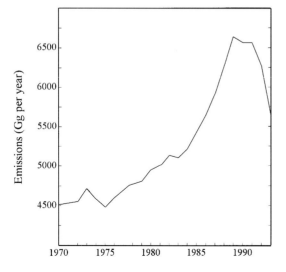

Ozone

There are no important emission sources of low-level ozone so its presence in air is the result of **photochemical reactions** involving the nitrogen oxides and **volatile organic compounds** (VOCs), along with a fraction transported down to the surface from the stratosphere. The presence of anthropogenic hydrocarbons and nitrogen oxides encourages the formation of ozone. These reactions take place, however, over substantial distances from source areas, making high ozone concentrations a widely distrib-

Figure 12.7 Trends in carbon monoxide emissions for Great Britain from 1970 to 1993 (data derived from the Digest of Environmental Statistics, 1994, 1995).

uted phenomena. Oddly enough, urban concentrations of ozone can be lower than rural values because of reactions involving nitric oxide that consume ozone (see Box 12.1). Given the importance of sunlight for ozone creation, it is understandable that the highest ozone concentrations are observed during the hot anticyclonic days of summer (Figure 12.8).

The Expert Panel on Air Quality Standards[16] recommended an eight-hour air-quality standard of 50 ppb ozone. Such a standard would be exceeded at some sites on more than eighty days each year. Achieving this standard would therefore require a reduction in the emissions of volatile organic compounds by between 75 and 80 per cent and a 95 per cent reduction in levels of nitrogen oxides across Europe. This goes far beyond reductions envisaged with simple controls. Most rural sites appear to show significant upward trends in ozone concentrations (Figure 12.9).[17]

Organic compounds

There are many sources of both volatile and less volatile organic compounds being emitted into the atmosphere. Pesticides, dioxins, polycyclic aromatic hydrocarbons and polychlorinated biphenyls (PCBs) are frequently grouped as toxic organic micropollutants (TOMPS). Many are **carcinogenic** and often involatile enough to be associated with airborne particulate matter. These are monitored on a much more limited basis than the volatile hydrocarbons.[18] They do have combustion sources, but the pesticides and polychlorinated biphenyls may simply evaporate from the ground surface.

Volatile organic compounds are also human carcinogens and air quality standards have been recommended for both benzene and 1,3 butadiene.[19] Other organic compounds may not be as significant as carcinogens, but the alkenes, toluene, xylene, aldehydes and hydrocarbons with more than five carbon atoms are important in aiding the formation of photochemical smog since these compounds have a high Photochemical Ozone Creation Potential.[20] Volatile organic compounds are produced during combustion, but also evaporate from many industrial operations.

There are now ten sites that monitor a range of volatile organic compounds in the air over the United Kingdom, but this network has been in operation only a short time. The site at Middlesbrough, which has been running since 1992, gives a most detailed picture of hydrocarbon behaviour at this site. This suggests that their relative concentrations are sensitive to the balance of industrial emissions and transport sources.[21] Sharp increases in styrene and 1,3 butadiene concentrations, which were found in the air from time to time, seem to be the result of industrial emissions (Figure 12.10).

Metals

Metals are frequently found in the polluted atmosphere, associated with the **particulate phase**. These may arise from industrial activities and combustion processes. A multi-element survey commenced in 1976 has monitored the concentrations of sixteen elements at five sites through to the present. In the early years of the survey the concentrations of most metals declined to a fairly stable level.[22] Lead has been of special concern because of its toxicity and the large amounts mobilised from leaded fuels. Lower lead concentrations in fuels, and more recently unleaded fuels, have caused a reduction in emissions (Figure 12.11).

INDOOR AIR POLLUTION

Most of us spend a large part of our lives indoors, particularly young children and the elderly who are known to be most susceptible to the effects of air pollution. The composition of the air we breathe indoors is, therefore, just as important as the air we breathe outdoors. Most people probably feel that their home provides a refuge from the often polluted air that they have to breathe in the outdoor environment. A comparison of typical concentrations of some important pollutants indoors and outdoors shows that in some respects they are right.[23] Concentrations of important pollutants such as sulphur dioxide and ozone are typically much lower indoors

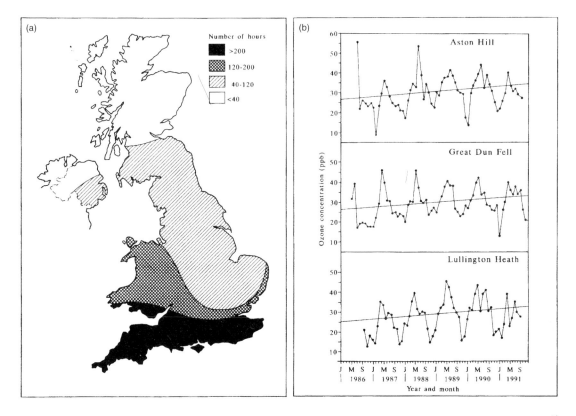

Figure 12.9(a) Number of hours with low-level ozone concentrations above 80 ppb during the period 1987 to 1990.[16]
Figure 12.9(b) Monthly average low-level ozone concentration trends from 1986 to 1991 at three rural sites.[17] See Figure 12.3 for locations.

than out, as is exposure to pollen and airborne lead. Indoors, however, we may be exposed to much higher concentrations (and for longer) of many other substances. These include nitrogen oxides which may exacerbate asthma and increase susceptibility to infections, and carbon monoxide which is lethal at high doses and at lower concentration can impair our ability to concentrate. The list also includes tobacco smoke where there is growing concern about the effects of passive smoking. Indoor air can also contain potentially carcinogenic substances such as formalde-

hyde and the radioactive gas radon where, in the worst affected parts of south-west England, doses in dwelling houses can exceed those permitted for workers in the nuclear industry. There can also be severe problems, particularly for asthmatics, from fungal spores and **allergens** such as those related to house-dust mites.

Unlike the situation for outdoor air pollution, there is little information on trends in indoor concentrations. There are strong reasons to believe, however, that exposures to indoor air pollution have probably

Figure 12.8 Summer haze over London, 1981, due to photochemical pollution. The top picture is a view from the roof of the old London County Hall taken on 28 August 1981, a day when ozone concentrations reached 220 μg/m³. The bottom picture was taken a month later when ozone concentrations reached only 30 μg/m³.

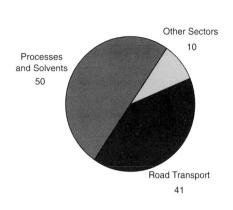

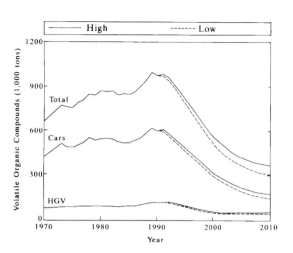

Figure 12.10a The relative importance for the United Kingdom of volatile organic compounds sources in the 1990s.[20]

Figure 12.10b Measured and estimated United Kingdom road transport sources of volatile organic compounds from 1970 to 2010.[18]

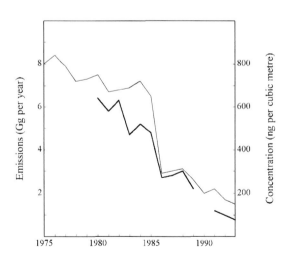

Figure 12.11 Emissions of lead from automotive sources, from 1975 to 1993, for the United Kingdom and mean lead concentrations, from 1980 to 1993, measured in central London.

increased in recent decades. One factor has been the emissions from new materials in the home. For example, formaldehyde resins are present in particle board which has been widely used as a building material and in furniture. There have also been problems of emissions from some installations of urea-formaldehyde foam insulation, although tighter regulations are helping in this area. Another factor pointing to increased exposure is that in the search for lower building costs and greater energy efficiency the ventilation rates of buildings have almost certainly decreased over time. Many modern houses are now built without chimneys, and draught-proofing of doors and windows is much more widespread. This means that there is a reduced flow of outside air with which to dilute chemical emissions from domestic materials and activities, with a resulting increase in their indoor concentration. Even more severe are the problems of so-called 'sick-building syndrome'. This has been experienced in many modern schools, hospitals and offices built to achieve high levels of energy efficiency to relatively airtight designs.

HEALTH ISSUES

A typical adult breathes more than 10,000 litres of air per day. Because this air is often polluted there is ample opportunity for the sensitive tissue of the respiratory tract to come into contact with significant quantities of potentially harmful substances. There has been long-term concern that air pollution may have contributed to the extremely high rates of chronic bronchitis that have, until recently, been characteristic of British cities. There has also been a persistent suspicion that smoke pollution, which can contain known carcinogens, may increase the risk of lung cancer – although there can be no doubt that smoking cigarettes is the main cause. More recently, there has been growing concern that air pollution, particularly that related to motor vehicles, may be an important contributor to the large increase in asthma that has occurred in the British Isles and in many other countries.

The potential health effects of air pollution, however, are not restricted to respiratory diseases. There is also growing evidence that, in individuals with other risk factors such as clogged up arteries (atherosclerosis), exposure to air pollution can increase the risks of heart attacks and stroke, which are two of the commonest causes of death in the United Kingdom. In recent years there seems to have been a growing public perception, fuelled by media interest, that air pollution poses a growing threat to health. As is often the case, the real picture is more complicated than this, with divergent trends in different areas. The prevailing gloomy view often ignores the major progress that has been made in reducing urban air pollution by smoke and sulphur dioxide that were formerly major problems; there seems little doubt that substantial benefits to health have ensued. It is undoubtedly true, however, that there is cause for concern about adverse trends of some other air pollutants, particularly those associated with motor vehicles.

Areas of progress

There is ample evidence that the high levels of smoke and sulphur dioxide that were characteristic of many British cities until the 1960s posed serious threats to health. The most dramatic of these occurred when cold weather and stagnant air associated with winter anticyclonic conditions led to the build-up over several days of extremely high levels of smoke and sulphur dioxide. Undoubtedly the most notorious of these incidents of extreme air pollution was the London smog of December 1952[24] (see also Chapter 13). For several days, concentrations of smoke and sulphur dioxide were hugely in excess of current World Health Organisation guidelines. The first signs of the health impact of the pollution crisis was a sharp increase in the number of patients being admitted to hospital with respiratory and circulatory problems. An official inquiry was later to show that there was also a marked increase in the death rate, with the main period of pollution experiencing about 4,000 deaths more than would normally be expected at that time of the year in London. The sharpest rise was in deaths from bronchitis but there were also large increases from other causes including pneumonia, heart attacks and strokes. The overwhelming majority of these deaths were of elderly people, suggesting strongly that the main impact of the pollution episode was on the section of the population who were particularly susceptible because of existing health problems.

The proportion of elderly people in the population has grown substantially during the twentieth century because of general improvements in the health status of the population and the consequent increases in life expectancy. Before the Second World War the elderly formed a much smaller fraction of the population than has been the case since, perhaps explaining why episodes of severe air pollution in earlier years do not seem to have been associated with as great an impact on mortality rates as they were in 1952. In spite of the continuing expansion of the susceptible, elderly population in the British Isles there has been no repetition of anything approaching the scale of the pollution-related health crisis represented by the London smog of 1952. This underlines the enormous benefits that have been associated with the improvements in smoke and sulphur dioxide levels in British cities since the 1950s.

As well as the acute health effects related to extreme episodes of air pollution, there is evidence that long-term exposure to relatively high levels of smoke and sulphur dioxide also posed serious risks to health. It has been shown that in the early 1950s there was a close association between levels of air pollution and mortality rates (especially for chronic bronchitis) in British cities.[25] This suggests that long-term exposure to air pollution was having a damaging effect on health, particularly in the cities with the worst air quality. By the early 1970s, however, by which time there had been major reductions in smoke and sulphur dioxide levels, there was little evidence of any association between mortality rates and air pollution. By the 1970s it therefore seems that urban smoke and sulphur dioxide concentrations had been reduced to levels where they no longer posed the significant threats to health that they had done only twenty years earlier.

Emerging problems

The good news on urban smoke and sulphur dioxide needs to be seen in the context of the possible health effects of some other air pollutants where trends are less favourable, particularly those arising from motor vehicles (Figure 12.12). A particular cause for concern is that increasing exposure to air pollution from traffic may be a factor in the major increase in the incidence of asthma that has occurred in the United Kingdom, Ireland and other western countries in recent decades. In support of the possibility of such a link it can be pointed out that during the relevant period both asthma and traffic-related air pollution have shown rising trends. Further support comes from experimental evidence that nitrogen oxides and ozone can increase bronchial responsiveness to allergens and viruses which are common triggers for asthma.

While there seems little doubt that exposure to air pollution can incite asthma attacks in some existing asthmatics it is, however, much less clear that this is a major factor in the induction of asthma in previously unaffected individuals.[26] For example, the incidence of asthma has been shown to be high in several rural

Figure 12.12 Heavy traffic on the M25 motorway around London. Air pollution from road vehicles has increased in recent years, as pollution from solid fuel burning has decreased.

areas, including some Scottish islands, where there is little air pollution. This has led some observers to suggest that the rise in asthma may have little to do with trends in air pollution. One suggestion is that the increase in asthma may result from changes in the indoor environment, especially increased exposure to the house-dust mite allergen. Another possibility is that the population may have become more susceptible to asthma because of changes in diet which have reduced **anti-oxidant** status, for example. It is, therefore, clear that there are several different possible causes for the current epidemic of asthma and, until there is further research, it is not possible to draw any firm conclusion on whether or not air pollution from traffic is a major factor.

The other area that is currently causing a great deal of concern is the possibility that the small

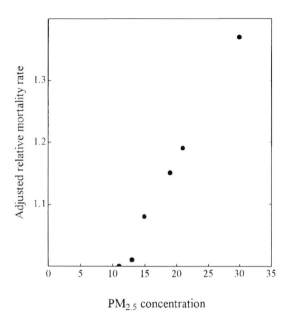

Figure 12.13 Adjusted relative mortality rates and average PM$_{2.5}$ concentrations in six cities in the United States.[8]

particles emitted from motor vehicles, especially those with diesel engines (and from some other sources), may have a significant and widespread detrimental impact on health. There is much interest in the effects of particles with an aerodynamic radius of less than ten microns (PM$_{10}$) and even more concern about smaller particles (PM$_{2.5}$). One of the principal reasons for concern is that the small size of such particles allows them to penetrate deep into the lungs where they can cause damage to sensitive tissue.

The health effects of exposure to small particles has been studied extensively in the United States. Research in several American cities has shown that hospital admissions for respiratory diseases rise significantly on days when PM$_{10}$ levels are high.[27] Increases in the number of deaths from circulatory as well as from respiratory causes on more polluted days have also been found. It has also been shown that, after controlling for other factors, mortality rates tend to be higher in cities with high levels of

PM$_{10}$, and especially PM$_{2.5}$, than in cities with cleaner air. Figure 12.13 shows the striking relationship between mortality rates and PM$_{2.5}$ in six American cities after adjusting for other factors such as cigarette smoking that affect the risk of death.[28] In most cases the proportional rise in mortality or morbidity that has been found is relatively small but, since it applies to very common health conditions, the absolute numbers of people affected can be very large.

There is, therefore, growing concern that exposure to small particles may pose a real threat to the health of the population, even if the increased risk for any particular individual is relatively small. Unfortunately, there is a dearth of British studies on this important subject and it is not known whether the United States findings would apply in the same way to the rather different conditions in the British Isles. Levels of PM$_{10}$ in some British cities, however, lie within the range at which effects have been observed in North America. One study has attempted to produce an estimate of small particle-related deaths in England and Wales by combining local measurements of PM$_{10}$ with the United States-based risk estimates.[29] Depending on the assumptions that are made this leads to an estimate of between 3,000 and 10,000 PM$_{10}$-related deaths per year in England and Wales. These findings, and the limitations of the information on which they are based, underline the need for further research on the possible health impacts of particulate pollution in the British Isles.

THE FUTURE

The future will see increasing regulation of a widening range of air pollutants. The Council of the European Union directive on ambient air quality assessment and management will be influential in changing air pollution policy. The directive examines the density of monitoring sites and the need to set concentration standards and alert thresholds. It seeks to widen interest beyond sulphur dioxide, nitrogen dioxide, fine particles, suspended particulate matter, lead and ozone, to include pollutants such as benzene, the polycyclic aromatic hydro-carbons, carbon monoxide, cadmium, arsenic, nickel and mercury. Emissions will no doubt decline over the next decades, but the secondary pollutants will continue to tax our regulatory approaches. The control of diesel particle emissions may be an important area for technological development. There is much political pressure to issue alerts when pollutant concentrations become extremely high, but the effectiveness of alerts in ameliorating the intensity of episodes is far from clear. Nevertheless, winter smogs are very sensitive to the concentration of nitric oxide and their occurrence may represent a good target for emission reduction alerts under stagnant winter conditions.

The future could also see some interesting effects of global environmental changes on air pollution. In particular, there is a possibility that the environment in Britain and Ireland will become more favourable for the development of summer ozone pollution. It has already been stressed that ozone at the surface is produced by the action of sunlight on hydrocarbons and nitrogen oxides. Episodes of ozone pollution in the British Isles are strongly associated with summer anticyclonic conditions when the relatively still air leads to a build up of the precursor pollutants and the high temperatures and strong sunlight favour the production of ozone. If climate change, as a result of the enhanced **greenhouse effect**, were to lead to warmer, sunnier summers in the region (and there are obvious uncertainties about this, see Chapter 15) this could lead to worse problems of ozone pollution. Rather paradoxically, there is also a possibility that the depletion of stratospheric ozone could lead to ozone *increases* in the lower atmosphere. This is because depletion of stratospheric ozone would lead to an increased penetration of the ultra-violet radiation which is involved in the photochemical production of ozone from its precursors. Any effects would be likely to be greatest in polluted urban areas, whereas in cleaner rural areas such reactions would be likely to be limited by nitrogen oxides. The effects would therefore be greatest where there are most people and where there is the greatest potential for harm.

NOTES

1 P. Brimblecombe and F.M. Nicholas, 'History and ethics of clean air', in R.J. Berry (ed.), *Ethical Dilemmas*, London, Chapman and Hall, 1993, pp. 72–85.
2 P. Brimblecombe, *The Big Smoke*, London, Methuen, 1987, 185 pp.
3 P. Brimblecombe and C. Bowler, 'The history of air pollution in York, England', *Journal of the Air and Waste Management Association*, 1992, vol. 42, pp. 1562–6.
4 Brimblecombe, op. cit.
5 It is important to distinguish between the different roles of ozone in the lower and upper atmosphere. In the upper atmosphere, the stratosphere, ozone is a valuable gas since it absorbs harmful ultra-violet radiation from the sun. In the lower atmosphere, ozone is a health hazard as well as being a greenhouse gas. In this chapter we are concerned primarily with low-level ozone. See Chapter 15 for a discussion about stratospheric ozone.
6 *This Common Inheritance: Britain's environmental strategy*, London, HMSO, 1990.
7 Commission of the European Communities, Council Directive, *Ambient Air Quality Management and Monitoring* 9514/95 Annex II, 1995.
8 AEA, *UK Smoke and Sulphur Dioxide Monitoring Networks*, Culham, AEA, 1995.
9 D.W. Dockery, C.A. Pope, X.P. Xu, J.D. Spengler, J.H. Ware, M.E. Fay. B.G. Ferris and F.E. Speizer, 'An association between air pollution and mortality in six United States cities', *New England Journal of Medicine*, 1993, vol. 329, pp. 1753–9.
10 Committee on the Medical Effects of Air Pollution, *Non-biological Particles and Health*, UK Department of Health, 1995.
11 Quality of Urban Air Review Group, *Airborne Particulate Matter in the United Kingdom*, London, Department of the Environment, 1996.
12 Expert Panel on Air Quality Standards, *Particles*, Department of the Environment, London, HMSO, 1995, 38 pp.
13 Quality of Urban Air Review Group, 1996, op. cit.
14 G.W. Campbell, J.R. Stedman and K. Stevenson, 'A survey of nitrogen dioxide concentrations in the UK using diffusion tubes, July–December 1991', *Atmospheric Environment*, 1994, vol. 28, pp. 477–86.
15 H.R. Anderson, E.S. Limb, J.M. Bland, A.P. de Leon, D.P. Strachan and J.S. Bower, *The Health Effects of an Air Pollution Episode in London, December 1991*, Culham, AEA Technology, 1995.
16 Expert Panel on Air Quality Standards, *Ozone*, London, HMSO, 1994.

17 Photochemical Oxidants Review Group, *Ozone in the United Kingdom 1993*, London, Department of the Environment, 1993.

18 Quality of Urban Air Review Group, '*Urban Air Quality in the United Kingdom*, London, Dept of the Environment, 1993.

19 Expert Panel on Air Quality Standards, *Benzene*, London, HMSO, 1994; Expert Panel on Air Quality Standards, *1,3-Butadiene*, London, HMSO, 1994.

20 N.R. Passant, *Emissions of Volatile Organic Compounds from Stationary Sources in the UK*, Stevenage, Warren Spring Laboratory, 1993.

21 J. Derwent, P. Dumitrean, J. Chandler, T.J. Davies, R.G. Derwent, G.J. Dollard, M. Delaney, B.M.R. Jones and P.D. Nason, *A Preliminary Analysis of Hydrocarbon Monitoring Data from an Urban Site*, AEA CS 18358030/005/Issue 2, Culham, AEA Technology, 1994.

22 Quality of Urban Air Review Group, 1993, op. cit.

23 J.D. Spengler and K. Sexton, 'Indoor air pollution: a public health perspective', *Science*, 1983, vol. 221, pp. 9–17.

24 Brimblecombe, op. cit.

25 S. Chinn, C. du V. Florey, I.G. Baldwin and M. Gorgol, 'The relation of mortality in England and Wales 1969–73 to measurements of air pollution', *Journal of Epidemiology and Community Health*, 1981, vol. 35, pp. 174–9.

26 Department of Health, *Committee on Medical Aspects of Air Pollution: Asthma and outdoor air pollution*, London, HMSO, 1995.

27 J. Schwartz, 'Air pollution and daily mortality: a review and meta analysis', *Environmental Research*, 1994, vol. 64, pp. 36–52.

28 Dockery *et al.*, op. cit.

29 D. Pearce and T. Crowards, *Assessing the Health Costs of Particulate Air Pollution in the UK*, CSERGE Working Paper GEC 95–27, Norwich, Centre for Social and Economic Research on the Global Environment, 1995.

GENERAL READING

P. Brimblecombe, *The Big Smoke*, London and New York, Routledge 1987, 185 pp.

P. Brimblecombe, *Air Composition and Chemistry*, Cambridge Environmental Chemistry Series 6, Cambridge, Cambridge University Press, 1996 (2nd edn), 253 pp.

D.M. Elsom, *Atmospheric Pollution: a Global Problem*, Oxford, Blackwell, 1992 (2nd edn).

C. Kirby, 'Urban air pollution', *Geography*, 1995, vol. 80, pp. 375–92.

R. Read and C. Read, 'Breathing can be hazardous to your health', *New Scientist*, 1991, 23 February, pp. 34–7.

13

'PHEW! WHAT A SCORCHER'
Weather Records and Extremes

Michael Dukes and Philip Eden

There are two kinds of statistics; the kind you look up and the kind you make up.
Rex Stout, *Death of a Doxy*

INTRODUCTION

The blistering summer of 1995 was drawing to a close and the conversation in the 'Red Lion' had once again turned to the heat and the drought. In the corner seat of the public bar old Fred was holding forth, as was his habit when the weather was discussed. After all, his memory went back so much further than everyone else's.

> Of course, you young fellers ain't old enough to remember '47. Now there was a year. We wuz snow-bound for months, two feet deep it were, drifts up to the top o' the houses even in the middle of March, last remnants didn't vanish till April. An' when the snow went we wuz flooded for two weeks, and to cap it all there was an 'urricane which took the roof off the village school. And then we 'ad the 'ottest summer in living memory, in the nineties by May, not a drop of rain through the whole of August, and that was the summer Denis Compton hit nigh on four thousand runs with eighteen 'undreds.

The young professionals listened politely. They were used to old Fred's vivid imagination. They all knew that the summer of 1995 had been the hottest and driest for centuries; after all, the water company spokesmen and even government ministers had been quoting the statistics almost daily.

In truth, old Fred's memory was rather more reliable than the water company statistics. All the statements attributed to him above are essentially correct. In 1995, the heat and drought were not as prolonged as in a number of other years, and the winter of 1994–5 had been one of three wettest of the twentieth century. Never has the capacity of those with vested interests to pick and choose statistics to suit their arguments been more apparent than in the 1990s.

PITFALLS IN THE STUDY OF EXTREMES

The British are obsessed with the weather, especially in its more extreme forms. In spite of a generally equable climate, we complain about what we consider to be excessive heat or abnormal cold, and we seek scapegoats during floods and droughts. Fortunately we have an excellent climatological network with a history of organisation and continuity second to none (see Chapter 7); thus unusual individual events, noteworthy months and seasons, and exceptional cumulative phenomena can readily be put into some sort of historical context. But those coming new to research into climatological extremes and climate history should be aware of some of the pitfalls that await; in particular they should never take published records for granted – even the most widely quoted ones.

In the British Isles our knowledge of climate fluctuations over the last three or four centuries is underpinned by several composite records. Some of the best known of these are the Central England Temperature series of monthly mean temperature values for the period 1659 to 1973 compiled by Professor Gordon Manley (see Chapter 9) and subsequently updated by the UK Met. Office; a catalogue of daily Central England mean temperature values; two England and Wales Precipitation series beginning in 1727 and 1766 respectively (see Chapter 10); and a London precipitation series for 1697 to date.[1] Anyone using these series should ensure that they are familiar with their origins and development and do not fall into the trap of confusing precision with accuracy. Similarly, many continuous records from individual locations are liable to be affected by changes of site, of instrumentation, and of observational practice, as well as trends due to urbanisation. Researchers should always be aware of these before they seek to explain discontinuities or trends. There is one well-known example in the literature of a paper which used 'second-hand' data to illustrate the effects on temperature of urbanisation in the Greater London area between 1878 and 1969. The author concluded an increase in the temperature difference between Kew Observatory and Rothamsted (Hertfordshire) of about 1°C due to urbanisation of the Kew site, dismissing a site change at Rothamsted from a walled garden to an open field some 15 m higher as unimportant. A cursory examination of the overlapping records for the two Rothamsted sites reveals that this was not the case.[2]

The efficacy of **data quality control** by the various authorities has also varied considerably over time. For many decades it arguably erred on the side of generosity towards doubtful readings, but since the late 1970s the quality control process has been automated and some people believe that it is now more likely to exclude genuine extreme values than to include doubtful ones.[3]

Accounts of past weather events should also be treated with caution. Dubious descriptions when repeated by reputable climatologists and meteorologists can acquire a false pedigree. One example is the widely documented account of widespread snow showers over low ground in England on 11 July 1888. In some publications the validity of these observations is queried, in others the account is repeated with no qualification, and very rarely has it been seriously challenged, yet the weather data available for the time suggests it was too warm for snow over low ground in much of England and Wales.[4]

Newspaper accounts of interesting weather events should always be taken with a large degree of caution (see Figure 13.1) – even those penned by well-known meteorological writers in the serious papers. This is not their fault; it is because newspaper articles are subject to the whim of a sub-editor. The newspaper sub-editor is a strange breed with two goals. The first is to force contributions to fit the space available, which may necessitate removing text, no matter its importance, or creating new text without reference to the original author. The second is to make these same contributions fit the style of the newspaper, which may involve extensive re-writing, again without reference to the author. Furthermore, eyewitness reports of unusual weather are notoriously unreliable.

One other source of error in extreme values, particularly with reference to temperature, may be mentioned. This is the usually taboo subject of invention. Those involved in quality control have always been aware of (very isolated) cases of observers who 'massage' their figures – usually maximum temperatures – to make them appear more exceptional. The reason is probably no more sinister than a desire for the observer's station to appear at the top of published lists or to be quoted regularly in the national news media. An off-the-record check with the quality-controlling authority is advised if such suspicions are aroused. Exaggeration of sunshine records by health or tourist resorts also falls into this category.

The above list of possible pitfalls is an extensive one but, in fact, the vast majority of weather observations are trustworthy. Investigations into extreme events necessarily involve examining data at the tails

Figure 13.1 Headline from the *Daily Mirror* newspaper, 4 August 1990.

of the frequency distribution, however, and it is vital to be aware that real weather may be only one of many possible explanations of extreme data.

A single chapter could not hope to provide an exhaustive coverage of abnormal weather events over the British Isles. Instead we have selected some prominent examples of extreme weather (extreme in British terms, that is), backed up by extensive lists of other notable events. The tables included in this chapter tend to show a bias towards England and Wales simply because of the wealth of observational data extending over a very long period in this part of the British Isles. In addition, extremes since 1900 are given more prominence because temperature and precipitation observations from individual stations prior to this century were less reliable (due to variations in instrumentation, exposure and observational techniques), while sunshine and wind measurements only became common towards the end of the last century. There are, however, a few outstanding exceptions and the very valuable long composite series – such as the Central England Temperature and England and Wales Precipitation – are widely used to give insight into how the more recent extremes compare to those from several hundred years ago.

SUMMER HEAT WAVES AND WINTER WARMTH

The media machine rouses itself at the merest hint of a new high temperature record. During the 1990s warmth has been dominant with a seemingly endless succession of summer heat waves and winter warmth, including the phenomenal hot spell in August 1990, the near-record warm summer of 1995 and the unusual mildness of November 1994 and October 1995 – as measured by the Central England Temperature.

The causes of exceptionally high temperatures in the British Isles vary from season to season.[5] Between mid-November and mid-February high temperatures usually occur on the lee side of a range of hills or mountains when a warm, moist tropical-maritime air mass is further warmed by the **Föhn effect**. Tropical maritime air usually arrives on our shores in a stiff south-westerly airflow, or occasionally even a southerly one, with low pressure south of Iceland and high pressure over Europe. This is ideal for high winter temperatures on the Devon and Somerset coast, the north coast of Wales, the coast of northeast England, and the southern shore of the Moray Firth. There was such a **synoptic** situation when the highest-ever January temperature was measured on 27 January 1958 and again on 10 January 1971 (see Figures 13.2 and 13.3). This January record goes to a weather station run by the University College of Wales, Bangor, near the little North Welsh village of Aber in the lee of Snowdonia, where on each occasion 18.3°C was logged.

There have been seven January days this century on which the temperature has reached or exceeded 17°C somewhere in the British Isles. On each occasion the highest temperature was measured in North Wales, five times at Aber. These include the period 8–11 January 1971. This was not only abnormally warm at Aber, but also at many places in North Wales, northern England and Scotland, several of which exceeded previous January site records. A new Scottish record for January of 16.7°C was set at Lairg in the Highlands.

In early spring and late autumn high temperatures require a combination of warm advection and solar heating, but between April and September a prolonged spell of calm anticyclonic conditions can bring near-record values without the **advection** of warm air. For example, on 20 June 1995 an **anticyclone** began to develop across the country in what was essentially a cool polar maritime **air mass**. By the 30th, a combination of warming due to subsidence and prolonged solar heating pushed daytime temperatures to between 32°C and 34°C, even though there had been no apparent change of air mass.

The five warmest years in the Central England Temperature series for each season are listed in Table 13.1. The newest entry is the summer of 1995. It will be long remembered for continuous days of bright sunshine and exceptional warmth. The first hint that this was to be an unusual summer came in early May, with the second warmest first week of

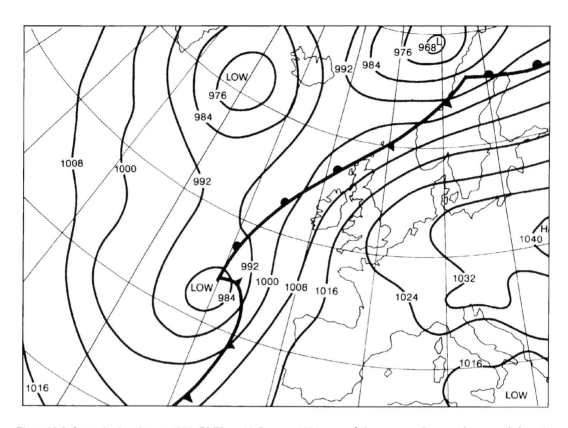

Figure 13.2 Synoptic situation at 1200 GMT on 10 January 1971, one of the warmest January days recorded in the British Isles. The temperature at Aber in North Wales reached 18.3°C.

Table 13.1 The five warmest, or equal warmest, years and seasons in the Central England Temperature series. (See Appendix D for a full listing of these data.)

Spring	°C	Summer	°C	Autumn	°C	Winter[a]	°C	Year	°C
1893	10.2	1976	17.8	1730	11.8	1869	6.8	1990	10.6
1945	10.1	1826	17.6	1731	11.8	1834	6.5	1949	10.6
1992	9.9	1995	17.4	1729	11.6	1989	6.5	1995	10.5
1952	9.9	1846	17.1	1818	11.6	1975	6.4	1989	10.5
1959	9.8	1983	17.1	1995	11.5	1686	6.3	1959	10.5
				1978	11.5			1921	10.5
				1959	11.5			1834	10.5
				1949	11.5			1733	10.5

[a] Winter refers to the year in which January fell.

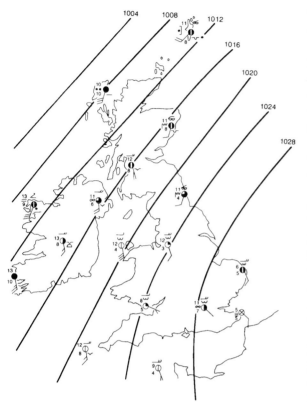

Figure 13.3 Synoptic observations and pressure pattern over the British Isles at 1200 GMT on 10 January 1971.

May on record. After a cool and cloudy first half of June, especially in the east, the real summer weather began. Temperatures were generally lower than during the hot spell of early August 1990, but there were a few new station records, including 28.1°C at Valentia, Co. Kerry on 27 June, 29.3°C at Clones in Co. Monaghan on 29 June, 31.8°C at Aberporth in west Wales on 2 August and 29.7°C at Aberdeen on 21 August. The temperature exceeded 32°C somewhere or other in the British Isles on twelve days during the summer, including a consecutive run of five days between 30 July and 3 August, the second longest such run on record. The highest temperature of the summer was 34.9°C at Kew Royal Botanic Gardens on 1 August. The 1995 summer ranked third in the Central England Temperature list (see

Table 13.1), thanks to a combination of the fifth warmest July and the hottest August on record.

There is no strict definition of a heat wave in the British Isles, but the accolade for the longest heat wave must go to a spell of weather between 23 June and 7 July 1976. On each of those fifteen days the temperatures exceeded 32°C somewhere in the country, with a reading of 35.9°C at Cheltenham on 3 July – the highest of that famous summer. But while 1976 holds the record for the longest heat wave, the most intense spell of heat happened in August 1990. It was short-lived, lasting only from the 1st to the 4th, but in that time many places in England and Wales and a few in Scotland established their all-time high temperature records. The heat reached a peak in Scotland, Northern Ireland, north-west England and west and North Wales on the 2nd with a Welsh record of 35.2°C at Hawarden Bridge in Clwyd, beating the previous record of 33.9°C at Newport in Gwent on 12 July 1923.[6] For most of England it was 3 August 1990 that was the hottest day in recorded history; Cheltenham's 37.1°C set a new all-time British Isles temperature record, exceeding the previous record of 36.7°C established at Raunds in Northamptonshire on 9 August 1911. There was also record night-time warmth; Brighton reported a minimum temperature of 23.9°C on the night of the 2nd–3rd, an unprecedented overnight value.

On 4 August 1990, Kew Royal Botanic Gardens reached 35.2°C making it the third successive day to exceed 35°C. There have been only twenty-one occasions since 1900 with a recorded temperature of 35°C or over and these are listed in Table 13.2. Such high values have never been reached in the Irish Republic, but Table 13.3 lists some of this country's warmest days on record.

Temperatures of 25°C or over have been measured in eight of the twelve months of the year. Table 13.4 lists the highest temperature in the British Isles for each month. One of the most notable values is the 35.6°C recorded at Bawtry in South Yorkshire on 2 September 1906 – a remarkable occurrence less than three weeks before the autumn equinox and a temperature only exceeded in the United Kingdom in four other years since 1900.

Table 13.2 Days with a maximum temperature of 35°C or over since 1900 in the British Isles. (Table gives highest value per date only.)

Year	Date	°C	Location
1900	20 July	35.2	Cambridge
1906	01 September	35.0	Collyweston, Leicestershire
1906	02 September	35.6	Bawtry, S. Yorkshire
1911	08 August	35.0	Kingston-on-Soar, Nottinghamshire
1911	09 August	36.7	Raunds, Northamptonshire
1911	13 August	35.6	Wilton, Wiltshire
1923	12 July	35.0	Bristol; Reading; Hitchin, Hertfordshire
1923	13 July	35.6	Camden Square, London
1932	18 August	35.0	St Helier, Jersey
1932	19 August	36.1	Tottenham, Enfield, Camden Square, Regent's Park, all London; Halstead, Essex.
1948	28 July	35.0	Milford, Surrey
1952	01 July	35.1	St Helier, Jersey
1957	29 June	35.6	Camden Square, London
1976	26 June	35.4	North Heath, W. Sussex; East Dereham, Norfolk
1976	27 June	35.5	Mayflower Park, Southampton
1976	28 June	35.6	Mayflower Park, Southampton
1976	02 July	35.7	Cheltenham, Gloucestershire
1976	03 July	35.9	Cheltenham, Gloucestershire
1990	02 August	36.6	Barbourne, Worcestershire
1990	03 August	37.1	Cheltenham, Gloucestershire
1990	04 August	35.2	Kew Botanic Gardens, London

Table 13.3 Some of the hottest days in the Republic of Ireland

Year	Date	°C	Location
1876	16 July	33.4[a]	Phoenix Park, Dublin
1887	26 June	33.3[b]	Kilkenny Castle, Co. Kilkenny
1921	12 July	32.2	Killarney, Co. Kerry
1934	09 July	32.2	Foynes, Co. Limerick
1975	03 August	30.5	Casement, Co. Dublin, and Kilkenny, Co. Kilkenny
1976	29 June	31.6	Shannon, Co. Clare
1983	13 July	31.4	Kilkenny, Co. Kilkenny
1983	14 July	32.2	Oak Park, Co. Carlow
1989	18 July	30.0	Shannon, Co. Clare
1990	02 August	31.0	Glasnevin, Dublin
1995	29 June	30.2	Shannon, Co. Clare
1995	02 August	31.5	Oak Park, Co. Carlow

[a] Value probably not recorded in a Stevenson Screen
[b] Officially accepted Irish temperature record

FREEZING WINTERS AND COOL SUMMERS

The year 1987 saw the media interest in global warming accelerate, with front page stories in the serious papers and extravagant sensationalism in the tabloids. Yet, ironically, in mid-January the British Isles experienced a spell of severe cold that rivals any in the instrumental record. It began on the 10th as

Table 13.4 Highest daily maximum temperature since 1900 for each month in the British Isles

Month	°C	Location	Date
January	18.3	Aber, Gwynedd	27th, 1958 and 10th, 1971
February	19.2	March, Cambridgeshire	23rd, 1990
March	25.0	Wakefield, W. Yorkshire	29th, 1965
		Cromer and Santon Downham, Norfolk	29th, 1968
April	29.4	Camden Square, London	16th, 1949
May	32.8	Camden Square and Regent's Park, London;	
		Tunbridge Wells, Kent; Horsham, W. Sussex	29th, 1944
June	35.6	Camden Square, London	29th, 1957
		Mayflower Park, Southampton	28th, 1976
July	36.0	Epsom, Surrey	22nd, 1911
August	37.1	Cheltenham, Gloucestershire	3rd, 1990
September	35.6	Bawtry, S. Yorkshire	2nd, 1906
October	29.4	March, Cambridgeshire	1st, 1985
November	21.7	Prestatyn, Clwyd	4th, 1946
December	18.3	Achnashellach, Highland	2nd, 1948

an intense anticyclone, centred over Scandinavia, ushered in an extremely cold easterly airflow across the British Isles. The severe weather lasted until about the 20th with abnormally heavy snowfalls across much of Essex, Kent, Sussex, Surrey and south London, but the cold was most intense on the 12th and 13th.

With large amounts of cloud and a stiff wind it was the low daytime maxima rather than night-time minima which were exceptional. The highest recorded value of the 12th in the United Kingdom was +0.1°C at Butt of Lewis in the Western Isles, but most of the country remained well below freezing all day. Typical maximum temperatures were between −5°C and −8°C, and on the 13th the temperature at Okehampton in Devon (elevation 372 m) did not exceed −8.5°C.[7] For many places these were record low maximum temperatures and, taking England and Wales as a whole, 12 and 13 January 1987 were probably the coldest days since January 1740,[8] the strong winds adding to the chill. Substantially lower maxima have occurred in the past at individual locations, but these have been invariably the result of persistent freezing fog, stagnant air over a deep snow cover, or inversion-trapped cold air in frost hollows, and have therefore been rather localised.

Strong easterly winds may be ideal for widespread

low maxima in winter, but Britain's lowest minima almost always occur in clear calm conditions, usually with a stagnant cold air mass across the country and a fresh snow cover. Some of the coldest nights of the twentieth century are listed in Table 13.5.[9] The winter of 1981–2 appears several times in the list, including Braemar's −27.2°C on 10 January 1982. This equalled its own all-time British low temperature record, set previously on 11 February 1895. More recently, Altnaharra (Scottish Highlands) came close to challenging this record with a reading of −27.0°C on 30 December 1995. By comparison, the lowest temperature ever recorded in southern Ireland is −19.1°C. This occurred on 16 January 1881 at Collooney, Co. Sligo.

The winter of 1981–2 was dominated by two exceptional cold spells, one in December and the other in January. The December cold weather began on the 8th and only abated after Christmas. Heavy snow fell widely, and several nights were ideal for rapid radiation cooling. On the 12th, Shawbury in Shropshire achieved a high of only −12°C, and, with clear skies, light winds and a deep snow cover, the night of the 12–13th saw the temperature plunge to −25.2°C – a new English record. The severe conditions returned on 6 January 1982 with 'cold' anticyclones over both Scandinavia and Greenland.

Table 13.5 Coldest nights since 1900 (Table gives lowest value per date only)

° C	Location	Date
−27.2	Braemar, Grampian	10 January 1982
−27.0*	Altnaharra, Highland	30 December 1995
−26.8	Grantown-on-Spey, Highland	08 January 1982
−26.6	Bowhill, Borders	11 January 1982
−25.9	Grantown-on-Spey, Highland	09 January 1982
−25.2	Shawbury, Shropshire	13 December 1981
−25.0	Braemar, Grampian	23 February 1955
−24.6	Carnwath, Strathclyde	13 January 1979
−23.6	Grantown-on-Spey, Highland	20 January 1984
−23.3	Rhayader, Powys	21 January 1940
−23.3	Balmoral and Logie Coldstone, Grampian	28 January 1910
−23.3	Braemar, Grampian	14 November 1919
−23.7	Altnaharra, Highland	28 December 1995
−22.8	Logie Coldstone, Grampian	14 March 1958
−22.8	Braemar, Grampian	15 November 1919
−22.7	Braemar, Grampian	27 January 1985
−22.6	Shawbury, Shropshire	12 December 1981
−22.6	Braemar, Grampian	07 January 1982
−22.2	Braemar, Grampian	22 February 1955
−22.2	Grantown-on-Spey, Highland	12 March 1958
−22.2	Grantown-on-Spey, Highland	18 February 1960
−22.2	Cannich, Grampian	30 December 1961
−22.2	Braemar, Grampian	18 January 1963
−22.0	Keith, Grampian	20 February 1978

* The support thermometer measured −27.2°C and this may become the accepted value.

Again, conditions became ideal for rapid radiative cooling by night, and localised exceptional daytime cold. The more noteworthy readings included a high of just −13°C at Tummel Bridge in Tayside on the 7th, and lows of −26.8°C at Grantown-on-Spey, Highland, on the 8th, and −26.1°C at Newport in Shropshire on the 10th – again breaking the English record. Braemar's British record minimum value of −27.2°C early on the 10th was followed by a new all-time low maximum of −19.1°C.

Persistent cold throughout a whole month is rare in the British Isles. Cold spells usually come in short sharp bursts with several intervening milder 'westerly' spells, but occasionally a winter month will pass without a thaw. The most recent example came in 1986 when persistent high pressure anchored over Scandinavia produced one of a very few totally 'easterly' months. As a consequence the month's Central England Temperature was below freezing (see Figure 13.4). This was only the eighth time since 1659 that February mean temperature has been below zero. Other notable bitterly cold months this century with a sub-zero Central England Temperature include January 1940, February 1947, February 1956, January and February 1963 and January 1979. These months were also very cold in much of Scotland, Wales and Ireland.

Icy cold is not always confined to the winter months, as the record low temperatures for each month in Table 13.6 indicate. Perhaps the most remarkable of all cold spells to hit the British Isles occurred in mid-November 1919. Overnight minima below −10°C were widespread and early on the 15th Braemar in Grampian registered a temperature of −23.3°C – easily the lowest temperature on record in November and a value only exceeded in five winters since. Other notable minima in this unprecedented early taste of winter included −21.7°C at

Figure 13.4 Frozen sea at Pegwell Bay, near Ramsgate, Kent, on the morning of 17 February 1986. This month was dominated by easterly airflow, and the Central England mean temperature was below zero.

Perth and −21.1°C at Balmoral (Grampian) and West Linton (Borders). Indeed, the maximum at Balmoral on 14 November 1919 was a mere −10°C. The cold was a little less severe in the rest of the country, but even so −12.7°C occurred at Scaleby, near Carlisle, −9.4°C at Rhayader in Wales, −12.2°C at Lisburn in what is now Northern Ireland, and −11.1°C at Markree Castle in Co. Sligo.

The temperature has fallen below freezing in every month of the year at some locations, including as far south as the Thames valley and also in southern Ireland where Mostrim in Co. Longford scraped

−0.3°C on 8 July 1889.[10] A cold summer's night under clear calm anticyclonic conditions is, however, often followed by a long sunny day and hence some of the more remarkable diurnal ranges in temperature have occurred after notable low summer minima. These include the greatest 12-hour range in temperature ever recorded in the British Isles which occurred at Rickmansworth in Hertfordshire on 29 August 1936. The overnight temperature dropped to 1.1°C, but within nine hours the air temperature had risen to 29.4°C, a range of 28.3°C. One of Scotland's greatest twenty-four hour ranges in temperature occurred

Table 13.6 Lowest daily minimum temperature for each month since 1900 in the British Isles

Month	°C	Date	Location
January	−27.2	10th, 1982	Braemar, Grampian
February	−25.0	23rd, 1955	Braemar, Grampian
March	−22.8	14th, 1958	Logie Coldstone, Grampian
April	−15.0	2nd, 1917	Newton Rigg, Cumbria
May	−9.4	4th and 11th, 1941	Lynford, Norfolk
		15th, 1941	Fort Augustus, Highland
June	−5.6	9th, 1955; 1st, 1962	Santon Downham, Norfolk
		3rd, 1962	Dalwhinnie, Highland
July	−2.5	15th, 1977	St Harmon, Powys
		9th, 1986	Lagganlia, Highland
August	−4.5	21st, 1973	Lagganlia, Highland
September	−6.7	26th, 1942	Dalwhinnie, Highland
October	−11.7	28th, 1948	Dalwhinnie, Highland
November	−23.3	14th, 1919	Braemar, Grampian
December	−27.0	30th, 1995	Altnaharra, Highland

Table excludes high-level stations such as Ben Nevis observatory and Cairngorm.

Table 13.7 Five coldest, or equal coldest, years and seasons in the Central England Temperature series. See Appendix D for a full listing of these data

Spring	°C	Summer	°C	Autumn	°C	Winter*	°C	Year	°C
1837	5.6	1725	13.1	1786	7.5	1684	−1.2	1740	6.8
1770	6.0	1695	13.2	1740	7.5	1740	−0.4	1695	7.3
1695	6.0	1816	13.4	1676	7.5	1963	−0.3	1879	7.4
1799	6.1	1860	13.5	1782	7.7	1814	0.4	1698	7.6
1782	6.1	1907	13.6	1692	7.7	1795	0.5	1694	7.7
1701	6.1	1823	13.6	1688	7.7			1692	7.7
				1675	7.7				

* Winter refers to the year in which January fell

in very different circumstances, between 24 and 25 January 1958 at Kincraig (Highland), where the temperature rose from −21.1°C to 4.3°C thanks to the advection of tropical air behind a **warm front**.

The five coldest years for each season in the Central England Temperature series are shown in Table 13.7. Cold summers include 1816, which became known as the 'year without a summer' (see Figure 13.5). This is often attributed to the volcanic eruption of Tambora in April 1815.[11] Cool and wet summers seem to have become rarer in the last twenty years, but that does not mean we have escaped unpleasant cold summer days. In July 1980 the daytime temper-

ature remained below 15°C at Manchester until the 9th, and an unusually active summer **cold front** on 9 July 1993 led to some particularly low lunch-time temperatures, including 7.5°C at Whipsnade in Bedfordshire and 9.2°C at Benson in Oxfordshire (the lowest since 5 July 1920), while the day maximum was just 8.7°C at Cape Wrath in northern Scotland.[12]

DELUGE AND DROUGHT

Probably the wettest inhabited places in the British Isles are Peny-Gwrhyd Hotel (Snowdonia), Seathwaite

Darkness

I had a dream, which was not all a dream.

The bright sun was extinguish'd, and the stars

Did wander darkling in the eternal space,

Rayless, and pathless, and the icy earth

Swung blind and blackening in the moonless air;

Morn came and went – and came, and brought no day,

And men forgot their passions in the dread

Of this their desolation; and all hearts

Were chill'd into a selfish prayer for light:

And they did live by watchfires – and the thrones,

The palaces of crowned kings – the huts,

The habitations of all things which dwell,

Were burnt for beacons; cities were consumed,

And men were gather'd round their blazing homes

To look once more into each other's face;

Figure 13.5 Lord Byron's poem 'Darkness' written at Geneva during 1816 and inspired by the very cold, cloudy and wet summer of that year following the eruption of the Tambora volcano in April 1815.

Farm (Borrowdale, Cumbria) and Kinlochquoich Lodge (Highland), each with an average of close to 3,120 mm of precipitation per year. At the other extreme, Gray's Thurrock (Essex) and Sheerness (Kent), with averages over the 1961 to 1990 period of about 510 mm and 530 mm per year respectively, have the dubious distinction of being the driest towns. Averaging precipitation in this way hides huge day-to-day, week-to-week and year-to-year variations, and it is at the extremes of temporal variability that precipitation intrudes on our comfortable, modern daily living, whether it be through a prolonged drought, a sudden violent downpour or a lengthy wet spell.

For those living in Lynmouth on 15–16 August 1952 there was such an intrusion, and it was devastating. The inhabitants of this small town on the

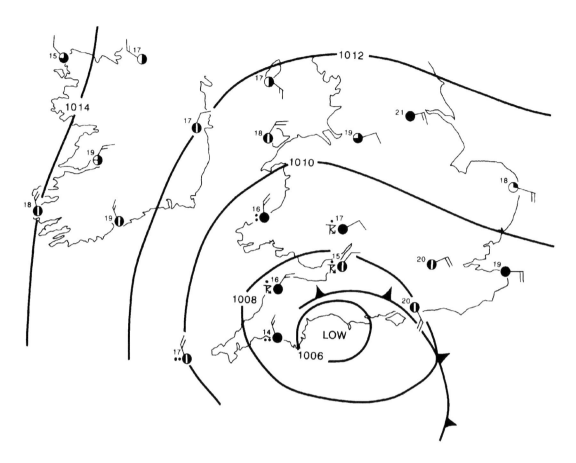

Figure 13.6 Synoptic observations and pressure pattern over southern Britain at 1500 GMT on 15 August 1952 – the Lynmouth flood.

northern flank of Exmoor had little warning of the disaster about to happen. The weather in the past month had been barely noteworthy in this part of Devon. A dry, warm spell in the second half of July was followed by an unsettled and thundery start to August. At midday on 12 August a small **depression** was located in the North Atlantic at 47°N, 34°W. Initially it moved east-south-east with an unremarkable central pressure of 1,007 hPa, but on the 14th and 15th it rounded an upper **trough** and swung north-east towards the West Country (see Figure 13.6). Associated thundery rain began to fall on the adjacent moors at 11.30 a.m. on 15 August. The depression was slow moving and as a conse-

quence the rain continued to fall with no real respite for twenty-two hours. Although much of the rain was not unusually heavy, it came down in torrents for a couple of hours in the late afternoon, and then again from about 9 p.m. to 11 p.m. Rain-gauges at Chivenor and at Wootton Courtenay, on Exmoor, indicate a peak six-minute rainfall intensity approaching 50 mm per hour during these periods.[13] The gauge at the remote spot of Longstone Barrow had been filled by 230 mm of rainwater between 4 p.m. on the 14th and 11.30 a.m. on the 16th, of which 229 mm was credited to the rainfall day of the 15th by the subsequent official investigation. At the time this was the third highest daily rainfall

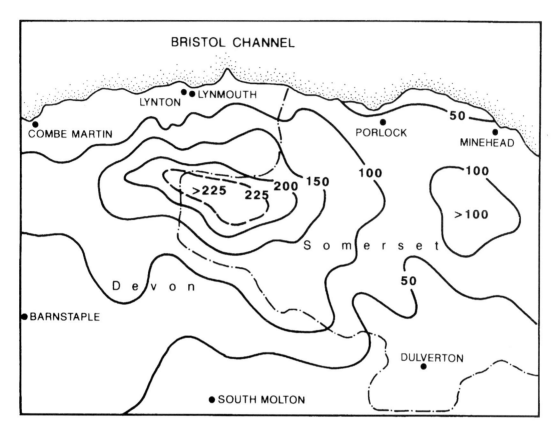

Figure 13.7 The distribution of rainfall over Exmoor (in mm), 14–15 August 1952 – the Lynmouth flood.

total ever recorded in the British Isles, and it has been speculated that at Simonsbath, in the heart of Exmoor, close to 300 mm of rain could have fallen during the storm.[14] The distribution of rainfall over Exmoor during this event is shown on Figure 13.7.

Lynmouth's geographical location is rather precarious; the northern slopes of Exmoor descend rapidly to the coast – by 450 m over a distance of 7 km – and the moor's drainage in this area is achieved by the East and West Lyn rivers which meet at Lynmouth just before reaching the sea. Given these circumstances it is hardly surprising that the consequences of this storm were catastrophic. During the night of 15–16 August enormous amounts of water surged down the two rivers sweeping a wave of water and debris through the town, which was devastated

(see Figure 13.8). Twenty-three buildings were demolished with seventy others seriously damaged; 420 people lost their homes; twenty-nine bridges in the area were swept away; 130 cars were washed out to sea, and thirty-four people lay dead.[15] This was the highest British Isles death toll from a flash flood in modern times.

Three years later, on 18 July 1955, another part of the country had a rainstorm to rival any these islands have seen in modern times, albeit without the tragedy Lynmouth experienced. The small village of Martinstown in Dorset (formerly Winterbourne St Martin) had a cloud-burst of truly tropical intensity. It began in mid-afternoon and continued with a short evening lull until about midnight, before tailing off during the early hours. An astonishing 279.4 mm

Figure 13.8 Damage in the centre of Lynmouth following the flood of 15 August 1952.

Table 13.8 Notable twenty-four hour precipitation totals

Amount (mm)	Location	Date
279.4	Martinstown, Dorset	18 July 1955
242.8	Bruton (Sexey's School), Somerset	28 June 1917
241.3	Upwey (Friar Waddon), Dorset	18 July 1955
238.8	Cannington, Somerset	16 August 1924
238.4	Loch Sloy Main Adit, Strathclyde	17 January 1974
228.6	Long Barrow, Devon	15 August 1952
228.6	Upwey (Higher Well), Dorset	18 July 1955
215.4	Bruton (King's School), Somerset	28 June 1917
213.1	Timberscombe, Somerset	28 June 1917
211.1	Rhondda (Lluest Wen Reservoir), Glamorgan	11 November 1929
211.1	Upwey (Elwell), Dorset	18 July 1955
208.3	Kinlochquoich, Highland	11 October 1916
204.0	Seathwaite, Cumbria	12 November 1897
203.2	Camelford, Cornwall	08 July 1957
200.7	Bruton (Pitcombe Vicarage), Somerset	28 June 1917
200.7	Wynford House, Dorset	18 July 1955

Table 13.9 Notable short-duration precipitation totals in the British Isles

Minutes	Amount (mm)	Location	Date
5	31.7[a]	Preston, Lancashire	10 August 1893
10	50.0[a]	Wisbech, Cambridgeshire	28 June 1970
15	55.9[a]	Bolton, Gtr Manchester	18 July 1974
15	38.1	Guernsey (St. Martin's Road)	14 August 1914
18	45.2	Arundel, W. Sussex	06 August 1956
20	63.5[a]	Hindolveston, Norfolk	11 July 1959
20	53.8	Canterbury (Harbledown), Kent	07 August 1875
25	67.0	Pershore, Worcestershire	11 June 1970
25	63.5	Silchester House, Hampshire	24 June 1933
30	80.0	Eskdalemuir, Dumfries	26 June 1953
30	73.7	Cambridge	22 July 1880
50	76.2	Acton, Gtr London	07 August 1960
55	89.9	Eskdalemuir, Dumfries	26 June 1963
60	92.2	Maidenhead, Berkshire	12 July 1901
70	80.0	Littleover, Derbyshire	09 July 1981
75	102.0	Wisley, Surrey	16 July 1947
75	94.0	Ilkley, W. Yorkshire	12 July 1900
105	152.4	Hewenden Reservoir, W. Yorks[b]	11 June 1956
120[a]	130.6	Knockholt, Kent	05 September 1958
120[a]	193.2	Walshaw Dean, W. Yorkshire[c]	19 May 1989
150	171.0	Hampstead, Gtr London	14 August 1975
170	159.0	Revesby Reservoir, Lincolnshire	07 October 1960
180	116.6	Louth (Elkington Hall), Lincs	29 May 1920
270	216.0	Cannington, Somerset	16 August 1974

[a] Estimated value or time.
[b] Now accepted after a long period of doubt. See M. Acreman, 'Extreme rainfall in Calderdale 19 May 1989', *Weather*, vol. 44, pp. 438-46, and J.M. Nicholls, 'Letter to the editor', *Weather*, 1990, vol. 145, p. 156.
[c] Authenticity of observation doubted by UK Met. Office. See V.K. Collinge, J. Thielen, J.F.R. McIlveen, 'Extreme rainfall at Hewenden Reservoir, 11 June 1956', *Meteorological Magazine*, 1992, vol. 121, pp. 166-71.

of rain was measured, making it the highest twenty-four-hour rainfall total on record in the British Isles, and beating the previous record (set at Bruton, Somerset, 28 June 1917) by 36.6 mm. Other notable one-day rainfall totals are listed on Table 13.8.

The twenty-four-hour value for Martinstown is enormous, but over shorter periods the precipitation intensity can be even greater. Probably the most intense precipitation ever recorded in the British Isles occurred at Preston in Lancashire on 10 August 1893. In just five minutes 32 mm of rain fell – an equivalent rate of 384 mm per hour, and one which would be remarkable even in the Tropics. For comparison, a list of other noteworthy short-duration precipitation totals in the British Isles is given in Table 13.9.

In contrast to these storms, many a month has gone by with serious precipitation shortages.[16] This can be seen in Table 13.10, which shows the five driest and wettest summers in the England and Wales Precipitation record discussed in Chapter 10. It is one of those ironies our climate can provide that the rainfall Martinstown had in just that one memorable July day in 1955 was marginally more than parts of east Kent received in the whole of the drought year of 1921 (compared to the annual average of 600 mm). Memorable dry periods which resulted in serious water shortages also occurred from the spring of 1975 until early September 1976, and from 1988 to 1992.

In 1975–1976 persistent **blocking** patterns kept rain-bearing fronts and depressions at bay for long

Table 13.10 Five wettest and driest years and seasons in the England and Wales Precipitation record (1766 to date). See Appendix D for a full listing of these data

Spring	mm	Summer	mm	Autumn	mm	Winter*	mm	Year	mm
Wettest:									
1782	363.0	1912	409.7	1852	455.8	1915	423.0	1872[a]	1284.9
1818	326.2	1879	409.2	1960	438.5	1990	421.0	1768	1247.3
1947	313.8	1829	396.3	1935	424.0	1877	418.0	1852	1213.0
1979	316.1	1860	370.8	1770	402.0	1995	415.0	1960	1194.8
1981	309.8	1768	362.2	1772	400.6	1994	388.0	1903	1180.3
Driest:									
1785	54.8	1995	67.0	1978	128.7	1964	88.9	1788	614.0
1893	71.5	1976	73.9	1817	131.9	1858	92.2	1921	629.0
1990	85.0	1800	91.5	1805	138.6	1891	105.2	1887	669.3
1938	95.0	1869	98.9	1788	145.8	1814	114.8	1854	672.9
1956	96.4	1818	102.6	1784	147.2	1934	118.7	1780	699.0

* Winter refers to the year in which January fell.
a Wettest of any twelve-month period.

periods, resulting in the driest twelve- and eighteen-month periods in England and Wales since 1820 (see Chapter 10). The drought was exacerbated because the preceding five-year period (1971 to 1975) was the driest since the middle of the last century, with average annual precipitation for the British Isles about 20 per cent below normal. The situation became serious during the spring of 1976 following a dry winter; reservoirs were rapidly drying and a massive 'save water' campaign was set into motion. The government even felt it appropriate to appoint a Minister for Drought (Mr Dennis Howell) and a Drought Bill was rushed through Parliament. Many people in South Wales, Southern England and East Anglia lost their household water supply and had to queue at standpipes. Crops withered and winter fodder was used to feed livestock because 'grazing meadows' were scorched to bare earth. But just after Dennis Howell had taken up his new position and when Britons really thought their changeable climate had gone forever, the heavens opened during the August Bank Holiday weekend, and autumn 1976 was one of the wettest on record (see Figure 13.9).

Autumn and winter precipitation is vital in order to replenish groundwater and remove soil moisture deficits. The record books are full of some quite

remarkable cold season water deficits, even in the wettest parts of the country. The winter of 1946–7 is infamous because of the severity of the frost and snow, but it was an 'easterly' season. That meant that western Scotland, in the **rain-shadow** of the Highlands, was relatively sunny and very dry. The combination of brisk east winds, low **dew points** and abundant sunshine left this part of the British Isles tinder dry by the end of February, and extensive peat fires provided an odd contrast to snow-bound land further to the south and east. Much of north-west Scotland had less than 5 per cent of normal February precipitation, and no measurable precipitation fell at Glencoe, Ardgour and Glenquoich – usually notoriously wet sites.[17] An 'easterly' February also occurred in 1963; western Scotland was again very dry with just a trace of precipitation at Fort William, situated at the foot of Ben Nevis.

Some of the longest periods with no precipitation are listed in Table 13.11. The outstanding event is the spring of 1893. No precipitation was measured at Mile End in east London for seventy-three consecutive days between 4 March and 15 May, while Twickenham in Middlesex was not far behind with a seventy-two day drought.[18] These figures assume that the sites had efficient observation, something now

Figure 13.9 Chew valley reservoir, Gloucestershire, in August 1976, following the driest eighteen months on record in England, and a year later in September 1977.

probably impossible to check. More recently, in the spring of 1990, parts of Buckinghamshire and Hertfordshire went thirty-nine days without any precipitation, and from mid-July to late-August 1995 mid- and east Kent were rainless for forty-two days.

SNOWFALL, BLIZZARDS AND ICE-STORMS

For a country situated at latitudes above 50°N it is remarkable the disruption even a slight snowfall can induce, particularly in southern Britain. Perhaps the reason lies in the great variability in amount and frequency of snow from year to year. Anyone in southern Ireland, southern England and South Wales born around 1970 saw little if any snow till they were seven or eight years old. Yet some winters produce snowstorm after snowstorm.

Deep snowfalls are commonplace in the hillier parts of the British Isles, but few can be as remarkable as the Dartmoor snowfall of 16 February 1929. The worst of it affected south-eastern slopes at an altitude of about 300 m, just to the west of Holme Chase. Snow fell solidly for fifteen hours and was

Table 13.11 Long rainless[a] periods in the British Isles (the table has one entry per event)

No. of days	Date	Location
73	04 March–15 May 1893	Mile End, Gt. London
59	14 August–11 October 1959	Finningley, S. Yorkshire
58	27 August–22 October 1969	Threadbare Hall, Suffolk
50	20 June–07 July 1976	Dale Fort, Dyfed[b]
49	01 April–19 May 1980	Warrington, Cheshire
48	19 September–05 November 1972	Langstamps, Essex; Pentlow, Suffolk
45	23 May–06 July 1975	Poole, Dorset[b]
44	29 July–10 September 1947	Wellingborough, Northamptonshire
43	30 September–01 November 1978	Flitwick, Bedfordshire
42	19 January–01 March 1947	Glenquoich, Highland
42	17 July–27 August 1995	Margate, Kent
41	12 February–24 March 1943	Scarborough, N. Yorkshire
40	29 March–07 May 1957	Ross-on-Wye, Herefordshire
40	31 March–09 May 1984	Witney, Oxfordshire
39	07 August–14 September 1973	Canterbury, Kent
39	22 April–31 May 1990	Little Chalfont, Bucks; Batchworth, Herts
38	09 June–16 September 1921	Patcham, W. Sussex
38	03 April–10 May 1938	Port William, Galloway; Limerick
37	23 August–28 September 1929	St Pancras, London
37	08 June–14 July 1949	Ventnor, Isle of Wight
37	24 January–01 March 1959	Herne Bay, Kent
37	11 May–16 June 1970	Bardney, Lincolnshire

[a] A rainless day is defined as one with ≤ 0.2 mm of measurable precipitation.
[b] And other stations in the same county.

accompanied by comparatively little wind, hence there was no drifting. Level snow by the end of the storm was close to 180 cm, probably making it the single largest one-day fall of snow ever seen at such a low elevation in these islands.[19] Some evidence of the validity of this measured snow depth comes from the nearest recording station, Dean Prior Vicarage, 8 km south of Holme Chase, where 62 mm of water equivalent was measured that day.[20] Local residents described the snow as the 'worst in living memory', and coming down as if 'shovelled'.

Although the term 'blizzard' has strict definitions which are rarely met in lowland Britain, in popular parlance it has come to mean a severe snowstorm or the combination of strong winds and heavy snow. Throughout the climate literature there are subjective definitions of a blizzard.[21] Some of the most disruptive blizzards are listed in Table 13.12. The winter of 1962–3, although often fine and very frosty,

had several extremely snowy spells. The really severe weather began over Christmas in what was an unusual weather pattern (see Figures 13.10 and 13.11). On Christmas Eve pressure was high over Denmark and easterly winds were feeding very cold polar continental air across much of England and Wales. Early on Christmas Day the temperature fell to −12.2°C at Warsop in Nottinghamshire. But developments were taking place to the north as a very strong build of pressure over Greenland and Iceland forced a cold **occlusion** southwards across Scotland, with Arctic air following. This front pushed slowly south across England and Wales on Boxing Day and, with pressure continuing to fall along the front, heavy snow fell widely. In London and the Home Counties snow began falling around mid-afternoon and did not stop until early on 28 December 1962, by which time it lay 30 to 45 cm deep in Kent, Surrey, Sussex and Hampshire, and

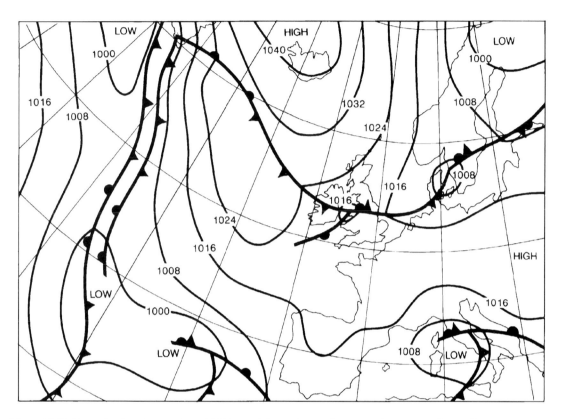

Figure 13.10 Synoptic situation at 1200 GMT on 26 December 1962, at the beginning of one of the coldest and snowiest winters in the British Isles since records began.

even on the Channel Islands 5 cm was recorded.

After a very brief respite the weather deteriorated on the 29th. With bitterly cold easterly winds covering the country, a depression moving slowly north from the Bay of Biscay provided a classic pattern for heavy snow (see Figures 13.12 and 13.13). This time the snow was accompanied by strong to gale force winds and by 31 December virtually the whole of England, Wales and southern Scotland was affected by badly drifted snow. The south was worst hit and many towns and villages were cut off by drifts 5 m high or more. Unlike so many British snowstorms, the snows of that particular winter were not followed by thaws and a return to 'westerly' weather. Freezing days and nights continued with the sort of persistent snow cover which is usually

associated with continental climates (see Figure 13.14). The more notable observations of 'number of days with snow lying at 9 a.m.' during the winter of 1962–3 included 86 at Braemar in Grampian, 70 at York, 58 at Regent's Park in London, 51 in Cardiff, 38 at Armagh in Northern Ireland, 30 at Cork in the Republic of Ireland and 17 on Guernsey.

More recent snowstorms to enter weather folklore include the Highland blizzard of January 1978, the 'Great West Country blizzard' of 18–19 February 1978, the often forgotten heavy snow which fell at intervals across north-east England and southern Scotland from 22 to 26 January 1984 with well over 30 cm in many places, the snowstorms which hit much of southern and eastern Britain in memorable 'easterly' spells in January 1987 and February 1991,

Table 13.12 A selection of disruptive snowstorms and blizzards in the British Isles

Date	Area worst affected	Maximum level snow depths
18–20 January 1881	Devon, Dorset, Hants, Isle of Wight	120 cm, Dartmoor; 80–100 cm, Isle of Wight.
4–8 December 1882	N. England, S. Scotland	80–100 cm, Berwickshire
9–13 March 1891	Devon, Cornwall, S. Wales	125–150 cm, Dartmoor
29 March 1901	N. England, N. Wales	100 cm, north Welsh hills
1–10 March 1909	SE England	50–70 cm, Kent
11–13 January 1913	N. England, S. Scotland	70 cm, Perthshire, Southern Uplands
1 April 1917	Ireland	100–120 cm, Cos Mayo, Galway and Clare
23–26 February 1933	Much of England, Wales, Ireland	70–80 cm, S. Wales, Yorkshire, Derbyshire
27–30 January 1940	Much of England, Wales and W. Scotland	60–100 cm, Yorkshire (120 cm, Sheffield), Derbyshire
18–20 February 1941	NE England, SE Scotland	100–120 cm, Co. Durham
22 January–16 March 1947	Most areas except W. Scotland	Numerous heavy falls; worst perhaps 4–6 March, 150 cm Clawddnewydd, Clwyd
14–18 January 1955	N. and NE Scotland	90–95 cm Grampian region
26–29 December 1962	S. England	50–75 cm, Hants, Surrey, Sussex, Kent
8–9 February 1963	Wales, SW England	150–160 cm, Tredegar, Gwent
16–20 February 1978	Devon, Somerset, Dorset, Glamorgan	85 cm, Nettlecombe Somerset
8–11 December 1981	Much of England and Wales	90–100 cm, N. Yorkshire
8–10 January 1982	S. England, S. Wales	60–80 cm Glamorgan
12–14 January 1987	S. and E. England	50–60 cm Mid and N. Kent
7–9 December 1990	Midlands	Up to 40 cm West Midlands
24–28 December 1995	Highlands and Islands	36 cm, Lerwick, Shetland
5–6 February 1996	Wales, W. England, SW Scotland	50 cm Southern Uplands

Table 13.13 A selection of out-of-season snowfalls in the British Isles

Date	Area most affected	Notes
29–31 October 1836	E. England, E. Anglia	25 cm, Newmarket, Suffolk
19–20 October 1880	E. England	20 cm, E. Sussex
12 May 1886	Most hilly areas	18 cm, Shropshire hills; 25 cm, Cumbria
10 June 1888	Scotland	15 cm, Highlands
7–11 July 1888	Hills in Scotland, N. England, Wales, N. Ireland	5 cm, Antrim hills
17–18 May 1891	Midlands, E. Anglia	15–18 cm, Norfolk
20–29 October 1895	Wales, W. Midlands	10–15 cm, Shropshire
23–25 April 1908	S. England, Midlands, E. Anglia	50–60 cm, Oxfordshire, Berkshire, Hampshire
27 April 1919	E. England	20–25 cm, London; 30–35 cm, E. Anglia
20 September 1919	Most hilly regions	5–10 cm, Dartmoor; 10–15 cm, Welsh hills

Table 13.13 continued

Date	Area most affected	Notes
15 May 1926	S. and C. England, Wales	10 cm, Cotswolds
25–29 October 1926	Most areas	5 cm, Hampstead; 30 cm, Perthshire
29 Sept–2 Oct 1927	Scotland, Cumbria	5–10 cm, Lake District
25–28 October 1933	Most areas	5–10 cm, Dartmoor and Lincs
31 October 1934	Most areas	21 cm, West Linton, Lothian
16–17 May 1935	Most areas including Scilly	12 cm, Tiverton, Devon; 60–90 cm, Pennines
8–9 May 1943	Scotland, N. England, N. Ireland	15 cm, Douglas, Isle of Man; 15–20 cm, Highlands
24–26 April 1950	Wales, S. England	40–45 cm, Wiltshire to Kent
6 October 1974	Scotland, E. England	Snow showers, Home Counties
2 June 1975	Most areas	Snow showers to south coast; 15 cm, Grampian
23–26 April 1981	Midlands northwards	60 cm, Derbyshire Hills
2–4 June 1991	Scotland, Cumbria	Frequent snow showers
13–14 May 1993	S. Scotland, N. Ireland	30 cm, Southern Uplands

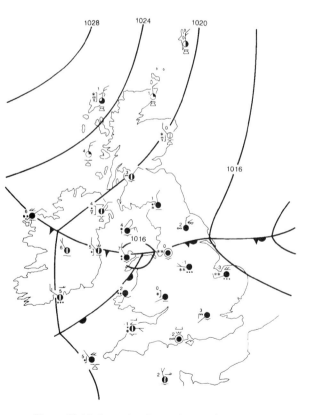

Figure 13.11 Synoptic observations and pressure pattern over the British Isles at 1200 GMT on 26 December 1962.

and the 1995 Christmas snowstorms in northern and eastern Britain which were unusually severe in the Northern and Western Isles.

Memorable snowfalls are not restricted to the winter·months. Some of the more spectacular out-of-season snowfalls are listed in Table 13.13. Perhaps the most outrageous of these happened early in June 1975. A blast of unseasonably cold air swept down from the Arctic, the northerly wind bringing hail, sleet and snow showers all the way to the south coast. This is the latest date in the season snow has fallen so widely across southern Britain since reliable records began. Snow actually lay on low ground in parts of Scotland, northern England and the Midlands, although it did not last long and just four days later the first of what turned out to be many summer heat waves began.

In North America one type of winter weather is feared even more than a snowstorm, but in the British Isles its rareness means few are aware of the dangers it can bring. This is freezing rain, which at its worst may constitute an ice-storm. When a snowflake can no longer be supported by updraughts within a cloud it begins to fall to earth. The flake will melt on descent if it passes through a sizeable layer of air with a temperature above freezing point, and will therefore reach the ground as rain or sleet.

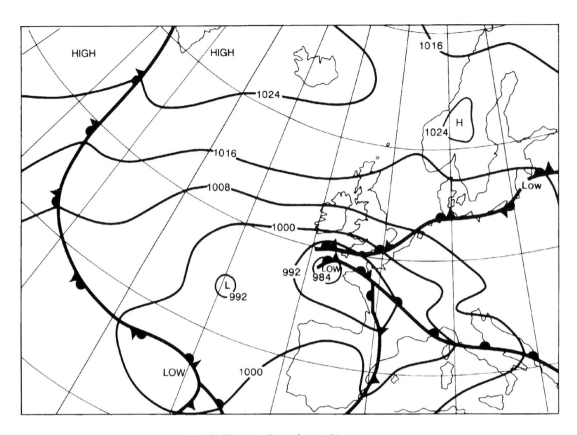

Figure 13.12 Synoptic situation at 0000 GMT on 30 December 1962.

If an inversion exists beneath the 'warm' layer the air near to the surface can still be well below 0°C. In this case the rain or sleet can freeze instantly on contact with any exposed surface, creating what is known as **glaze**.

Undoubtedly the most severe ice-storm to affect the British Isles occurred in January 1940, but, because of wartime censorship of the media, few people outside the affected area knew about it. Between the 26th and 29th much of the country was hit by a very bad snowstorm. Level snow depths were commonly over 30 cm with drifts between 2 and 4 m. In some parts of the country, however, freezing rain rather than snow was the main cause for concern. The areas hit by the ice-storm stretched across Kent, Essex, Dorset, Hampshire and Sussex and as

far north as the west Midlands and North Wales. On 27 January, many people watching the rain falling outside believed a thaw had arrived, but with temperatures mostly between –2°C and –4°C the rain was freezing on contact with exposed surfaces, encasing power lines, tree branches, roads and footpaths in clear ice. Eyewitness reports tell us that travelling by foot or vehicle became extremely hazardous, and as the volume of ice increased, power lines and tree branches came crashing down. A strong south-east wind led to many south- and east-facing windows becoming completely frozen shut. People living in Hampshire and Sussex described the night of the 27th–28th as very eerie, with an almost constant crashing of tree branches fooling some into believing it was a bombing raid.

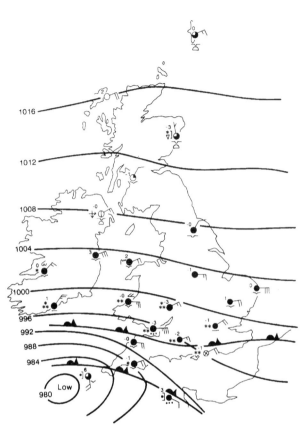

Figure 13.13 Synoptic observations and pressure pattern over the British Isles at 0600 GMT on 30 December 1962.

Conditions like January 1940 are rare over the British Isles, but a rather less dramatic episode on 20 January 1966 led to 245 road accidents, thirty-six seriously hurt, 201 slightly hurt and four fatalities; indeed the London Metropolitan Police reported accidents up 70 per cent on normal levels.[22] In March 1969 an ice-storm affected the southern Pennines. This event hit the news because the sheer weight of ice caused the Emley Moor television transmitter to collapse, creating serious difficulties for Yorkshire Television.[23] More recently, the winter of 1995–6 had several mostly minor freezing rain events, but severe enough to close the M4 motorway in South Wales on 30 December.

SEVERE GALES AND RECORD GUSTS

Deep depressions from across the Atlantic regularly track close to or over the British Isles, bringing frequent gales to coastal areas in the north and west and also to hilly regions (see Chapter 11). For many parts of central and eastern England a wind speed of more than 34 knots for at least a ten-minute period (i.e., a gale) is comparatively rare. London, for example, averages about two gale days a year compared with nearly fifty days at Lerwick in Shetland.[24] This means many stormy days come and go without note in the windier and least populated areas, but on those occasional days when storm-force winds affect our major towns and cities, disruption ensues.

The infamous 'Great Storm' which struck most of southern England and East Anglia with insufficient warning during the early hours of 16 October 1987 is a prime example. By popular consent this was the worst wind storm to hit southern Britain since 1703, not because of the number of storm-related deaths (comparatively low at eighteen), but because the wind speeds across the affected area were quite unprecedented in modern times. It is the strongest gusts which do much of the damage and some of the more notable peak gust speeds include 100 knots at Shoreham in Sussex, 94 knots at Langdon Bay near Dover, 92 knots at Ashford in Kent and even 82 knots in the centre of London – nearly 17 knots higher than any previous gust measured in the capital. Calculations made on the basis of this storm suggest such wind speeds in southern Britain have a **return period** of over 200 years.[25]

The damage in the three or four hours the storm was at its peak was quite unprecedented.[26] An estimated fifteen million trees were lost, cars and lorries were crushed by falling trees, walls were blown down, power and telephone lines came crashing down leaving thousands of homes in south-east England without electricity. Rail and road links were badly disrupted, the biggest problem being the inability of staff to travel to work. On the south coast of England hundreds of boats were smashed and a bulk carrier capsized in Dover harbour. All in

Figure 13.14 Drifting snow at Annfield Plain, County Durham, February 1963.

all, at 1995 prices, about £2,000 million worth of damage was brought about by an Atlantic depression which was not unprecedentedly intense, but just happened to be on an unusual track (see Figures 13.16 and 13.17, pp. 290, 291).

Table 13.14 catalogues some of the more notable wind storms to strike the British Isles, while Table 13.15 lists the top gust speeds on record. The highest death rate of any documented storm to hit the country occurred on 7–8 December 1703 (Gregorian calendar). The storm hit southern England and South Wales; approximately 9,000 people lost their lives, many of them at sea. Indeed it is believed that England lost more ships to this storm than in any battle up to that time.[27] Estimated surface wind speeds may have been as much as 90 knots and stronger than that in gusts and squalls.

In comparison, the highest gust speed ever recorded in the British Isles was 150 knots, measured automatically on the summit of Cairn Gorm on 20 March 1986. The record for a low-level station is 126 knots at Kinnaird Head lighthouse at Fraserburgh, Grampian, on 13 February 1989. Other exceptional peak gusts occurred on 16 February 1962 as a deep depression tracked eastwards to the north of Scotland. North-east England and Scotland were worst affected with registered gusts including 103 knots at Lowther Hill, Strathclyde. But this date is remembered most by the residents of Sheffield. Mean wind speeds reached 65 knots in Sheffield with a peak gust of 83 knots; part of the city suffered extensive structural damage with close on 7,000 houses damaged, three people killed and 250 injured. Strangely, while Sheffield was being battered, sur-

Figure 13.15 North Fleet, Kent, after the 'Burns' Day' storm of 25 January 1990.

October 1987, affected most of the country and struck during daylight hours (see Figure 13.15). Although peak wind speeds were generally lower than those during the October storm, the damage was greater and forty-seven deaths were attributed to the severe winds. In what was a remarkably windy winter, another severe windstorm hit the country on 26 February – a further fourteen people were killed.

The second half of the twentieth century has produced a surprising revelation about the climate of the British Isles – the relatively high frequency of tornadoes (see Figure 13.18, p. 292). The Tornado and Storm Research Organisation have shown that there are an average of about twenty days during the year when tornadoes are reported in the United Kingdom. Fortunately, most of these intense, localised storms are weak by comparison to those in the Midwest of the United States, but occasional substantial tornadoes have occurred causing considerable damage. For example, on 21 May 1950 a single tornado left a 100 km trail of damage stretching from Buckinghamshire through Bedfordshire and Cambridgeshire and ending in north Norfolk. Worst hit was Linslade, now in Bedfordshire, where fifty houses were demolished, a bakery destroyed and farm outhouses lifted bodily and dumped some distance away. Four years later on 8 December 1954, it was the turn of north-west London to be hit by a tornado. At Gunnersbury tube station half the roof was ripped off, injuring six people. At nearby Acton a car was reportedly lifted 15 feet off the ground, and with debris flying through the air a further six people were hurt.

The most startling tornado outbreak on record in the British Isles happened on 23 November 1981 when 105 separate tornadoes were reported, and this came during a twelve-day spell in which 152 tornadoes were observed.

Statistics show that a tornado can occur in any month of the year, but autumn is the favoured season and the afternoon the most likely time of day. The most susceptible areas are eastern England, with Bedfordshire having the greatest reported frequency of tornadoes per unit area.

rounding areas had substantially lower wind speeds. Investigations revealed that exaggerated wave motion induced by uplift over the Pennines was to blame. Airflow was compressed into the bottom of a wave trough right over Sheffield, leading to increased wind flow through the city.[28]

Widespread high wind speeds during the so-called 'Burns' Day' storm of January 1990 included a maximum gust of 94 knots at Aberporth, an exposed site in west Wales.[29] This storm, unlike that of

Table 13.14 A selection of notable wind storms in the British Isles

Date	Area worst affected	Notes
7–8 December 1703	S. England and Wales	≅ 9,000 dead, most at sea.
6–7 January 1839	Ireland and NW Britain	≅ 400 dead, v. severe in Ireland. 'Night of the big wind'.
28 December 1879	Most of the British Isles	Train destroyed as Tay Bridge collapses, much loss of life.
14–15 October 1881	E. Scotland and E. England	129 dead in the Eyemouth, Berwickshire, fishing fleet disaster.
28–29 October 1927	W. Ireland, W. and N. Wales, NW England	Sea surge flooding W. Wales, NW England. Five drowned at Fleetwood. Ten lost at sea off coast of Co. Mayo.
17 December 1952	Scotland, N. and E. England	A 96-knot gust at Cranwell Lincs; 91 knots at Speke, Liverpool.
31 January–1 February 1953	North Sea coasts of England and Scotland	Severe North Sea storm surge with NW–NNW gales. 109-knot gust on Orkney. 350 drowned in English coastal floods.
16–17 February 1962	NE England and Scotland	WNW–NW gales. Exaggerated wave-motion caused huge damage in Sheffield. 103-knot gust at Lowther Hill, Strathclyde.
11–12 January 1974	Ireland	Sw'ly gust of 94 knots at Cork. Co. Cork and Co. Sligo affected by severe sea-surge flooding.
2–3 January 1976	Ireland, England and Wales	28 dead, much damage. 91-knot gusts at Wittering and Coventry in the Midlands. £176 million insurance cost at 1976 prices.
16 October 1987	S. England and E. Anglia	90–100-knot gusts caused enormous damage to buildings and trees in the early hours. 18 dead.
8–9 February 1988	Ireland, W. England	6 dead in severe w'ly gale.
25 January 1990	Much of British Isles	'Burns' Day' storm. Severe daytime sw'ly gales led to 47 deaths. Widespread damage with huge insurance claims.
26 February 1990	Much of British Isles	More severe damage and 14 dead.

Notes: Largely extracted from Lamb.[25]
Gregorian (new style) calendar only.

SMOGS, FOGS AND RECORD SUNSHINE

Human intervention contributed the key element to Britain's most deadly meteorological event since the 'Great Storm' of 1703. It happened during a quiet spell of anticyclonic weather in early December 1952; no great energetic weather systems crashing across the country were to blame. Instead it came in the form of a fog – a thick pungent acidic fog, often called a 'pea-souper' by Londoners and now usually referred to in the British Isles as smog.

Until the 1950s, the end result of industrial and residential coal burning was the flux of tonnes of soot and large concentrations of sulphur dioxide straight into the atmosphere at or near street level (see Chapter 12). During spells of strong westerly

Table 13.15 Highest peak gust speeds recorded in the British Isles since 1900 (at sites below 300 m above mean sea-level). Single highest gust speed in any one event is shown

Knots	Location	Date
126	Fraserburgh, Grampian	13 February 1989
118	Kirkwall, Orkney	07 February 1969
109	Fair Isle	16 January 1990
101	Tiree, Inner Hebrides	26 February 1961
100	Fair Isle	14 November 1978
100	Shoreham, Sussex	15 October 1987
98[a]	Foynes Island, Co. Limerick	18 January 1945
98	Tiree, Inner Hebrides	31 January 1957
98	Malin Head, Co. Donegal	16 September 1961
98	Stornoway, Western Isles	12 December 1962
97[a]	Quilty, Co. Clare	27 January 1920
96	Scilly, Isles of Scilly	06 December 1935
96	Cranwell, Lincolnshire	17 December 1952
96	Claremorris, Co. Mayo	27 January 1974
96	Butt of Lewis, Western Isles	31 December 1991

[a] Top limit of station's **anemometer** reached; higher speeds may have gone unrecorded.

Table 13.16 Sunniest and dullest months at Kew and Stornoway. The Kew record comprises Kew Observatory from 1876 to 1980 and Kew Royal Botanic Gardens from 1981 to date. Stornoway's record is from 1881 to date

	Kew				Stornoway			
	Hours	Year	Hours	Year	Hours	Year	Hours	Year
January	92	1984	15	1885	66	1963	12	1907
February	115	1988	19	1947	106	1975	22	1943
March	183	1907[a]	57	1888	172	1911	55	1938
April	245	1984[b]	79	1920	230	1962	95	1944
May	320	1989	114	1932	294	1994	93	1944
June	302	1975	105	1909	278	1970	92	1942
July	334	1911	103	1888	227	1917	57	1939
August	279	1947[c]	109	1912	262	1899	75	1942
September	224	1911	64	1945	175	1903	70	1965
October	160	1959	51	1894	135	1898	34	1921
November	104	1889	22	1897	73	1968	19	1881
December	72	1986	0	1890	54	1935	6	1884

[a] Provisional figures indicate March 1995 may have been sunnier.
[b] April 1990 equally as sunny.
[c] Provisional figures indicate August 1995 may have been sunnier.

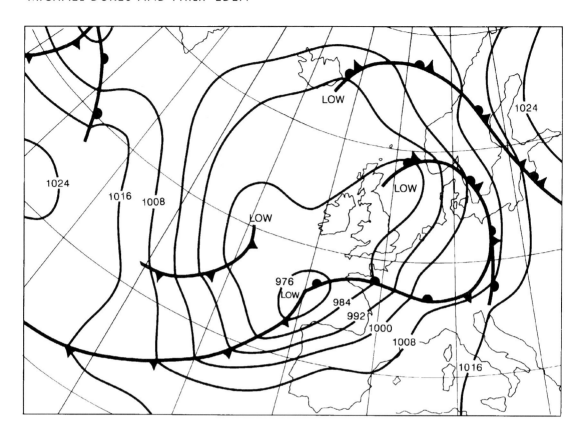

Figure 13.16 Synoptic situation at 1200 GMT on 15 October 1987 – the day before the 'Great Storm'. Although causing extensive damage over southern England, this October storm comes quite a long way down the list of storms affecting the British Isles when ranked by absolute severity (see Chapter 11).

winds and frequent rains this was a manageable problem, at least in the British Isles. But during episodes of quiet high pressure weather a combination of stagnant air and a strong temperature inversion trapped the gases and soot close to ground level and concentrations would increase day by day until the weather changed. Particulate matter, including soot, is **hygroscopic**; the particles encourage water vapour to condense on them before the air is truly saturated. Periods of anticyclonic weather in winter with associated light winds and lack of cloud already provide excellent conditions for the formation of fog. Add to this the huge concentrations of artificial hygroscopic nuclei such as soot and conditions are ideal for thick, persistent smog.

The period 5–9 December 1952 provided perfect conditions for smog across south-east England and the London area. During this spell of weather, continuous thick smog affected large parts of south-east England, and it was mostly very cold with persistent frost. At its worst the smog contained enormous quantities of black smoke, and sulphur dioxide combined with water droplets and oxygen to produce sulphuric acid. Visibility was officially reported below 20 m within the Greater London area, and unconfirmed reports of visibility below 10 m were common. Smoke and sulphur dioxide concentration levels were greatest in Lambeth, Westminster, Southwark and in the City, where daily values averaged as much as fifteen times normal early 1950s background levels.[30]

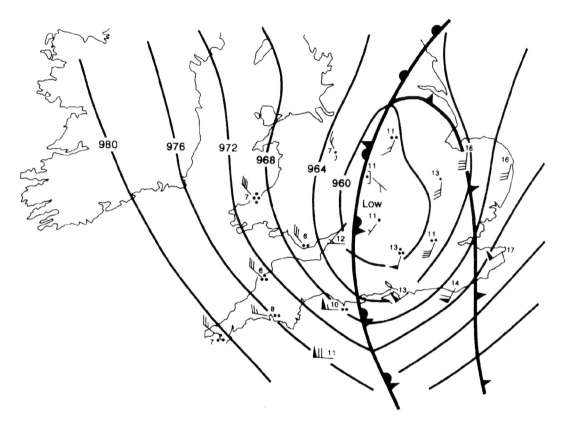

Figure 13.17 Synoptic observations and pressure pattern over southern Britain at 0400 GMT on 16 October 1987 – the 'Great Storm'.

Throughout the smog, there was much disruption to traffic, and football matches were postponed, but it was the elderly and chronically ill who suffered most. The death rate from bronchitis was ten times the average and that from influenza seven times the average. In all, an estimated 4,000 people died in the old London County Council area as a result of the smog, with a further 2,000 deaths in Middlesex, Essex, Kent, Surrey and Hertfordshire. Many of those who died were already weak, and with temperatures close to or a little below freezing throughout the event, hypothermia may also have been a contributing factor. On 5 July 1956 the Clean Air Act reached the statute books and the effects were quickly beneficial. In early December 1962, another smog enveloped the London area, and lasted even longer. Sulphur dioxide levels were actually even higher than the 1952 event, but smoke concentrations were 60 per cent lower. This time the smog-related death rate was reduced to 700 in the London County Council area.

The longest well-documented London smog lasted four and a half days between 26 November and 1 December 1948. But 'pea-soupers' were not confined to Britain's largest city; Glasgow and Clydeside suffered in November 1925 and November 1977, and Birmingham, Manchester and Liverpool experienced a very severe smog episode between the 19th and 28th of November 1936.[31] More recently, Dublin endured serious smog episodes in November 1988 and November/December 1989. Since the introduction of the various Clean Air Acts, water-

Figure 13.18 A newspaper engraving of a waterspout seen off Worthing on Sunday, 21 August 1864. The engraving is based on a sketch drawn by the Rev. F. Piggott of Worthing College, Sussex.

based smogs and fogs have become a much rarer visitor to urban areas (although, conversely, photochemical smog has increased markedly, especially since 1976; see Chapter 12). The worst fogs in recent years, therefore, have been mostly confined to suburban and rural areas; the great urban fogs are hopefully a thing of the past.

If the frequency of autumn and winter fogs has been reduced in urban districts, it is reasonable to expect the number of sunshine hours (and perhaps also temperature) to have risen. Nowhere is this more true than in London. Prior to 1950 the average number of hours of bright sunshine in November at Kew Observatory was about 52 hours. In the period 1961 to 1990, the average shot up to 69 hours. The figures are even more impressive further towards central London. At Kingsway in Central London, the

increase over the same period has been a staggering 32 hours, nearly doubling the 1921 to 1950 average of 38 hours. If evidence were needed that this is mainly due to the Clean Air Act and not to a significant change in the character of the November weather, then at Rothamsted in the Hertfordshire countryside the increase in sunshine hours over the same period is just eight hours.

Sunnier autumns and winters in the large cities mean sunnier years, and this change is worth bearing in mind when reviewing sunshine records from urban areas. It is also worth noting that the measurement of sunshine is not an exact science. Most observers use a Campbell–Stokes recorder (Figure 13.19). This method has its critics because it over-records during periods of intermittent bright sunshine, under-records hazy sunshine in winter, and fails to record

CONCLUSIONS

The weather in the British Isles will continue to be an endless source of fascination and will doubtless provide us with further extremes and new records over the years to come. Continued global warming may mean more hot summers and at some stage the absolute maximum temperature record for the islands may again be broken.

It is often mistakenly believed, however, that a warming of our climate will mean no more cold winters and an end to chilly summers. Winters with a high 'northerly' or 'easterly' component are still bound to occur, bringing widespread snow and very low temperatures. A perfect example of a warm year with temperature extremes from each end of the spectrum occurred in 1995. The summer was hot, dry and very sunny and October was on a par with the mildest ever measured. Yet the first half of December brought an easterly outbreak with some snow and sub-zero maximum temperatures, especially in southern Britain. Then the last week of December produced a memorable spell of severe weather. Snowstorms caused havoc in northern and eastern Britain (especially Shetland, north Highland and Lewis) around Christmas. After the snow ceased, clear and calm anticyclonic conditions allowed intense radiative cooling over the deep powdery snow-cover. This produced some exceptionally low minima including −27.0°C at Altanaharra, Sutherland, on the 30th − a new December extreme for the United Kingdom and only 0.2°C above the record for any month. Daytime temperatures were also very low with a maximum of −15.0°C at Braemar, Grampian, on the 29th. And all this occurred in the third warmest year in the Central England Temperature record!

NOTES

1   B.G. Wales-Smith, 'Revised monthly and annual totals of rainfall representative of Kew, Surrey for 1697–1870 and an updated analysis for 1697–1976', *Hydrological Memorandum*, Meteorological Office, vol. 43, 1980.

2   B.J. Moffitt, 'The effects of urbanisation on mean

Figure 13.19 A Campbell–Stokes sunshine recorder, the standard measuring instrument for hours of sunshine.

for at least twenty minutes after sunrise and before sunset even on the clearest of days. A world-wide change in the method of recording sunshine, however, would make historical comparisons very difficult, even with a substantial overlap period.

The dramatic contrasts in amounts of sunshine that occur from one month to another is a notable feature of the climate of the British Isles. This can be seen to an extent from Table 13.16, which shows the sunniest and dullest occurrences for each month at Kew (London) and Stornoway (Western Isles).

temperatures at Kew Observatory', *Weather*, vol. 27, 1972, pp. 121–9.

3   See G.P. Northcott, 'A reassessment of the highest temperature during July 1959', *The Meteorological Magazine*, 1991, vol. 120, pp. 230–2; M. Acreman, 'Extreme rainfall in Calderdale 19 May 1989', *Weather*, 1989, vol. 44, pp. 438–46; P. Eden, 'Letter to the editor', *Weather*, 1991, vol. 146, pp. 330–1.

4   P. Eden, *Weatherwise*, London, Macmillan, 1995, pp. 114–18.

5   J.D.C. Webb and G.T. Meaden, 'Britain's highest temperatures by county and by month', *Weather*, 1993, vol. 48, pp. 282–91.

6   See G.T. Meaden and J.D.C. Webb, 'Britain's highest temperatures for every day of the year, 1 January to 31 December', *The Journal of Meteorology*, 1984, vol. 9, pp. 169–76, and S. Burt, 'The exceptional hot spell of early August 1990 in the United Kingdom', *International Journal of Climatology*, 1992, vol. 12, pp. 547–67.

7   R. Brugge, 'Low daytime temperatures over England and Wales on 12 January 1987', *Weather*, 1987, vol. 42, pp. 146–52.

8   G. Manley, 'The great winter of 1740', *Weather*, 1958, vol. 13, pp. 11–17.

9   J.D.C. Webb, 'Britain's lowest temperatures for every date from 1 January to 31 May and 1 October to 31 December', *The Journal of Meteorology*, vol. 10, pp. 45–51.

10 P.K. Rohan, *The Climate of Ireland*, Dublin, Meteorological Service, 1986, 2nd edn, p. 141.

11 See M.R. Rampino and S. Self, 'Historic eruptions of Tambora (1815), Krakatau (1883), and Agung (1963), their stratospheric aerosols and climatic impact', *Quaternary Research*, 1982, vol. 18, pp. 127–43, and H. Stommel and E. Stommel, 'The year without a summer', *Scientific American*, 1979, vol. 240, pp. 176–86.

12 M.V. Young, 'An Exceptional summer cold front – 9 July 1993', *Weather*, 1994, vol. 49, pp. 249–53.

13 See A. Bleasdale and C.K.M. Douglas, 'Storm over Exmoor on August 15, 1952', *The Meteorological Magazine*, 1952, vol. 81, pp. 353–67, and W.A.L. Marshall, 'The Lynmouth flood', *Weather*, 1952, vol. 7, pp. 338–42.

14 Bleasdale and Douglas, op. cit.

15 R. Stirling, *The Weather of Britain*, London, Faber and Faber Ltd, 1982, pp. 69–70.

16 Prior to 1961 the UK Met. Office defined an absolute drought as a fifteen-day period in which no day records more than 0.2 mm of rain. In fact, drought is also influenced by potential evapotranspiration, and a cool dry misty couple of weeks of 'anticyclonic gloom' in November, for example, hardly constitutes a real drought, so the definition was dropped.

17 Meteorological Office, *Monthly Weather Report February 1947*, London, HMSO, 1947.

18 R. Bushell, 'Droughts in 1893', *British Rainfall 1893*, 1894, vol. 33, pp. 137–47.

19 E.G. Bilham, *The Climate of the British Isles*, London, Macmillan and Co. Ltd, 1938, p. 241.

20 Meteorological Office, 'Heavy falls on rainfall days in 1929', *British Rainfall 1929*, 1930, vol. 69, pp. 54–82.

21 See L.C.W. Bonacina, 'Snowfall in the British Isles during the half century, 1876–1925', *British Rainfall 1927*, 1928, vol. 67, pp. 260–87; L.C.W. Bonacina, 'Snowfall in the British Isles during the decade 1925–1936', *British Rainfall 1936*, 1937, vol. 76, pp. 272–92; L.C.W. Bonacina, 'Snowfall in the British Isles during the decade 1936 to 1945', *British Rainfall 1948*, 1949, vol. 88, pp. 209–17; L.C.W. Bonacina, 'Snowfall in the British Isles during the decade 1946 to 1955', *British Rainfall 1955*, 1956, vol. 95, pp. 219–30; L.C.W. Bonacina, 'Chief events of snowfall in the British Isles during the decade, 1956–1965', *Weather*, 1966, vol. 21, pp. 42–6; M.C. Jackson, 'A classification of the snowiness of 100 winters – a tribute to the late L.C.W. Bonacina', *Weather*, 1977, vol. 32, pp. 91–7.

22 See G. Parker and A.A. Harrison, 'Freezing drizzle in south-east England on 20 January 1966', *Meteorological Magazine*, 1967, vol. 96, pp. 108–12, and T.H. Kirk, 'The synoptic situation attending an occurrence of freezing drizzle', *Meteorological Magazine*, 1967, vol. 96, pp. 112–15.

23 J.K. Page, 'Heavy glaze in Yorkshire, March 1969', *Weather*, 1969, vol. 24, pp. 486–95.

24 Stirling, op. cit., p. 209.

25 H.H. Lamb, *Historic Storms of the North Sea, British Isles and Northwest Europe*, Cambridge, Cambridge University Press, 1991, pp. 189–91.

26 See G. Hill, *Hurricane Force*, London, Collins, 1988; B. Ogley, *In the wake of the Hurricane*, Westerham, Kent, Froglets Publication Ltd, 1988; B. Ogley, I. Currie and M. Davison, *The Kent Weather Book*, Westerham, Kent, Froglets Publication Ltd, 1991; *Weather* (Special Issue), 1988, vol. 43, pp. 65–142.

27 Lamb, op. cit., pp. 59–72.

28 C.J.M. Aanensen, 'Gales in Yorkshire February 16, 1962', *Geophysical Memoir*, vol. 108.

29 E. McCallum, 'The Burns' Day storm, 25 January 1990', *Weather*, 1990, vol. 45, pp. 166–73. This storm is incorrectly referred to as the 'Burns' Day' storm since the Scottish festival being used as a marker is actually 'Burns' Night'.

30 C.K.M. Douglas and K.H. Stewart, 'London fog of December 5–8 1952', *The Meteorological Magazine*, 1953, vol. 82, pp. 67–71.

31 C.W.G. Daking, 'Unusual persistence of fog', *The Meteorological Magazine*, 1942, vol. 71, pp. 252–3.

GENERAL READING

P. Eden, *Weatherwise*, London, Macmillan, 1995.

H.H. Lamb, *Historic Storms of the North Sea, British Isles and Northwest Europe*, Cambridge, Cambridge University Press, 1991.

P.K. Rohan, *The Climate of Ireland*, Dublin, Meteorological Service, 1986, 2nd edn.

R. Stirling, *The Weather of Britain*, London, Faber and Faber Ltd, 1982.

Part 4

FORECASTING THE FUTURE

If you can look into the seeds of time,
And say which grain will grow and which will not,
Speak then to me.
William Shakespeare, *Macbeth*

14

FORECASTING THE BRITISH ISLES WEATHER

Clive Pierce, Michael Dukes and Graham Parker

Some are weatherwise, some are otherwise.
Benjamin Franklin, *Poor Richard's Almanac*, February 1735

INTRODUCTION

The weather has enormous impact on our lives, from the relatively insignificant effects on our daily routine, to the catastrophic destruction of whole communities during severe storms, floods and droughts. It is therefore not surprising that from the very earliest times humans have tried to understand and predict the changing moods of the weather with, it must be said, varying degrees of success.

In the first of three sections in this chapter, some milestones in our understanding of science and meteorology are reviewed in the context of their impact on weather forecasts. Written evidence from southwest Asia shows that the ancient Egyptian and Babylonian civilisations were interested in weather prediction as long ago as 3000 BC. With a few exceptions, little headway was made in our understanding of meteorology, however, until the rebirth of science in western Europe during the Renaissance period. Indeed, the greatest advances in the field have come during the last one hundred years or so with the formulation of a complete set of equations to describe the physical behaviour of the atmosphere.

In the second part of the chapter the modern-day preparation of weather forecasts is examined in some detail. Today, weather forecasting is an enor-

mously complex and costly exercise, involving the co-ordinated effort of many tens of thousands of people world-wide. A huge number of weather observations from the land, sea, air and space, are collected and distributed globally on a daily basis. These data are assimilated into elaborate, computer-based models of the Earth's atmosphere that generate forecasts of atmospheric behaviour for hours, days or weeks ahead. A review of data collection and processing is followed by a look at approaches to modelling of the atmosphere. A discussion of the range of techniques employed in forecasting weather from a few minutes to one month ahead is followed by a summary of the range of services supplied by the United Kingdom Met. Office and private weather consultancies in the British Isles.

In keeping with the chronological organisation of the chapter, the third and final section briefly explores the future of weather forecasting from a number of different perspectives. The issue of atmospheric predictability and its implications for Numerical Weather Prediction are reviewed in the light of projected developments in science and technology, and the theoretical limitations imposed by chaos theory. A look to the future of weather forecasting would not be complete without some reference to economic issues. Perhaps most noteworthy is the rising cost of collection and distribution of weather data world-wide, and its potential implications for weather service provision on both national and international scales.

A HISTORICAL PERSPECTIVE

The evolution of weather forecasting, both in theory and practice, is intimately linked to the growth of meteorology as a science (see Table 14.1 for a chronology of some key landmarks). The foundations of the modern science of meteorology were laid by the Greek civilisation in Europe between about 650 and 300 BC (see Figure 14.1). Aristotle (384–322 BC) was one of the first Greek philosophers to apply systematic observation to the study of meteorology, work which culminated in the oldest comprehensive treatise on the subject. In his book, *Meteorologica*,[1] written in 340 BC, Aristotle demonstrated rudimentary understanding of the hydrological cycle, clouds and the vertical structure of the atmosphere.

Unfortunately, this highly productive period in the early history of meteorology and other natural sciences was relatively short-lived. With the fall of the ancient Greek civilisation in the first century BC, scientific investigation generally declined. Nevertheless, the wealth of knowledge accumulated by the Greeks was not lost. During the centuries of comparative intellectual darkness that followed, the Muslims translated Aristotle's *Meteorologica* into Arabic, whence it could later be translated into Latin by western scholars. Remarkably, about 1,500 years after the fall of the ancient Greeks, *Meteorologica* was still the unquestioned authority on meteorology in western Europe. Early in the Renaissance period, meteorology was not a formally recognised discipline. Rather, it was the concern of individuals whose livelihoods, in one way or another, were dependent upon the vagaries of the weather. As such, it had more to do with accumulated weather lore, astrology and superstition than any scientifically based understanding of atmospheric behaviour. With the development of a scientific method founded on quantitative, rather than qualitative observation, Aristotle's *Meteorologica* was gradu-

Figure 14.1 The Tower of Winds (Athens), built about 40 BC by the astronomer Andronikos of Kyrrhos.

ally superseded, and the study of weather slowly transformed from a philosophy to an applied physical science.

Experiments with primitive thermometers and barometers during the first half of the seventeenth century demonstrated that variations in atmospheric temperature and pressure accompanied changes in the weather. This inspired a good deal of conjecture regarding the behaviour of the atmosphere. The idea of a weather observing network to study these changes soon followed. As early as 1653, Grand Duke Ferdinand II of Tuscany organised a weather observing network in northern Italy. Despite the tremendous vision of this early scientific patron, it was not until the end of the eighteenth century that such networks were more widely introduced (see Chapter 7).

Table 14.1 Some landmarks in the evolution of the science of weather forecasting.

Date	Description
3000 BC	The Babylonians use astrological signs to forecast the weather.
650 BC	The ancient Greeks begin systematic observations of the weather.
340 BC	The Greek philosopher, Aristotle (384–322), publishes *Meteorologica*.
1250 AD	An English scholar, Roger Bacon (1214–94), advocates experiment-based scientific method.
1337	William Merle, an English rector, begins the earliest known systematic record of the weather.
1550	An Italian mathematician, Girolamo Cardano (1501–76), disputes Aristotle's work.
1600	Galileo Galilei (1564–1642) demonstrates the use of a thermoscope, the forerunner of the thermometer.
1637	A contemporary of Galileo, René Descartes (1596–1650), publishes a philosophy of scientific method.
1643	An Italian mathematician, Evangelista Torricelli (1608–47), devises the barometer.
1653	The renowned scientific patron, Grand Duke Ferdinand II of Tuscany (1610–70), establishes the first weather observing network in northern Italy.
1663	The English physicist, Robert Hooke (1635–1703), proposes a *Method for Making a History of the Weather*.
1684	The German mathematician, Gottfried Wilhelm Leibnitz (1646–1716), publishes a description of a branch of mathematics known as calculus.
1687	Isaac Newton (1642–1727), an English mathematician and physicist, publishes *Philosophiae Naturalis Principia Mathematica* which includes the three natural laws of motion. He also demonstrates the use of calculus.
1688	The English astronomer, Edmund Halley (1656–1742), draws a map showing wind direction between 30°N and 30°S.
1735	George Hadley (1685–1758) publishes a theory of atmospheric motion involving a cyclic circulation, later known as the Hadley circulation.
1746	The first mathematical study of atmospheric motion by the Frenchman, Jean Le Ronde d'Alembert (1717–83).
1755	The German mathematician, Leonhard Euler (1707–83) develops equations of fluid motion using Newton's second law, and partial differential equations.
1771	Johann Heinrich Lambert (1728–77) proposes the world-wide taking of weather observations.
1813	An American, Amerigo Avogadro (1776–1856) uses the Ideal gas equation relating pressure, volume and temperature of a gas.
1823	Siméon Denis Poisson (1781–1840), student of Pierre Simon LaPlace, derives an adiabatic equation of volume changes for gases.
1827	The German meteorologist, Wilhelm Heinrich Dove (1803–76), formulates a theory of mid-latitude storm development ('Law of Gyration'), based upon the concept of conflicting equatorial and polar air currents.
1835	Gustave-Gaspard Coriolis (1792–1843) describes the absolute acceleration of moving bodies in a system rotating about a vertical axis.
1840	The electric telegraph, conceived by Samuel Morse in 1832, comes into operational use.
1841	James Pollard Espy (1785–1860), an American meteorologist, publishes his convective theory of storms, based upon adiabatic cooling, and latent heat release.
1843	William C. Redfield (1789–1857) explains the rotary motion of storms in terms of centrifugal force. The American, Elias Loomis (1811–89), draws the first synoptic chart.
1848	James Glaisher (1809–1903) organises the first synoptic weather observing network in England. The first weather reports are published in the *Daily News*.
1854	Rudolph Clausius (1822–88) introduces the concept of entropy. Admiral Robert FitzRoy (1805–65) is appointed head of the Meteorological department of the Board of Trade.
1857	Professor Christophorus Buys Ballot (1817–90), a Dutch meteorologist, discovers that low atmospheric pressure is on the left of an observer standing with their back to the wind in the northern hemisphere. The opposite is true in the southern hemisphere.

Table 14.1 Continued

Date	Description
1858	A German meteorologist, Hermann von Helmholtz (1821–94), introduces the concept of *vorticity*.
1859	The American scientist, William Ferrel (1817–91), formulates the first mathematical equations of atmospheric motion on a rotating earth. Urbain Le Verrier (1811–77), a Frenchman, draws isobars on a weather map.
1860	The British Meteorological Office publish the first weather summaries in *The Times*. Storm warning cones are erected in British ports early in 1861.
1864	Theodore Reye (1838–1919), a Swiss mathematician, publishes a paper on dry adiabatic and saturated adiabatic processes. Müller draws isallobars on weather charts.
1868	An American engineer, H. Peslin, constructs a thermodynamic diagram for use in tracing the motions of imaginary parcels of air undergoing dry and saturated adiabatic changes.
1883	Ralph Abercromby (1842–97), an English meteorologist, publishes his book, *Principles of Weather Forecasting*, in which he describes the characteristic weather associated with the passage of a depression.
1888	Wladimir Köppen (1846–1940), a German meteorologist, uses the thermal wind relationship to derive fields of pressure at different altitudes.
1898	A Norwegian mathematician and physicist, Vilhelm Bjerknes (1862–1951), describes his Circulation theorem, in which rotary atmospheric motion is caused by horizontal temperature contrasts (baroclinicity).
1906	Sir William Napier Shaw (1854–1945), director of the British Meteorological Office, describes the concept of air trajectory analysis. The invention of the Tephigram is also attributed to him.
1911	Sir William Napier Shaw publishes *Forecasting Weather* in which he summarises the current state of weather forecasting.
1921	The Norwegian, Vilhelm Bjerknes, and a colleague, Halvor Solberg (1895–1974), publish the wave or polar front theory of cyclone development.
1922	An English meteorologist, Lewis Fry Richardson (1881–1953), expounds the theory and practice of numerical weather prediction in his book, *Weather Prediction by Numerical Process*. H. Jeffreys simplifies the Primitive equations using *scale analysis*.
1929	A Frenchman, Robert Bureau, makes the first successful atmospheric sounding with a radiosonde.
1939	Carl-Gustav Rossby (1898–1957), a Swedish meteorologist, relates long (Rossby) wave motions in the mid-latitude troposphere to conservation of absolute vorticity.
1946	Purpose-built weather radars are used by meteorologists for the first time.
1950	Sutcliffe and Forsdyke demonstrate a relationship between jet streams and the development of cyclones and anticyclones. The UK Met. Office begins research into numerical weather prediction. The World Meteorological Organisation is born.
1960	The first meteorological satellite, TIROS 1, is launched from Cape Canaveral, Florida in North America.
1965	Numerical weather prediction in the UK Met. Office takes over from manually produced weather forecasts.
1968	The World Weather Watch programme, formulated by the WMO in 1963, commences.
1974	The plotting of synoptic weather charts is automated in the UK Met. Office. Weather observations are fed automatically into its numerical models.
Early 1980s	Numerical weather forecast accuracy at the UK Met. Office is improved significantly with the introduction of a more powerful computer and a ten-layer model.
1988	Nine-member Ensemble numerical weather forecasts begin at the UK Met. Office.
1994	The UK Met. Office commence NWP operations on the Cray C90 supercomputer. This is capable of performing 16,000,000,000 floating point calculations per second.

Towards the end of the seventeenth century, two vital scientific breakthroughs, one in physics, the other in mathematics, were made by the great English scientist, Isaac Newton. The first of these was the invention of mechanics, that branch of physics concerned with bodies in motion. The second was the development of a mathematical tool known as calculus. Together, these later formed the basis for the quantitative study of fluids in motion and thus were necessary prerequisites to the emergence of meteorology as an applied physical science.

One of the earliest attempts to explain the motions of the atmosphere was made by the English astronomer, Edmund Halley. In 1688, he summarised the wind regime in the tropical and sub-tropical oceans in the form of a chart illustrating the direction of the air flow with wind arrows. This was based upon the accumulated weather observations of numerous ships plying established trade routes. Halley believed this general circulation to be driven by the sun's heat and this led him to consider the role of thermal **convection** in atmospheric motion. Perhaps his most important contribution to meteorology, however, was the introduction of the concept of winds as a general circulation of air over the Earth's surface. Almost fifty years later, Halley's ideas were taken a step further by another English scientist, George Hadley. In 1735, Hadley published the results of his study of the atmosphere in *Philosophical Transactions*.[2] His explanation of the atmospheric circulation outlined by Halley not only involved thermal convection, but also took account of the Earth's rotation. He envisaged a thermally direct 'cell' in which warm air rising on the equator moved polewards at high altitude, later to descend in the sub-tropics and return to the equator as a cooler, near surface air current.

In retrospect, Hadley's work on the general circulation was one of the milestones in dynamical meteorology. It generated considerable interest amongst the early meteorological theoreticians. One of these, the German mathematician, Leonhard Euler, presented the first correct mathematical explanation of fluid flow in 1755. This involved Newton's second law of motion and the concept of partial differential equations. Euler's studies, and later research by

Gustave-Gaspard Coriolis on the effects of the Earth's rotation in 1835, paved the way for William Ferrel's derivation of the equations of atmospheric motion on a rotating Earth. First published in 1859, these formulae represented another important advance in the theory of meteorology. Even so, they received relatively little attention until some decades later. Ferrel's work was overshadowed by developments in other areas of meteorology which were of some practical worth to meteorologists of the day.

First, the growth in popularity of weather observing from the late eighteenth century onwards led to the discovery that periods of bad weather in mid-latitudes were, more often than not, associated with cyclonic disturbances originating over the Atlantic Ocean. Consequently, most theoretical exploration in meteorology during the nineteenth century was focused on identifying the causes of cyclone development. In particular, the discovery of the first law of thermodynamics led to the realisation that adiabatic changes (those in which sensible heat energy is neither gained nor lost) played an important role in atmospheric motion. The American meteorologist, James Espy, recognised this fact in formulating his convective theory of cyclone development during the 1830s.[3]

Second, the introduction of the electric telegraph early in the 1840s, made the generation of near real-time weather charts possible. The Englishman James Glaisher (see Figure 14.2), was the first person to demonstrate this when he organised the collection of weather reports for the *Daily News* in 1848.[4] With this facility for rapid data collection came the potential for producing the first observation-based short-range weather forecasts. This could be achieved by simple linear extrapolation of current weather, the assumption being that weather patterns would move in the direction of the prevailing wind. By the early 1860s, when the recently formed Meteorological Department of the Board of Trade (the forerunner of the United Kingdom Met. Office) began the first daily weather report and gale warning services, weather forecasting, still in its infancy, was essentially an observation-based practice. It had no means of exploiting the complicated theoretical work of Ferrel.

Figure 14.2 James Glaisher, FRS (1809–1905) who established the first network of reliable weather observatories in the British Isles.

In the years following 1861, the pioneering weather predictions made by the Met. Office under Admiral Robert FitzRoy received a mixed reception. Indeed, they came in for an increasing amount of criticism for their failing to provide timely warnings of bad weather. Consequently, weather forecasts were abandoned altogether for a number of years. Their unreliability is understandable in view of the extremely limited information and knowledge then at the disposal of meteorologists. Despite these setbacks however, the work of Christophorus Buys-Ballot, Ralph Abercromby, Napier Shaw and others, together with improvements in weather observing networks, saw that the impetus towards an improved weather forecast service was maintained through the

closing decades of the Victorian era and into the twentieth century. In 1911, Shaw published his book, *Forecasting Weather*, in which he succinctly summarised the practice of weather forecasting as follows, 'The business of the forecaster is to determine what type of barometric distribution is to be expected within the next twenty-four hours and to assign to it the appropriate weather.'[5] The current surface pressure distribution was predicted forward in time using a combination of **pressure tendency** information and forecaster experience. The weather associated with this new pressure distribution was forecast using Abercromby's conceptual models relating weather to high and low pressure areas.[6]

From Shaw's description it is apparent that weather forecasters still made little use of the growing body of meteorological theory. Yet the importance of both thermodynamic and dynamic influences on atmospheric motion was understood in quantitative terms well before the end of the nineteenth century. What was missing was a coherent theory of atmospheric motion embodying both thermodynamic and dynamic principles. Only then could realistic conceptual models of mid-latitude cyclones evolve, and these formed an essential part of practical weather forecasting.

A breakthrough came in 1898 when the Norwegian mathematical physicist, Vilhelm Bjerknes, made the link between thermodynamically and dynamically forced atmospheric motions in a powerful theorem later know as the Bjerknes circulation theorem.[7] The success of Bjerknes' approach lay in his ability to recognise that the density of the air could vary when atmospheric pressure was constant. This recognition led him to distinguish between two atmospheric states. These he described as **barotropic** and **baroclinic**. In short, the former represents an atmosphere in which there are no horizontal variations in temperature, whilst in the latter the converse applies.

Some time later, Bjerknes, and another Norwegian, Halvor Solberg, working at the Bergen School of Meteorology, put forward an entirely new and innovative conceptual model of cyclone development.[8] This had its foundations in the earlier theoretical work of Bjerknes and the analysis of detailed weather observations. Their model treated cyclones as transient features of the general atmospheric circulation, formed in baroclinic zones on the **polar front**. As such, they were associated with two distinct lines of **convergence** and weather discontinuity called **cold** and **warm fronts**. The adoption of the term 'front' in this context has its origins in the conceptual similarity between the movement of weather discontinuities in the atmosphere and the troop formations used in the First World War.

It would be difficult to exaggerate the importance of Norwegian polar front theory in terms of its subsequent impact on mid-latitude weather forecasting. Not only did it revolutionise the way meteorologists viewed the growth and decay of weather systems, but it also transformed practical weather forecasting. One of the key tasks of the weather forecaster soon became the identification of frontal systems on weather charts. Allied to this, the concept of the **air mass** was used to identify and distinguish broad categories of weather. Somewhat surprisingly, the United Kingdom Met. Office was slow to adopt these new techniques and it was not until 1935 that frontal and air mass analyses became two of the key tasks of British weather forecasters.[9]

During the 1920s and 1930s, the aviation-led growth in **radiosonde**-based upper air measurements enabled meteorologists to examine the three-dimensional structure of entire weather systems for the first time. Wind observations from the middle and upper troposphere revealed unforeseen wave-like motions with wavelengths of the order of 2,000 km. These so called **long waves** were not static features. They appeared both to move and amplify or contract over time. Furthermore, the positions of individual **ridges** and **troughs** were shown to be intimately related to the surface pressure distribution.

The recognition of a clear relationship between long waves in the upper air and the growth and decay of transient surface weather systems was another major step forward for twentieth-century meteorology. Evidently, if the behaviour of these waves could be forecast then, through well-established dynamical theory, so could corresponding changes in the underlying pressure fields. It was a Swedish

meteorologist, Carl-Gustaf Rossby, who finally demonstrated that the propagation of long waves was dependent on three factors: their wavelength, the poleward rate of change of the **Coriolis parameter**, and the mean **zonal flow** rate.

By 1945, knowledge of upper air motion had altered weather forecasting practice significantly. Predictions of changes in the contour patterns of **atmospheric thickness** and constant pressure charts were used, in conjunction with surface weather maps, to generate forecasts for up to forty-eight hours ahead.[10] Unfortunately, these techniques were very time-consuming, involving the manual analysis of large quantities of data. Furthermore, there was still a large subjective component to the forecasting process. Not surprisingly, weather forecasts were frequently wrong. Upon being told that the Met. Office got its forecasts right 60 per cent of the time, Winston Churchill is reputed to have said that, if he believed the opposite of the weather forecast to be true, he would be right almost as often!

The Second World War also saw the first use of ground-based radar for meteorological purposes. Through the operation of military radar during periods of bad weather, it was quickly discovered that areas of moderate or heavy precipitation generated detectable radar echoes. Later, this feature was employed to track belts of precipitation and identify convective storms. Since the 1950s, radars designed specifically for the detection of precipitation have come into widespread use (see Plate 8).

Numerical weather prediction

One of the most significant breakthroughs in weather forecasting this century has undoubtedly been the development of computer-based Numerical Weather Prediction (NWP) models. The potential of such an approach was first recognised by an English meteorologist, Lewis Richardson. In 1922, Richardson published a book, *Weather Prediction by Numerical Process*,[11] in which he demonstrated a method for predicting changes in surface atmospheric pressure over west Germany from an initial distribution of pressure over western Europe. This involved a numerical solution of the primitive equations: a set of fundamental equations governing atmospheric motion, described earlier by Bjerknes (see Box 14.1). Many thousands of calculations and months of hard work were necessary to produce a six-hour forecast – there were no computers or electronic calculators – yet the final answer proved to be completely wrong. Unknowingly, Richardson had used flawed mathematical techniques. Not to be deterred however, he envisaged a time when a co-ordinated army of 64,000 human computers would perform similar calculations quickly enough to keep ahead of the evolving weather.

Richardson was well ahead of his time. In fact, it was not until the 1950s that the advent of the electronic computer made the realisation of his dream possible. The first operational NWP model was introduced by the Swedish Military Weather service in 1954 and, with the invention of faster computers in the 1960s, such models became both more widely used and progressively more sophisticated. In the United Kingdom Met. Office, computer-based NWP forecasts were compared for accuracy and reliability against human forecasters before their full introduction in 1965.[12] Today, highly complex baroclinic NWP models, run on extremely powerful supercomputers, are used by National Meteorological Centres all over the world.

This computer-driven revolution finally freed meteorologists from the tedium of manual data analysis, a task which had previously been an integral part of weather forecasting. Nowadays, the role of the weather forecaster is more concerned with the assessment and interpretation of computer-based weather predictions and the preparation of intelligible forecasts for the lay person. These tasks draw on his or her experience and knowledge of local weather – human skills which are, at present, mostly beyond the capabilities of a programmed machine.

The Space Age Revolution

Prior to the launch of the first meteorological satellites in the early 1960s, weather forecasting for the large, data-sparse regions of the Earth was almost impossible. In particular, very few weather reports were available from the Pacific and Atlantic oceans, yet these were some of the most active areas in terms of weather. During the 1950s, North American experiments with rocket-mounted cameras demonstrated the enormous potential of high altitude weather observations for supplying information on the movement of hurricanes. This provided the impetus for the world's first meteorological satellite programme.

Between 1960 and 1965, North America launched ten Television and Infra-red Observation Satellites (TIROS I–X) into low altitude, quasi-polar orbits.[13] These carried television cameras designed to capture high resolution pictures of the Earth's cloudy atmosphere. On board, a facility called Automatic Picture Transmission relayed the images directly back to a Communications and Data Acquisition centre on the ground. For the first time, forecasters had access to real-time pictures of the atmosphere over hundreds of thousands of square kilometres.

This had an enormous impact on weather forecasting in low and mid-latitudes. Between July 1961 and December 1964, 118 tropical storms were tracked by TIROS satellites. Many of these would previously have gone unnoticed until they reached populated land areas. Now, evasive action to save life and property could be taken several days, rather than several hours, ahead. Over Europe and the North Atlantic, satellite pictures of frontal wave **depressions** confirmed the validity of many aspects of the Bjerknes–Solberg conceptual model. They also brought to light certain smaller-scale features of the atmospheric circulation hitherto unimagined. For example, the now familiar 'open cell' structure of shower clouds in polar maritime airstreams would probably not have been identified were it not for such pictures.

Today, global coverage of the Earth's surface and atmosphere is afforded by a host of **geostationary** and **sun-synchronous** satellites (see Figure 14.3). These are maintained by various member countries of the United Nations for the benefit of the World Meteorological Organisation's World Weather Watch

BOX 14.1 THE PRIMITIVE EQUATIONS

The primitive equations are a set of fundamental equations governing large-scale atmospheric motions, and form the theoretical basis for NWP models. They appear in a variety of configurations according to the choice of co-ordinate system and the mathematical notation used. Here, the equations are given in their component form. The time derivatives refer to changes at a point fixed relative to the rotating Earth. The three dimensions of the co-ordinate system are represented by x, y (the two orthogonal horizontal dimensions) and p (the vertical dimensions, atmospheric pressure).

$$\frac{\partial u}{\partial t} = -u\frac{\partial u}{\partial x} - v\frac{\partial u}{\partial y} - w\frac{\partial u}{\partial p} - \frac{\partial \phi}{\partial x} + fv + F_x \quad \text{(1a)}$$

$$\frac{\partial v}{\partial t} = -u\frac{\partial v}{\partial x} - v\frac{\partial v}{\partial y} - w\frac{\partial v}{\partial p} - \frac{\partial \phi}{\partial y} - fu + F_y \quad \text{(1b)}$$

$$\frac{\partial \phi}{\partial p} = -\frac{RT}{p} \quad \text{(2)}$$

$$\frac{\partial T}{\partial t} = -u\frac{\partial T}{\partial x} - v\frac{\partial T}{\partial y} + w\left(\frac{R}{c_p}\frac{T}{p} - \frac{\partial T}{\partial p}\right) + \frac{H}{c_p} \quad \text{(3)}$$

$$\frac{\partial w}{\partial t} = -\left(\frac{\partial u}{\partial x} + \frac{\partial v}{\partial y}\right) \quad \text{(4)}$$

$$\frac{\partial q}{\partial t} = w\frac{\partial q}{\partial p} + \left(u\frac{\partial q}{\partial x} + v\frac{\partial q}{\partial y}\right) + C \quad \text{(5)}$$

Equations 1a–b represent the local time rate of change of horizontal wind speed in the west–east (u) and south–north (v) directions. The first two terms on the right-hand side of each equation constitute the horizontal advection of the wind field; w is the vertical wind speed; ϕ is the geopotential, a measure of altitude; f is the Coriolis parameter; F_{x-y} represent the effects of friction.

Equation 2 is the hydrostatic equation. It represents the vertical component of atmospheric acceleration as a balance between the downward acceleration due to gravity and the upward acceleration due to the decrease of atmospheric pressure with height. R is the gas constant for dry air.

Equation 3 is the thermodynamic energy equation. The local time rate of change of temperature, T, is given in terms of advection of the temperature field, adiabatic energy transformations, and 'in situ' diabatic energy changes; c_p is the specific heat capacity of dry air at constant pressure; H is the diabatic (sensible) heat term.

Equation 4 is the continuity equation. It implies the conservation of mass. The sum of the terms inside the brackets on the right-hand side represent the horizontal stretching or squeezing of the air. This is balanced by a vertical expansion or contraction term on the left-hand side.

Equation 5 represents the conservation of moisture. The local time rate of change of specific humidity, q, is described in terms of convection, advection of the humidity field, and 'in situ' changes in the moisture content of the air (C).

programme (see below and Table 14.2a). Sophisticated remote sensing devices, including **radiometers** and **atmospheric sounders**, afford a wealth of information on land and sea surfaces, clouds, and the three-dimensional distribution of atmospheric temperature, humidity and composition (see Tables 14.2b–c). Such observations provide an increasingly important supplement to Earth-based weather measurements and are now routinely integrated in NWP models.

WEATHER FORECASTING TODAY

In this section, the modern preparation of weather forecasts is explored in some detail, including the gathering of weather observations, the use of computer-based weather models, the role of the meteorologist, and the range of forecast services available in the British Isles. Throughout, the reader should refer to Figure 14.4, which provides an overview of the weather forecasting process in the form of a flow diagram.

Figure 14.3 An infra-red image from NOAA 11, taken at 0340 GMT on 24 January 1990. A westerly air flow affects the British Isles on this day, categorised as 'W' in the Lamb Catalogue (see Appendix B). Supplied courtesy of the University of Dundee.

Table 14.2a Current operational meteorological satellites as of May 1995

Orbit	Satellites
Sun-synchronous	NOAA 12, 14
Geostationary	Meteosat 5 (0°W), INSAT 2 (70°E), GMS 4 (140°E), GOES 7 (135°W), GOES 8 (75°W)

Table 14.2b Sun-synchronous meteorological satellites

Launch	Satellite	Features	Measurements
1960–65	TIROS[a] 1–10 North American	TV camera, Automatic Picture Transmission, five-channel radiometer.	Daylight cloud cover. Cloud top and Earth surface temperatures. Water vapour.
1964–78	NIMBUS I–VII North American	Three TV camera system, temperature–humidity, microwave radiometers.	Day and night cloud cover. Vertical profiles of temperature, pressure and water vapour.
1966–69	ESSA[b] 1–9 North American	Amalgam of features from TIROS and NIMBUS series.	Daylight cloud cover. Earth–atmosphere radiation balance.
1970–76	NOAA[c] 1–5 North American	Very High Resolution Radiometer. Vertical Temperature Profile Radiometer.	Day, night cloud cover. Water vapour. Vertical temperature profiles every 12 hours.
1979–81	NOAA 6, 7	Two satellites in the same orbit, ninety degrees apart, giving two passes per day. Advanced Very High Resolution Radio-meter. TIROS Operational Vertical Sounder. Data Collection System.	Global cloud, ice and snow cover, day and night. Cloud top, Earth and sea-surface temperatures. Vertical profiles of temperature and humidity. Receipt and distribution of Earth-based observations.
1983–86	NOAA 8–10	Similar to NOAA 6, 7. Earth Radiation Budget Experiment.	Similar to NOAA 6, 7. Energy exchange measurements for Earth–atmosphere system.
1988–91	NOAA 11, 12	Enhanced versions of NOAA 8–10. The AVHRR on NOAA 11 failed in 1994.	Similar to NOAA 8–10.
1993–94	NOAA 13, 14	Enhanced versions of NOAA 11, 12. NOAA 13 failed after launch.	Similar to NOAA 11, 12.
2000	METOP 1 European	Enhanced imaging instruments, and a high precision sounder.	Global cloud cover, day and night. Land and sea-surface temperatures. Cloud-top temperatures. Vertical profiles of temperature and humidity. Cloud motion winds. Estimates of precipitation.

Notes:[a] Television and Infra-red Observation Satellite.
[b] Environmental Science Services Administration.
[c] National Oceanic and Atmospheric Administration.

Table 14.2c Geostationary meteorological satellites

Launch	Satellite	Features	Measurements
1966–74	ATS[a] 1–6 North American	Amalgam of features from TIROS and NIMBUS series.	Day, night cloud cover over visible Earth disc. Cloud top, earth surface temperatures. Water vapour.
1974–78	SMS[b] 1–2/GOES[c] 1–3 North American	Two-channel Visible-Infra-red Spin-Scan Radiometer. Data Collection System.	Day, night cloud cover over visible Earth disc. Cloud top, Earth surface temperatures. Cloud motion winds. Receipt and distribution of Earth-based observations.
1977–95	GMS[d] 1–5 Japanese	Two-channel Visible-Infra-red Spin-Scan Radiometer Addition of a water vapour channel on GMS 5. Data Collection System.	Similar measurements to those made by GOES 1–3 satellites.
1977–96	Meteosat 1–7 European	Automatic Picture Trans-mission. Three-channel imaging radiometer. Data Collection System.	Day, night cloud cover over visible Earth disc. Cloud motion winds. Cloud top, sea-surface temperatures. Water vapour. Receipt and distribution of Earth-based weather observations.
1980–87	GOES 4–7	VISSR Atmospheric Sounder. Data Collection System.	Similar to GOES 1–3. Vertical temperature and humidity profiles.
1983–94	INSAT 1–2 Indian	Multi-purpose: tele-communications, weather forecasting, search and rescue.	Data not freely available.
1994 – ?	GOES NEXT 8–12 North American	Eighteen-channel radiometer.	Similar to GOES 4–7. Vertical temperature and humidity profiles measured at forty levels.
1994–95	GOMS[e] 1–2 Russian	Two-channel radiometer. Addition of a water vapour channel on GOMS 2.	Similar measurements to those made by GOES 1–3 satellites.
2000	MSG[f] 1 European	Radiometers with 16 channels: 4 high resolution visible, 6 infra-red, 2 water vapour, and 4 pseudo-sounding.	Cloud cover and type. Fog and snow cover. Cloud top temperatures. Cloud motion winds. Sea- and land-surface temperatures. Detailed vertical profiles of temperature and humidity. Earth–atmosphere radiation budget data.

Notes: [a] Applications Technology Satellite.
[b] Synchronous Meteorological Satellite.
[c] Geostationary Operational Environmental Satellite.
[d] Geostationary Meteorological Satellite.
[e] Geostationary Operational Meteorological Satellite.
[f] Meteosat Second Generation.

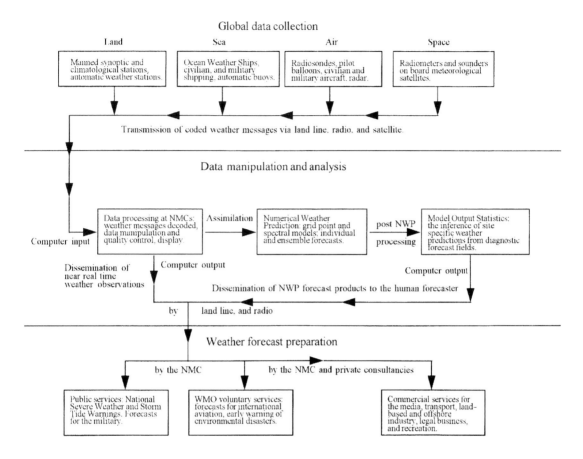

Figure 14.4 A schematic illustration of the modern-day preparation of weather forecasts.

The collection of weather data

With the introduction of the electric telegraph in the 1840s came the first opportunity for the collection of near real-time weather reports from widely separated observers and the possibility of generating up-to-date weather maps. A group of weather watching stations connected by telegraph was first established in the British Isles and France around 1860 and other developed countries soon followed suit. Thus began the weather monitoring network that now covers a large proportion of the Earth's surface. This so-called synoptic network is co-ordinated by the World Meteorological Organisation (WMO), a specialised agency of the United Nations, as part of an operation termed World Weather Watch.[14] The latter encompasses not just the collection of weather data, but their processing and dissemination and the provision of regional and global services by individual member countries of the United Nations.

The primary aim of World Weather Watch is to provide a regular and coherent record of the state of the atmosphere. This must be sufficiently detailed and accurate to allow the generation of useful weather forecasts for weather sensitive operations on land, sea and in the air. Consequently, the standardisation of observational practice is of the utmost importance.

The WMO provide detailed guidelines on such matters and these include the location of meteorological instruments and their use, times of observation and the order in which instruments should be read.[15]

Surface weather observations

At land-based synoptic weather stations, measurements of dry-bulb and wet-bulb temperature (the latter for calculating humidity), cloud amount and type, height of cloud base, weather type, visibility, wind direction and speed, atmospheric pressure, pressure tendency and past weather are made at regular intervals. Frequency of observation varies according to the type of station in question. Airports, for example, report the weather hourly, whilst others make measurements only at what are termed the main synoptic hours: 0000, 0600, 1200 and 1800 GMT.[16] In addition, there are also climatological weather stations, usually run by amateur meteorologists, and these make one observation a day at 0900 GMT.

Over the sea, the weather is monitored by merchant shipping, a few specialised weather ships, and an increasing number of unmanned, instrumented buoys with radio transmitters. Not surprisingly, the interpretation of marine weather reports is somewhat different to that applied to land-based reports. For example, an observation of wind speed and direction made on board a moving ship must be corrected for the ship's motion. The scarcity and uneven distribution of weather observations from sea areas have long posed problems for the weather forecaster – major weather systems such as tropical storms form over the sea, not over the land. The introduction of meteorological satellites, however, has done much to improve the situation.

Upper air weather observations

Weather reports above ground level come from a variety of sources. Radiosonde balloons make measurements of atmospheric pressure, temperature, relative humidity and wind up to altitudes of between 20 and 30 km, twice daily at 0000 and 1200 GMT. Smaller balloons, called pilot balloons, are utilised to measure cloud base height and high altitude winds. In common with surface weather reports, these so-called upper air observations are also reduced to compact coded messages for ease of communication.

The spatial coverage of upper air observations, whilst relatively dense over the British Isles and other wealthy countries, is very poor or non-existent over substantial areas of land and over the sea. These large gaps in the upper air network are partly filled with measurements made by automatic sensors attached to civilian aircraft. Unfortunately, the latter tend to be concentrated along the flight paths of the major airlines and are prone to errors. Today, the main source of weather data in remote parts of the world is the meteorological satellite. During the past three decades, satellite-based remote sensing instrumentation has become evermore sophisticated. As a consequence, weather observations from space are now of sufficient accuracy and reliability to be routinely used in NWP models. Furthermore, weather satellites form an integral part of the Global Telecommunications Network, facilitating the exchange of weather data around the Earth.

Weather data dissemination and processing

It should be apparent by now that an enormous quantity of data is potentially available to the meteorologist from the array of sources outlined above. The task of collating these data is evidently a huge one. At the point of observation – whether it be on land, at sea, or in the air – weather reports must be reduced to compact coded messages. These are then electronically transmitted, via satellite, radio or land line, to centres that co-ordinate their storage and subsequent processing. Box 14.2 demonstrates how a coded numerical weather message can be deciphered and plotted in symbolic form, and Figure 14.5 shows

Figure 14.5 A plotted surface weather chart for 0800 GMT, 4 October 1995. This day is categorised as 'CS' – cyclonic southerly air flow – in the Lamb Catalogue (see Appendix B). Reproduced courtesy of the UK Met. Office.

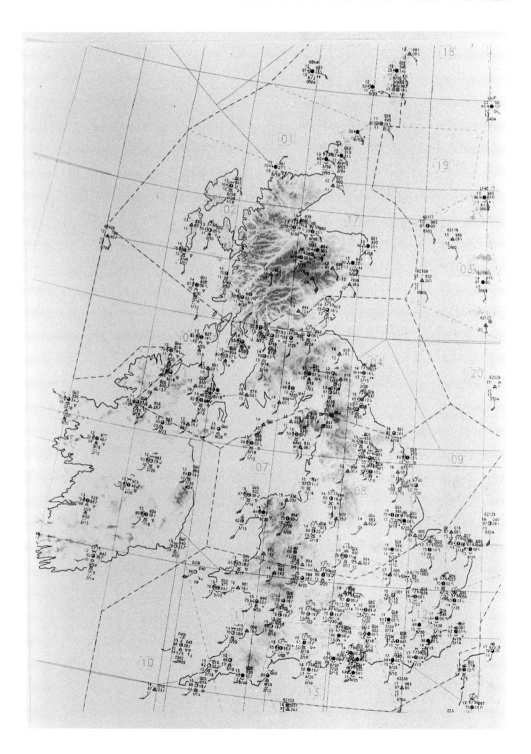

BOX 14.2 PLOTTING WEATHER CHARTS

Weather observations are transmitted to National Meteorological Centres in the form of coded messages comprising strings of five figure numbers. The structure of these messages is internationally agreed so that the information they contain can be decoded by anyone with a knowledge of the relevant codes. An excerpt from a coded message compiled by a land-based weather station in the United Kingdom is shown below. A brief explanation of each five digit group is provided beneath.

03853 41480 62213 10106 20041 40007 52020 70381 84901

03853   International index number 03 – British Isles, 853 – station number.

41480   4 – No rainfall data included in message, 1 – present weather included in message, 4 – coded height of lowest cloud 300–600 m, 80 – visibility 80–50 = 30 km.

62213   6 – Total amount of cloud observed = 6 eighths, 22 – wind direction 220°, 13 – speed 13 mph.

10106   1 – Indicator, 0 – sign is positive, 106 – dry-bulb temperature 10.6°C.

20041   2 – Indicator, 0 – sign is positive, 41 – dew-point temperature 4.1°C.

40007   4 – Indicator, 0007 – sea-level pressure 1000.7 hPa.

52020   5 – Indicator, 2 – pressure characteristic 'increasing', 020 – pressure tendency = 2.0 hPa.

70381   7 – Indicator, 03 – present weather = clouds developing, 81 – past weather = showers, partly cloudy.

84901   8 – Indicator, 4 – 4 eighths of cloud at 300–600 m, 9 – low-level cloud = cumulonimbus with anvil top, 0 – no medium-level cloud, 1 – high-level cloud = cirrus or 'mares' tails'.

These coded weather reports are normally displayed on charts in symbolic form. The above message would be drawn like this:

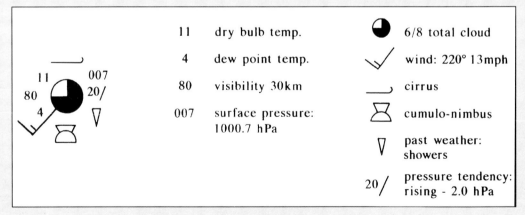

Figure 14.5 shows a plotted surface weather chart for 4 October 1995. The meteorologist would normally draw isobars, and weather fronts on this. In addition, significant features such as precipitation may be highlighted in colour, time permitting.

a plotted surface weather map for 0800 GMT on 4 October 1995. Each station circle and its associated symbols represents a separate weather observation.

Data collection tends to be co-ordinated by National Meteorological Centres and these, in turn, pass their information via the Global Tele-communication Network to World Meteorological Centres. Under WMO guidelines, the three World Meteorological Centres – Moscow, Washington and Melbourne – have the responsibility of issuing weather analyses and forecasts on a global scale. These products are used by a larger number of Regional and National Meteorological Centres in the preparation of weather forecasts for their designated regions of the Earth.[17]

NUMERICAL WEATHER PREDICTION MODELS

Meteorologists first experimented with computer-based numerical weather prediction models in the 1940s. Since then, computer technology has advanced to such an extent that these hitherto scarce and costly tools are now becoming more commonplace. They are used to make predictions of atmospheric behaviour on a wide range of space and time-scales, from the minute to minute growth and decay of individual clouds, to global changes in climate over millennia (see Chapters 15 and 16).

The majority of operational NWP models used in day-to-day weather forecasting are what are termed grid-point models. In these, the atmosphere is repre-sented by a number of vertical layers, each of which contains a network of grid points fixed relative to the Earth's surface. Each grid point is assigned certain values describing the state of the atmosphere at that location and these values are assumed to be representative of the rectangular box of air or parcel surrounding the grid point. When the model is run, time advances in a series of discrete steps. At each step the complete set of physical equations governing atmospheric behaviour is used to determine a new state of the atmosphere.

In the real atmosphere, motion occurs on a wide range of scales from the molecular to the synoptic

and the spacing between successive grid points, both in the horizontal and the vertical, will determine the scales of motion that can be resolved. Evidently, the issue of model resolution is of fundamental impor-tance to the weather forecaster. If the spacing of grid points is too coarse, then atmospheric disturbances having a significant impact on daily weather may not be modelled effectively. On the other hand, if the resolution is too fine, then the model may be too slow or too costly to run given the limitations on computing power. In reality, the choice of resolution is always a compromise, arrived at by considering such factors as the area covered by the model, the minimum scale of features that must be resolved and the power and affordability of available computer resources.

The behaviour of synoptic scale features of the weather, such as the mid-latitude depression, is gen-erally well represented by NWP models. Smaller-scale phenomena, for example **polar lows**, are treated less well, however, because they cover fewer grid points. It follows that atmospheric disturbances whose charac-teristic dimensions are less than the model grid spacing will be unresolvable. Such features may, none the less, be important to the weather forecaster. Showers fall into this category. The effects of these, and other so-called sub-grid scale features of the atmosphere, are simulated using an approach called parameterisation. In short, this is a means of estimat-ing the average effects of these processes using large-scale model variables. Some of the most important parameterisations are those concerned with **boundary layer** processes, the development of cloud and precip-itation, convection, radiation, and **gravity waves**.

In order to represent the physical state of the atmosphere, a NWP model must be able to repre-sent three basic properties: atmospheric mass, motion and moisture. The way these quantities are measured in the real world differs from the way they are repre-sented in the model. For example, the weather observer measures moisture in the atmosphere with a wet-bulb thermometer or hygrometer. From this, a **dew-point temperature** can be derived and this is an indirect measure of the amount of water vapour present in the air. In NWP however, moisture content is represented explicitly as specific humidity, the mass of water per unit mass of dry air.

To complicate matters further, a distinction must be drawn between variables directly represented in a model and those diagnosed for use within it. Continuing with the example of moisture, cloud water and precipitation are diagnosed from the primary model variable, specific humidity. Consequently, the former are said to be diagnostic variables, whilst the latter is referred to as a primary prognostic variable. The basis of numerical weather forecasting is the prediction of changes in model prognostic variables over time. From these predictions, cloud, precipitation, surface temperature and other quantities necessary for the preparation of weather forecasts may be diagnosed. The complete set of prognostic and diagnostic formulae coded into a NWP model represent a synthesis of the equations governing large-scale atmospheric motions. These equations are often referred to as the primitive equations and are summarised in Box 14.1.

Before a NWP forecasting model is run, it must be supplied with information on the current state of the atmosphere, a process referred to as data assimilation. Weather observations collected as part of the WMO World Weather Watch programme are used for this purpose. Despite the improvements in data coverage afforded by weather satellites, geographical 'holes' in model-assimilated observations still exist. These are filled in with so-called background fields from earlier model runs. The process of data assimilation can be envisaged as the blending of received observations of the real atmosphere with these background fields in such a way that an objective best estimate of the true state of the atmosphere is achieved.

The United Kingdom Met. Office unified model

The UK Met. Office runs a suite of grid-point-based NWP models collectively known as the Unified Model. This comprises an operational forecasting model, and ocean and climate models, and can be used in a variety of modes. The software runs on an extremely powerful super-computer – the Cray C90 – which is capable of performing sixteen billion floating point calculations every second. For daily weather forecasting, the operational forecasting model can run in two different formats: Global and Limited Area. These are normally referred to as the Global and Limited Area Models (LAM) respectively.

As its name implies, the LAM confines its representation of the atmosphere to a small fraction of the globe: the north Atlantic and western Europe. This is the area of relevance to weather forecasters in the British Isles since the majority of our weather originates over the Atlantic. Unlike the Global Model, the LAM has clearly defined boundaries beyond which the state of the atmosphere is undefined. To prevent it from behaving unpredictably at the margins of the model area, it must be provided with what are termed boundary conditions from the latest Global Model run.

Table 14.3 presents some summary statistics on the two models. Both divide the atmosphere into nineteen horizontal layers. The vertical resolution of these layers is greatest at the surface for the reason that most of the rapid variations in atmospheric processes occur within the boundary layer. Horizontal resolution is defined by a latitude–longitude co-ordinate system. In the Global Model this varies markedly with latitude, between about 70 km at 60°N and 140 km on the Equator. Consequently, it can only resolve the largest scales of atmospheric motion with any degree of accuracy. In the LAM however, the co-ordinate system is artificially rotated so that the more evenly spaced lines of latitude and longitude found on the Equator are positioned over the model area. This feature, coupled with a finer horizontal resolution (about 50 km), permits the LAM to simulate the day-to-day development and decay of synoptic-scale weather systems more effectively.

Weather observations from the World Weather Watch are input into the Unified Model using a continuous data assimilation cycle. This involves repeatedly inserting the latest reports into the evolving model. The benefit of such an approach is that the model atmosphere is regularly nudged

Table 14.3 Summary statistics for the UK Met. Office Unified and Mesoscale Models[a]

Model characteristic	Global	Limited Area	Mesoscale
Coordinate system	Lat./Long.	Rotated pole Lat./Long.	Equatorial Lat./Long. fine mesh
Resolution:			
Rows/columns	217/288	132/229	92/92
Grid points per level	62,496	30,228	8,464
Levels/levels with moisture	19/16	19/16	31/28
Total grid points	1,187,424	574,332	262,384
Grid point spacing W–E	1.25° longitude	0.4425° longitude	0.15° longitude
Grid point spacing S–N	0.83° latitude	0.4425° latitude	0.15° latitude
Grid length at 60°N/140°W	70 km/140 km	Average: 50 km	Average: 16.8 km
Time-step: Physics[b]	20 min.	15 min.	5 min.
Dynamics[c]	10 min.	5 min.	1 min. 40 s

Notes: [a] All information on this page appears courtesy of the UK Met. Office.
[b] The time-step between model evaluation of parameterisations, e.g., convection, precipitation, etc.
[c] The time-step between successive solutions of the fundamental equations.

towards the state of the real atmosphere, whilst maintaining its internal consistency. In addition, assimilation of data on a continuous cycle obviates the need for a separate initialisation time-step. The Global Model runs on a six-hourly assimilation cycle, the LAM on a three-hourly cycle.

Apart from the Unified Model, the Met. Office also runs a number of other weather models, the most noteworthy of which is the Mesoscale Model (see Table 14.3). This short range, high resolution NWP model is designed to provide detailed weather predictions for the British Isles for up to thirty hours ahead. Its use in practical weather forecasting will be mentioned later.

The ECMWF and spectral numerical models

The European Centre for Medium-range Weather Forecasting (ECMWF) based near Reading in Berkshire is an intergovernmental organisation whose responsibility is to undertake research into NWP, the primary aim being to improve weather forecasts in the medium range (four to ten days ahead). The Centre runs an operational weather forecasting model which uses a spectral representation of the atmosphere, unlike the grid-point-based Unified Model of the United Kingdom Met. Office. One obvious drawback with grid point models is their representation of continuously varying fields of pressure, temperature or humidity by discrete points. Spectral models get round this problem by depicting these same horizontal fields in terms of finite sums of certain functions, called **spherical harmonics**.

Much of the research into medium range weather forecasting done at the ECMWF is concerned with ensemble numerical weather prediction.[18] This approach involves the multiple running of an NWP model for the same forecast period. In each run, the model's starting conditions are slightly modified and, as a result, a range of different forecasts – called an ensemble – is produced. In this way the meteorologist may get some idea of how sensitive is the current weather situation to subtle shifts in atmospheric state. It is a peculiarity of the atmosphere that in certain situations its behaviour is inherently more predictable than in others. Ensemble forecasts can help to identify these more predictable states and afford some means of assessing the reliability of individual model predictions.

PRACTICAL WEATHER FORECASTING: THE HUMAN–MACHINE MIX

From the preceding discussion one might be under the false impression that in this age of numerical weather prediction the human weather forecaster is largely redundant. Far from it. Computer models, for all their sophistication, are still generally unable to produce weather forecasts in a form that is both meaningful and relevant to the lay person. One of the principal tasks of meteorologists is to use their knowledge and experience of weather, particularly with regard to its idiosyncrasies in their local area, to provide an intelligible forecast tailored to customers' needs. Thus, ideally, weather forecasting is an amalgam of the quantitative predictive skills of NWP models and the accumulated wisdom of the trained meteorologist. This is sometimes referred to as the human–machine mix.

When a weather forecaster arrives for work, his or her first objective is to build up a comprehensive, mental picture of current weather conditions and their recent evolution in the relevant geographical region. This will involve close examination of surface (e.g., Figure 14.5) and upper air weather charts to identify areas of significant weather, for example, cloud and rain or strong winds. These features may then be placed within the broader weather picture afforded by satellite pictures and numerical model forecasts. The detailed methodology employed by a meteorologist in the preparation of a weather forecast on a given day will vary considerably according to the current weather situation, the range of the forecast, the customer for which the forecast is intended and the time available. For example, a short range weather forecast for holiday-makers on a dry, sunny day in summer, is likely to be much easier, and of less significance, than a forty-eight-hour prediction for the offshore oil industry during a wet and stormy winter. There is insufficient space here to give a proper explanation of the plethora of forecasting techniques used. Instead, a general description of methods employed in making weather predictions for one hour to one month ahead are discussed.

Nowcasts

Weather forecasts for the period ranging from a few minutes to about six hours ahead are often referred to as nowcasts. Here, meteorologists tend to rely heavily upon a detailed analysis of current weather observations (see Box 14.2 and Figure 14.5) and the visual information afforded by satellite imagery and rainfall radar (if this is available). The essence of the nowcast is an extrapolation-based prediction that assumes areas of significant weather will continue to move at more or less the same speed and with little change in their extent or severity in the near future. Such an assumption is valid unless certain features of the current weather, or short range numerical model guidance, indicate otherwise. For example, the pattern of pressure tendencies occurring over the last three hours is a valuable indicator of likely future changes in the distribution of pressure and therefore the development or decay of weather systems. Such indicators can be compared with detailed, short range weather predictions made by mesoscale models. If either suggest the rapid development or decay of a weather system, then the linear extrapolation forecast may have to be amended accordingly.

Short range weather forecasts

Other NWP models, such as the Met. Office Unified Model, are employed in the preparation of weather forecasts for between one and five days ahead. The extent of the Earth's atmosphere represented by these models must be significantly larger to account for the greater distances moved by weather systems during the period of the forecast. For example, a depression forming off the coast of Newfoundland can be over the British Isles two to three days later. The need to model a larger volume of atmosphere poses constraints on model resolution. The latter tends to be coarser in an effort to limit the run time and cost of model operation. The result is that model output is less readily transposed into useful forecasts. It is here that the knowledge and experience of the human forecaster comes into play. For example, given forecast fields of atmospheric pressure, temperature and humidity, and an indication of the extent and severity of any atmospheric instability, a trained meteorologist will be able to infer the sort of weather that can be expected over the period of the model forecast.

Medium range weather forecasts

The reliability of individual NWP model forecasts falls off with increasing lead time. Beyond about three days their worth may only be limited to general guidance on future weather trends, although this is not always the case. In the last ten years or so some improvement in the quality of weather forecasts between about four and ten days ahead (medium range forecasting) has been achieved with the aid of ensemble NWP techniques. The multiple running of a numerical model to reflect a plausible range of initial atmospheric states allows a skill score to be assigned to individual members in the ensemble. If the forecasts so produced are seen to converge, then some confidence can be placed in the value of individual model predictions. On the other hand, if they appear to diverge, then this may indicate a less predictable atmospheric state. In this latter case, individual medium range forecasts are likely to be viewed with some scepticism and weather forecasts will be worded to reflect this uncertainty.

Long range weather forecasts

Given optimal atmospheric conditions, the limit of predictability of daily weather is theoretically about two weeks.[19] Beyond this, objective tests indicate that skill is no longer present in daily forecasts. At first glance, it might seem somewhat contradictory to suppose that the atmosphere is at all predictable beyond this deterministic threshold. The limit of atmospheric predictability for certain types of circulation, however, is significantly longer than for others. Consequently, on occasions, this threshold may be much greater than the average value of two weeks. At a more fundamental level, slow fluctuations in **sea-surface temperature** may impart a longer range predictability to the atmosphere on time-scales of a season or more, particularly in tropical latitudes.

Until quite recently, the most common long range (up to one month ahead) weather forecasting technique involved the use of weather analogues. Initially, this entailed trying to find periods of past weather that were analogous to the current weather over western Europe and the north Atlantic. The forecast for the coming weeks would then be based upon the weather following the historical analogue. The sorts of data used in making the comparison included mean sea-level pressure and **500 hPa heights**. Given the vast quantities of archived data, and the availability of high-powered computers, this approach could generate a range of good analogues for a variety of current weather situations.

During the 1970s analogue techniques were improved by the use of what were termed predictors.[20] These were groups of weather analogues for particular types of extreme event. For example, if one wanted to forecast whether July was going to be wet or dry, data for ten wet and ten dry Julys would be selected from a data archive. Composite 500 hPa charts of the preceding Junes were then drawn for each case. Subtraction of the two composites gave a difference chart and these differences were tested for statistical significance. The June chart for the current year could then be examined in the significant areas to see whether it was more like a June preceding a wet or a dry July.

WEATHER SERVICE PROVISION

Many of our work and leisure activities are sensitive to the weather, and this susceptibility is reflected in the wide range of forecast products and services available to the public and commerce. In the United Kingdom, public sector weather forecasts and related utilities are provided by the Met. Office, whose headquarters are located in Bracknell, Berkshire. The Irish Met. Office assumes a similar role in Ireland. Public services can be grouped into three broad categories.

First, there are duties performed for the 'public good'. These include essential forecasts connected

Table 14.4 The range of commercial weather services provided by the UK Met. Office

Service		Description
Media	Television	Regional and national weather forecasts on BBC and ITV. A business unit, International Weather Productions (IWP), is involved in the selection and training of ITV presenters, and the development of the advanced computer graphics used in the presentations.
	Radio	Television weather broadcasters also read forecasts on the radio for the BBC and Independent stations.
	Newspapers	The Press & Distribution Unit (PDU) at the Met. Office is responsible for the issuing of weather forecasts to the Press. Most daily Newspapers publish a weather forecast of some description, often accompanied by a weather chart.
	Telephone/Fax	The PDU also has responsibility for preparing Telephone- and Fax-based weather forecasts. Weathercall and Marinecall are premium rate telephone services. MetFAX is a dial-up Fax forecast aimed at private aviators, mariners, and educational establishments.
	MIST	The Meteorological Information Self-briefing Terminal (MIST) is a PC-based weather information system for downloading and viewing up-to-the-minute actual and forecast meteorological information.
Transport	Air	AIRMET is a weather forecast aimed at aviators, balloon operators, gliding and parachute clubs. Other products include Terminal Aerodrome Forecasts (TAFs), Meteorological Aerodrome Reports (METARs), and Significant Weather charts (SIGWX). The last of these provides guidance on the location and strength of jet streams, cloud, and severe weather to international aviation.
	Sea	METROUTE is a ship routeing service intended to help shipping operators avoid damage to marine craft caused by severe weather.
	Land	OpenRoad, OpenBridge, OpenRoadFreight, OpenRail, and OpenRunway are services designed to help maintain the free flow of traffic during bad weather. Advanced warning of severe weather can help the relevant authorities minimise the effects on transportation. OpenRoad makes use of a Road Ice Prediction model that forecasts minimum road surface temperatures.
Primary industry	Offshore	The extraction of oil and gas in the North Sea is highly weather sensitive. Here the requirement is for the forecast of wind, wave height and frequency, and 'weather windows' when seas will be calm enough for major operations such as the installation of rigs.
	Water	Real time rainfall radar images, and forecasts (Nimrod) are used by the Environment Agency in river flow forecasting, and pollution control. Five day precipitation forecasts and heavy rainfall warnings can help minimise the risks of flooding.
	Agriculture and forestry	The Rainfall and Evaporation Calculation System (MORECS) can be used to assess water balance. Forecasts of crop yields, crop disease risk, growing days, and Lamb wind-chill are available. Other services include FARMPLAN, FARMCALL.
Utilities	Construction/Engineering	SITEPLAN, SITESUPPORT are consultancy services geared to facilitating the design and construction of buildings. FORESITE and SHORESITE supply site-specific weather forecasts for inland and coastal engineering projects respectively.
	Energy	Electricity generators use forecasts of electricity demand up to ten days ahead. These are based upon predicted maximum and minimum temperatures, wind speed and direction, and rainfall. Early warnings of lightning, snow and ice accretion, and strong winds are produced to help organise the maintenance of power lines.

Table 14.4 The range of commercial weather services provided by the UK Met. Office

Service		Description
	Manufacturing and retail distribution	The Weather Initiative (TWI) is a business unit concerned with the provision of tailored weather forecasts and consultancy services for those involved in the sale and distribution of goods. Weather Sensitivity Analysis (WSA) can identify retail products whose sales are weather dependent. This information may be used as an objective basis for stock control management.
Professional	Legal	BasicProof, ExtraProof and TotalProof are services designed to produce site-specific weather information and consultancy for the legal profession. For example, the weather is often implicated in insurance claims e.g., storm damage to buildings.
Recreation	Sport, travel and tourism	World and European city weather forecasts. WSA based demand forecasts for caterers. Weather forecasts, warnings, and climate statistics for those planning outdoor events. MORECS Irriplan Weatherfax for the maintenance of courses and pitches.

with public safety, such as the National Severe Weather Warning and Storm Tide Warning services. In addition, the Office is required to notify the public of potentially dangerous atmospheric pollution incidences. Enquiries and complaints can be addressed to an Enquiries Officer based at the Met. Office head-quarters. Other utilities considered to be in the 'national interest' also fall into this category. For example, the Met. Office provides meteorological support to the armed forces. The Mobile Meteorological Unit supplies weather forecasts for military exercises, both in the British Isles and overseas, and is regularly involved in the support of operations under the direction of the United Nations.

A second category of services is concerned with the Met. Office's role as a member of the WMO. One of its responsibilities here is the issuing of warnings in the event of international environmental disasters, such as the nuclear accident that occurred at Chernobyl in the Ukraine in 1986. Others include preparing weather forecasts for international civil aviation. The Central Forecast Office at the Met. Office is a World Area Forecast Centre. One of its main duties in this capacity is to produce significant weather (SIGWX) charts for international flights. These enable aircraft to avoid severe weather – for example, thunderstorms – or to make use of strong winds to save fuel on long distance flights.

Finally, there is a wide, and expanding, range of services available to customers on a commercial basis. Excluding revenue from national and international aviation, the most lucrative commercial services are those produced for the media, in particular television and radio. Next in order of revenue-earning potential are weather forecasts for the so-called utilities – electricity generators, the construction industry, manufacturers and retail distributors. Services for land transport, the offshore oil and gas industries and agriculture are also valuable income generators. A brief description of the sorts of weather information supplied by the United Kingdom Met. Office to these and other customers can be found in Table 14.4.

One cannot present a balanced view of weather service provision in the British Isles without some mention of the increasingly important contribution made by the private sector. The largest private weather consultancy in the British Isles is Ocean-routes (UK) Ltd.[21] Based in Aberdeen, this American-owned company is ideally located to service the needs of the offshore oil and gas industries in the North Sea. This work, together with its ship routeing business, provides the majority of its revenue, although in recent years Oceanroutes has broadened its operations to include the provision of land-based services such as winter road ice prediction. Here, it is in direct competition with the Met. Office. The Weather

Department specialises in weather forecasting for television and holds a number of important contracts, including weather presentations for Central Television, the Midlands-based division of the Independent Television (ITV) company. Its success has led it into competition with the United Kingdom Met. Office in a bid to win important contracts such as national television weather broadcasts. Nobel Denton Weather Services is another important competitor in the commercial weather forecasting sector, although much of its work is based outside Britain.

Although the remaining private sector organisations have much smaller annual turnovers, they too have their niches in the sphere of weather consultancy. For example, Weather Action specialises in long range weather prediction; British Weather Services and the Philip Eden Weather Consultancy provide weather forecasts for the media; and Weather Watchers' expertise lies in the supply of road weather information in Scotland.

It should be apparent by now that weather forecasts play a crucial role in many aspects of our daily lives. Without them, our advanced civilisation, dependent as it is on rapid, efficient communications, could not function as well as it does. In recent decades increasing pressure on National Meteorological Centres to provide value for money has encouraged attempts to quantify the economic and social benefits of reliable weather forecasts. In the United Kingdom, it has been estimated that weather forecasts save commerce as much as £600 million per year.[22] About two-thirds of this is attributed to savings in the construction and transport sectors of the economy, the remaining third to savings in agriculture and energy production.

WEATHER FORECASTING: THE FUTURE

The last three decades have seen enormous advances in remote sensing, telecommunications and electronic computing. These developments have radically changed the practice of weather forecasting. What was once a manually intensive, technologically primitive process has now become a highly automated one, relying upon evermore sophisticated weather satellites and radars, global telecommunications and some of the world's fastest supercomputers. What used to be a relatively low key operation is now a high profile business affecting millions of people in today's affluent and increasingly mobile society. The ever-growing demands on weather services reflect the increasing costs to commerce and industry of the weather forecaster's failure to provide timely and accurate weather predictions. In looking to the future of weather forecasting, one cannot just be concerned with issues such as the limits to atmospheric predictability and the changing role of the human forecaster. From an economic standpoint, it is important to consider the future of an industry whose rapidly rising costs dictate its increasing involvement in competitive commercial ventures, but whose very existence depends upon a fragile gentlemen's agreement allowing the free exchange of weather data around the world.

Figure 14.6 illustrates how errors in the United Kingdom Met. Office's global forecasting model have changed over the period. It is apparent that today's 72-hour numerical weather predictions are about as accurate as their 24-hour counterparts were in the early 1970s. The question arises whether future technological progress will continue to bring accompanying improvements in the accuracy of numerical weather forecasts. The United Kingdom Met. Office aims to reduce present errors in 24-hour NWP by a further 10 per cent in the next few years.[23] This improvement is expected to be a consequence of progress in three distinct areas. First, current work on parameterisation schemes promises to refine the representation of sub-grid scale processes – such as radiation balance and cloud and precipitation formation – in numerical models. Second, the introduction of higher resolution imagers and sounders on weather satellites launched during the remaining years of the 1990s and early next century is anticipated to improve the accuracy and spatial coverage of weather observations assimilated into numerical models. Finally, the computational capacity of supercomputers is expected to increase with the transition from conventional computer processors to Massively

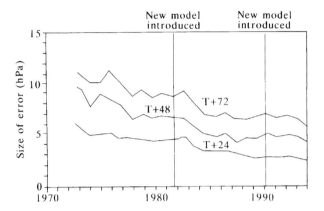

Figure 14.6 Errors in the UK Met. Office's global forecasting model, 1970 to 1994. T+24, T+48 and T+72 refer to forecast made for 24 hours, 48 hours and 72 hours ahead, respectively. Reproduced courtesy of the UK Met. Office.

Parallel Processors. At the moment, supercomputers used in NWP can do as many as ten thousand million (10^{10}) arithmetic operations per second. By early next century, it has been suggested that this figure could increase by five orders of magnitude to around one thousand billion (10^{15}) operations per second.[24] Meteorologists believe that this growth in processing power will allow routine runs of higher resolution grid-point models and the extension of ensemble-based forecasting methods to consider a greater range of initial atmospheric conditions.

Despite these foreseen advances, some scientists have argued that our ability to forecast the weather by numerical methods is fast approaching the theoretical limits of atmospheric predictability. In the past few decades the concept of chaos and its implications in this area have come very much to the fore. Mathematicians have demonstrated that, even in deterministic models such as those used in NWP, chaos theory dictates that the accuracy with which the current state of the atmosphere must be known to predict all future states exactly is infinite.[25] Consequently, no matter how great the improvements in the quantity and quality of weather observations, a perfect forecast will always be impossible.

Nevertheless, the question remains how the technological advances described above can be used to optimise their beneficial effects on weather forecast accuracy. Here meteorologists are divided. Some believe that the most fruitful way forward is to continue the current trend towards running higher resolution, evermore complex grid-point models. Others think that this would be a waste of technological resources and that weather forecasting would be better served if the emphasis was placed on providing an indication of the reliability of NWP forecasts. This might be achieved by further developing the ensemble approach. Given that both the above approaches have their limitations it would seem that an amalgam of the two methods is likely to prove the best way forward.

A related issue concerns the future role of the meteorologist in the human–machine mix. The last two decades have seen steady improvements in the reliability of NWP. Should these continue apace, it has been suggested that the role of the meteorologist will gradually decline to the point where human input into weather forecasts is negligible. Certainly it is true that today's high resolution mesoscale numerical models can provide weather forecasts in sufficient detail to warrant minimal interpretation by meteorologists. These models are not always reliable, however, and the watchful eye of the meteorologist is important if weather forecasts, particularly in the short range, are to take account of any divergence between model predictions and the evolution of weather in the real world. Furthermore, the complex human skills involved in preparing weather forecast texts tailored to customers' specific needs are not easily acquired by an electronic computer.

Despite these difficulties, the North American National Weather Service began using simple Computer Worded Forecasts on an operational basis as early as the mid-1970s.[26] These utilised **Model Output Statistics** from NWP models to generate a series of forecast matrices. Each matrix contained sequences of forecast weather elements for a given location. The Computer Worded Forecast was generated by feeding these forecast matrices into an automatic text generator. The latter would then construct coherent sentences from a combination of the weather data supplied and a pre-programmed knowledge of the semantics of weather forecast scripts. Today, more sophisticated versions of these early Computer Worded Forecasts are used operationally, or experimentally, by National Meteorological Centres across the world. Notwithstanding suggestions that the role of the human forecaster will decline in the future, scientific and technological developments such as NWP and Computer Worded Forecasts have yet to remove the need for the human component in the human–machine mix; rather they have transformed the demands made on meteorologists. Assuming that current trends persist, one must conclude that, in the near future at least, the human forecaster will continue to play a crucial part in day-to-day weather forecasting.

In recent decades, the cost of the global weather industry, encompassing as it does the maintenance of an up-to-date infrastructure in the face of scientific and technological progress, has risen significantly. This has led some countries to question the validity of the gentleman's agreement permitting the free global exchange of weather data as part of the WMO World Weather Watch programme. In addition, the expense entailed in running a National Meteorological Centre has led a good many countries to encourage commercial activities as a means of offsetting their overall expenditure on weather services. This leads to a potential conflict of interest. On the one hand, a country expects to receive free weather data from its neighbours. On the other, it may increasingly be encouraged to charge for weather services provided to these same neighbours, the basis for which are the data provided freely to it. In Europe, the air of increasing commercialisation and competition has led member countries of the European Union to formulate an agreement on the provision of data and weather services known as ECOMET.[27] This came into effect in January 1996. The primary aim of ECOMET is to establish an agreed framework for commercial and non-profit making National Meteorological Centre operations within Europe. It is hoped this agreement will prevent future conflicts of interest such as that outlined above.

NOTES

1 Aristotle, *Meteorologica*, trans. H.D.P. Lee, Cambridge, Mass., Harvard University Press, 1926.
2 G. Hadley, 'Concerning the cause of the general trade winds', *Philosophical Transactions,* 1735, vol. 34, pp. 58–62.
3 G. Kutzbach, *The Thermal Theory of Cyclones, A History of Meteorological Thought in the Nineteenth Century*, Historical Monograph Series, American Meteorological Society, 1979, pp. 22–5.
4 W. Marriot, 'The earliest telegraphic daily meteorological reports and weather maps', *Quarterly Journal Royal Meteorological Society*, 1903, vol. 29, p. 124.
5 W. Napier Shaw, *Forecasting Weather*, London, Constable and Company Ltd, 1911.
6 O.M. Ashford, 'Development of weather forecasting in Britain 1900–1940: the vision of L.F. Richardson', *Weather*, 1992, vol. 47(10), pp. 394–402.
7 Kutzbach, op. cit., p. 159.
8 V. Bjerknes and H. Solberg, 'Meteorological conditions for the formation of rain', *Geofysiske Publikasjoner*, 1921, vol. 2, no. 3.
9 O.M. Ashford, op. cit.
10 R.A.S. Ratcliffe, 'Weather forecasting in Britain, 1939-80', *Weather*, 1993, vol. 48, no. 9, pp. 299–304.
11 L.F. Richardson, *Weather Prediction by Numerical Process*, Cambridge, Cambridge University Press, 1922.
12 Ratcliffe, op. cit., p. 302.
13 J. Fishman and R. Kalish, *The Weather Revolution, Innovation and Imminent Breakthroughs in Accurate Forecasting*, New York, London, Plenum Press, 1994, pp. 82–5.

14 G.J. Day *et al.*, 'Scientific statement on International organisation for co-operation in meteorology', *Weather*, 1992, vol. 47, no. 8, p. 310.

15 The Meteorological Office, *The Observer's Handbook*, London, HMSO, 1982.

16 GMT – Greenwich Mean Time – is used as the universal time zone for meteorological observations. GMT is sometimes replaced with UTC – Universal Time Clock; these time units are identical.

17 World Meteorological Organisation, *The WMO Achievement, 40 Years in the Service of International Meteorology and Hydrology*, Geneva, The WMO no. 729, 1990, p. 16.

18 M.J.S. Harrison, 'Ensembles, higher resolution models and future computing power – a personal view', *Weather*, 1994, vol. 49, no. 12, pp. 398–406.

19 T.N. Palmer and others, 'Scientific statement on extended-range atmospheric prediction', *Weather*, 1992, vol. 47, no. 8, pp. 306–9.

20 Ratcliffe, op. cit., p. 303.

21 N. Lynagh, 'Viewpoint: The range of weather forecast services provided by the private sector and a view on the services provided to the public', *Weather*, 1995, vol. 50, no. 6, pp. 226–8.

22 ESA, 'Meteosat: There in all weather', Paris, ESA, p. 23.

23 J.C.R. Hunt, 'Developments in forecasting the atmospheric environment', *Weather*, 1994, vol. 49, no. 9, pp. 312–18.

24 H.E. Brooks and C.A. Doswell, 'New technology and numerical weather prediction – a wasted opportunity', *Weather*, 1993, vol. 48, no. 6, pp. 173–7.

25 T.N. Palmer, 'A nonlinear dynamical perspective on climate change', *Weather*, 1993, vol. 48, no. 10, pp. 314–26.

26 H.R. Glahn, 'Computer worded forecasts', *Bulletin of the American Meteorological Society*, 1979, vol. 60, pp. 4–11.

27 A. Douglas, F. Duvernet and R. Hoerson, 'Potential economic benefits from the introduction of ECOMET', Conference on the Economic Benefits of Meteorological and Hydrological Services, WMO, Geneva, Switzerland, 19–23 September 1995, pp. 205–8.

GENERAL READING

R.G. Barry and R.J. Chorley, *Atmosphere, Weather and Climate*, London and New York, Routledge, 6th edn, 1995.

J.F.R. McIlveen, *Fundamentals of Weather and Climate*, London, Chapman and Hall, 1992.

D.H. McIntosh and A.S. Thom, *Essentials of Meteorology*, London, Wykeham Publications Ltd, 1981.

P.G. Wickham, *The Practice of Weather Forecasting*, London, HMSO, Third Impression, 1980.

15

GLOBAL WARMING AND THE BRITISH ISLES

Sarah Raper, David Viner, Mike Hulme and Elaine Barrow

Is it not the height of silent humour to cause an unknown change in the Earth's climate?
Robert Graves, *The Meeting*

INTRODUCTION

There is increasing evidence that global climate is changing and that at least some of these changes are related to pollution of the atmosphere resulting from human activities. This is one of the conclusions reached in the latest report from the Intergovernmental Panel on Climate Change published in 1996,[1] and over the last ten years this concern has led to the creation of the United Nations Framework Convention on Climate Change. This Convention has been ratified by most of the nations of the world and its objective is to prevent 'dangerous interference in the climate system'. Global climate change will have a variety of effects at continental and regional scales and, as has been shown in Chapter 9, there is evidence from the British Isles that there has been a general warming of climate in this region over the last two or three centuries and, in particular, over the last two or three decades. Is this climate change in the British Isles part of a larger scale trend and, if so, can it be clearly related to pollution of the global atmosphere by humans? How can our knowledge of how the climate system works be used to make predictions, not just of the weather of tomorrow or next week (see Chapter 14), but of climate over the next ten to one hundred years?

What do such predictions as have been made signify for the climate of the British Isles of the next century, the climate that our children and grandchildren will experience? These are some of the questions tackled in this chapter, but we begin by summarising some of the background to the **greenhouse effect** and how human activities can modify it.

THE GREENHOUSE EFFECT

The atmosphere contains naturally occurring gases which are very important for our climate because they maintain the Earth's temperature about 30°C warmer than it would be in their absence – this is the greenhouse effect. The most important of these so-called greenhouse gases is water vapour; others, such as naturally occurring carbon dioxide and methane, also contribute to the total effect. Greenhouse gases are transparent to the incoming short-wave (ultra-violet) radiation of the sun which heats the Earth's surface. In turn, the warmed surface emits longwave (infra-red) radiation, some of which is absorbed and scattered in all directions by the greenhouse gases, thus reducing the energy escaping to space (see Figure 15.1). For this reason the gases are referred to as **radiatively active**. The amount of

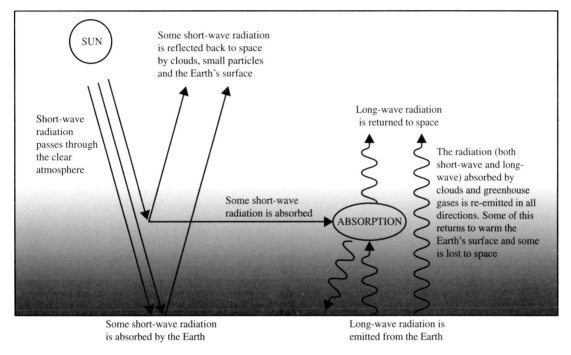

Figure 15.1 A general illustration of the greenhouse effect.

longwave radiation emitted from a surface increases with increasing temperature so a balance is established when the temperature is such that the long-wave radiation escaping to space is equal to the incoming shortwave radiation.

The **anthropogenic** emissions of greenhouse gases have altered the natural balance between incoming and outgoing radiation causing an enhanced (or anthropogenic) greenhouse effect. The most important of these gases are carbon dioxide (emitted from fossil fuel combustion and changing land use), methane (from ruminants, rubbish tips, rice paddies, coal mines and gas extraction), ozone (a chemical derivative from car pollution), halocarbons (from refrigeration and other industrial uses) and nitrous oxide (from car pollution).

Natural ozone is very important in the stratosphere because it absorbs ultra-violet radiation which is harmful for life. Over the past few decades stratospheric ozone depletion (the **Ozone Hole**), which is

attributed to anthropogenic chlorofluorocarbon emissions, has caused great concern.[2] For this reason, strong controls on the future emission of chlorofluorocarbons are being implemented as laid down in the latest amendment to the 1987 Montreal Protocol on ozone-depleting substances. Because ozone is a greenhouse gas, this ozone depletion in the **stratosphere** causes a small reduction in the greenhouse effect. Significant increases in tropospheric ozone, however, due largely to transport pollution, result in a significant positive contribution to the greenhouse effect from ozone concentration changes in the atmosphere as a whole.

In order to assess the climate change which might result from past and continued emissions of greenhouse gases there are a number of calculations which have to be made in a consistent manner as illustrated in Figure 15.2. These calculations begin with the reconstruction of past emissions and, for the future, the construction of **emissions scenarios**. In 1992,

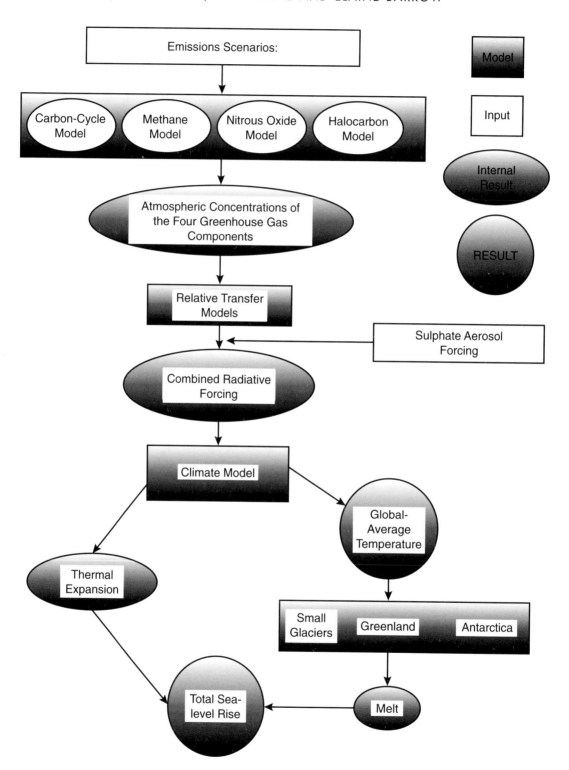

the Intergovernmental Panel on Climate Change defined a set of six greenhouse gas emissions scenarios for the world, scenarios which portrayed emissions through the whole of the twenty-first century.[3] These scenarios have since been modified to account for the rapid phase out of chlorofluorocarbons due to the concern over stratospheric ozone depletion. The scenarios represent possible projections of greenhouse gas emissions in the absence of new policies to reduce them and are therefore regarded as 'non-intervention' scenarios. They are referred to as IS92a–f. The IS92a scenario falls roughly in the middle of the range and adopts intermediate assumptions about future global population, economic growth, and the mix of conventional and renewable energy sources. The range of possible future emissions scenarios, even in the absence of new control policies, is the first of a series of uncertainties which enter at various stages in the process of trying to determine climate and sea-level projections for the next century.

The next step is to calculate future atmospheric concentrations of the greenhouse gases for a given emissions scenario. **Gas cycle models** are invoked to keep track of the sources, sinks and movements of the gases. For example, to estimate future carbon dioxide concentrations the uptake of carbon dioxide by the oceans and biosphere needs to be considered. The carbon cycle is complicated and involves many feedbacks, so estimating future carbon dioxide concentrations also involves some uncertainty.

The estimated change in the global-average **radiative forcing** due to increases in greenhouse gas concentrations which have occurred since the pre-industrial era – assumed to be 1765 – is about 2.5 watts per square metre (Wm^{-2}). For the IS92a emissions scenario the calculated change in forcing over the next century is about 5.1 Wm^{-2}, more than twice that observed to date.[4] Over the last few years, however, it has become evident that when averaged over the global atmosphere a substantial part of the anthropogenic greenhouse effect may have been offset by a negative forcing due to the anthropogenic emis-

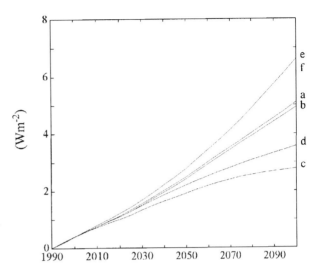

Figure 15.3 Projections of global-average radiative forcing from 1990 to 2100 due to greenhouse gas emissions and sulphate aerosols. The six curves represent the six IS92 emissions scenarios of the Intergovernmental Panel on Climate Change.

sion of **aerosols**. Sulphate aerosols, formed from sulphur dioxide emitted during the combustion of fossil fuels and partly responsible for producing **acid rain**, have been regarded as particularly important.

The size of this negative forcing is very uncertain and, unlike greenhouse gas forcing, is not geographically uniform. Thus, determining the pattern of future aerosol forcing is crucial for understanding the regional patterns of future climate change. For the global average, the aerosol forcing is estimated to be about –1.3 Wm^{-2} at present, thus potentially offsetting about half of the anthropogenic greenhouse effect. The uncertainty is large, however, and future patterns of aerosol forcing are of course unknown. A further increase in this global aerosol forcing of about –0.6 Wm^{-2} may accompany the emissions scenario IS95a. Using mid-estimates for the aerosol forcing, the future global-average total forcing projections for the six emissions scenarios IS92a–f are shown in Figure 15.3.

Figure 15.2 A schematic representation of the steps involved in estimating future global temperature and sea level change.

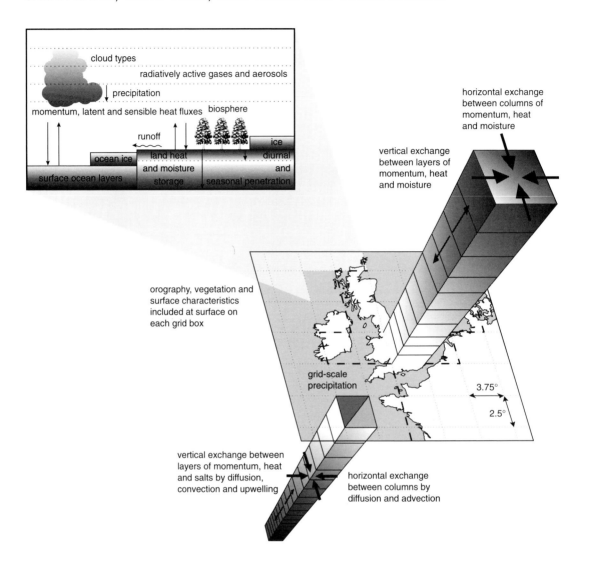

Figure 15.4 Schematic representation of the climate system as modelled by a global climate model, in this case HADCM2.

GLOBAL AND REGIONAL CLIMATE CHANGE

The effect of a change in radiative forcing on climate can be assessed using a hierarchy of climate models. These range from simple global-average box models to complex three-dimensional models of the coupled atmosphere–ocean system (see the schematic diagram in Figure 15.4), similar to the Numerical Weather Prediction models discussed in Chapter 14. The oceans play a major part in the climate system because they act like a flywheel by damping the response of the climate to a given forcing. Thus an essential element in climate models, whether simple or complex, is their ability to simulate the effect of

BOX 15.1 THE CLIMATE SENSITIVITY

The climate sensitivity can be defined as the equilibrium, or eventual, global-average temperature
change for a doubling of the carbon dioxide
concentration. Its value depends on the various
feedbacks in the climate system which may
enhance or suppress the direct response of the
climate system to forcing. One such example of
a positive feedback is the ice-albedo feedback.
Clean ice and snow have a high **albedo** which
means they reflect a large proportion of the
incoming solar radiation directly back to space.
Less snow and ice associated with warmer temperatures in the future means a lower albedo so that
more of the incoming solar radiation will be
absorbed at the surface, thus causing a further
increase in temperature. Another important positive feedback is the water vapour feedback. With
warmer temperatures and increased evaporation
the concentration of water vapour in the atmosphere will increase and water vapour is itself an
important greenhouse gas. Apart from the large
aerosol forcing uncertainties, the unknown value
of the climate sensitivity is perhaps the single
most important uncertainty in the whole process
of calculating future climate change. It is best
estimated from the complex three-dimensional
climate models which explicitly include the
known important feedbacks and is thought to
be in the range 1.5°C–4.5°C with a best estimate
of 2.5°C.

oceanic **thermal inertia**. If the model does not represent the deep ocean, only the hypothetical eventual
– or equilibrium – change can be assessed; no rate
of climate change, so-called 'transient' effects, can be
simulated in such models. Simple climate models are
a powerful tool for the rapid calculation of the global
temperature response to different forcing scenarios.
If regional details and other climate variables, such
as precipitation or wind speed, are required, however, it is necessary to use the much more complex
three-dimensional ocean–atmosphere global climate
models. Results from both these types of models are
given below.

Simple model projections

The global warming projections shown in this
chapter are produced by the simple climate model
used by the Intergovernmental Panel on Climate
Change.[5] In this model, separate boxes for land and
ocean areas in each hemisphere are distinguished.
The two ocean areas are each modelled with a single
ocean column split into forty layers so that the heat
penetration into the deep ocean can be represented.
The most important parameter which has to be set

in this simple model is the **climate sensitivity** (see
Box 15.1).

Before looking at the temperature projections for
the next century it is advisable first to look at how
well the simple climate model simulates the observed
global warming since pre-industrial times during the
mid-eighteenth century. This warming is estimated
to be about 0.5°C.[6] It turns out that the model tends
to underestimate the warming for all three recommended values of the climate sensitivity. Thus, for
climate sensitivities of 1.5°C, 2.5°C and 4.5°C, past
warming estimated by the model is, respectively,
about 0.2°C, 0.3°C and 0.4°C. To obtain an estimate
of 0.5°C from the model requires a climate sensitivity of about 6°C, which is larger than the values
estimated by most complex models. These results
may indicate that the value of the negative aerosol
forcing chosen in the model is too high, or they may
reflect the presence of natural variability in the
observed record which the model cannot simulate.[7]
The combined effect of uncertainties in the aerosol
forcing and the climate sensitivity, and the unknown
size of the trend caused by natural variability over
the last century, makes the unambiguous detection
of the anthropogenic greenhouse effect from the

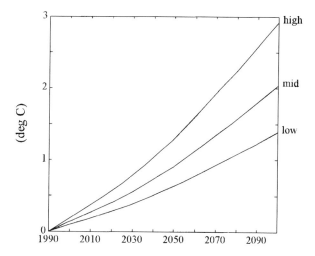

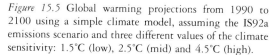

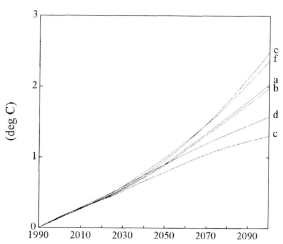

Figure 15.5 Global warming projections from 1990 to 2100 using a simple climate model, assuming the IS92a emissions scenario and three different values of the climate sensitivity: 1.5°C (low), 2.5°C (mid) and 4.5°C (high).

Figure 15.6 Global warming projections from 1990 to 2100 using a simple climate model, assuming a value of 2.5°C for the climate sensitivity and the six Intergovernmental Panel on Climate Change emissions scenarios: IS92a–f.

global-average temperature series alone not possible at the present time. Techniques which seek to detect in the observations the *geographical patterns* of temperature change predicted by complex climate models are, however, beginning to have some success. These results suggest that the observed climate change is unlikely to be entirely due to natural causes.[8]

Using the simple climate model, forced with radiative changes corresponding to emissions scenario IS92a, global warming projections from 1990 are shown in Figure 15.5 for the range of values for the climate sensitivity. The changes to the year 2100 are 1.4°C, 2.0°C and 2.9°C respectively. The effect of the six different emissions scenarios on global warming, using a fixed value of 2.5°C for the climate sensitivity, is shown in Figure 15.6. The differences between the scenarios are small before about 2050, after which time the results diverge substantially. By 2100 the range of warming resulting from the different emissions scenarios is 1.3°C to 2.5°C. To put the above results into recent perspective, the estimate using IS92a and a climate sensitivity of 2.5°C gives a rate of global warming over the next century

of about 0.2°C per decade – about four times that observed over the last 150 years. This estimated rate of future warming *is* consistent, however, with the global warming observed since the 1960s.[9]

Simple models can also be used to obtain estimates of future global-average sea-level rise.[10] Both **thermal expansion** of the oceans and the net melting of land ice are expected to contribute to an increase in the ocean volume with global warming. For the calculation of melting land ice, glaciers and small ice-caps and the large ice-sheets of Greenland and Antarctica are considered separately. Interestingly, Antarctica is expected to contribute a small fall in sea-level over the next century. This is because the very cold atmosphere over Antarctica would be able to hold more water vapour if it warmed, which may result in greater snowfall and hence an accumulation of ice. Although the main contributions to sea-level change are expected to be from thermal expansion and the melting of glaciers and small ice-caps, the uncertainties, especially surrounding the response of the ice-sheets to climate change, are very large. Taken together, the ice-melt model uncertainties produce

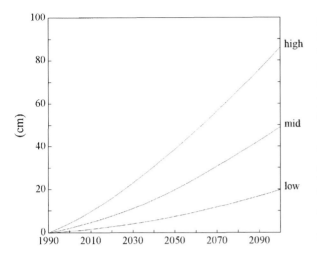

Figure 15.7 Global sea-level rise projections from 1990 to 2100 using a simple climate model, assuming the IS92a emissions scenario and low, mid and high settings for the climate and sea level model parameters.

uncertainties in the sea-level rise projections which are of a similar size to the uncertainties associated with the climate sensitivity. When these two sets of uncertainties are combined a wide range of sea-level projections results. This range is shown in Figure 15.7 for the IS92a emissions scenario. The central estimate for sea-level rise in 2100 is about 50 cm, but the range is from 20 to 86 cm. This result can be compared with the observed rise in sea-level over the last century of between 10 and 20 cm.

Complex model projections

Although simple climate models are powerful tools for assessing global-average temperature changes, in order to study the regional patterns of future climate change – such as for the British Isles – it is neces- sary to perform climate change experiments using complex three-dimensional global climate models. The Hadley Centre, and before that the United Kingdom Met. Office, has over the years developed and performed a series of climate change experiments with such models.[11] The latest model version is called HADCM2 and we show here some results

from one recent experiment.[12] The range of possible future global warming projections produced using the simple climate model, provides a context in which to place this example of a complex model experiment. HADCM2 is a nineteen-level atmos- phere model coupled to a twenty-level ocean model and has a climate sensitivity of about 2.5°C.

Commencing in 1860, the model was forced with estimates of the past and future (based on IS92a) greenhouse gas and aerosol forcing. This experiment was called the sulphate experiment, or SUL for short. The aerosol forcing used represents only the *direct* scattering effect on the energy budget of the atmos- phere and amounts to a global average of -0.65 Wm^{-2} in 1990. This is a much smaller negative forcing than the IPCC mid-estimate of -1.3 Wm^{-2} quoted earlier, which included an estimate of the *indirect* effects of aerosols on the radiative properties of clouds. Also, the effect of tropospheric ozone increases (estimated to be about 0.4 Wm^{-2} in 1990) is ignored in the HADCM2 SUL experiment. These, and other differences, result in a 1990 forcing that is about 0.44 Wm^{-2} greater than that which was used for the simple model results presented in Figure 15.5.

The resulting global warming projection from the HADCM2 SUL experiment is shown in Figure 15.8, together with the observed surface temperature changes from 1860 to 1995. In contrast to the smooth temperature projections obtained from the simple climate model, the complex climate model results have a year-to-year variation which is caused by the natural variability represented in this complex model. With the total forcing used in the HADCM2 exper- iment being larger than in the simple model results shown earlier, the complex model results shown here reproduce well the overall observed warming from 1860 to 1990. This result should not, however, be used to justify the lower aerosol forcing used in the HADCM2 experiment. The agreement shown in Figure 15.8 could be coincidental and some other, hitherto unknown, factors may be operating in the real world. Natural variability may also have affected the past trends in the observations and/or the model results. For this HADCM2 SUL simulation the

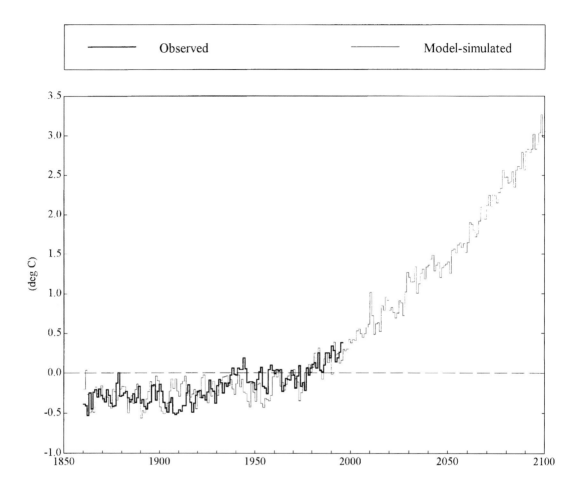

Figure 15.8 Global-average temperature change from 1860 to 2100 from the HADCM2 SUL experiment[12] and from 1860 to 1995[6] from the observations. The changes are plotted with respect to the average temperature of 1961 to 1990. The observed temperature data are listed in Appendix D.

overall warming from the 1961 to 1990 period to the end of next century is between 2.5°C and 3.0°C.

Unlike the anthropogenic greenhouse gases, aerosols only remain in the atmosphere for a few days before they are washed out by rain. The geographical distribution of the aerosol forcing is therefore uneven and depends on the location of the industrial sources of sulphur dioxide. The resulting patterns of regional anthropogenic climate change are quite sensitive to this distribution. The patterns of regional change for temperature and precipitation from the

HADCM2 SUL experiment for the years centred around 2050 are shown in Plates 9 and 10 for the winter and summer seasons. The global warming by this date is about 1.5°C with respect to the 1961 to 1990 average.

It is evident from the temperature change maps that the simulated warming is far from globally uniform. A striking feature is the strong winter warming in the Arctic of over 4°C, although this region cools slightly in summer. Generally, warming is simulated over the land masses in both seasons, although a small area of cooling is evident over eastern China in winter. The latter is due to a strong local influence of sulphur dioxide emissions. Cooling is also simulated in Southern Hemisphere summer over parts of the Southern Ocean, although this is due to changes in ocean processes rather than to aerosol effects. For the Earth as a whole, precipitation is expected to increase by about 2 per cent in both seasons, but regional differences are marked. For example, increases in precipitation occur over large parts of the equatorial oceans in both seasons, but decreases predominate elsewhere in the tropical and sub-tropical oceans. Over land areas the South Asian monsoon weakens and precipitation decreases over northern Australia during their (austral) summer monsoon season. At high northern latitudes, increases in precipitation tend to dominate, especially in the winter season.

CLIMATE CHANGE IN THE BRITISH ISLES

The British Isles form only a very small part of the Earth's surface and because the global climate models like HADCM2 contain only a coarse geographical representation, the climate changes simulated by such models are not very detailed at such a small regional scale. Furthermore, different experiments with different models yield different patterns of change even with the same forcing, so the results described here must be interpreted as only one possible scenario out of a range of several. The results from HADCM2 SUL for the thirty-year period centred on 2050 were extracted for the British Isles and are shown in Plates 11 and 12 for mean temperature and precipitation for the winter and summer seasons. The temperature changes are added to the current climate, defined as the average of 1961 to 1990, whereas the precipitation changes are shown as per cent change from current climate at the original HADCM2 resolution.

The increase in temperature is between 1.2°C and 1.6°C in both seasons and is distributed over the British Isles with greater warming in the east than in the west. In Scotland, the results show increased precipitation in both seasons – up to 5 per cent or

more in winter. Further south, in the summer, the warmer temperatures are accompanied by modest decreases in precipitation – about 5 per cent. The driest areas of the British Isles therefore experience the greatest decrease in summer precipitation. If, within one generation, a warmer and drier climate such as this ensued in the southern regions of the British Isles there would clearly be major consequences for the viability of a range of social and economic activities. The next section explores some of the implications of such a climate change for the region.

HOW CLIMATE CHANGE MIGHT AFFECT THE BRITISH ISLES

Many environmental assets and economic activities in the British Isles are highly sensitive to weather and climate – outdoor recreation, agriculture, water supply, transport and the construction industry are just a few. Any change in climate, or in its variability, will inevitably have consequences for such activities and considerable attention is now being given to defining the scope and magnitude of such impacts both now[13] and in the future.[14]

Changes in temperature and precipitation are not necessarily the most important climatic changes to affect the assets and activities of the British Isles. Reductions in snowfall will be important for transport or recreation, increases in wind speed will affect forestry and the construction industry, changes in radiation will alter agricultural production and the design of buildings, and increases in potential evapotranspiration will affect the agricultural and water resource base of the country. There are also likely to be important secondary effects of a general warming of the climate which will have implications for health and tourism. Higher air temperatures (particularly in summer), combined with increased local emissions of nitrogen oxides and hydrocarbons (particularly from the transport sector), will increase tropospheric ozone concentrations over urban areas, increase the formation of photochemical smogs and lead to changes in air quality (see Chapter 12).

Table 15.1 Some recent extreme annual and seasonal temperature anomalies from the 1961 to 1990 average and their approximate estimated return periods under current (1961–90) climate and under the HADCM2 SUL scenario for the years centred around 2050. The global warming by this date is about 1.5°C. Estimates derive from statistical analysis of the Central England Temperature record

| | Seasonal anomaly (°C) | | Return period (years) | |
	Temperature	Anomaly	1961–90	2050
Annual 1990	10.6	+1.1	65	1.6
Summer 1976	17.8	+2.5	310	5.5
Summer 1995	17.4	+2.1	90	3
Winter 1988/9	6.5	+2.4	30	4
Winter 1962/3	−0.3	−4.4	230	∞

Table 15.2 Average annual frequencies of daily temperature extremes for six locations around the British Isles for current climate (1961 to 1990 average) for the years centred around 2020 and 2050 under a global warming scenario.[a] Santon Downham is in Norfolk, Hillsborough is near Belfast and Fortrose is near Inverness

	Latitude (°N)	1961–90	2020	2050
		Average annual frequency of frost days (T_{min} <0°C)		
Plymouth	50.4	21	10	8
Oxford	51.8	42	26	18
Santon Downham	52.4	79	60	44
Hillsborough	54.6	41	27	19
Durham	54.8	56	39	28
Fortrose	57.6	31	22	15
		Average annual frequency of hot days (T_{max} >25°C)		
Plymouth	50.4	3	4	8
Oxford	51.8	12	15	20
Santon Downham	52.4	11	17	26
Hillsborough	54.6	1	2	3
Durham	54.8	3	5	7
Fortrose	57.6	0	0	0

[a] These results are from E.M. Barrow and M. Hulme, 'The changing probabilities of daily temperature extremes in the UK related to future global warming and changes in climate variability', *Climate Research*, 1996, vol. 6, pp. 21–31. The global warming by 2020 and 2050 in this example is slightly higher (about 0.1° and 0.2°C respectively) than the HADCM2 SUL scenario used in this chapter.

The likelihood of extreme seasonal temperatures will also change in a warming climate. Table 15.1 shows the estimated return periods for some recent extreme seasonal temperature anomalies over the British Isles under conditions of climate change. The estimated return period of the summer temperatures of 1995 changes from about ninety years at present to, on average, about once every three years by 2050. Assuming these seasonal extremes are independent events, the probability of two successive summers like 1995 occurring during a decade in the middle of next century is about 90 per cent (i.e., very likely). Such an event would have severe implications for agriculture and water resources in many parts of the British Isles.

Change in (boreal) winter mean temperature

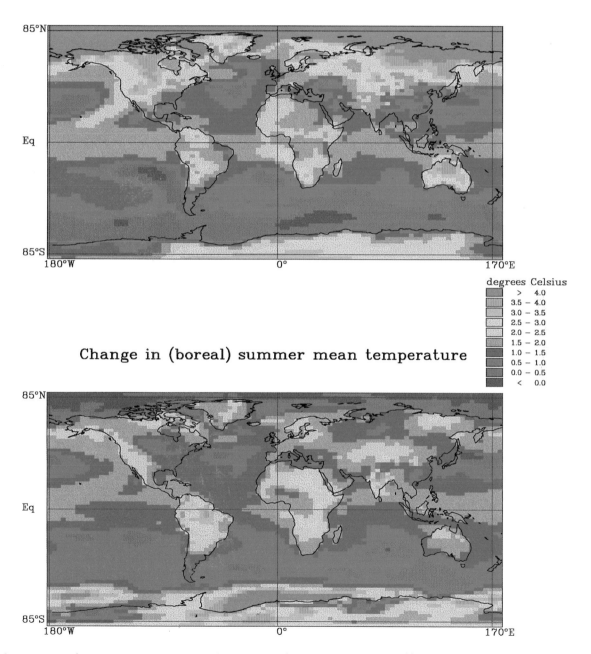

degrees Celsius
> 4.0
3.5 – 4.0
3.0 – 3.5
2.5 – 3.0
2.0 – 2.5
1.5 – 2.0
1.0 – 1.5
0.5 – 1.0
0.0 – 0.5
< 0.0

Change in (boreal) summer mean temperature

Plate 9 Mean surface air temperature change in degrees Celsius by the period 2035 to 2064 with respect to the average of 1961 to 1990 for the boreal winter (top) and summer (bottom) seasons. Results from the HADCM2 SUL global climate model experiment.

Change in (boreal) winter precipitation

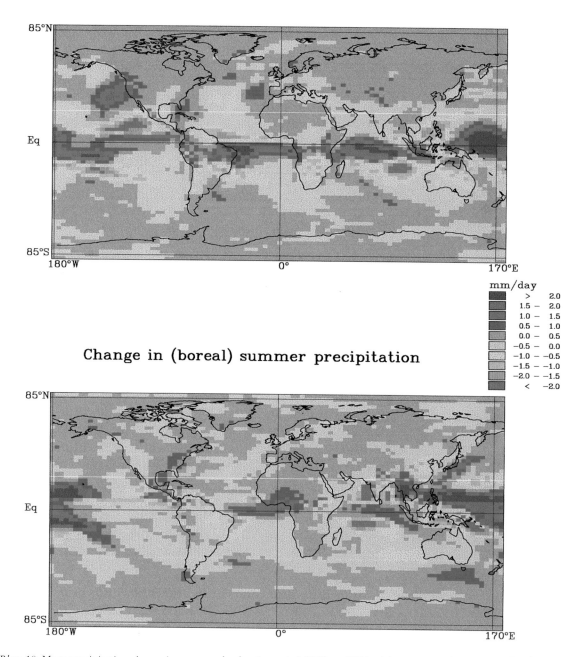

Change in (boreal) summer precipitation

Plate 10 Mean precipitation change in mm per day by the period 2035 to 2064 with respect to the average of 1961 to 1990 for the boreal winter (top) and summer (bottom) seasons. Results from the HADCM2 SUL global climate model experiment.

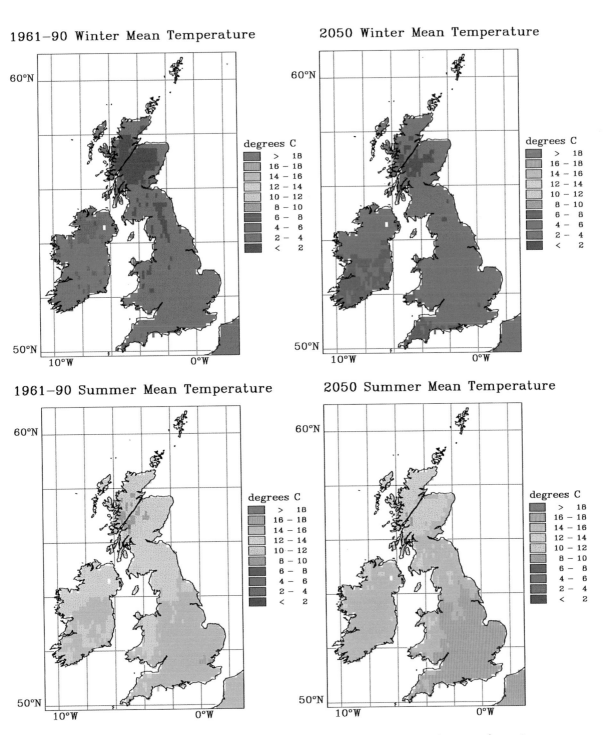

Plate 11 Mean observed surface air temperature in degrees Celsius for the winter (top) and summer (bottom) seasons for the period 1961 to 1990 (left), and for the period 2035 to 2064 (right) after adding the changes from the HADCM2 SUL global climate model experiment.

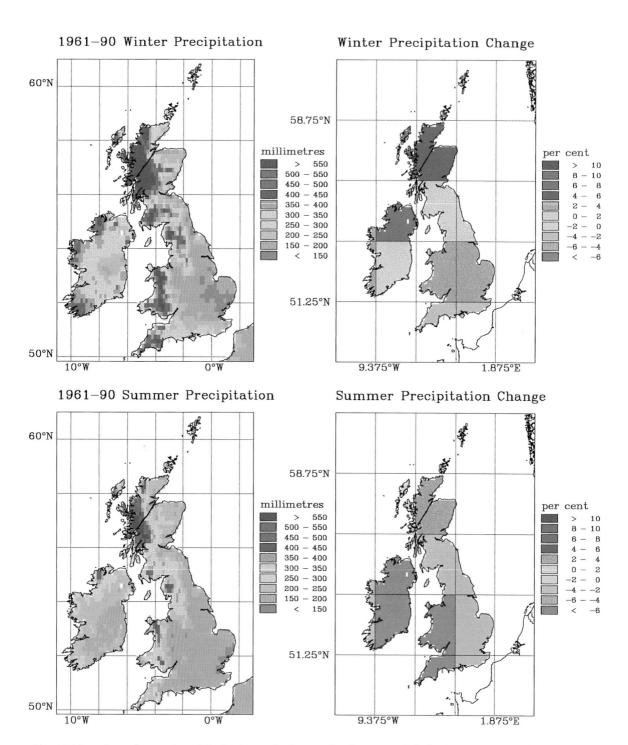

Plate 12 Mean observed seasonal precipitation in mm for the period 1961 to 1990 (left) for the winter (top) and summer (bottom) seasons, and the percentage change in this total for the period 2035 to 2064 (right) obtained from the HADCM2 SUL global climate model experiment.

Changes in the frequency of *daily* temperature extremes are also likely to accompany a climatic warming and such changes will have important consequences for species distributions, agriculture and human health. Table 15.2 presents one set of estimates of changes in the average annual frequencies of hot days and air frosts for six locations representing a range of British and Irish climates. By 2050, frost frequencies fall by about 50 per cent at all six locations, with the largest reduction at Plymouth where on average only eight air frosts are estimated to occur compared to twenty-one at present. The average frequency of hot summer days increases throughout the country (except at Fortrose where even by 2050 the climate is not warm enough to generate days warmer than 25°C), with the average frequency at Oxford nearly doubling from twelve at present to twenty by the year 2050.

A further consequence of climatic warming in the British Isles will be changes in the **degree days** above and below certain temperature thresholds. These thermal indices are used in a variety of climate impact sectors including agriculture and forestry (growing degree days) and energy and the built environment (heating and cooling degree days). Growing degree days (accumulated mean temperatures above 5.5°C) increase by between 24 and 34 per cent by 2050, with Durham, for example, recording as many growing degree days by 2050 as Plymouth does at the present time. **Heating degree days** (accumulated mean temperatures below 15.5°C) reduce by a smaller percentage (between 16 and 23 per cent) and the latitudinal shift of zones is also not as great as for growing degree days. Durham, for example, still records more heating degree days by 2050 than does Plymouth under average 1961 to 1990 conditions. **Cooling degree days** (accumulated mean temperatures above 18°C) increase everywhere, although owing to the small totals at present the percentage changes are not always meaningful. Nevertheless, cooling degree days nearly double by 2050 for locations such as Oxford and Santon Downham in the south of England.

For many environmental and economic processes it is changes in the frequency of the heaviest rain-

Figure 15.9 Flooded meadows, Shalford, near Guildford, Surrey, after very heavy winter rains. The climate change scenario presented in this chapter suggests that winter flooding episodes similar to this may increase in the future due to increased rainfall intensity.

fall events that will be of greatest significance – for example, for soil erosion, agriculture and flood estimation. Suggestions from complex model experiments such as HADCM2 are that such frequencies will increase in a warmer world, and that for the British Isles these increases would be greatest in the winter months (Figure 15.9). Similar analyses of average daily wind speeds also suggest an increase in frequencies of the stronger winds, although more so over the northern British Isles than in the south. The climate models from the Hadley Centre used here also suggest an increase in storminess over the British Isles with global warming. Future climate warming is likely to lead to a reduction in snow cover – for example, fewer days with snow lying – although reductions in the number of snowfalls, while likely, may not be so great or so certain. Changes in the synoptic origins of the winter precipitation may also be important in determining the effect of climate change on snowfall.[15]

Potential evapotranspiration (PET) is a calculated quantity usually derived from average temperature, vapour pressure, wind speed and **global radiation**. Each of these variables has a different relative influence on the eventual PET estimate, with temperature and relative humidity having the strongest influences. Although a warming climate is most likely to lead to increases in PET – some general estimates suggest about a 5 per cent increase for 1°C of warming – changes in relative humidity may greatly alter this value or even lead to reductions in PET. Changes in plant **stomatal conductivity** resulting from higher carbon dioxide concentrations may also be important for transpiration losses. Previous sensitivity work in the British Isles[16] has suggested that such water use efficiency gains could offset a substantial proportion of any climatically induced increase in PET.

CONCLUSIONS

Research is continuing into making better predictions of climate change at regional scales and into better understanding how environmental assets and economic activities are affected by climate change. Establishing for certain how climate change will affect the British Isles over the next few decades is very difficult. Not only is there uncertainty in the nature of the climate change which will affect the region, but the impacts will be conditioned by the nature of economic growth in this region and world-wide, by the advent of new technologies and by the changes in human aspirations and behav-

iour. What seems certain, however, is that managers will need to be more aware of the possible range and nature of climate changes and incorporate such possibilities into their medium- to long-term planning. This is already being adopted as good practice in a number of sectors such as the water and insurance industries.

NOTES

1 The exact wording from this report was that, 'the balance of evidence suggests that there is a discernible human influence on global climate'. From Intergovernmental Panel on Climate Change, *Climate Change 1995: the science of climate change*, J.T. Houghton, L.G. Meiro Filho, B.A. Callander, N. Harris, A. Kattenburg and K. Maskell (eds), Cambridge, Cambridge University Press, 1996, 572 pp.

2 R. Russell-Jones and T.M.L. Wigley (eds), *Ozone Depletion: health and environmental consequences*, Chichester, J. Wiley and Sons, 280 pp.

3 J. Leggett, W.J. Pepper and R.J. Swart, 'Emissions scenarios for the IPCC: an update', in J.T. Houghton, B.A. Callander and S.K. Varney (eds), *Climate Change 1992: the supplementary report to the IPCC scientific assessment*, Cambridge, Cambridge University Press, 1992, pp. 75–95.

4 These figures are taken from S.C.B. Raper, T.M.L. Wigley and R.A. Warrick, 'Global sea level rise: past and future', in J.D. Milliman and B.U. Haq (eds), *Sea-level Rise and Coastal Subsidence: causes, consequences and strategies*, Dordrecht, Kluwer, 1996, pp. 11–45.

5 Results from successive versions of this model are discussed in T.M.L. Wigley and S.C.B. Raper 'Implications of revised IPCC emissions scenarios', *Nature*, 1992, vol. 357, pp. 293–300; Raper *et al.*, op. cit.; and A. Kattenburg, F. Giorgi, H. Grassl, G.A. Meehl, J.F.B. Mitchell, R.J. Stouffer, T. Tokioka, A.J. Weaver and T.M.L. Wigley, 'Climate models – projections of future climate', in J.T. Houghton, L.G. Meiro Filho, B.A. Callander, N. Harris, A. Kattenburg and K. Maskell (eds), *Climate Change 1995: the science of climate change*, Cambridge, Cambridge University Press, 1996, pp. 285–357.

6 P.D. Jones, 'Hemispheric surface air temperature variability – a re-analysis and an update to 1993', *Journal of Climate*, 1994, vol. 7, pp. 1794–1802.

7 These results are from Raper *et al.*, op. cit.. Solar and volcanic influences are examples of causes of natural variability which are not simulated in the model.

8   B.D. Santer, K.E. Taylor, T.M.L. Wigley, J.E. Penner, P.D. Jones and U. Cubasch, 'Towards the detection and attribution of an anthropogenic effect on climate', *Climate Dynamics*, 1995, vol. 12, pp. 77–100.

9   From the observed global temperature record plotted in Figure 15.8, the warming from 1966–75 to 1976–85 was 0.12°C and from 1976–85 to 1986–95 it was 0.19°C.

10 See Raper *et al.*, op. cit.; and R.A. Warrick, C. Le Provost, M. Meier, J. Oerlemans and P. Woodworth, 'Changes in sea level', in J.T. Houghton, L.G. Meiro Filho, B.A. Callander, N. Harris, A. Kattenburg and K. Maskell (eds), *Climate Change 1995: the science of climate change*, Cambridge, Cambridge University Press, 1996, pp. 358–405.

11 There have been three main earlier experiments performed by the UK Met. Office; in 1986 (UKLO; C.A. Wilson and J.F.B. Mitchell, 'A doubled CO_2 climate sensitivity experiment with a global climate model including a simple ocean', *Journal of Geophysical Research*, 1987, vol. 92, pp. 13,315–43), in 1989 (UKHI; J.F.B. Mitchell, C.A. Senior and W.J. Ingram, 'CO_2 and climate: a missing feedback', *Nature*, 1989, vol. 341, pp. 132–4), and in 1991 (UKTR; J.M. Murphy and J.F.B. Mitchell, 'Transient response of the Hadley Centre coupled ocean–atmosphere model to increasing carbon dioxide. Part II, spatial and temporal structure of the response', *Journal of Climate*, 1995, vol. 8, pp. 57–80).

12 J.F.B. Mitchell, R.A. Davis, W.J. Ingram and C.A. Senior, 'On surface temperature, greenhouse gases and aerosols: models and observations', *Journal of Climate*, 1995, vol. 8, pp. 2364–86.

13 M.G.R. Cannell and C.E.R. Pitcairn (eds), *Impacts of the Mild Winters and Hot Summers in the UK in 1988–1990*, London, UK Dept of the Environment, HMSO, 1993, 154 pp.

14 Climate Change Impacts Review Group, *The Potential Effects of Climate Change in the United Kingdom*, London, Dept of the Environment, HMSO, 1991, 124 pp. Also Climate Change Impacts Review Group, *Review of the Potential Effects of Climate Change in the United Kingdom*, London, Dept of the Environment, HMSO, 1996, 247 pp.

15 S.J. Harrison, 'Differences in the duration of snow cover on Scottish ski-slopes between mild and cold winters', *Scottish Geographical Magazine*, 1993, vol. 109, pp. 37–44.

16 N.W. Arnell and N. Reynard, *Impact of Climate Change on River Flows in the UK*, Wallingford, Institute of Hydrology Report to the UK Department of the Environment, 1993, 129 pp.

GENERAL READING

Climate Change Impacts Review Group, *Review of the Potential Effects of Climate Change in the United Kingdom*, London, Dept of the Environment, HMSO, 1996, 247 pp.

Intergovernmental Panel on Climate Change, *Climate Change 1994: Radiative Forcing of Climate Change and an Evaluation of the IPCC IS92 Emissions Scenarios*', J.T. Houghton, L.G. Meira Filho, J. Bruce, H. Lee, B.A. Callander, E. Haites, N. Harris and K. Maskell (eds), Cambridge, Cambridge University Press, 1995, 339 pp.

Intergovernmental Panel on Climate Change, *Climate Change 1995: The Science of Climate Change*', J.T. Houghton, L.G. Meiro Filho, B.A. Callendar, N. Harris, A. Kattenburg and K. Maskell (eds), Cambridge, Cambridge University Press, 1996, 572 pp.

R.A. Warrick, E.M. Barrow and T.M.L. Wigley (eds), *Climate and Sea Level Change: Observations, Projections and Implications*, Cambridge, Cambridge University Press, 1993, 424 pp.

16

CLIMATE BEYOND THE TWENTY-FIRST CENTURY

Clare Goodess and Jean Palutikof

A first rate theory predicts; a second rate theory forbids; and
a third rate theory explains after the event.
A.I. Kitaigorodskii

INTRODUCTION

It makes good sense to consider climate change and its impacts on agricultural and economic activity over the next century in the planning and decision-making process (see Chapter 15). The need to consider climate change beyond the twenty-first century, and thousands of years into the future, is less obvious. The **Last Glacial Maximum** (about 21,000 years BP), for example, is considered relatively recent in geological terms (see Chapter 4), but is prehistoric (part of the **Palaeolithic** period) from a human perspective.[1] The dramatic changes in human technology since this prehistoric period make it difficult to envisage how society might change over the next 10,000 years, let alone the next 100,000 years, or to accept that climate change so far into the future is of anything other than academic interest.

There are two reasons, however, why it is becoming increasingly important to consider very-long-term climate change. First, the impacts of some human activities may persist much longer than the typical decade or century planning time-scales. The most obvious example of such an activity is the generation and disposal of radioactive waste. A number of countries have active research programmes which are attempting to foresee natural and human-induced climate changes over the next million years and to assess the safety implications for underground disposal. The authors of this chapter, for example, are involved in a project to assess future climate states in the United Kingdom which is focused on the potential low- and intermediate-level nuclear waste repository site at Sellafield.[2]

The second major justification for research into very-long-term climate change in the future concerns the potential impacts of human activity on climate itself. The evidence presented in the previous chapter indicates that human-induced global warming will persist through the twenty-first century. It is important to know what will happen beyond this time. Will the climate system return to 'normal'? If so, when, and what is the 'normal' pattern of change? Or will global warming itself bring about irreversible changes in the climate system? If research suggests that human-induced changes are likely to be irreversible (or may persist for an unacceptably long time), the pressure on the international community to take effective remedial action in the near-term may increase.

PREDICTING CLIMATE THOUSANDS OF YEARS INTO THE FUTURE

A common reaction to researchers such as ourselves investigating climate thousands of years into the future is 'weather forecasters can't even get tomorrow's weather right, so how can you possibly predict climate 5,000 years from now?' Clearly it is not possible to make long-term climate predictions in the same way that weather forecasts are made. It is, however, possible to say something about the range of conditions likely to be experienced in the British Isles over the next 100,000 years or so, and how this range of conditions compares with the range of conditions experienced in the past.

In order to make reliable weather forecasts, the meteorologist must look beyond the British Isles (see Chapter 14). The controversy over the forecasting of the October 1987 storm, for example, demonstrates the need for reliable and detailed information about conditions in the North Atlantic. For assessment of very long-term future climate change, what is happening elsewhere becomes even more important. Global changes in ice volume, sea-level, and in atmospheric and oceanic circulation had major impacts on the climate of the British Isles over the **Quaternary** period (the last 1.8 million years).[3] Future changes to the climate system of similar magnitude are also likely to have major impacts on British climate.

Weather forecasters have two major tools at their disposal – the instrumental record of immediate past and current weather conditions, and numerical weather prediction models (see Chapter 14). Similarly, climatologists have two major investigative tools. First, the reconstructed record of long-term past climate change and, second, a range of climate models of varying complexity. In the next section, we consider the extent to which the past climate record can be used as a guide to the future. We then discuss the use of climate models to investigate future climate changes over the next 100,000 years. We start by assuming that there will be no human intervention over this period, but go on to show that this is an unrealistic assumption and that the future

will not be exactly like the past. The use of climate models to investigate the relationship between **anthropogenic** forcing (e.g., the enhanced **greenhouse effect**) and **natural forcing** (e.g., orbital changes) is then discussed. Finally, we consider the contradictions and uncertainties associated with the prediction of future very-long-term climate change.

THE PAST AS A GUIDE TO THE FUTURE

Past climate can be used as a (partial) guide to future climate. The reconstructed climate record for the period since the appearance of permanent high-latitude ice-sheets (1.8 million years BP to the present day) tells us about the range of conditions in the past over the time-scales of interest (see Chapters 4 and 5). Analysis of this record should allow us to identify the forcing mechanisms operating on these very long time-scales and whether or not these natural mechanisms are likely to operate in the same, or similar, way in the future. If so, we can reasonably assume that the same, or similar, range of climate conditions will occur in the future.

Climate varies on all time-scales in response to a mix of random and periodic mechanisms or forcing factors (Figure 16.1).[4] Random variations caused by **stochastic** mechanisms can be thought of as 'white noise' and are, by their very nature, unpredictable. Of more interest to the climatologist are the peaks of variability associated with periodic forcing mechanisms which are, in theory, predictable. Periodic variations include the familiar annual temperature cycle and the more controversial 11-, 22- and 80-year-long cycles attributed to solar variability.[5] Very much longer scales of variability have also been identified. The most notable of these occur at about 100,000, 41,000, 23,000 and 19,000 years and are attributed to systematic and periodic variations in the Earth's orbit.[6] These peaks are strongly evident in ocean cores and other palaeoclimatic records of the Quaternary period.[7]

There are a number of reasons why it is not unreasonable to assume that, in the absence of human intervention, the major mechanisms of climate

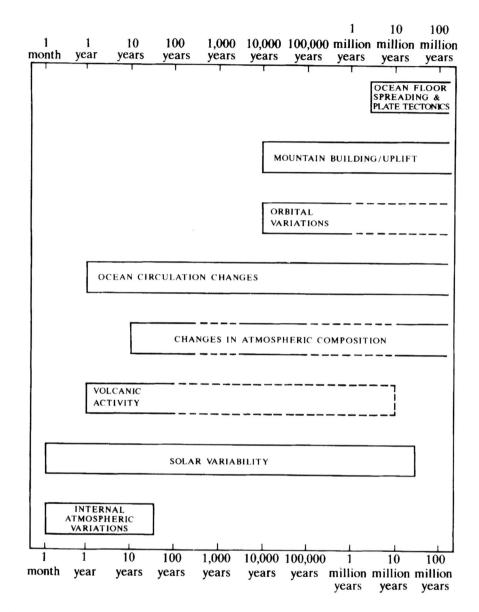

Figure 16.1 Major mechanisms of climate change operating on time-scales of one month to 100 million years (excluding feedback mechanisms). Solid lines indicate time-scales over which these mechanisms have the major impact. Dashed lines indicate time-scales over which they may have some impact.

change operating over the next million years or so will be similar to those operating over the Quaternary period. This period was characterised by a strong periodic pattern of variability: the succession of glacial–interglacial cycles. It is widely, although not universally, considered that **orbital variations** are a major cause of these glacial–interglacial cycles.[8] No major changes in geographical boundary conditions, due for example to **continental drift**, have occurred over this period. Thus, taking the Quaternary period as a guide to the next 100,000 or one million years, we have a characteristic pattern of change (glacial–interglacial cycles), a likely cause (orbital changes), and a characteristic time-scale for the forcing mechanism (from about 20,000 to 100,000 years).

There are, however, two major caveats to proceeding on this basis. First, **orbital forcing** can only ever explain part of the observed variability. Random mechanisms and shorter time-scale mechanisms (such as solar variability and volcanic eruptions) will also occur and will impose shorter-term fluctuations on the broad patterns of change. Second, if major changes in boundary conditions *do* occur in the future, it will not be legitimate to use the past as a guide to the future. It is unlikely that continental drift or **tectonic uplift** will have a major impact on the time-scales considered here. The retreat of the polar ice-caps due to global warming could, however, have a major impact. For now, we proceed on the assumption of no human intervention in the climate system, but return to this issue in a later section.

The simplest approach is to use the Quaternary period as an empirical guide to the future, although this causes some difficulty in the British Isles because of the lack of long continuous land-based palaeoclimatic records. Nonetheless, we can make some general statements based on an analysis of ocean core records, long pollen records from the European continent and fragmentary land-based records from the British Isles.[9] The present warm interglacial period, the **Holocene**, began about 10,000 years ago (see Chapter 5). Over the last four glacial–interglacial cycles, a period of about 500,000 years, we estimate

that interglacial conditions have characterised the British Isles for only about 9 per cent of the time. An empirical analysis of the British Isles record supports the conclusion reached over fifteen years ago by John and Katherine Imbrie: 'statistically speaking then, the present interglacial is already on its last legs, tottering along at the advanced age of 10,000 years and can be expected to end within the next 2,000 years'.[10]

It is, however, unnecessary to assume that the next glacial–interglacial cycle will be exactly like the last. Climatologists have sufficient understanding of the most likely causal mechanism (orbital changes) to run physically based models of the next glacial–interglacial cycle.

MODELLING THE NEXT GLACIAL–INTERGLACIAL CYCLE

The climate models which have been used to simulate future glacial–interglacial cycles were not designed specifically for this purpose, but rather to determine whether orbital forcing is a plausible cause of the Quaternary glacial–interglacial cycles. The first orbital-based models were developed in the mid-1970s and are relatively simple, statistical models.[11] Ice-sheet models, one-dimensional energy balance models and three-dimensional models, such as global climate models, have also been used to investigate orbital forcing. The large computing requirements mean that it is not possible to run global climate models for tens or hundreds of thousands of years into the future. It is, however, possible to use a sectorally averaged (essentially a 2.5-dimensional) model. A model of this type is under development at the University of Louvain-la-Neuve in Belgium by a team led by André Berger[12] and we refer to this model as the LLN palaeoclimate model.

At present only the Northern Hemisphere is simulated by the LLN palaeoclimate model. It consists of a climate model, coupled to models of the Greenland, North American and Eurasian ice-sheets and to a relatively simple ocean model. The model does not have a conventional geography. Instead, each 5°

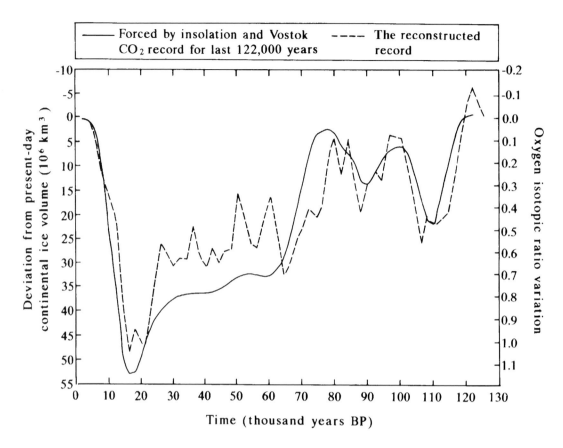

Figure 16.2 Global ice volume simulated by the Louvain-la-Neuve (LLN) model forced by insolation and the Vostok CO$_2$ record for the last 122,000 years (solid line) compared with the reconstructed record[15] (dashed line).

latitude zone is divided into seven sectors representing open ocean, sea-ice, ice-free land, snow-covered land, and three ice-sheets. The model is forced by daily insolation at each latitude band calculated from the orbital changes.[13] In some simulations it is also forced additionally by natural changes in the atmospheric concentration of carbon dioxide which are known to accompany global changes in ice volume.[14]

The LLN palaeoclimate model has been used to simulate ice volume over the last glacial–interglacial cycle (see Figure 16.2). Model performance was assessed by comparison with an oxygen isotope record of global ice volume.[15] The model and reconstructed

ice volumes agreed reasonably well. The ability of the model to reproduce the main features of the last two glacial–interglacial cycles, including the rapid melting of ice at the end of glacial periods, and the fact that it incorporates processes known to be important on these time-scales, makes it a very appropriate model for simulations of the next glacial–interglacial cycle.

One of the first simulations of long-term future climate performed with a physically based model was completed using an early version of the LLN palaeoclimate model. This simulation provided an estimate of Northern Hemisphere ice volume over the next 80,000 years (Figure 16.3). It indicated that, if we

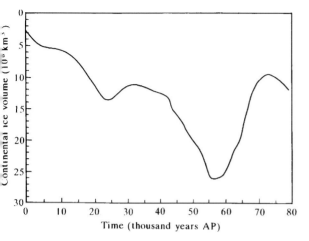

Figure 16.3 Future continental ice volume, including the Greenland ice sheet, as simulated by the Louvain-la-Neuve model with orbital forcing only.

ignore the potential for human-induced change, the world's climate over the next few thousand years would begin a slow deterioration towards glacial conditions. The final cooling extreme at 60,000 years after present (AP) is of similar magnitude to that at 21,000 years BP, the Last Glacial Maximum. Evidence from this model and earlier, simpler models indicates that conditions as warm as the present would not again be reached until 120,000 years AP.

The reliability of the output in Figure 16.3 is affected by limitations in the model, including those related to the representation of topography, clouds and deep-ocean circulation, and the omission of the Southern Hemisphere. It should also be noted that this simulation does not incorporate the effects of future natural changes in atmospheric carbon dioxide concentrations.[16] In simulations which incorporate past natural changes in carbon dioxide concentrations, about 50 per cent of the temperature change and about 30 per cent of the ice-volume change from the Last Glacial Maximum to the Holocene inter-glacial maximum is attributed to carbon dioxide changes.[17] The carbon dioxide forcing for the past is known but, for simulations of the future, natural carbon dioxide variations have to be predicted. This

is not an easy task and is likely to introduce additional uncertainties to long-term future simulations of climate.

Output from models such as the LLN palaeoclimate model tells us something about global and hemispheric conditions, but does not tell us much about conditions in the British Isles over the next glacial–interglacial cycle.

The next glacial–interglacial cycle in the British Isles

The simplest assumption that can be made for this purpose is that periods of similar global ice volume will be associated with similar climatic conditions in the British Isles. On this basis, the model results discussed in the previous section indicate that conditions at 60,000 years AP (the next glacial maximum) will be similar to those observed at 21,000 years BP (the Last Glacial Maximum), and that conditions at 120,000 years AP (the next interglacial maximum) will be similar to those at 125,000 years BP (the height of the last, **Ipswichian**, interglacial before the present one). If we represent climate change in a stylised form as a succession of climate states we can use this approach to interpret the entire global ice-volume record simulated for the next 125,000 years as the succession of climate states likely to be experienced in the British Isles.

We use four climate states to describe the range of conditions experienced in the British Isles over the Quaternary period. These are defined by a widely used climate classification system and can, therefore, be associated with present-day **climate analogue regions**.[18] The four climate states and their analogue regions are: temperate (present-day British Isles); **boreal** (present-day north Norway and north Sweden); **periglacial** or tundra (present-day south-west Alaska and north Russia); and glacial (present-day south-west Greenland).

Output from the LLN palaeoclimate model for the next 125,000 years, and from other simpler climate models, provides the basis for the climate index shown here (Figure 16.4). This index represents the succession of climate states likely to occur in

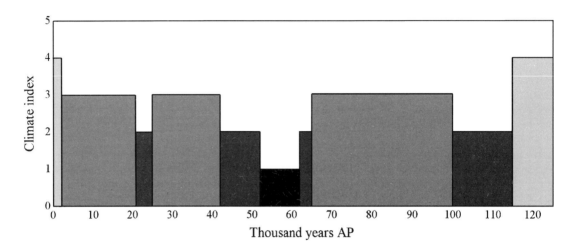

Figure 16.4 A climate index showing the succession of major climate states likely to be experienced in the British Isles over the next 125,000 years constructed from orbitally forced climate model output. The potential effects of enhanced greenhouse gas warming are *not* included.

northern parts of the British Isles over the next 125,000 years. The succession will be somewhat different in the southern parts of the British Isles not covered by ice-sheets during the last (or next) glacial maximum. This climate index suggests that the next 125,000 years will be dominated by cooler and drier, boreal and periglacial conditions. Conditions as severe as the Last Glacial Maximum (with average annual temperatures in the British Isles 10°C to 20°C lower than present), or as warm as the present day, are relatively rare and are not likely to occur until at least 52,000 and 115,000 years AP respectively.

Like the last glacial–interglacial cycle, the next glacial–interglacial cycle is likely to be accompanied by major changes in sea-level. At the height of the last glaciation, for example, global sea-level was about 130 m lower than present because of the huge amount of water locked up in the continental ice-sheets (Chapter 4). During the next glacial period, indicated at 52,000 to 62,000 years AP in our climate index, it is likely that much of the area covered by the North Sea at present will again be dry land, and that the Irish Sea will be reduced to a small channel. It is also probable that the warming influence of the Gulf Stream will be absent as the **thermohaline circulation** of the ocean shuts down again.[19] These changes will bring far more continental conditions to the British Isles. During this future glacial period, it is likely that the atmospheric circulation over the British Isles will be dominated by easterly airflow, associated with the re-establishment of a glacial **anticyclone** system over the Fennoscandian ice-sheet.[20]

Our interpretation of the model output suggests that major climate changes are to be expected in the

British Isles over the next few tens of thousands of years. At present, human communities exist in northern Scandinavia, Russia and Alaska, the analogue regions for the boreal and periglacial climate states. Permanent communities also exist in coastal Greenland, our glacial analogue region. So, even without major technological advances, some human settlement should be possible in the British Isles throughout the next 125,000 years, although it is likely that present-day population densities will be unsustainable in the more northern areas during the coldest periods. Technological changes may, however, radically alter our perception of habitable and uninhabitable climates.

Knowledge of past climate and the range of forcing mechanisms (see Figure 16.1) tells us that the real climate is more variable than implied by output from models such as the LLN palaeoclimate model or by the climate index shown here. Evidence from pollen, beetle and ocean sediment records indicates that warmer periods – **interstadials** – occurred throughout the last glaciation.[21] The causes of these events are not fully understood, but their rapid onset and short duration suggests that they are related to mode changes in the thermohaline circulation. It is likely that similar events will occur during future periods of glaciation. Thus, the colder climate states indicated in the climate index may be interrupted by warmer periods, lasting from a few hundred to possibly a thousand years. During the deglaciation period leading up to the next interglacial period, it is possible that at least one minor glacial re-advance may occur, similar in magnitude to the **Younger Dryas** at about 11,000 years BP (see Chapter 5).

ANTHROPOGENIC EFFECTS: WHY THE FUTURE WON'T BE LIKE THE PAST

So far, we have implied that natural mechanisms of climate change will operate over the next one million years or so in the same way that they have operated over the last million years or so. We have, however, stressed that it is only legitimate to use the past as a guide to the future if it is assumed that there will be no human intervention and no major changes in boundary conditions. The evidence presented in Chapter 15 indicates that changes in the **cryosphere** (i.e., changes in the extent of ice-sheets, sea-ice, glaciers and snow) are likely to accompany anthropogenic global warming. Such changes may be critical for the future response of the climate system to orbital forcing.

The polar ice-sheets are considered to play a particularly important role in transforming the relatively weak **insolation** changes associated with orbital forcing into global glacial–interglacial cycles. They are, for example, involved in an ice–**albedo** feedback effect. If some external mechanism, such as a decrease in incoming radiation, initiates ice-sheet growth, then the surface albedo increases. More long-wave radiation is reflected back to space and the surface cools further. Thus the ice–albedo feedback is a positive feedback mechanism, reinforcing the initial forcing mechanism. It has been suggested that the impact of global warming on the global cryosphere may seriously limit the effectiveness of the ice–albedo feedback and other positive feedback mechanisms. At one extreme, it has been proposed that these mechanisms could be weakened to such an extent that the initiation of further glaciation will be prevented. This is the so-called 'irreversible greenhouse effect' and is one of three possible patterns which can be used to describe the relationship between global warming and orbital forcing.

The simplest assumption that can be made is of a relatively brief (say, 1,000-year) period of global warming followed by a return to the 'natural pattern' of glacial–interglacial cycles. The second possibility is that, following a longer period of global warming (up to 10,000 years in length), the next glaciation will be delayed and will be less severe. The third possibility is the irreversible greenhouse effect. In order to determine which of these patterns is most likely, we need to address two major issues. First, how high will atmospheric greenhouse gas concentrations rise, and for how long will they remain at enhanced levels? Second, what can climate models tell us about the relationship between the enhanced greenhouse effect and orbital forcing?

How high will greenhouse gas concentrations rise?

There are still major uncertainties concerning the operation of the global carbon cycle and concentrations of the main greenhouse gas, carbon dioxide.[22] These uncertainties are reflected in the range of greenhouse gas **emissions scenarios** produced by the Intergovernmental Panel on Climate Change, although this panel has not attempted to produce scenarios beyond AD 2100.[23] Without a better understanding of the global carbon cycle it will be difficult to produce reliable estimates of future greenhouse gas concentrations for the longer-term future or to determine the likely persistence of the enhanced greenhouse effect.

Despite the large uncertainties, some limited attempts have been made to investigate atmospheric carbon dioxide concentrations thousands of years into the future using simple global carbon cycle models.[24] One model has been run over one million years, assuming that all recoverable fossil fuel reserves will be burnt.[25] In the 'save fuel' scenario, reserves are exhausted in about AD 2400, while in the 'burn fuel' scenario reserves run out after about one hundred years. Two other scenarios reflect continued deforestation (the 'burn forests' scenario), and an immediate halt to deforestation (the 'save forests' scenario). The response of atmospheric carbon dioxide concentrations to these four scenarios over the next 5,000 years is shown in Figure 16.5. The highest simulated atmospheric carbon dioxide concentration of about 2,100 ppmv occurs in about AD 2400 when the 'burn fuel' and 'burn forests' scenarios are combined. In this case, the model takes over one million years to return to present-day carbon dioxide concentrations. These model results provide a rare and tentative look at how carbon dioxide concentrations may evolve thousands of years into the future.

Carbon dioxide is not the only greenhouse gas, but no attempts so far have been made to investigate long-term changes in other greenhouse gases such as methane. Recently published results from global climate models indicate that the global warming observed over the last century is less than expected due to greenhouse gases alone because of the offsetting cooling effect of sulphate **aerosols** (see Chapter 15). Estimates of the present-day sulphate aerosol concentrations, the spatial variations of these concentrations, their atmospheric lifetime, and the magnitude of their direct and indirect radiative effects, are all highly uncertain.[26] It is not, therefore, possible to predict how the relative balance of sulphate aerosol forcing and greenhouse gas forcing may change in the longer-term future.

Clearly it is very difficult to make judgements about energy use over the next few thousand years and thus predict how greenhouse gas forcing will operate. Nevertheless, the results presented here indicate that, unless action is taken to reduce emissions and to stop deforestation, peak carbon dioxide (or **carbon dioxide equivalent**) concentrations of 1,000 to 2,000 ppmv may be reached within a few centuries. Greenhouse gas concentrations could remain at enhanced levels for hundreds of thousands of years into the future.

The relationship between the enhanced greenhouse effect and orbital forcing

Three-dimensional global climate models

One of the longest published climate change experiments comprised a set of three 500-year simulations performed with a coupled atmosphere–ocean global climate model (Figure 16.6).[27] The first simulation was of climate with current carbon dioxide concentrations. In the second simulation, carbon dioxide concentrations increased by 1 per cent per annum, until they stabilised after 140 years at carbon dioxide levels four times those at present. In the third simulation, carbon dioxide concentrations again increased by 1 per cent per annum, but stabilised after seventy years at levels twice those of today. Over the first 140 years of the carbon dioxide quadrupling simulation, global-average surface air temperature increased by about 5°C (Figure 16.6). Although carbon dioxide concentrations then stabilised, global temperature increased by a further 1.5°C by the end of the simulation. This residual warming was due

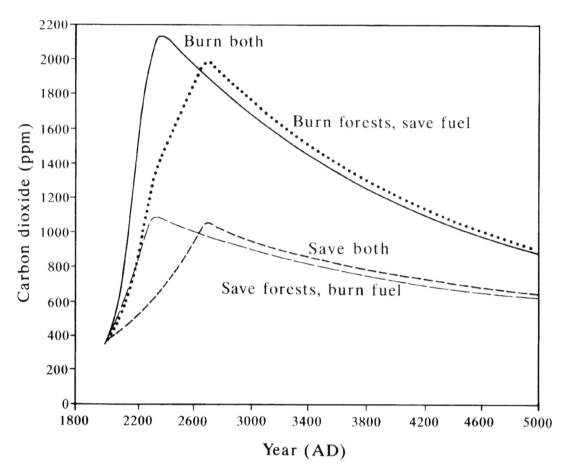

Figure 16.5 Atmospheric carbon dioxide concentrations under various long-term scenarios for fossil fuel combustion and forest clearance (simulated using Walter and Kasting's 1992 global carbon cycle model[25]).

mainly to the **thermal inertia** of the deep oceans. It implies that global temperature is likely to continue rising for hundreds of years after greenhouse gas concentrations stabilise.

One–dimensional climate models

A one-dimensional upwelling-diffusion energy-balance model (similar to that described in Chapter 15) has been used to investigate the global temperature response to 'unrestricted' carbon dioxide emis-sions over the next 10,000 years, and to compare this with the climate response to orbital forcing.[28] The global average temperature response to an emissions scenario in which carbon dioxide concentrations six times those at present (i.e., about 2,000 ppmv) are reached by between AD 2100 and 2300 and in which subsequent carbon dioxide concentrations remain just above present-day levels up to 50,000 years AP, is shown in Figure 16.7.[29] Depending on the assumed **climate sensitivity**, the maximum increase in global-average temperature is between 4° and 13°C and occurs between AD 2200 and

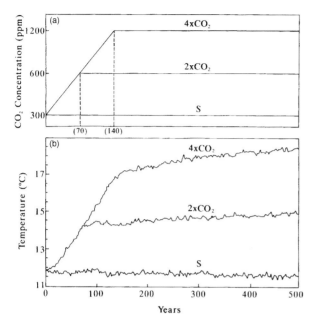

Figure 16.6 Forcing scenarios and output from three 500-year simulations with the coupled atmosphere-ocean GFDL GCM[27]: S, standard integration; $2 \times CO_2$: CO_2 increased by 1 per cent per annum until the concentration reaches two times the present value; $4 \times CO_2$: CO_2 increased by 1 per cent per annum until the concentration reaches four times the present value. (a) CO_2 concentration (ppm), (b) global temperature.

2400. After 10,000 years, global temperature is still between 2° and 5°C warmer than present. This compares with a global temperature change simulated by the same model of only 0.04°C at 10,000 years AP due to orbital forcing alone. The authors, Kwang-Yul Kim and Thomas Crowley, conclude that anthropogenic warming may well dominate orbital cooling for at least the next 10,000 years.[30]

This one-dimensional model has also been used to investigate the response of global temperature to the long-term carbon dioxide emissions scenarios shown in Figure 16.5, and to compare this response with temperature changes reconstructed from ocean cores (Figure 16.8).[31] With unrestricted emissions and a high or medium climate sensitivity, the simulated warming leads to a global temperature higher than

any 'natural' temperatures found over the last 600 million years. Even with restricted carbon dioxide emissions (the combined 'save fuel' and 'save forests' scenario of Figure 16.5) and a low climate sensitivity, the simulated global temperature is higher than any 'natural' temperatures found over the last 150 million years. These model results suggest that future global warming may generate temperature changes over the next few centuries which are very large even on the geological time-scale.

2.5-dimensional Models: The LLN Palaeoclimate Model

The LLN palaeoclimate model does not have a carbon cycle. The effects of anthropogenic carbon dioxide must, therefore, be simulated indirectly, or future carbon dioxide concentrations must be prescribed. The first approach was adopted in one of the first tests of the sensitivity of orbital forcing to global warming performed by André Berger in 1991.[32] In this 80,000-year sensitivity experiment, the effects of anthropogenic carbon dioxide were prescribed by removing the Greenland ice-sheet and running the model with present-day carbon dioxide concentrations. Orbital-related insolation forcing was used in both the control (Figure 16.3) and perturbed (Figure 16.9) simulations. The main results were that, as a consequence of anthropogenic carbon dioxide in the atmosphere, ice-sheets do not reappear in the Northern Hemisphere until about 15,000 years AP, and the next glaciation is delayed by 2,000 to 3,000 years and is less extensive than it would otherwise be (Figure 16.9).

The representation of global warming in the experiment described above is crude, although not entirely unrealistic. Other model-based studies and observational studies of the past behaviour of the Greenland ice-sheet suggest that it *will* be much reduced in extent, and may eventually disappear completely, during a prolonged period of global warming.[33]

The implications for the British Isles can be explored by constructing a climate index which

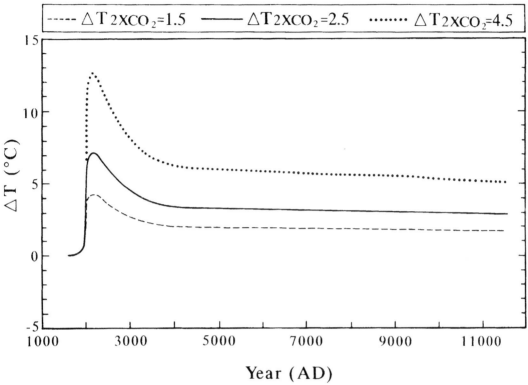

Figure 16.7 Global-average temperature change over the next 10,000 years as a response to elevated CO_2 concentrations of 2,000 ppmv between AD 2100 and 2300, and concentrations slightly higher than today thereafter (using Kim and Crowley's 1994 energy balance model[28]). The three simulations result from different values of the climate sensitivity being applied: 1.5°C, 2.5°C and 4.5°C.

incorporates the effects of anthropogenic global warming, using the same method employed to construct Figure 16.4, and based on the evidence discussed above. The new index – Figure 16.10 – commences with a 1,000-year period of subtropical or Mediterranean conditions representing greenhouse warming, followed by a prolonged, 24,000-year, period of temperate conditions. The glacial period from 52,000–62,000 years AP, seen in Figure 16.4, is replaced by a period of periglacial conditions, from 50,000–65,000 years AP. This period would be comparable with the Younger Dryas, with ice-caps present only in the highest, northern mountain areas of the British Isles. The conclusion is that global warming may restrict and delay, but not prevent, the re-advance of ice-sheets over the British Isles.

In an as yet unpublished set of sensitivity experiments for the next 150,000 years, Paul Burgess used the LLN model to compare the climate response to two different future carbon dioxide concentrations.[34] The first case of 280 ppmv of carbon dioxide (the average interglacial concentration) is compared with the climate response to a stepwise reduction in carbon dioxide concentration from 350 ppmv to 280 ppmv over the first 13,000 years of the model simulation and thereafter a constant 280 ppmv. Although carbon dioxide concentrations are equal (280 ppmv) in both experiments after 13,000 years, differences in ice volume persist until 65,000 years AP (Figure 16.11), and occur despite a period

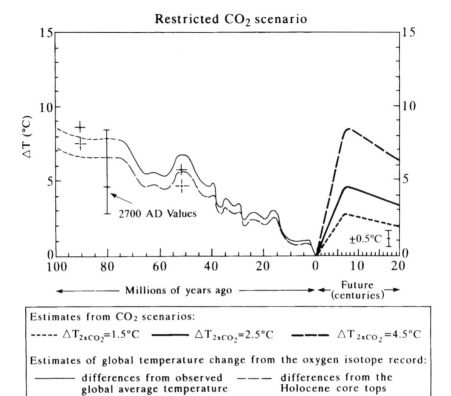

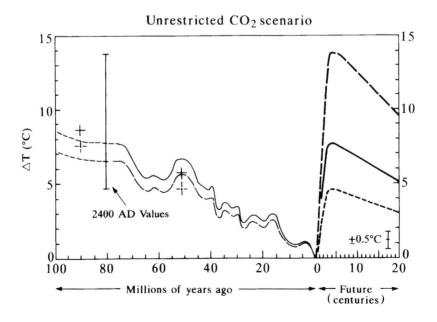

of about 18,000 years when Northern Hemisphere ice-sheets are absent from both simulations. It is concluded that the simulated climate system shows a long-term memory response to these quite modestly different carbon dioxide concentration scenarios.

Sensitivity experiments performed with the LLN palaeoclimate model indicate that the climatic effects of enhanced greenhouse gas forcing may persist for tens of thousands of years. Figure 16.11 shows that on these long time-scales the climate system is very sensitive to different magnitudes of greenhouse gas forcing, and that even relatively modest increases in carbon dioxide forcing (in comparison with the scenarios shown in Figure 16.5) may severely restrict the future long-term development of Northern Hemisphere ice-sheets.

The future pattern of change

On the basis of the model evidence discussed here, and assuming that greenhouse gas concentrations will eventually return to levels close to those of the present day, we consider that, of the three possible outcomes describing the relationship between enhanced greenhouse warming and orbital forcing, the second possibility – delayed and weakened glaciation – is the most likely. The third possibility – the irreversible greenhouse effect – is considered to have a low, but non-zero, probability of occurrence. The model results which have become available over the last few years imply that the probability of an irreversible greenhouse effect is higher than we might have thought, say, five years ago.

The evidence also indicates that the next few thousand years will be very warm in geological terms.

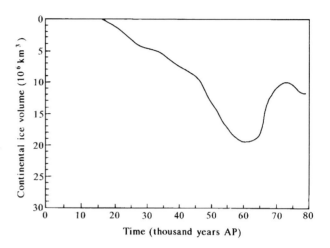

Figure 16.9 Future continental ice volume, without the Greenland ice sheet, as simulated by the Louvain-la-Neuve model, assuming orbital *and* greenhouse gas forcing (cf. Figure 16.3).

Indeed, it is likely that the British Isles will be warmer than at any time during the Quaternary period by the end of the twenty-first century.[35] The most recent model results indicate that the period of greenhouse gas-induced warming will be longer than suggested by earlier studies. On the basis of the latest evidence we could, for example, justify the extension of the 1,000-year period of subtropical or Mediterranean conditions shown in Figure 16.10 by up to 10,000 years.

CONTRADICTIONS AND UNCERTAINTIES

Climatologists have sometimes been accused of making contradictory predictions with regard to

---

Figure 16.8 Comparison of future greenhouse projections against the geologic record (using Crowley and Kim's 1995 energy balance model[31]). Curves to the left of zero represent estimates of past global temperature changes from the oxygen isotope record: dashed lines represent differences from the Holocene core tops, and continuous lines represent differences from the observed global average temperature. Crossbars indicate fitting points for calibration of oxygen isotope curves in terms of global temperature. Labelled scale on the left-hand side represents calculated values for peak warming on the right-hand side of the figure. Curves to the right of zero represent global warming estimates assuming restricted (top) and unrestricted (bottom) CO_2 emissions scenarios, using three standard values for the climate sensitivity. 'Error bars' represent a generous estimate of the range of natural variability based on records of the last 1,000 years.

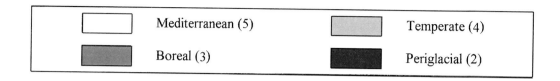

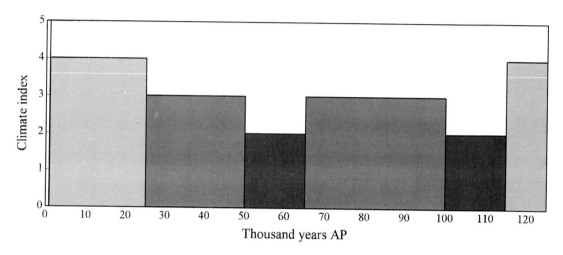

Figure 16.10 A climate index showing the succession of major climate states likely to be experienced in the British Isles over the next 125,000 years constructed from orbitally forced model output. The potential effects of enhanced greenhouse gas warming *are* incorporated (cf. Figure 16.4).

future climate. Twenty years ago, some climatologists were predicting the onset of the next ice age and now, most, although not all, climatologists are predicting global warming. In this chapter we have attempted to show that another 'Ice Age' *is* likely to follow a period of global warming, but that this Ice Age is likely to be delayed and to be less extreme than we might expect in the absence of global warming. Its onset is also many human generations into the future. Greenhouse warming and another Ice Age are not, therefore, incompatible predictions, but depend on the future time-scale considered.

There are clearly major uncertainties in any attempt to predict climate over such a long time horizon. Some of these, such as the uncertainty concerning future atmospheric carbon dioxide concentrations, have already been noted. Our understanding of exactly how the relatively weak insolation forcing associated with orbital changes is translated into glacial–interglacial cycles is still incomplete, and some researchers continue to argue that orbital changes cannot be a major cause of these cycles.[36] None of the models described in this chapter are fully realistic. The LLN palaeoclimate model, for example, does not have a Southern Hemisphere, nor does it include the deep ocean circulation. At best, the models described here incorporate two major forcing mechanisms, anthropogenic greenhouse gas forcing and orbital forcing. They do not employ the full range of forcing mechanisms known to operate over different time-scales (see Figure 16.1). And there may be other, as yet undiscovered, anthropogenic effects which may counteract or intensify the enhanced greenhouse effect.

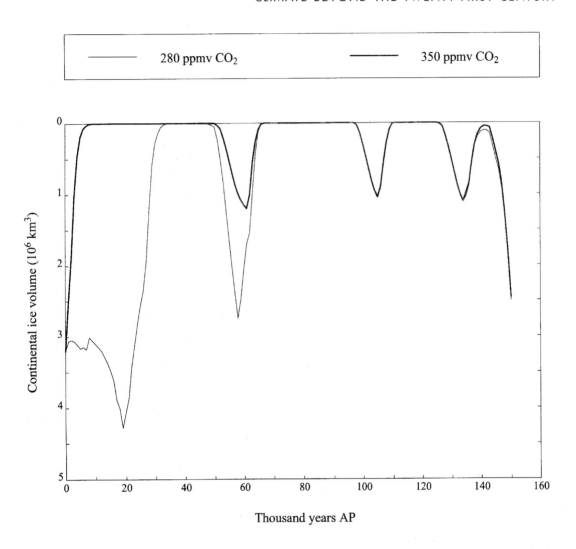

Figure 16.11 Northern Hemisphere ice volume, 0 to 150,000 years AP, simulated by the Louvain-la-Neuve model for experiments with future atmospheric CO_2 concentrations of 280 and 350 ppmv.

CLIMATE RESEARCH: THE NEXT TWENTY-FIVE YEARS

Throughout this chapter we have argued that enhanced greenhouse warming and orbital forcing will be the two major, though not the only, forcing mechanisms over the next 100,000 to one million years. Twenty-five years ago, neither mechanism was widely recognised or accepted by the international research community. The growing acceptance of these mechanisms is a reflection of the rapid advances in climate reconstruction and modelling over the last twenty-five years. It is now possible, for example, to reconstruct a high-resolution 200,000-year record of past changes in atmospheric carbon dioxide concentrations from air bubbles trapped in an Antarctic ice-core.[37] Improvements in our understanding of the climate system, more data for model validation, and greater computing power, make it possible to run complex three-dimensional models for up to 1,000 years and to run one- and two-dimensional models for hundreds of thousands of years.

Despite these advances, there are still some major uncertainties to be resolved. As our understanding of the climate system improves, it sometimes seems that more questions are raised than answered. This means that the next twenty-five years of climate research should be as exciting as the last twenty-five years have been, and that we can look forward to equally rapid advances in our knowledge.

NOTES

1 The conventional dates and names for all the geological periods discussed in this chapter are given in Table 4.1 in Chapter 4. The Last Glacial Maximum is generally assigned a ^{14}C date of 18,000 years BP, which, on the basis of the best present-day evidence, is equivalent to a calibrated age (in calendar years) of about 21,000 years BP. See Box 4.1, Chapter 4, and Box 5.1, Chapter 5, for discussions of dating techniques and problems. Climate conditions in the British Isles during the Last Glacial Maximum are also described in Chapter 4.

2 C.M. Goodess, J.P. Palutikof and T.D. Davies, *Studies of Climatic Effects and Impacts Relevant to Deep Underground Disposal of Radioactive Waste*, Nirex Safety Series, NSS/R267, 1992, 398 pp., available from UK Nirex Ltd, Harwell; S.T. Adcock, M.D.G. Dukes, C.M. Goodess and J.P. Palutikof, *A Critical Review of the Climate Literature Relevant to the Deep Disposal of Radioactive Waste*, Nirex Science Report, 1996 (in press), 292 pp., available from UK Nirex Ltd., Harwell.

3 See Chapter 4. For a discussion of the links between global and regional British Isles climate over the last glacial–interglacial cycle see C.M. Goodess, J.P. Palutikof and T.D. Davies, *The Nature and Causes of Climate Change: Assessing the long term future*, London, Belhaven Press, 1992, pp. 157–73.

4 J.M. Mitchell, 'An overview of climatic variability and its causal mechanisms', *Quaternary Research*, 1976, vol. 6, pp. 481–94; A. Berger, 'Spectrum of climatic variations and their causal mechanisms', *Geophysical*

Surveys, 1979, vol. 3, pp. 351–402; T.M.L. Wigley, 'Climate and paleoclimate: what we can learn about solar luminosity variations', *Solar Physics*, 1981, vol. 74, pp. 435–71.

5 A.B. Pittock, 'Solar variability, weather and climate: an update', *Quarterly Journal of the Royal Meteorological Society*, 1983, vol. 109, pp. 23–55; P.M. Kelly and T.M.L. Wigley, 'Solar cycle length, greenhouse forcing and global climate', *Nature*, 1992, vol. 360, pp. 328–30.

6 See Box 4.2, Chapter 4; J. Imbrie, 'A theoretical framework for the Pleistocene ice ages', *Journal of the Geological Society*, 1985, vol. 142, pp. 417–32; A. Berger and C. Tricot, 'Global climatic changes and astronomical theory of paleoclimates', in A. Cazenave (ed.), *Earth Rotation: Solved and Unsolved Problems*, Dordrecht, Reidel, 1986, pp. 111–29.

7 J. Imbrie, J.D. Hays, D.G. Martinson, A. McIntyre, A.C. Mix, J.J. Morley, N.G. Pisias, W.L. Prell and N.J. Shackleton, 'The orbital theory of Pleistocene climate: support from a revised chronology of the marine [18]O record', in A. Berger, J. Imbrie, J. Hays, G. Kukla and B. Saltzman (eds), *Milankovitch and Climate*, Dordrecht, Reidel, 1984, pp. 269–306; P.L. de Boer and D.G. Smith (eds), *Orbital Forcing and Cyclic Sequences*, Oxford, Blackwell, 1994, 559 pp.

8 See Box 4.2, Chapter 4 and reviews in Goodess *et al.*, *Nature and Causes of Climate Change*, pp. 11–50, and Adcock *et al.*, 1995, op. cit., pp. 7–81.

9 Goodess *et al.*, *Studies of Climatic Effects*, pp. 228–46.

10 J. Imbrie and K.P. Imbrie, *Ice Ages, Solving the Mystery*, London, Macmillan Press, 1979, p. 178.

11 For reviews of the range of orbital-based models see the following. A. Berger, J. Imbrie, G. Kukla and B. Saltzman (eds), *Milankovitch and Climate, Parts 1–2, Understanding the Response to Orbital Forcing*, Dordrecht, Reidel, 1984, two volumes; J. Imbrie, 1985, op. cit; A. Berger, 'Milankovitch theory and climate', *Reviews of Geophysics*, 1988, vol. 26, pp. 624–57; Goodess *et al.*, *Nature and Causes of Climate Change*, pp. 25–34.

12 H. Gallée, J.P. van Ypersele, T. Fichefet, C. Tricot and A. Berger, 'Simulation of the last glacial cycle by a coupled, sectorially averaged climate–ice sheet model. 1. The climate model', *Journal of Geophysical Research*, 1991, vol. 96, pp. 13,139–61; H. Gallée, J.P. van Ypersele, T. Fichefet, I. Marsiat, C. Tricot and A. Berger, 'Simulation of the last glacial cycle by a coupled, sectorially averaged climate–ice sheet model. 2. Response to insolation and CO_2 variations', *Journal of Geophysical Research*, 1992, vol. 97, pp. 15,713–40.

13 A. Berger, 'Long-term variations of caloric insolation resulting from the Earth's orbital elements', *Quaternary Research*, 1978, vol. 9, pp. 139–67.

14 The model has been forced by the record of atmospheric carbon dioxide reconstructed from air bubbles trapped in the Vostok ice core from Antarctica, as described by A. Berger, H. Gallée and C. Tricot, 'Glaciation and deglaciation mechanisms in a coupled 2-dimensional climate–ice sheet model', *Journal of Glaciology*, 1993, vol. 39, pp. 45–9. It has also been forced by a 200,000-year long CO_2 record reconstructed from ocean core data, as described by A. Berger, C. Tricot, H. Gallée and M.F. Loutre, 'Water-vapour, CO_2 and insolation over the last glacial–interglacial cycles', *Philosophical Transactions of the Royal Society of London B*, 1993, vol. 341, pp. 253–61.

15 L.D. Labeyrie, J.C. Duplessy and P.L. Blanc, 'Variations in the mode of formation and temperature of oceanic deep waters over the past 125,000 years', *Nature*, 1987, vol. 327, pp. 477–82.

16 Natural variations in atmospheric carbon dioxide concentrations accompanied past glacial–interglacial cycles and are one of the most plausible mechanisms for reinforcing the relatively weak insolation forcing associated with the orbital changes. See Box 4.2, Chapter 4, and reviews in Goodess *et al.*, *Nature and Causes of Climate Change*, pp. 35–40, and Adcock *et al.*, 1995, op. cit., pp. 40–53.

17 Berger *et al.*, 'Glaciation and deglaciation mechanisms'.

18 We use a modified version of the Köppen–Trewartha classification system. G.T. Trewartha, *An Introduction to Climate*, McGraw-Hill, 4th edition, 1968; W. Rudloff, *World Climates*, Stuttgart, Wissenschaftliche Verlagsgesellschaft, 1981, 632 pp.

19 W.S. Broecker, 'Unpleasant surprises in the greenhouse?', *Nature*, 1987, vol. 328, pp. 123–6; J.C. Duplessy, L. Labeyrie, N. Kallel and A. Juillet-Leclerc, 'Intermediate and deep water characteristics during the Last Glacial Maximum', in A. Berger, S. Schneider and J.C. Duplessy (eds), *Climate and Geo-Sciences*, Dordrecht, Kluwer, 1989, pp. 105–20.

20 COHMAP Members, 'Climatic changes of the last 18,000 years: observations and model simulations', *Science*, 1988, vol. 241, pp. 1,043–52.

21 For evidence from the British Isles, see the following: G.R. Coope, 'Climatic fluctuations in northwest Europe since the last interglacial, indicated by fossil assemblages of Coleoptera', in A.E. Wright and F. Moseley (eds), *Ice Ages: Ancient and Modern*, Liverpool, Seel House Press, 1975, pp. 153–68; T.C. Atkinson, K.R. Briffa and G.R. Coope, 'Seasonal temperatures in Britain during the past 22,000 years, reconstructed using beetle remains', *Nature*, 1987, vol. 325, pp. 587–92. For evidence from the North Atlantic and Greenland see the following: G. Bond, W. Broecker, S. Johnsen, J. McManus, L. Labeyrie, J. Jouzel and G.

Bonani, 'Correlations between climate records from North Atlantic sediments and Greenland ice', *Nature*, 1993, vol. 365, pp. 143–7.

22 R.A. Kerr, 'Global change – fugitive CO_2: it's not hiding in the ocean', *Science*, 1992, vol. 256, p. 35; C.B. Field, 'Carbon cycle: Arctic chill for CO_2 uptake', *Nature*, 1994, vol. 371, pp. 472–3.

23 J.T. Houghton, B.A. Callander and S.K. Varney (eds), *Climate Change 1992: The Supplementary Report to the IPCC Scientific Assessment*, Cambridge, Cambridge University Press, 1992, 200 pp.

24 E.T. Sundquist, 'Long-term aspects of future atmospheric CO_2 and sea level changes', in R. Revelle (ed.), *Sea-level Change*, Natural Research Council Studies in Geophysics, National Academy Press, 1990, pp. 193–207; J.C.G. Walker and J.F. Kasting, 'Effects of fuel and forest conservation on future levels of atmospheric CO_2', *Palaeogeography, Palaeoclimatology, Palaeoecology (Global and Planetary Change Section)*, 1992, vol. 97, pp. 151–89.

25 Walker and Kasting, 1992, op. cit.

26 A. Jones, D.L. Roberts and A. Slingo, 'A climate model study of indirect radiative forcing by anthropogenic sulphate aerosols', *Nature*, 1994, vol. 370, pp. 450–3.

27 S. Manabe and R.J. Stouffer, 'Century-scale effects of increased atmospheric CO_2 on the ocean atmosphere system', *Nature*, 1993, vol. 364, pp. 215–18.

28 K.Y. Kim and T.J. Crowley, 'Modeling the climatic effect of unrestricted greenhouse emissions over the next 10,000 years', *Geophysical Research Letters*, 1994, vol. 21, pp. 681–4.

29 H. Perry and H.H. Landsberg, 'Projected world energy consumption', in Natural Research Council (ed.), *Energy and Climate*, Washington, DC, National Academy Press, 1977, pp. 35–50; Sundquist, 1990, op. cit.

30 Kim and Crowley, 1994, op. cit.

31 T.J. Crowley and K.Y. Kim, 'Comparison of long-term greenhouse projections with the geologic record', *Geophysical Research Letters*, 1995, vol. 22, pp. 933–6.

32 A. Berger, H. Gallée and J.L. Melice, 'The Earth's future climate at the astronomical timescale', in C.M. Goodess and J.P. Palutikof (eds), *Future Climate Change and Radioactive Waste Disposal*, Nirex Safety Series NSS/R257, Harwell, UK Nirex Ltd, 1991, pp. 148–65.

33 A. Letréguilly, P. Huybrechts and N. Reeh, 'Steady-state characteristics of the Greenland ice sheet under different climate states', *Journal of Glaciology*, 1991, vol. 37, pp. 149–57; T.J. Crowley and S.K. Baum, 'Is the Greenland Ice Sheet bistable?', *Paleoceanography*, 1995, vol. 10, pp. 357–63.

34 These simulations have been performed by Paul Burgess, a Ph.D. student in the Climatic Research

Unit, in collaboration with colleagues from the University of Louvain-la-Neuve. The work is funded by UK Nirex Ltd under Contract SCRS 0268.

35 The Ipswichian and Holocene thermal maxima were no more than 2°C warmer than present in the British Isles (see Chapters 4 and 5). Warming of this magnitude may occur in the British Isles by the end of the next century (see Chapter 15).

36 I.J. Winograd, T.B. Coplen, J.M. Landwehr, A.C. Riggs, K.R. Ludwig, B.J. Szabo, P.T. Kolesar and K.M. Revesz, 'Continuous 500,000-year climate record from vein calcite in Devils-Hole, Nevada', *Science*, 1992, vol. 258, pp. 255–60.

37 J. Jouzel, N.I. Barkov, J.M. Barnola, M. Bender, J. Chappellaz, C. Genthon, V.M. Kotlyakov, V. Lipenkov, C. Lorius, J.R. Petit, D. Raynaud, G. Raisbeck, C. Ritz, T. Sowers, M. Stievenard, F. Yiou and P. Yiou, 'Extending the Vostok ice-core record of palaeoclimate to the penultimate glacial period', *Nature*, 1993, vol. 364, pp. 407–12.

GENERAL READING

A. Berger and M-F. Loutre, 'Modelling the climate response to astronomical and CO_2 forcings', *Comptes Rendus de l'Academie des Sciences*, 1996, vol. 323, pp. 1–16.

T.J. Crowley and K.Y. Kim, 'Comparison of long-term greenhouse projections with the geologic record', *Geophysical Research Letters*, 1995, vol. 22, pp. 933–6.

C.M. Goodess, J.P. Palutikof and T.D. Davies, *Studies of Climatic Effects and Impacts Relevant to Deep Underground Disposal of Radioactive Waste*, Nirex Safety Series, NSS/R267, 1992, 398 pp., available from UK Nirex Ltd, Harwell.

C.M. Goodess, J.P. Palutikof and T.D. Davies, *The Nature and Causes of Climate Change: Assessing the long term future*, London, Belhaven Press, 1992, 248 pp.

K.Y. Kim and T.J. Crowley, 'Modeling the climate effect of unrestricted greenhouse emissions over the next 10,000 years', *Geophysical Research Letters*, 1994, vol. 21, pp. 681–4.

APPENDICES

Appendix A

CLIMATE MAPS OF THE BRITISH ISLES

Chapter 3 described the climatology of the British Isles using statistics from the period 1961 to 1990 and showed a series of colour maps depicting average seasonal conditions for temperature, precipitation and a few other climate variables. In this appendix, we present a further selection of colour maps showing average annual conditions over the British Isles for a more comprehensive selection of fifteen surface climate variables. These climate maps are plotted using data on a ten minute latitude/longitude grid and are derived from 1961 to 1990 station averages. The construction of the climatologies is described in the following publication: E.M. Barrow, M. Hulme and T. Jiang, *A 1961–90 Baseline Climatology and Future Climate Change Scenarios for Great Britain and Europe. Part I: 1961–90 Great Britain baseline climatology*, A Report Prepared for the Landscape Dynamics and Climate Change TIGER IV Consortium, Climatic Research Unit, Norwich, 34 pp. The variables plotted are as follows:

Plate A.1: Mean surface air temperature, diurnal temperature range, mean maximum and mean minimum surface air temperature. Diurnal temperature range is the difference between mean maximum and minimum temperature, and mean temperature is the average of mean maximum and minimum temperature.

Plate A.2: Precipitation total (liquid plus solid), daily precipitation intensity (the average depth of precipitation on 'raindays'), 'rainday' frequency (days with greater than 0.1 mm precipitation) and 'snow-days' frequency (days with snow lying on the ground at 0900 GMT).

Plate A.3: Daily sunshine intensity (average number of hours of sunshine per day), daily solar radiation intensity (derived from sunshine hours and expressed as mega Joules per square metre, MJm^{-2}), cloud cover (proportion of sky covered) and wind speed (10 m equivalent).

Plate A.4: Vapour pressure, relative humidity and 'frostday' frequency (days with grass minimum temperature below $0°C$).

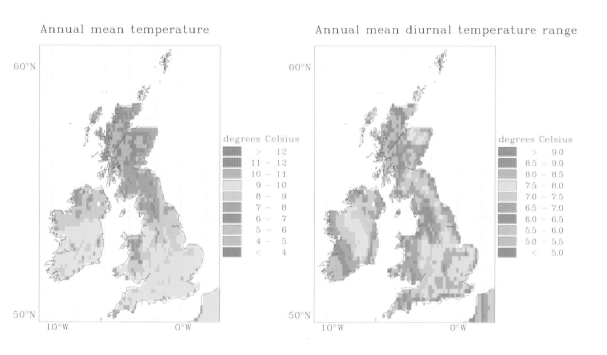

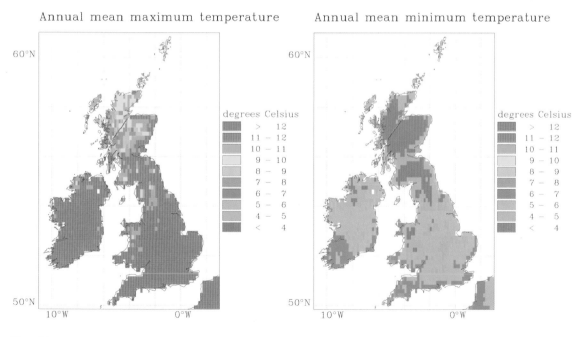

Plate A1

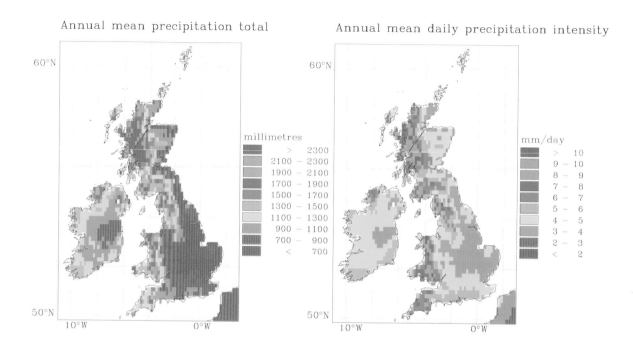

Annual mean precipitation total

Annual mean daily precipitation intensity

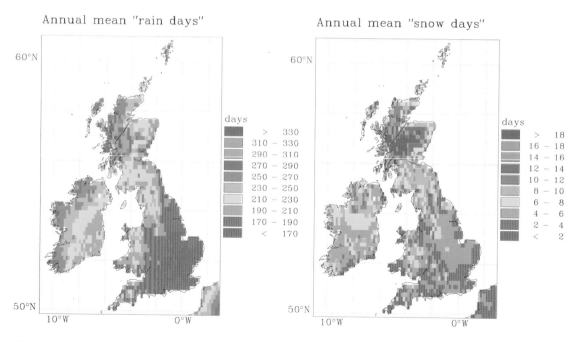

Annual mean "rain days"

Annual mean "snow days"

Plate A2

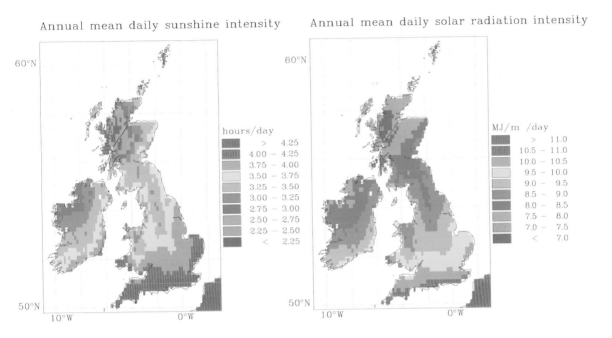

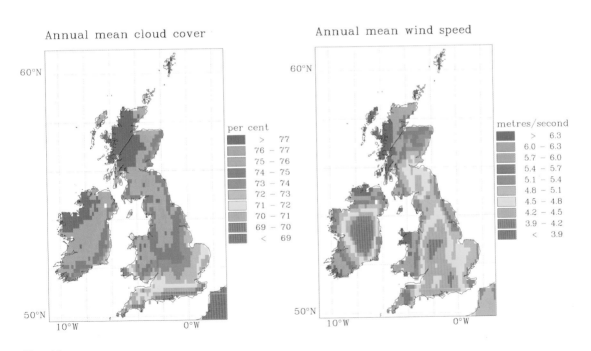

Plate A3

Annual mean vapour pressure

Annual mean relative humidity

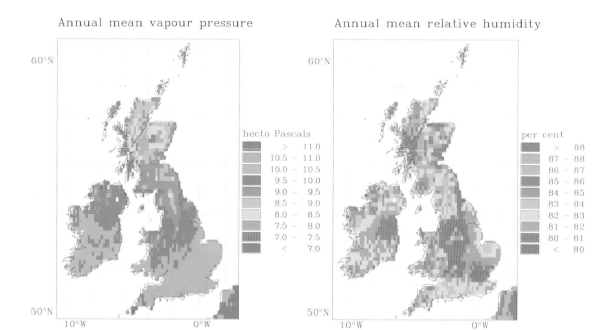

Annual mean "frost days"

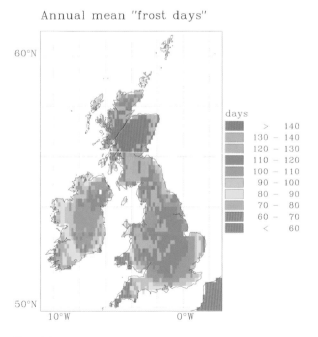

Plate A4

Appendix B

THE LAMB CATALOGUE, 1972–95

One of the most widely used daily synoptic classifications for the British Isles is that developed by Professor H.H. Lamb. The full Catalogue runs from 1 January 1861 to the present day, and its application to the understanding of weather in the British Isles has been discussed in Chapter 8. The Catalogue was published in full in 1972 in *Geophysical Memoirs (London)*, vol. 16 (116), 85 pp., and is updated in *Climate Monitor*, the quarterly publication of the Climatic Research Unit. We publish here the Catalogue from 1972 to 1995, ensuring continuity from the 1972 publication. In addition to the eight directional components (NE, E, SE, S, SW, W, NW and N), there are the two vorticity components (A, anticyclonicity, and C, cyclonicity) and also an unclassifiable type (U).

Lamb Catalogue for 1972

	Jan	Feb	Mar	Apr	May	Jun	Jul	Aug	Sep	Oct	Nov	Dec	
1	AE	S	C	W	C	W	W	CN	A	C	U	W	1
2	E	S	SW	W	C	W	W	N	A	E	AW	CW	2
3	E	S	C	W	N	W	W	A	AN	A	AW	W	3
4	E	S	SW	CW	NE	S	C	C	A	A	AW	CW	4
5	U	SE	C	CW	CE	C	CW	W	A	A	W	CW	5
6	SE	C	C	CW	C	CS	W	SW	A	E	W	SW	6
7	SE	C	C	C	C	C	W	CSW	W	S	W	C	7
8	S	CSW	E	CN	S	C	W	C	U	ASW	AW	C	8
9	SE	SW	SE	W	CS	C	W	CW	C	W	W	W	9
10	S	SW	AE	C	W	CN	W	AW	NW	C	W	CW	10
11	CS	SW	A	C	C	C	AW	A	NW	ANE	NW	SW	11
12	S	C	A	NW	C	C	A	S	ANW	A	C	SW	12
13	C	C	ASE	CW	NE	N	A	A	CNW	A	CN	SW	13
14	S	AW	S	C	NE	A	A	A	ANE	A	N	SW	14
15	S	C	AS	ANW	N	AN	A	A	A	A	AW	S	15
16	CS	C	ASE	A	A	ANW	A	AW	ANW	A	C	S	16
17	C	E	ASE	ANW	A	W	A	NW	AN	ANE	N	SE	17
18	U	E	ASE	AN	C	W	AE	ANW	AN	A	A	AS	18
19	W	A	U	ANE	CS	W	ANE	A	ANE	ANE	CS	A	19
20	W	U	A	A	S	W	A	ANW	A	N	C	A	20
21	W	U	A	AN	S	W	ANE	ANW	AW	ANW	C	A	21
22	W	A	A	ANE	S	W	AE	A	A	NW	NW	A	22
23	W	A	A	ANE	S	NW	C	A	A	NW	N	W	23
24	W	AE	A	AE	SW	W	N	A	AE	AW	A	A	24
25	W	SE	U	A	W	SW	AN	A	A	ASW	ANW	S	25
26	C	S	W	AN	C	SW	A	A	A	S	A	S	26
27	N	S	NW	AN	N	C	A	AE	A	S	W	SE	27
28	AE	SW	NW	W	W	C	A	AE	A	W	W	C	28
29	AE	SW	W	C	CW	AS	A	AE	A	SW	W	U	29
30	E		CW	C	C	SW	C	A	S	SW	W	A	30
31	SE		CW		NW		C	A		U		ASW	31

Lamb Catalogue for 1973

	Jan	Feb	Mar	Apr	May	Jun	Jul	Aug	Sep	Oct	Nov	Dec	
1	ASW	ANW	W	W	NW	W	SW	A	W	A	S	E	1
2	W	A	W	C	A	SW	ASW	W	W	A	S	A	2
3	A	A	W	AW	E	CW	A	CW	W	A	S	NW	3
4	A	AW	W	W	C	A	A	CW	SW	AE	C	NW	4
5	A	W	W	W	C	A	A	C	A	E	N	W	5
6	A	W	NW	CNW	C	A	C	C	AW	C	ANW	NW	6
7	A	W	AW	N	C	A	ANW	W	A	U	AW	C	7
8	A	CW	A	N	NW	A	A	AW	A	AW	W	N	8
9	A	NW	A	N	W	NW	A	W	A	W	W	A	9
10	A	NW	A	N	NW	NW	W	CW	ANE	U	C	W	10
11	AS	W	A	N	ANW	AW	NW	A	A	ANE	NW	W	11
12	AS	W	AE	N	AW	CW	A	A	A	A	W	NW	12
13	CSE	NW	A	NW	W	NW	W	A	A	E	NW	NW	13
14	S	C	A	ANW	AN	A	C	A	A	CE	NW	N	14
15	C	C	A	ANW	A	AS	CE	AE	ASE	CE	NW	A	15
16	U	A	A	ANW	ASE	AS	C	U	S	NE	N	NW	16
17	A	U	A	A	E	W	C	NW	C	N	A	NW	17
18	A	A	AN	AN	SE	A	NW	A	C	NW	W	W	18
19	S	NW	A	N	CE	CW	C	U	C	C	AN	SE	19
20	C	AW	A	N	E	CNW	C	A	N	C	A	CE	20
21	C	NW	A	CN	E	A	CN	AE	NW	ANW	A	C	21
22	W	NW	A	CNE	CSE	A	CNW	ASE	N	AW	U	C	22
23	W	NW	ASW	E	C	A	CNW	A	N	A	AW	C	23
24	ASW	CN	SE	E	U	A	NW	A	U	A	CNW	E	24
25	AW	U	CW	ANE	A	A	N	A	U	A	N	A	25
26	CW	U	AW	A	AS	A	AN	A	W	A	AN	W	26
27	NW	A	AS	N	SE	A	AN	W	CW	S	A	W	27
28	ANW	AW	U	CN	C	U	AN	A	CW	CW	C	W	28
29	AW		A	C	CS	W	A	W	CNW	A	N	W	29
30	W		W	C	C	ASW	A	NW	N	A	U	W	30
31	NW		W		N		A	W		ASE		A	31

Lamb Catalogue for 1974

	Jan	Feb	Mar	Apr	May	Jun	Jul	Aug	Sep	Oct	Nov	Dec	
1	A	CS	C	E	C	AW	C	NW	C	AN	C	W	1
2	S	CSW	NW	CSE	C	W	C	NW	C	CN	SW	W	2
3	S	CW	N	A	E	AW	CNW	A	C	NE	C	ASW	3
4	S	W	E	A	E	A	CW	U	W	N	U	W	4
5	S	C	E	A	NE	U	C	A	W	N	A	NW	5
6	SW	C	S	A	AE	CNW	U	AS	W	C	U	W	6
7	SW	U	AS	AE	A	W	AW	S	C	CN	SW	W	7
8	C	W	SE	AE	AS	NW	ASW	C	W	N	W	W	8
9	U	CW	E	AE	S	NW	W	C	W	N	CW	W	9
10	SW	SW	E	C	CS	NW	W	C	ASW	N	W	W	10
11	SW	C	E	CE	CS	NW	CW	W	S	N	CW	N	11
12	SW	CW	CE	E	S	A	W	C	S	ANW	W	NW	12
13	SW	C	E	AE	S	A	CN	ASW	W	A	S	NW	13
14	W	W	U	A	A	A	W	CSW	ASW	A	C	W	14
15	W	C	W	ANE	ASE	ANE	C	CSW	C	U	CW	W	15
16	C	C	C	A	S	A	C	W	AW	C	U	W	16
17	W	U	W	A	S	C	NW	NW	W	U	U	NW	17
18	W	A	W	A	S	W	NW	AN	A	W	CNE	NW	18
19	AW	A	C	A	A	ASW	W	A	A	CW	A	AW	19
20	AW	A	C	A	AW	AS	NW	A	AW	NW	C	W	20
21	ASW	W	A	A	W	ASE	W	A	W	N	C	SW	21
22	SW	W	ASE	A	CNW	AE	W	AW	W	N	C	SW	22
23	W	ANW	E	ANE	N	E	W	SW	C	AN	C	SW	23
24	W	A	E	ANE	N	E	NW	W	C	NW	C	W	24
25	W	A	E	ANE	AN	E	W	W	C	NW	NW	W	25
26	W	A	E	AE	A	NE	W	W	NW	NW	W	W	26
27	SW	A	E	E	NW	NE	W	AW	C	NW	NW	W	27
28	SW	S	E	E	CNW	U	W	AS	N	N	NW	W	28
29	SW		A	C	AN	C	W	CSE	N	N	W	NW	29
30	SW		A	U	A	CSW	CW	CE	N	AN	W	AW	30
31	SW		A		U		CW	CSE		C		AW	31

Lamb Catalogue for 1975

	Jan	Feb	Mar	Apr	May	Jun	Jul	Aug	Sep	Oct	Nov	Dec	
1	W	AW	S	N	W	N	A	A	A	CSW	CW	C	1
2	ASW	A	S	N	AW	C	A	A	A	C	CW	CN	2
3	W	A	CS	N	A	CN	A	A	NW	CNW	W	NW	3
4	AW	ASE	C	N	A	A	NE	SE	NW	W	AW	ANW	4
5	AW	A	C	CN	A	S	AN	S	NW	W	CW	ANW	5
6	W	AE	W	A	E	S	A	S	C	ANW	NW	A	6
7	NW	AE	W	N	AE	A	AE	A	A	A	A	ANW	7
8	W	E	C	N	CNE	ASE	CE	CE	W	A	A	A	8
9	W	ASE	C	N	CN	E	C	C	W	A	A	A	9
10	SW	SE	CN	CNW	N	A	C	U	W	NE	E	A	10
11	SW	SE	ANE	NW	U	A	C	A	C	A	SE	A	11
12	SW	C	AE	NW	C	A	CSW	C	C	A	E	N	12
13	SW	C	AE	W	C	ANW	S	A	N	A	E	A	13
14	SW	SE	ANE	C	C	NW	S	U	CN	CS	A	A	14
15	CSW	ASE	N	C	NE	CN	C	C	ANW	C	C	A	15
16	SW	S	AE	W	U	CN	U	C	U	C	N	AN	16
17	C	W	AE	SW	A	CNW	C	CNW	W	A	N	ANE	17
18	C	A	AE	S	A	SW	A	A	C	AE	ANW	A	18
19	W	A	AE	C	A	SW	S	SW	AS	ASE	NW	A	19
20	W	A	AE	W	ANE	A	W	W·	W	SE	AN	A	20
21	W	A	A	SW	ANE	AE	W	W	AW	E	A	AW	21
22	W	S	CW	U	ANE	A	W	NW	ASW	SE	AW	AW	22
23	W	AS	NW	A	AN	A	CW	W	W	S	SW	W	23
24	W	AS	N	A	ANE	A	NW	W	W	ASW	W	NW	24
25	C	ASE	ANW	A	AE	A	AW	A	C	A	W	NW	25
26	W	ASE	U	A	AE	AN	AW	A	U	ASW	W	ANW	26
27	C	A	CNE	A	E	ANE	A	A	C	ASW	C	AW	27
28	C	AS	CN	A	NE	A	A	A	W	AS	CW	W	28
29	W		N	W	NE	A	A	U	S	S	W	W	29
30	SW		N	W	N	A	A	NE	CS	S	W	W	30
31	W		N		NE		A	A		S		W	31

Lamb Catalogue for 1976

	Jan	Feb	Mar	Apr	May	Jun	Jul	Aug	Sep	Oct	Nov	Dec	
1	C	AE	A	C	AW	C	ASE	NW	N	CE	W	C	1
2	C	E	A	CN	W	AN	AE	NW	N	C	W	CNW	2
3	NW	E	AS	W	CW	A	E	ANW	NW	W	W	C	3
4	AW	AE	S	W	U	A	AE	ANW	NW	CW	W	CN	4
5	W	E	SE	NW	E	AW	A	A	A	SW	W	W	5
6	AW	SE	SE	NW	A	AW	E	A	A	W	CS	SW	6
7	W	S	E	N	A	A	E	A	A	CW	CSW	C	7
8	W	W	E	A	A	S	E	A	C	A	S	CW	8
9	W	W	A	A	A	W	C	A	CN	U	U	NW	9
10	NW	W	ASW	AW	NW	AW	A	A	C	S	C	ANW	10
11	NW	NW	W	U	CW	ASW	AS	A	C	C	C	A	11
12	AW	C	C	A	C	AW	S	A	CN	C	CNE	A	12
13	AW	U	CE	U	W	AW	CS	A	CNW	W	A	A	13
14	ANW	A	U	CN	ASW	AW	SW	A	C	C	U	ASE	14
15	A	A	C	A	SW	A	SW	AE	CNE	C	ASW	ASE	15
16	AW	A	C	A	SW	A	C	A	U	C	ANW	SE	16
17	A	SE	A	A	SW	A	AW	A	C	CS	U	SE	17
18	AW	S	U	A	C	W	U	A	SE	C	U	E	18
19	W	S	ASE	AE	C	CW	W	A	S	C	A	SE	19
20	W	S	S	AE	C	U	NW	A	S	C	A	E	20
21	NW	S	S	AE	W	A	NW	A	S	SW	AN	C	21
22	CNW	S	U	E	ASW	AS	ANW	E	CSE	S	N	SE	22
23	NW	SW	A	ANE	A	ASW	ANW	SE	E	C	AN	SE	23
24	N	W	A	NE	S	A	NW	E	E	S	ANW	E	24
25	N	W	W	ANE	C	A	NW	A	E	U	AW	ANE	25
26	NW	AW	W	NE	U	A	A	ANE	SE	C	CW	A	26
27	U	A	W	ANE	A	A	A	ANE	SE	E	W	N	27
28	CS	SW	AW	A	SE	A	AN	AE	CSE	E	W	N	28
29	S	SW	W	A	C	A	ANW	E	C	E	W	U	29
30	E		W	A	C	A	N	C	CE	ANE	C	CS	30
31	E		CW		U		N	NW		A		C	31

Lamb Catalogue for 1977

	Jan	Feb	Mar	Apr	May	Jun	Jul	Aug	Sep	Oct	Nov	Dec	
1	C	W	S	CW	CNE	A	AW	A	W	NW	W	A	1
2	A	SW	SW	NW	NE	A	AS	AW	W	NW	C	ASE	2
3	A	W	W	N	C	A	A	W	AW	W	CW	SE	3
4	AW	W	W	AN	C	NW	ASE	C	AW	W	W	SE	4
5	W	W	AW	ANW	C	N	AE	N	W	C	W	SE	5
6	A	W	AS	N	W	C	AE	A	CW	CE	W	CE	6
7	ANW	C	SW	N	C	CN	ANE	U	AW	E	SW	CE	7
8	NW	U	S	CN	U	U	NE	A	U	CSE	W	C	8
9	CN	S	W	N	C	NE	NE	A	AW	C	W	CS	9
10	CN	C	S	NW	CW	C	ANE	A	W	W	SW	S	10
11	CN	C	W	NW	C	C	ANE	AE	W	AS	W	C	11
12	C	C	SW	NW	C	C	NE	E	ANW	SW	NW	W	12
13	C	CS	CS	NW	N	NE	N	E	A	AS	NW	W	13
14	C	C	U	AN	A	NE	ANE	E	A	S	NW	AW	14
15	CN	U	S	A	E	ANE	A	E	AN	S	NW	A	15
16	A	C	S	U	NE	ANE	W	CE	NE	S	N	A	16
17	A	C	S	NE	A	ANE	C	E	NE	S	ANW	U	17
18	S	C	C	A	A	N	C	CNE	AE	S	W	A	18
19	SW	C	C	A	A	N	CW	C	ANE	S	W	A	19
20	CS	C	CNE	AW	AE	ANE	NW	C	ANE	S	CN	S	20
21	CS	C	E	W	A	AN	W	NE	ANE	S	N	S	21
22	CS	C	E	C	A	AE	W	NE	AE	S	NW	S	22
23	C	CE	E	C	AE	A	W	A	ASE	SW	NW	CSW	23
24	S	CE	U	CW	AE	U	CW	CS	SE	W	N	W	24
25	C	U	A	W	E	W	N	C	S	AW	AN	AW	25
26	C	A	ASE	W	E	NW	N	C	W	ASW	A	W	26
27	C	AE	CNE	W	AE	W	NW	N	U	W	A	NW	27
28	U	A	AN	CW	A	W	N	A	CW	W	A	N	28
29	A		A	C	N	NW	A	W	W	SW	A	NW	29
30	A		A	C	ANE	W	ANW	C	W	SW	A	NW	30
31	U		SW		A		A	W		W		NW	31

Lamb Catalogue for 1978

	Jan	Feb	Mar	Apr	May	Jun	Jul	Aug	Sep	Oct	Nov	Dec	
1	AW	C	C	C	CE	S	W	CE	AN	A	SW	SE	1
2	W	U	CS	E	E	A	W	C	A	A	W	SE	2
3	NW	W	C	E	E	ASE	CNW	C	A	W	SW	SE	3
4	A	SW	AN	E	E	SE	N	C	A	AW	SW	U	4
5	W	W	A	AE	NE	S	CN	C	U	W	SW	SE	5
6	SW	CNW	AS	A	CNE	W	N	C	CS	ASW	S	SE	6
7	W	U	ASW	A	U	W	NW	N	U	S	CS	SE	7
8	A	E	W	A	E	NW	NW	N	W	S	W	C	8
9	W	AE	AW	N	ANE	NW	W	N	W	C	A	S	9
10	W	NE	AW	N	A	NW	A	A	W	C	U	S	10
11	C	CE	ASW	N	A	NW	AE	A	W	CS	AS	S	11
12	A	NE	W	C	CNW	AN	E	U	AW	S	W	S	12
13	A	N	SW	C	CNW	AE	A	W	AW	ASE	W	C	13
14	A	U	C	N	C	A	AN	SW	W	U	W	C	14
15	A	A	C	A	C	C	AN	CW	W	C	C	C	15
16	C	CE	C	A	A	CE	AN	W	AW	NW	W	N	16
17	C	E	N	A	ASE	NE	A	AW	ANW	NW	SW	A	17
18	CW	AE	W	U	E	A	NW	AS	A	NW	W	A	18
19	C	E	W	CS	E	NW	NW	S	A	AW	W	A	19
20	C	E	W	C	A	CW	N	W	A	ANW	W	E	20
21	SW	SE	NW	C	A	C	W	AW	A	AW	W	CE	21
22	W	SE	W	AS	A	C	SW	W	AW	NW	W	SE	22
23	C	S	CNW	ASE	N	C	SW	A	AW	AW	W	SE	23
24	C	S	W	E	AN	N	W	A	W	W	W	C	24
25	W	S	W	E	A	NW	S	AN	W	AW	NW	C	25
26	CNW	CS	CW	CE	A	NW	SW	A	W	ANW	A	C	26
27	C	S	W	C	A	W	S	AN	NW	A	ANE	SE	27
28	C	S	SW	CE	A	W	SW	A	W	A	A	C	28
29	CN		SW	E	A	NW	A	AN	CNW	A	A	C	29
30	A		SW	CE	ASE	CW	CNE	AN	CNW	ASW	A	CE	30
31	CW		S		AE		CNE	NW		W		E	31

Lamb Catalogue for 1979

	Jan	Feb	Mar	Apr	May	Jun	Jul	Aug	Sep	Oct	Nov	Dec	
1	A	C	W	W	CN	A	NW	C	S	CS	W	W	1
2	A	A	W	C	N	AE	ANW	C	W	SE	AW	W	2
3	S	ANW	W	C	N	E	A	N	W	C	W	W	3
4	U	U	W	CN	NW	C	A	A	AW	C	W	W	4
5	A	A	W	N	N	C	A	SW	AS	C	NW	CW	5
6	U	A	W	C	CW	C	AW	SW	SW	SE	NW	ASW	6
7	W	C	W	C	C	C	W	CW	S	S	CW	C	7
8	W	ANE	W	C	C	C	W	CW	C	S	CW	S	8
9	W	E	W	E	U	A	AW	C	U	S	NW	C	9
10	C	E	W	E	A	A	A	W	A	S	W	C	10
11	N	E	W	C	U	A	A	A	AW	S	C	C	11
12	A	CE	ANW	C	AW	U	A	SW	W	S	U	U	12
13	A	CE	CN	CS	ASW	NW	NW	SW	NW	E	U	CSW	13
14	AS	E	CNE	E	ASW	CNW	W	C	ANW	CNE	C	C	14
15	S	AE	NE	NE	A	CNW	A	W	A	NE	C	C	15
16	U	E	CNE	A	C	ANW	AW	C	ASW	AN	C	W	16
17	ANE	E	C	A	C	A	AW	C	W	A	W	C	17
18	E	U	CE	A	C	A	NW	U	W	AW	NW	NW	18
19	E	A	U	AW	A	A	NW	N	W	W	U	N	19
20	C	SE	CW	W	CS	A	NW	CW	C	A	A	AN	20
21	SE	U	C	NW	C	CW	NW	NW	N	A	A	A	21
22	E	A	NW	W	S	W	NW	CW	A	U	AW	A	22
23	CNE	A	A	C	C	W	NW	C	NW	SE	W	C	23
24	N	A	AS	CN	CSW	C	AW	C	A	S	AW	W	24
25	C	A	C	N	C	NW	U	C	W	S	CSW	ASW	25
26	CN	A	C	N	C	W	A	NE	W	CE	AW	SW	26
27	W	W	C	ANW	C	W	AS	A	A	E	A	C	27
28	C	W	CN	NW	S	W	CSW	A	A	U	SW	C	28
29	C		CN	NW	CSW	NW	CW	A	A	W	SW	CNW	29
30	C		N	N	C	NW	C	A	SE	W	W	N	30
31	CW		NW		C		C	S		W		A	31

Lamb Catalogue for 1980

	Jan	Feb	Mar	Apr	May	Jun	Jul	Aug	Sep	Oct	Nov	Dec	
1	ANE	U	A	CW	E	W	CN	S	A	W	S	A	1
2	A	C	AN	ANW	E	AW	A	ASE	U	A	E	N	2
3	SW	C	A	A	E	A	W	S	U	A	E	N	3
4	W	C	A	A	E	AS	C	C	SW	A	AE	N	4
5	CNW	C	C	A	E	U	C	C	W	AW	AE	NW	5
6	CN	CE	C	A	E	C	C	C	W	CW	E	N	6
7	U	C	C	ANW	CNE	CN	C	CE	SW	CW	E	A	7
8	SE	S	CN	N	A	C	CNE	AE	W	NW	E	A	8
9	C	C	W	NW	A	E	N	A	W	NW	NE	SW	9
10	A	W	C	ANW	ASE	E	N	ASW	C	C	A	SW	10
11	A	AW	W	A	SE	CNE	NW	C	W	CNE	N	W	11
12	A	S	W	S	CSE	CNE	NW	C	C	A	N	SW	12
13	A	CSW	CNW	U	SE	CE	C	SW	CW	U	AW	W	13
14	N	W	U	A	SE	C	C	S	U	C	SW	C	14
15	AN	A	E	U	ASE	C	N	C	S	NE	C	CNW	15
16	A	W	A	A	A	CSW	A	U	SW	CNE	C	W	16
17	A	SW	C	NW	AE	C	W	SW	S	CN	C	W	17
18	A	C	CE	NW	A	W	C	NW	W	N	U	W	18
19	CS	U	E	N	A	W	C	ANW	S	A	S	W	19
20	CS	C	NE	N	CN	CNW	CN	W	U	A	SW	C	20
21	C	S	CNE	ANW	NE	CW	AW	NW	C	S	SW	U	21
22	C	C	C	NW	AN	CW	ASW	ANW	CS	C	SW	W	22
23	C	U	SE	NW	AN	C	SW	A	SW	C	SW	CW	23
24	C	A	CS	A	AN	C	A	A	AW	C	C	CW	24
25	U	U	S	A	NW	N	S	A	A	A	C	W	25
26	ANE	A	S	A	C	N	C	ASE	A	S	NW	CNW	26
27	A	A	C	A	C	NW	C	U	U	SW	NW	A	27
28	S	A	C	A	CN	C	SE	SW	A	CW	N	W	28
29	S	A	C	ANE	CN	N	CE	C	A	AW	A	W	29
30	C		AW	AE	C	C	C	C	A	A	A	W	30
31	C		SW		C		S	A		SE		W	31

Lamb Catalogue for 1981

	Jan	Feb	Mar	Apr	May	Jun	Jul	Aug	Sep	Oct	Nov	Dec	
1	NW	AW	C	A	N	AS	W	AN	AE	C	W	A	1
2	NW	W	C	A	N	S	C	A	A	C	ASW	ANW	2
3	NW	CW	CNE	AE	C	SW	C	A	A	C	W	ANW	3
4	N	NW	A	A	CN	W	SW	A	A	CN	A	N	4
5	NW	NW	A	AE	U	SW	SW	A	A	C	A	CNW	5
6	CNW	W	SW	A	C	CW	SW	A	A	C	A	NW	6
7	A	W	W	A	S	CSW	ASW	A	CS	CW	A	NW	7
8	AW	C	CW	A	CS	C	U	A	W	C	A	CN	8
9	NW	C	W	A	E	C	U	A	ASW	C	A	CN	9
10	AN	N	SW	A	CE	C	U	A	S	CW	AW	W	10
11	A	A	W	CS	CE	C	C	A	SW	NW	CW	CNE	11
12	N	AS	W	U	C	A	C	A	S	NW	NW	A	12
13	NW	A	C	ANE	C	ASW	W	U	C	NW	ANW	U	13
14	NW	A	CN	AE	U	AW	NW	ANW	A	U	AW	C	14
15	NW	A	N	AE	S	NW	W	ANW	U	N	ASW	C	15
16	CW	A	N	AE	CSE	NW	W	AN	AW	A	W	CE	16
17	C	A	N	AE	C	N	NW	A	SW	A	W	E	17
18	U	A	NW	AE	CS	A	NW	AW	C	U	CW	A	18
19	NW	A	C	NE	CS	CNW	W	CW	C	W	W	A	19
20	U	A	C	ANE	S	N	W	N	C	NW	CW	S	20
21	W	U	C	N	CS	A	W	NW	W	N	W	CSE	21
22	AW	C	C	N	CS	A	NW	NW	W	N	W	CE	22
23	A	CE	C	U	S	A	C	A	W	NW	W	E	23
24	AW	A	C	C	SW	N	CNW	A	SW	W	NW	U	24
25	ANW	A	S	C	C	N	W	A	CS	CNW	AW	A	25
26	AW	SE	W	CNE	C	N	W	A	C	W	W	S	26
27	A	SE	S	NE	C	N	ANW	A	C	W	W	S	27
28	A	CS	S	NW	C	N	A	A	C	CW	NW	SE	28
29	A		U	NW	S	NW	A	U	SW	CW	AW	CSE	29
30	A		U	NW	S	NW	A	E	C	W	C	SE	30
31	A		U		S		A	AE		CW		SE	31

Lamb Catalogue for 1982

	Jan	Feb	Mar	Apr	May	Jun	Jul	Aug	Sep	Oct	Nov	Dec	
1	C	ASW	CW	E	NW	S	SW	C	W	S	CSW	A	1
2	S	S	W	SW	W	A	C	CE	AW	SW	A	A	2
3	U	S	C	S	C	ASE	CNW	CE	W	U	A	A	3
4	C	S	NW	SE	NW	SE	CW	C	U	C	S	U	4
5	C	S	AW	SE	C	SE	W	C	CSE	C	S	AW	5
6	U	SW	W	C	C	C	C	C	U	CNE	S	A	6
7	A	W	U	C	U	A	AW	A	W	N	SE	U	7
8	CE	W	SW	N	A	A	A	W	AW	NE	C	C	8
9	A	SW	SW	ANW	A	E	C	W	A	N	CW	C	9
10	A	SW	C	NW	A	E	U	AW	A	C	W	C	10
11	ANE	SW	W	N	AE	C	A	ASW	A	C	SW	C	11
12	A	SW	C	AN	A	C	AE	W	W	C	W	C	12
13	A	SW	W	A	ASE	N	E	W	A	C	W	U	13
14	A	ASW	W	A	SE	ANE	CE	W	A	CN	C	W	14
15	S	A	CSW	A	CSE	U	C	C	A	W	NW	W	15
16	S	A	C	A	S	C	W	W	A	S	W	W	16
17	SW	AE	CNW	A	U	AE	AW	W	A	S	W	CNW	17
18	S	SE	U	A	U	C	A	C	A	SW	W	W	18
19	S	ASE	C	A	S	U	A	W	C	SW	W	W	19
20	S	SE	C	A	C	A	A	NW	CS	W	W	CW	20
21	CSW	U	U	A	S	E	A	W	C	W	CSW	N	21
22	C	U	AW	A	SW	CE	AN	CW	W	C	SW	NW	22
23	W	A	A	ANW	C	C	A	CW	SW	U	C	W	23
24	W	U	A	A	W	U	AN	W	SW	W	C	W	24
25	W	W	A	A	SW	SE	ANW	W	S	A	C	W	25
26	U	SW	A	A	U	C	ANW	W	C	SW	C	W	26
27	A	ASW	A	A	U	C	A	W	C	ASW	C	W	27
28	NW	W	A	ANW	AW	C	A	W	SW	AS	CNW	AW	28
29	NW		N	NW	A	W	A	ASW	U	S	A	ASW	29
30	NW		A	NW	A	W	U	W	A	S	A	ASW	30
31	AW		A		U		E	W		CS		W	31

Lamb Catalogue for 1983

	Jan	Feb	Mar	Apr	May	Jun	Jul	Aug	Sep	Oct	Nov	Dec	
1	W	C	A	NE	C	C	W	CN	C	S	W	S	1
2	W	NW	S	N	C	C	W	NW	C	SW	A	ASW	2
3	CW	A	W	W	C	U	AW	A	C	SW	A	A	3
4	W	W	A	C	U	U	A	AW	C	SW	A	SW	4
5	W	A	AW	C	SE	ANE	A	A	C	W	A	W	5
6	W	N	AW	C	C	A	A	A	AW	AW	A	A	6
7	W	N	AW	C	C	SE	E	A	A	W	S	A	7
8	W	A	A	C	C	C	SE	A	C	W	CS	CW	8
9	W	A	A	C	C	W	A	A	C	W	E	C	9
10	W	N	AW	C	C	AW	AE	A	C	W	E	U	10
11	SW	AN	A	C	C	W	AE	A	N	W	E	A	11
12	W	A	S	N	C	W	ANE	NW	N	SW	SE	U	12
13	W	A	S	AW	C	W	AN	ANW	C	SW	A	SW	13
14	NW	A	SW	AW	CS	C	A	A	S	W	A	SW	14
15	NW	AE	W	A	S	A	AW	SW	C	SW	A	CS	15
16	NW	E	W	W	SE	A	U	U	C	C	AN	SE	16
17	W	ASE	W	CW	CSE	A	C	A	W	W	AN	SE	17
18	NW	A	C	CN	C	A	NW	AS	SW	W	A	SE	18
19	ANW	A	U	CN	C	A	A	SE	W	W	A	C	19
20	AW	AE	W	C	C	AE	A	CE	W	A	N	C	20
21	AW	ASE	W	C	N	AE	A	C	C	A	ANE	C	21
22	ASW	SE	CNW	C	W	A	A	A	AS	A	A	C	22
23	AS	SE	C	C	U	N	C	C	S	A	A	U	23
24	SW	S	N	SE	U	A	C	A	U	A	S	S	24
25	W	S	NW	CSE	A	A	U	A	A	A	SW	C	25
26	W	U	U	E	AN	NW	A	A	ASW	AW	C	W	26
27	W	CW	C	E	N	NW	AN	A	A	W	C	W	27
28	W	W	N	C	CN	NW	A	A	E	AN	C	AW	28
29	W		W	C	CN	NW	ANW	A	E	A	U	AW	29
30	W		W	E	E	AN	A	A	E	AW	A	AW	30
31	C		C		E		A	S		W		W	31

Lamb Catalogue for 1984

	Jan	Feb	Mar	Apr	May	Jun	Jul	Aug	Sep	Oct	Nov	Dec	
1	W	CW	W	AE	E	SE	A	U	U	C	SW	C	1
2	W	C	CN	AE	E	CS	AN	C	C	CS	C	C	2
3	CW	W	ANW	AE	U	C	AN	C	C	C	CN	S	3
4	W	W	AW	SE	U	CSE	A	CE	N	C	N	S	4
5	W	W	A	A	ANE	C	A	W	AN	CN	ANE	S	5
6	W	W	A	CN	ANE	C	A	W	A	W	E	SW	6
7	NW	W	A	NE	ANE	NE	A	CNW	A	W	E	SW	7
8	NW	CNW	A	AN	A	N	S	A	W	W	C	AW	8
9	W	AW	A	A	A	CN	SE	A	C	W	C	A	9
10	W	AW	AN	A	CN	A	CS	AN	NW	W	U	A	10
11	C	A	C	W	ASE	A	S	A	W	AW	S	A	11
12	W	ASE	C	AW	A	AW	W	ANE	AW	ASW	SW	S	12
13	C	A	E	ASW	A	W	W	-	SW	ASW	S	C	13
14	C	A	E	CW	A	AW	W	A	C	ASW	C	C	14
15	W	A	E	CNW	U	A	NW	A	A	A	C	C	15
16	C	AS	E	ANW	CN	A	ANW	A	U	ASW	E	C	16
17	CW	S	A	A	CNE	A	ANW	A	C	SW	CE	C	17
18	C	S	A	ASW	N	A	AN	ASW	U	C	CE	W	18
19	U	SE	A	ASW	A	A	A	A	W	C	C	W	19
20	A	C	A	ASW	C	A	A	ASE	CW	CNW	C	W	20
21	SE	C	SE	AW	N	W	A	E	C	W	C	AW	21
22	W	C	SE	A	C	CW	A	E	C	CW	C	SW	22
23	C	U	CSE	A	CNE	NW	A	C	CN	W	CSW	CSW	23
24	C	A	C	ASE	N	NW	A	E	N	C	CW	W	24
25	U	A	C	AE	N	NW	A	AE	U	C	CW	W	25
26	E	A	C	AE	AN	ANW	A	A	U	W	AW	C	26
27	CE	A	C	ASE	AN	ANW	ANW	A	S	ASW	SW	U	27
28	C	A	C	ASE	N	N	AW	AW	S	S	SW	AS	28
29	SW	A	C	SE	A	N	SW	W	CS	SW	S	A	29
30	W		CN	SE	A	AN	C	W	C	SW	CS	U	30
31	W		ANE		S		C	W		S		ANW	31

Lamb Catalogue for 1985

	Jan	Feb	Mar	Apr	May	Jun	Jul	Aug	Sep	Oct	Nov	Dec	
1	AN	W	C	C	NW	A	AW	W	C	SE	N	SE	1
2	A	W	C	U	CN	A	A	W	C	SE	N	SE	2
3	ANE	AS	C	SE	N	AE	A	W	C	SE	A	CSW	3
4	ANE	A	C	SE	C	U	U	C	C	SW	CS	SW	4
5	NE	SE	W	SE	C	NE	C	C	U	SW	C	C	5
6	N	E	SW	C	C	NE	W	W	A	SE	W	C	6
7	ANE	E	AW	C	A	N	A	W	AW	SW	C	C	7
8	AN	E	A	C	U	W	U	W	A	W	SE	C	8
9	N	E	ASW	C	N	CW	ANW	SE	A	W	C	C	9
10	A	E	A	C	AN	NW	AW	SW	A	SW	N	U	10
11	A	E	A	C	A	C	W	C	A	A	N	A	11
12	A	E	A	W	AE	C	W	C	U	A	A	SW	12
13	E	E	NW	C	E	N	C	SE	W	A	A	W	13
14	AE	AE	NW	NW	C	AN	C	C	W	A	U	SW	14
15	AE	A	CNW	AW	C	ANW	W	C	W	A	C	W	15
16	E	A	CN	AW	ASE	A	W	C	W	A	W	ASW	16
17	E	A	A	A	A	AW	SW	C	W	A	A	W	17
18	E	A	A	A	CE	W	W	SE	ASW	A	A	W	18
19	E	U	AE	U	E	C	W	SE	W	A	A	W	19
20	SE	AW	SE	N	E	U	W	CSW	ASW	A	AE	SW	20
21	C	A	C	ANE	CE	C	W	CW	C	A	AE	SW	21
22	C	A	CE	ANE	CNE	C	C	W	U	AE	ANE	SW	22
23	NW	SW	C	A	C	C	AW	C	U	SE	ANE	CSW	23
24	C	A	C	A	C	C	A	C	A	A	NE	C	24
25	C	SE	C	AN	U	CW	CSE	C	AS	AE	ANE	CSE	25
26	U	A	CNE	ANW	CS	C	C	W	A	AE	N	CE	26
27	U	A	N	N	C	W	C	SW	A	A	CN	N	27
28	C	ASE	AW	W	AW	CW	C	ASW	ASE	A	C	AN	28
29	W		U	C	A	W	C	A	AS	A	A	A	29
30	W		C	W	A	SW	C	CS	SE	A	SE	U	30
31	W		C		A		NW	C		CNE		C	31

Lamb Catalogue for 1986

	Jan	Feb	Mar	Apr	May	Jun	Jul	Aug	Sep	Oct	Nov	Dec	
1	C	E	A	C	A	W	A	C	W	A	U	W	1
2	C	E	A	C	C	W	A	C	CW	A	A	SW	2
3	NE	E	A	N	CE	W	AW	C	U	A	AW	SW	3
4	U	E	W	AN	CE	NW	W	C	AW	ASE	AW	SW	4
5	C	CE	W	ANE	CSE	N	W	SW	AW	ASW	W	C	5
6	C	CE	W	AE	CSE	A	NW	C	AW	AW	A	ASW	6
7	C	AE	AW	AE	AE	NW	W	C	A	AW	W	S	7
8	C	ANE	ASW	E	CS	W	NW	A	AN	AW	W	C	8
9	U	A	W	NE	S	CSW	W	A	A	SW	W	U	9
10	SW	A	AW	NE	SW	C	NW	E	AN	W	SW	SW	10
11	W	A	S	AN	CW	A	A	A	ANE	A	C	SW	11
12	W	SE	SE	N	CS	A	A	A	A	A	S	S	12
13	W	SE	SE	C	C	A	A	U	U	S	S	SW	13
14	CNW	E	S	C	CSW	A	AW	CS	CNE	W	C	SW	14
15	NW	E	S	C	CW	AE	AW	C	NE	A	SW	W	15
16	W	E	S	C	A	E	SW	C	AN	A	SW	W	16
17	W	E	S	C	SE	C	W	N	AN	A	SW	CW	17
18	W	E	C	W	SW	A	AW	N	A	W	CW	CW	18
19	CW	E	W	W	AS	AE	AW	N	A	W	C	W	19
20	W	A	CSW	C	S	E	W	A	A	CW	CW	NW	20
21	W	U	W	C	W	AE	W	U	AW	W	C	N	21
22	W	C	W	C	SW	E	NW	C	AW	W	C	AN	22
23	C	NE	W	C	SW	E	NW	CN	A	W	CW	A	23
24	N	ANE	C	C	SW	C	W	AN	A	CSW	W	ASW	24
25	A	AE	NW	U	SW	S	W	C	A	C	SW	W	25
26	A	A	W	A	SW	ASE	W	C	A	W	W	W	26
27	W	A	W	C	C	AE	SW	N	AW	W	SW	W	27
28	W	A	CW	W	W	AE	C	N	AW	W	SW	W	28
29	C		CW	AW	ANW	E	C	N	A	W	ASW	U	29
30	CE		C	ASW	AW	A	C	NW	A	CW	SW	C	30
31	E		C		W		C	A		C		C	31

Lamb Catalogue for 1987

	Jan	Feb	Mar	Apr	May	Jun	Jul	Aug	Sep	Oct	Nov	Dec	
1	C	S	SW	C	U	W	AW	W	U	SE	A	A	1
2	U	S	U	C	CNW	S	A	W	A	E	A	AE	2
3	A	W	A	E	AN	C	A	N	S	E	A	A	3
4	W	A	U	CE	ANW	C	A	AN·	S	SE	A	A	4
5	W	SW	A	E	A	C	A	AN	W	S	ASE	E	5
6	N	C	S	CSE	A	C	A	N	CW	W	A	NE	6
7	A	AW	S	C	A	C	NW	N	W	C	A	A	7
8	A	SW	SE	C	A	N	ANW	N	AW	C	U	A	8
9	C	C	E	C	A	N	A	N	W	C	C	A	9
10	U	C	E	CW	ANW	CNE	U	A	W	C	U	A	10
11	E	C	E	C	CNW	CN	C	S	SW	U	C	A	11
12	E	C	A	A	N	N	A	S	W	C	C	ASE	12
13	E	CE	A	AW	NW	N	A	SW	W	C	C	A	13
14	E	NE	A	A	C	CNE	S	AW	W	CS	W	A	14
15	AE	A	NW	A	CN	N	CS	AW	AW	C	S	SE	15
16	AE	ANE	NW	A	W	U	C	A	A	C	W	S	16
17	ASE	AN	W	S	C	NW	C	W	SW	S	W	S	17
18	ASE	AN	CNW	S	N	C	C	W	A	S	SW	W	18
19	AS	AN	N	SW	N	C	C	AW	S	CS	W	W	19
20	A	A	NW	W	AN	U	NE	U	C	C	NW	W	20
21	A	A	W	AW	N	C	NE	C	S	C	NW	AW	21
22	A	ANE	SW	AS	NE	C	ANE	C	SW	C	NW	A	22
23	A	A	C	AS	NE	C	AN	C	SW	A	N	AS	23
24	A	ASE	C	ASE	E	C	N	NE	W	A	N	S	24
25	A	SE	C	E	E	C	N	C	NW	A	ANE	SW	25
26	A	SE	CSW	A	E	U	CNW	N	AN	S	A	SW	26
27	A	SW	C	A	CNE	U	NW	A	A	C	A	CSW	27
28	ANE	W	N	A	U	SW	W	W	A	A	A	S	28
29	A		NW	S	AW	SW	CW	A	A	ASE	U	SW	29
30	A		AW	U	SW	W	W	A	SE	E	A	SW	30
31	A		W		W		W	ASE		U		SW	31

Lamb Catalogue for 1988

	Jan	Feb	Mar	Apr	May	Jun	Jul	Aug	Sep	Oct	Nov	Dec	
1	W	C	N	W	SE	C	C	A	C	A	A	E	1
2	W	CW	W	C	CE	C	C	A	C	A	A	E	2
3	W	C	N	A	C	C	C	AW	W	A	AS	C	3
4	C	C	N	A	C	N	C	W	SW	CS	ASE	C	4
5	CS	CW	NW	A	AW	A	C	AW	ASW	C	A	NW	5
6	C	NW	NW	A	A	A	C	A	AS	W	A	ANW	6
7	W	W	NW	A	E	CNE	C	A	S	NW	SE	AW	7
8	S	W	A	AN	CE	NE	C	SE	W	C	SE	AW	8
9	SW	C	W	A	A	NE	W	S	A	C	S	W	9
10	W	CNW	ANW	A	A	ANE	SW	U	AW	U	SW	AW	10
11	W	W	NW	A	E	AE	CW	CSW	W	CE	ASW	AW	11
12	S	W	NW	ANE	E	A	CSW	C	CW	C	AW	A	12
13	S	S	U	A	E	A	C	SW	N	CE	A	A	13
14	S	S	U	AS	E	A	CN	C	AN	A	A	A	14
15	A	SW	C	S	E	A	A	AW	A	A	A	A	15
16	A	A	C	CS	E	A	W	A	ANW	A	AS	A	16
17	AW	A	A	C	N	ANE	W	A	ANW	E	C	A	17
18	U	AW	AS	SE	N	A	AW	CS	A	E	AN	W	18
19	CS	A	SW	SE	N	A	W	C	A	E	N	W	19
20	SW	A	CSW	C	A	A	A	C	A	CSE	N	W	20
21	W	A	C	A	A	W	CSW	NW	AS	SE	A	W	21
22	C	A	C	AE	ASE	A	C	W	SW	SE	A	W	22
23	U	AN	C	AE	U	A	C	W	CW	SE	A	W	23
24	C	N	C	A	S	A	CSW	W	CW	SE	A	W	24
25	C	N	CW	U	S	A	SW	W	W	S	A	SW	25
26	C	ANW	C	C	CSE	A	W	W	W	S	A	W	26
27	NE	N	W	AE	S	AE	W	W	SW	C	A	W	27
28	E	N	W	E	S	N	C	CW	AW	AN	W	ASW	28
29	C	N	C	CE	C	C	CW	W	NW	A	U	ASW	29
30	C		C	E	C	C	C	SW	A	A	C	ASW	30
31	C		U		C		W	CSW		A		A	31

Lamb Catalogue for 1989

	Jan	Feb	Mar	Apr	May	Jun	Jul	Aug	Sep	Oct	Nov	Dec	
1	A	ASW	U	SE	ASW	N	C	ANW	NW	A	SW	A	1
2	A	ASW	CSE	CSE	ASW	C	A	A	ANW	A	SW	A	2
3	S	SW	C	E	A	CN	A	ANW	A	A	W	A	3
4	W	W	SW	E	A	N	A	A	A	ASE	CW	A	4
5	CW	W	S	E	A	C	A	A	AW	W	CNW	A	5
6	C	W	S	C	A	C	A	CSW	A	NW	W	A	6
7	W	W	AW	C	A	N	C	W	A	NW	W	A	7
8	W	ASW	SW	U	A	A	U	W	A	NW	C	ANE	8
9	W	S	SW	S	N	CS	A	W	ANE	NW	W	A	9
10	AW	AW	SW	C	A	S	AW	W	AE	NW	CSW	A	10
11	SW	AW	ASW	C	C	S	AW	W	AE	W	CS	A	11
12	W	W	U	C	C	S	ANW	W	U	W	A	U	12
13	SW	W	C	C	W	S	ANW	SW	W	W	A	C	13
14	W	W	C	CN	W	A	NW	SW	CW	W	A	C	14
15	W	W	C	U	ASW	A	ANW	SW	W	AW	ASE	C	15
16	W	A	CNE	E	AW	A	A	SW	W	SW	SE	SE	16
17	A	SW	A	N	A	A	A	W	SW	ASW	SE	C	17
18	A	SW	W	AN	A	ASE	A	AW	SW	S	SE	C	18
19	ASW	W	W	ANE	A	A	A	ASW	SW	S	E	U	19
20	CSW	W	W	NE	ASE	U	ASE	SW	SW	SW	E	SE	20
21	CW	A	W	NE	SE	A	SE	ASW	C	CSW	E	C	21
22	AW	W	W	N	E	A	U	AW	C	SW	NE	W	22
23	SW	W	W	N	U	ANW	C	AW	SW	W	A	W	23
24	S	C	W	N	A	AW	A	AW	ASW	W	ANE	S	24
25	S	C	SW	N	A	W	U	U	AW	W	A	SW	25
26	SW	C	S	C	A	W	A	C	AW	A	A	A	26
27	S	NW	C	C	A	C	AW	N	A	SE	A	A	27
28	AW	CN	A	C	A	CW	W	A	A	C	A	A	28
29	AW		A	U	AN	A	W	AW	A	C	A	SE	29
30	AW		AW	SW	N	C	NW	W	A	W	A	SE	30
31	A		A		N		NW	A		W		SE	31

Lamb Catalogue for 1990

	Jan	Feb	Mar	Apr	May	Jun	Jul	Aug	Sep	Oct	Nov	Dec	
1	S	C	CNW	U	ASE	SW	C	A	U	W	NW	A	1
2	S	C	ANW	C	ASE	C	NW	A	AW	W	N	A	2
3	S	W	AW	N	A	CW	W	A	W	W	N	ANW	3
4	S	S	W	A	A	CW	C	AW	W	W	N	AN	4
5	S	S	W	AW	U	SW	C	NW	W	W	ANE	A	5
6	SW	C	W	A	C	C	W	NW	C	C	A	A	6
7	W	C	W	AE	C	C	W	ANW	NW	N	A	C	7
8	W	W	W	A	C	C	W	AW	ANW	A	ASE	C	8
9	W	W	W	A	C	N	W	AW	A	W	SE	U	9
10	W	SW	SW	W	N	N	AW	AW	A	W	C	A	10
11	SW	C	AW	W	AN	N	AW	W	A	SW	U	N	11
12	W	CW	A	CW	ANE	AN	A	W	A	S	S	N	12
13	W	W	S	C	AE	AN	A	W	A	S	CS	A	13
14	W	W	ASW	W	U	A	ASE	W	A	S	CW	A	14
15	W	U	SW	W	SE	A	A	C	A	C	W	A	15
16	W	U	ASW	W	C	AS	A	C	A	S	W	A	16
17	W	SW	S	W	U	S	A	W	NW	CE	W	ASE	17
18	W	SW	S	W	AE	CS	AS	U	W	CE	W	A	18
19	SW	SW	SW	C	AE	S	A	C	NW	E	W	ANW	19
20	SW	SW	SW	E	ANE	C	A	W	W	E	C	C	20
21	SW	ASW	W	E	N	C	A	AW	NW	E	N	W	21
22	SW	S	W	A	A	C	A	AW	C	E	A	SW	22
23	W	SW	W	A	NW	CW	A	A	C	E	W	SW	23
24	W	C	NW	A	N	CW	A	U	CN	CSE	C	SW	24
25	C	SW	A	A	ANE	W	ASE	A	N	CSE	CE	W	25
26	W	C	A	AW	A	C	SE	A	AN	C	NE	CW	26
27	C	W	U	A	A	C	C	ASW	A	C	A	W	27
28	C	C	A	A	A	U	S	S	ASW	C	A	W	28
29	S		A	A	ASW	U	S	W	U	C	A	W	29
30	S		A	A	W	C	SW	W	C	C	A	W	30
31	SW		A		AS		A	W		C		CW	31

Lamb Catalogue for 1991

	Jan	Feb	Mar	Apr	May	Jun	Jul	Aug	Sep	Oct	Nov	Dec	
1	W	SE	A	W	N	NE	C	C	C	W	CSW	A	1
2	W	SE	S	W	N	U	CE	S	NE	AW	W	A	2
3	W	SE	S	W	N	CN	E	ASW	A	W	CW	A	3
4	W	E	SE	C	N	U	AE	ASW	A	CW	C	A	4
5	C	E	S	C	N	C	E	SW	ANE	W	N	AE	5
6	W	E	E	SW	NW	C	E	C	A	AW	W	A	6
7	C	E	E	W	CN	NE	E	A	A	S	W	AS	7
8	C	E	CE	W	A	A	C	A	A	S	W	A	8
9	SW	E	C	SW	A	C	U	ASW	A	E	NW	A	9
10	SW	N	C	S	A	C	AS	W	A	E	W	ASE	10
11	C	U	S	S	W	W	S	W	ANE	E	W	A	11
12	U	CN	S	C	W	C	W	AW	A	E	C	A	12
13	A	A	S	ANE	C	C	W	AW	S	C	W	A	13
14	ASE	NW	SW	A	ANW	C	CW	A	SW	CN	CN	A	14
15	SE	C	S	AE	AN	C	W	W	W	U	C	S	15
16	SE	N	C	NE	N	N	W	W	W	W	C	SW	16
17	S	U	C	AN	A	N	AW	W	SW	NW	C	W	17
18	S	A	SW	N	AW	CNW	C	A	W	N	C	W	18
19	SW	A	W	N	W	C	C	AW	AW	AN	CNE	W	19
20	AW	SW	C	ANW	W	C	U	AW	A	ANW	A	W	20
21	A	C	NW	NW	AW	C	A	S	SW	ANW	A	C	21
22	A	C	N	U	A	C	ASE	CE	W	ANW	ASW	W	22
23	A	SW	N	ASW	A	C	C	C	W	ANW	S	NW	23
24	A	W	A	S	A	C	C	W	W	A	S	A	24
25	A	A	A	C	A	C	N	A	C	A	S	A	25
26	A	S	AE	A	A	C	NW	A	N	SE	SW	W	26
27	A	S	A	ASE	A	C	ASW	A	U	SE	SW	A	27
28	AS	A	A	AE	A	ANW	A	A	E	CE	SW	AW	28
29	C		A	C	ANE	AW	E	A	CNE	S	S	A	29
30	U		A	C	ANE	C	E	A	U	S	A	A	30
31	U		AW		NE		C	ASE		S		W	31

Lamb Catalogue for 1992

	Jan	Feb	Mar	Apr	May	Jun	Jul	Aug	Sep	Oct	Nov	Dec	
1	W	A	W	CE	NW	C	CE	W	CW	C	W	SW	1
2	W	A	W	NE	ANW	C	AE	W	C	C	C	C	2
3	W	W	W	N	ASW	C	C	CW	NW	C	CW	W	3
4	U	W	ASW	A	AW	U	C	W	NW	E	W	W	4
5	SW	W	CSW	W	AW	NE	AN	W	A	AE	W	W	5
6	SW	AW	CSW	CS	W	NE	A	A	S	AN	AW	CS	6
7	C	SW	C	C	W	E	A	A	W	A	AW	C	7
8	C	S	A	U	W	E	A	A	W	A	A	U	8
9	U	W	SW	A	C	CE	A	C	SW	AN	W	A	9
10	A	SW	W	A	C	E	W	W	S	AN	W	A	10
11	A	CSW	W	A	CSW	AE	C	SW	C	NE	C	AW	11
12	A	S	W	W	SW	A	CW	C	CSW	A	NW	W	12
13	A	W	W	W	S	A	CW	C	CW	A	A	W	13
14	A	SW	NW	C	S	AW	A	A	W	N	SE	W	14
15	A	W	W	C	A	NW	A	ASW	CW	N	C	C	15
16	A	NW	W	W	A	A	S	W	A	N	C	W	16
17	A	A	ASW	W	A	A	SW	W	CE	N	NW	S	17
18	A	U	CSW	W	AE	AN	SW	A	C	CN	W	W	18
19	A	A	W	A	A	A	SW	A	C	C	W	A	19
20	A	A	W	AS	A	A	SW	U	CS	CNE	AW	A	20
21	A	AW	W	U	E	A	C	C	E	E	C	A	21
22	A	SW	NW	A	AE	A	W	CS	C	CNW	C	A	22
23	ASE	W	N	S	E	A	SW	CSW	C	CNW	S	A	23
24	AS	SW	N	S	E	A	SW	SW	C	W	SW	A	24
25	A	A	NW	SW	E	AW	SW	C	U	C	CSW	A	25
26	A	SW	N	CSW	E	AW	SW	C	AE	W	W	A	26
27	A	SW	N	CW	E	A	W	C	E	C	W	A	27
28	A	A	W	NW	E	A	A	C	SE	C	A	A	28
29	A	SW	C	W	E	U	A	CSW	S	CN	S	A	29
30	A		CE	W	CE	C	A	C	C	NW	CSW	ASE	30
31	A		C		C		A	C		AW		ASW	31

Lamb Catalogue for 1993

	Jan	Feb	Mar	Apr	May	Jun	Jul	Aug	Sep	Oct	Nov	Dec	
1	A	A	E	C	CNW	U	A	SW	A	C	A	W	1
2	A	A	E	C	NW	C	AW	C	AN	C	SE	W	2
3	A	A	AE	C	A	A	W	C	AN	C	SE	W	3
4	SW	A	A	C	A	A	NW	C	ANE	C	SE	W	4
5	W	A	ANW	CS	A	A	ANW	NW	A	C	SE	W	5
6	W	A	A	C	A	A	ANW	AW	ASE	CE	E	W	6
7	W	A	A	U	A	AW	ANW	W	SE	C	U	W	7
8	W	A	A	S	A	A	W	W	CE	CE	A	C	8
9	W	AS	ASE	C	AE	SE	CNW	NW	C	C	W	C	9
10	W	ASE	S	CS	E	C	NW	NW	C	C	C	C	10
11	W	AS	S	CE	E	C	NW	C	C	CNE	W	CNW	11
12	W	AS	S	CE	NE	C	AW	NW	CE	CNE	W	C	12
13	CW	AS	S	E	C	C	C	ASW	E	CNE	C	C	13
14	W	AS	S	A	C	C	C	C	NE	NE	U	U	14
15	W	A	W	A	C	AW	C	C	CNE	N	A	C	15
16	W	ANW	W	W	CSW	W	C	ANW	C	AN	AS	NW	16
17	W	A	W	W	C	W	C	A	A	A	SE	W	17
18	CW	NW	W	W	C	C	C	A	ASE	A	S	W	18
19	W	NW	AW	U	C	C	C	AW	S	A	SE	NW	19
20	W	NW	AW	CS	U	ANW	CNW	AW	S	N	ASE	C	20
21	W	N	W	W	C	NW	W	ANW	CSW	AN	E	CNW	21
22	W	A	W	SW	SE	N	W	N	U	A	CE	C	22
23	W	A	W	CS	E	N	C	N	A	A	E	CNW	23
24	W	AW	ANW	C	E	AN	CW	N	U	A	CS	CNW	24
25	NW	W	A	CNE	E	A	CNW	N	NW	A	ASW	C	25
26	W	N	A	NE	E	AW	W	A	AN	A	A	AE	26
27	W	N	ASW	AE	E	A	CW	A	A	A	ASE	A	27
28	CW	ANE	S	AE	C	A	W	A	C	AE	SE	W	28
29	U		S	A	C	A	CW	A	C	A	S	W	29
30	A		S	C	C	A	CW	A	C	A	W	W	30
31	A		CS		C		W	A		AE		W	31

Lamb Catalogue for 1994

	Jan	Feb	Mar	Apr	May	Jun	Jul	Aug	Sep	Oct	Nov	Dec	
1	W	W	C	C	A	U	A	C	CNE	A	W	S	1
2	C	W	W	W	AS	SW	SE	C	A	C	AS	S	2
3	C	C	W	W	S	C	SE	C	U	N	CS	SW	3
4	C	C	W	W	C	CNW	C	C	W	AN	CS	C	4
5	C	U	W	W	W	W	C	NW	W	A	C	CSW	5
6	C	S	W	C	SW	W	C	A	SW	ASW	S	SW	6
7	AN	SW	W	C	U	W	C	A	S	ASW	S	SW	7
8	S	W	W	C	C	NW	U	A	C	A	SE	CSW	8
9	S	A	W	C	U	NW	ASW	A	C	A	SE	W	9
10	S	U	AW	U	SE	ANW	SW	NE	C	AS	CE	W	10
11	S	U	W	A	E	A	A	NE	C	A	U	W	11
12	C	ASE	W	ANW	E	A	U	NW	C	A	C	CW	12
13	CW	SE	W	AN	E	A	A	ANW	C	A	SW	N	13
14	C	SE	W	AN	E	ANW	A	A	CE	A	W	A	14
15	AE	E	W	ANE	NE	NW	W	AW	CNE	ANE	CW	ASW	15
16	N	AS	W	A	NE	NW	A	SW	N	AE	W	AW	16
17	A	S	NW	A	NE	W	ASE	C	AN	AE	SW	SW	17
18	W	S	C	A	ANE	W	ASE	W	A	SE	ASW	W	18
19	W	SE	CNW	U	AE	W	A	CW	CW	SE	SW	W	19
20	W	E	U	C	E	W	A	W	C	CSE	SW	NW	20
21	W	E	U	C	CE	W	A	A	A	CS	ASW	A	21
22	W	E	W	C	C	W	A	SE	A	C	SW	A	22
23	NW	E	W	SE	ANE	A	A	C	A	C	W	A	23
24	W	E	W	CS	E	C	ASE	C	U	CW	A	ASW	24
25	W	SE	W	SW	E	U	S	C	A	C	AW	W	25
26	W	SE	A	SW	E	W	S	W	A	C	AW	W	26
27	W	C	S	SW	ANE	ASW	SW	W	ANW	C	A	U	27
28	AW	C	W	ASW	A	S	A	CW	U	CW	A	W	28
29	W		W	ASW	A	W	A	ANW	A	C	A	SW	29
30	W		SW	A	A	A	CS	A	U	C	AS	C	30
31	AW		C		A		C	E		C		N	31

Lamb Catalogue for 1995

	Jan	Feb	Mar	Apr	May	Jun	Jul	Aug	Sep	Oct	Nov	Dec	
1	N	W	W	W	AS	NW	A	AE	A	W	ANE	S	1
2	A	A	CW	W	A	U	N	A	CN	SW	A	S	2
3	A	W	C	U	A	C	N	ANE	C	SW	A	U	3
4	S	AW	W	A	A	C	AN	ANE	N	CS	A	A	4
5	SW	AW	C	W	A	ANW	W	A	N	SW	AS	A	5
6	W	W	W	W	A	NW	W	AN	C	SW	A	E	6
7	AW	U	C	ANW	U	N	AW	A	E	SW	AW	E	7
8	W	U	C	A	CN	N	A	A	C	S	AW	CS	8
9	W	A	SW	A	N	N	ASE	A	C	CSW	CNW	A	9
10	NW	C	S	AW	ANE	N	E	ASE	C	ASW	C	A	10
11	N	C	AW	AW	NE	N	SE	S	C	A	CE	A	11
12	A	C	A	A	N	ANE	S	SW	C	A	E	A	12
13	AW	C	W	A	N	NE	S	W	C	ASE	SE	ANE	13
14	W	CSW	CNW	NW	N	N	C	AW	C	S	C	AE	14
15	SW	CW	CNW	NW	A	N	C	A	CN	SW	C	AE	15
16	SW	C	W	NW	CNE	W	C	A	CN	S	C	AE	16
17	CS	CW	CW	C	CNE	W	CSW	A	A	SW	N	E	17
18	SW	W	W	N	N	W	U	A	E	AW	A	NE	18
19	C	W	NW	N	CNW	SW	ASW	AE	NE	W	A	NE	19
20	CSW	W	A	N	A	AW	SW	A	C	ANW	S	NE	20
21	C	W	A	C	AS	A	W	A	CW	A	S	A	21
22	C	W	A	NE	S	A	A	A	W	SW	SW	CSE	22
23	CNW	W	ASW	C	S	A	AW	NW	W	S	SW	C	23
24	W	C	W	E	C	A	A	W	W	CS	C	C	24
25	C	CNW	NW	E	CS	A	ASE	CW	W	CSW	C	CN	25
26	CN	NW	N	NE	S	AE	CSE	C	W	SW	CS	N	26
27	U	W	N	A	S	AE	C	NW	NW	W	C	A	27
28	C	W	C	A	S	A	ASW	N	NW	A	CE	A	28
29	C		A	AE	C	A	S	NW	A	A	E	ASE	29
30	C		ASW	A	C	A	U	AN	U	AS	SE	E	30
31	A		W		NW		E	A		A		E	31

Appendix C

THE DAILY CENTRAL ENGLAND TEMPERATURE, 1961–95

The 'Central England' temperature was originally devised and published in 1953 by the late Gordon Manley and the record represents the mean monthly temperature of the English Midlands. The series commences in 1659. A *daily* Central England temperature record was constructed by the Hadley Centre, part of the UK Met. Office, a few years ago and this is now routinely updated by them. Some applications of the Central England temperature record have been discussed in Chapter 9, where plots of the seasonal data from 1659 to the present have been shown. These seasonal mean temperatures from 1659 to 1995 are listed in Appendix D. Here, we present graphs of the daily data from 1961 to the present in a form which allows easy visual interpretation. On each of the graphs that follow, the daily mean temperature during the respective year is shown as a histogram (the thin 'blocky' line), while the thirty-day average for that year is shown as the heavy black curve. To compare each year with the long-term average, also shown (the bolder dashed line) is the average daily temperature for the period 1961 to 1990.

Reference: D.E. Parker, T.P. Legg and C.K. Folland, 'A new daily Central England Temperature series, 1772–1991', *International Journal of Climatology*, 1992, vol. 12, pp. 317–42.

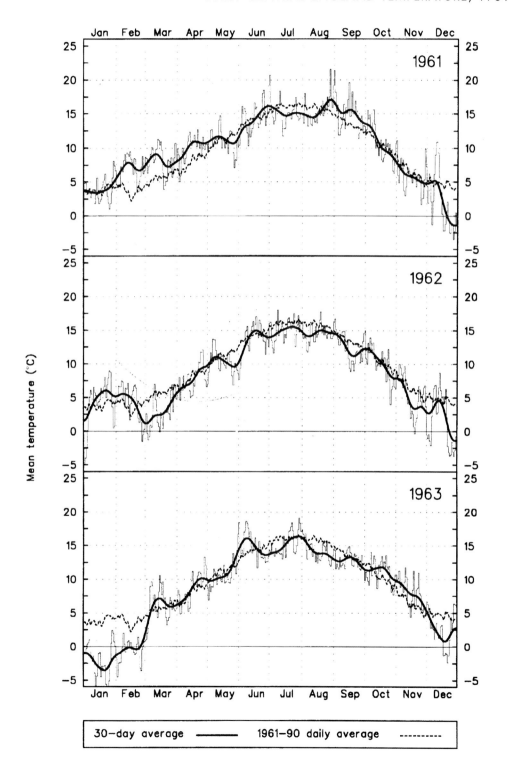

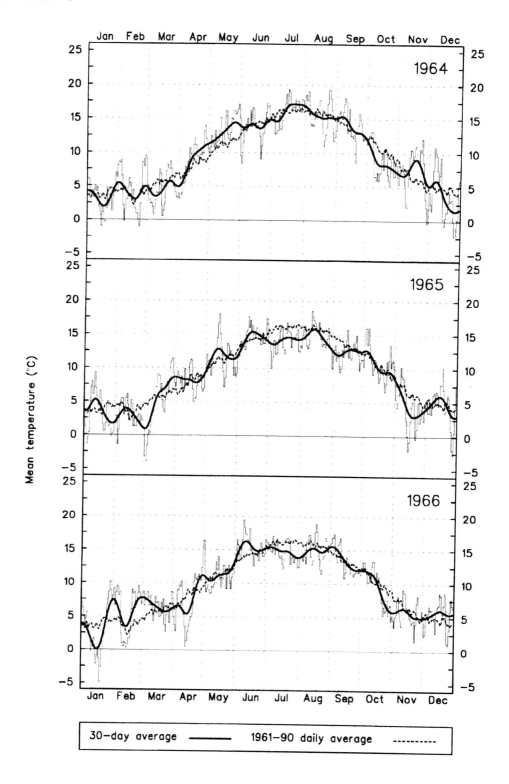

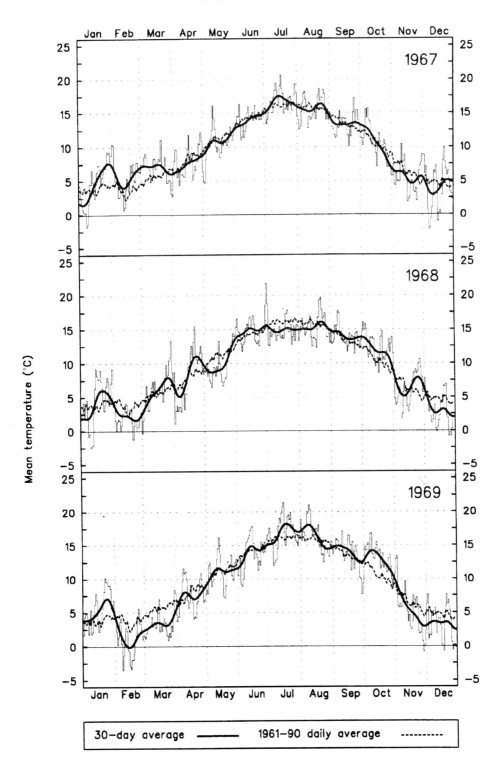

30-day average ———   1961–90 daily average ----------

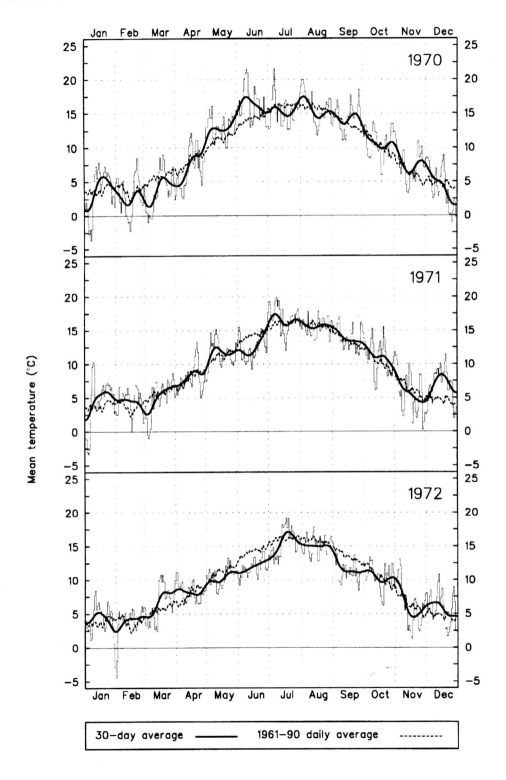

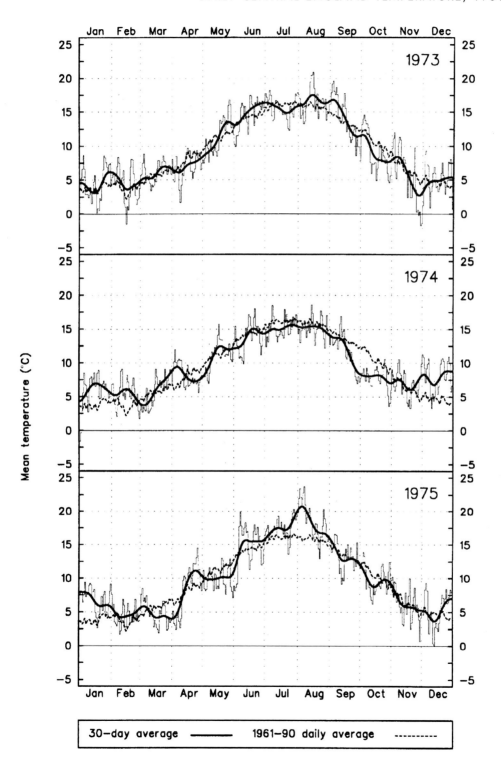

Mean temperature (°C)

| 30-day average | —————— | 1961-90 daily average | ---------- |

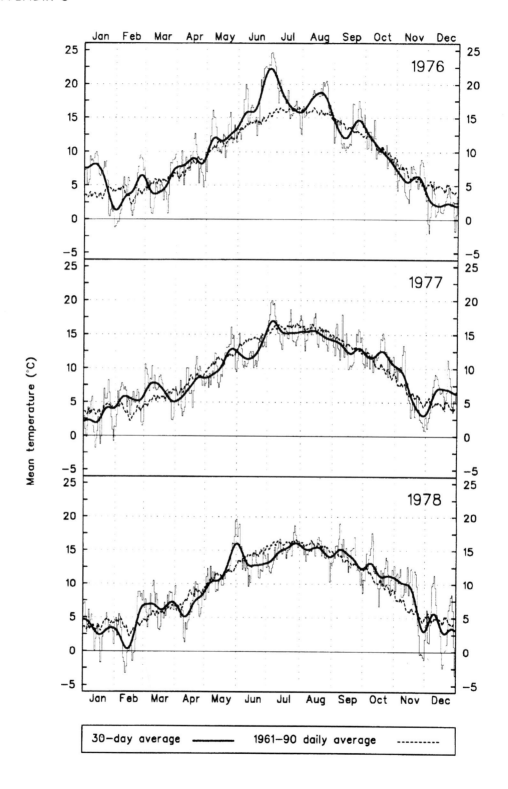

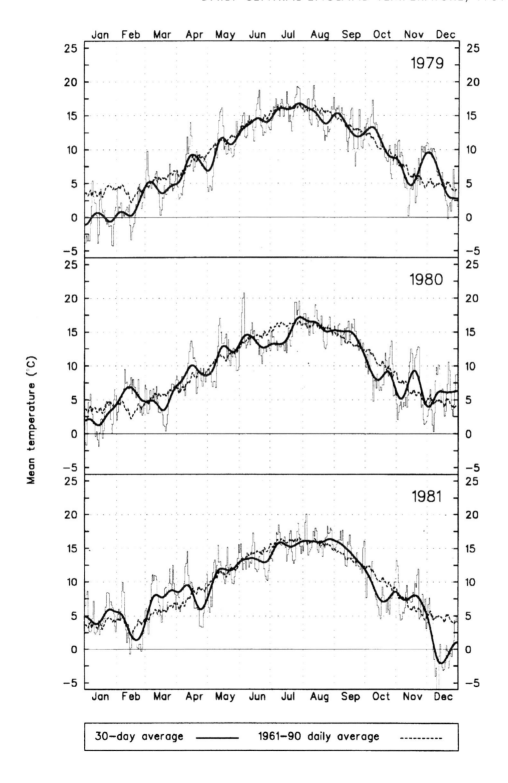

30-day average ————   1961–90 daily average ----------

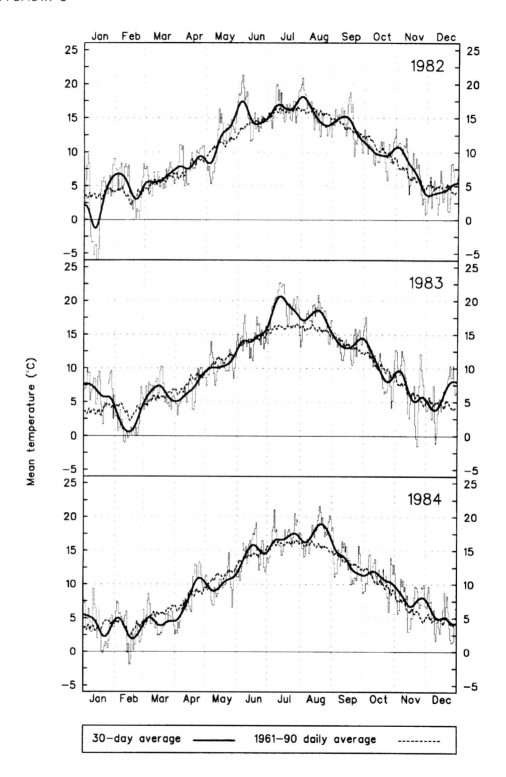

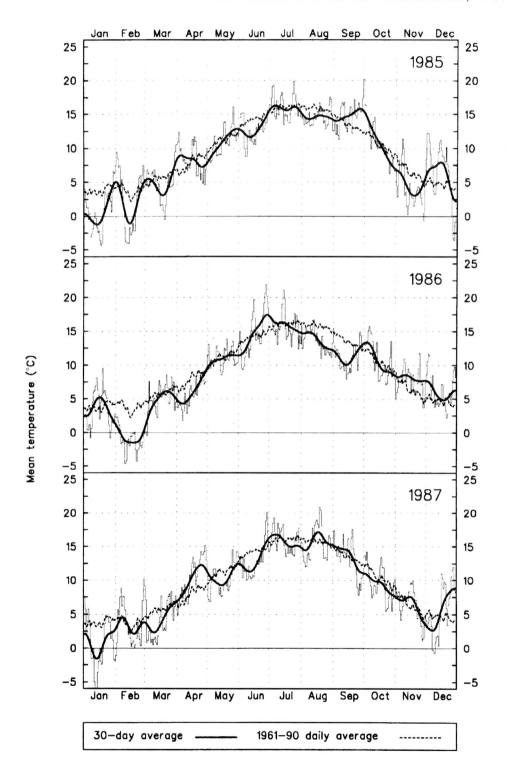

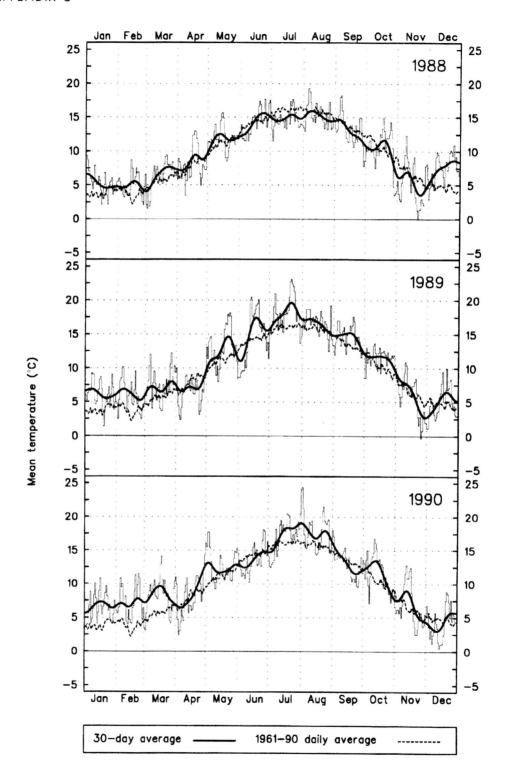

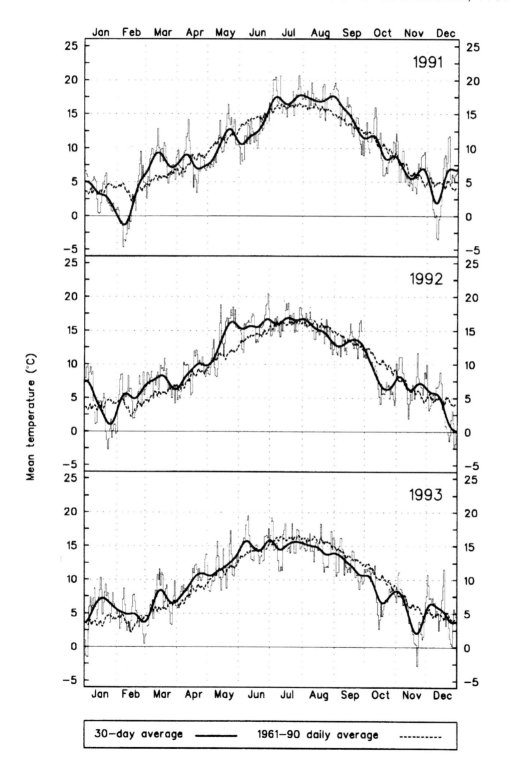

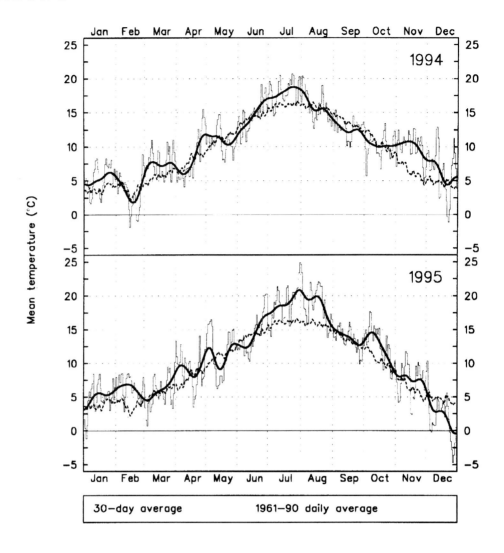

Appendix D

LISTINGS OF CLIMATE DATASETS

In this appendix are listed seasonal and annual values of five commonly used climate datasets, relevant to British Isles and global climate. The data are complete from the first year of the respective records, through 1996. The sources of the data are acknowledged and scientific papers are cited which contain a detailed discussion on how they are constructed.

Table D.1: 'Central England' temperature. These seasonal and annual mean temperatures are expressed in degrees Celsius from 1659 to 1996. The series was originally constructed by Professor Gordon Manley, but is now updated by the Hadley Centre. The data are discussed in the following two papers: G. Manley, 'Central England Temperatures: monthly means 1659 to 1973', *Quarterly Journal of the Royal Meteorological Society*, 1974, vol. 100, pp. 389–405; D.E. Parker, T.P. Legg and C.K. Folland, 'A new daily Central England Temperature series, 1772–1991', *International Journal of Climatology*, 1992, vol. 12, pp. 317–42.

Table D.2: England and Wales precipitation. These seasonal and annual precipitation totals are expressed in mm from 1766 to 1995. The source of the data is Dr Phil Jones of the Climatic Research Unit, and the construction of the record is discussed in the following papers: J.M. Gregory, P.D. Jones and T.M.L. Wigley, 'Precipitation in Britain: an analysis of area-average data updated to 1989', *International Journal of Climatology*, 1991, vol. 11, pp. 331–45; T.M.L. Wigley, J.M. Lough and P.D. Jones, 'Spatial patterns of precipitation in England and Wales and a revised homogeneous England and Wales precipitation series', *Journal of Climatology*, 1984, vol. 4, pp. 1–26.

Table D.3: North Atlantic Oscillation index. These seasonal and annual values represent the sea-level pressure difference between the Azores and Iceland, measured in hectopascals (hPa). Nearly all values are positive indicating higher pressure in the Azores than Iceland; the exception is the winter of 1880/1 when the NAO index was negative. The winter of 1962/3 also had a very small positive value. The two stations used to generate these data were Ponta Delgada, on the Azores, and Stykkisholmur, Iceland. The data are supplied by Dr Phil Jones of the Climatic Research Unit. See the following paper for a discussion about the North Atlantic Oscillation: J.W. Hurrell, 'Decadal trends in the North Atlantic Oscillation; regional temperatures and precipitation', *Science*, 1995, vol. 269, pp. 676–9.

Table D.4: Northern Hemisphere (land areas only) surface air temperature anomaly. These seasonal and annual temperature anomalies from 1851 to 1995 are expressed as differences in degrees Celsius from the 1961 to 1990 average. The source of the data is Dr Phil Jones of the Climatic Research Unit and the data set is discussed in the following paper: P.D. Jones, 'Hemispheric surface air temperature variability – a reanalysis and an update to 1993', *Journal of Climate*, 1994, vol. 7, pp. 1794–802.

Table D.5: Global-average (land and marine areas) surface air temperature anomaly. These seasonal and annual temperature anomalies from 1856 to 1995 are expressed as differences in degrees Celsius from the 1961 to 1990 average. The data are those used by the Intergovernmental Panel on Climate Change and are supplied by the Hadley Centre. The data are discussed in numerous papers, but see D.E. Parker, P.D. Jones, C.K. Folland and A. Bevan, 'Interdecadal changes of surface temperature since the late nineteenth century', *Journal of Geophysical Research*, 1994, vol. 99, pp. 14,373–99; N. Nicholls, G.V. Gruza, J. Jouzel, T.R. Karl, L.A. Ogallo and D.E. Parker, 'Observed climate variability and change', in J.T. Houghton, L.G. Meiro Filho, B.A. Callendar, A. Kattenburg and K. Maskell (eds), *Climate Change 1995: the science of climate change*, Cambridge, Cambridge University Press, 1996, pp. 133–92.

Table D.1 'Central England' mean surface air temperature (°C)

Year	Win	Spr	Sum	Aut	Ann	Year	Win	Spr	Sum	Aut	Ann
1659		8.0	15.0	9.3	8.8						
1660	2.0	8.7	15.0	9.7	9.1	1700	3.3	7.3	14.5	8.9	8.6
1661	5.0	8.3	14.7	10.7	9.8	1701	3.2	6.1	16.2	9.5	8.7
1662	5.7	8.3	15.0	10.0	9.5	1702	5.1	7.4	14.8	9.6	9.3
1663	1.7	7.3	14.7	10.0	8.6	1703	3.5	8.6	15.4	8.5	9.1
1664	4.7	8.0	15.7	9.3	9.3	1704	3.5	8.4	16.0	8.9	9.1
1665	2.0	7.3	15.0	9.3	8.3	1705	3.4	7.8	15.1	8.2	8.7
1666	3.7	8.3	16.7	10.3	9.8	1706	3.8	9.2	16.1	9.9	9.8
1667	2.3	6.3	16.0	9.3	8.5	1707	3.7	8.2	16.3	9.7	9.4
1668	4.3	7.7	15.3	10.0	9.5	1708	4.5	8.8	15.3	10.2	9.7
1669	3.3	7.7	16.0	10.0	9.0	1709	1.2	8.0	15.2	10.3	8.7
1670	2.0	8.0	15.3	10.0	8.9	1710	3.2	8.0	15.0	10.3	9.5
1671	3.5	8.0	15.0	9.3	9.0	1711	4.8	8.7	15.3	9.8	9.4
1672	2.3	8.0	15.0	9.7	8.8	1712	3.8	7.8	15.3	9.5	9.1
1673	3.7	7.8	15.2	8.0	8.3	1713	4.2	6.8	14.2	9.3	8.6
1674	2.5	6.5	13.7	8.7	8.1	1714	4.5	7.5	15.8	9.8	9.4
1675	2.8	6.8	13.7	7.7	7.8	1715	4.3	9.0	14.8	10.3	9.4
1676	5.0	7.7	16.8	7.5	8.8	1716	0.8	8.0	15.0	9.2	8.4
1677	2.0	8.2	15.3	8.7	8.8	1717	3.3	7.7	14.8	9.5	9.0
1678	1.8	7.2	15.3	9.8	8.4	1718	3.2	8.0	16.2	9.8	9.3
1679	1.0	7.8	16.2	9.2	8.7	1719	3.8	7.8	16.7	9.8	9.5
1680	3.2	7.7	14.7	10.5	8.9	1720	4.0	7.5	14.7	9.5	9.1
1681	1.0	7.5	15.0	10.7	8.7	1721	3.8	7.2	15.2	9.7	8.9
1682	3.7	7.5	14.5	9.3	9.0	1722	4.5	8.0	14.7	10.4	9.4
1683	3.8	8.8	15.0	8.0	8.5	1723	3.1	9.4	15.3	10.6	9.8
1684	-1.2	7.5	15.5	8.7	7.9	1724	5.2	7.8	15.5	9.4	9.3
1685	2.7	8.7	14.3	10.0	9.1	1725	3.7	8.0	13.1	9.7	8.7
1686	6.3	9.3	15.5	9.5	10.1	1726	3.1	8.7	16.0	10.3	9.3
1687	4.7	7.3	14.5	9.3	9.0	1727	3.7	9.3	16.2	10.0	9.9
1688	3.7	6.5	14.5	7.7	7.8	1728	3.3	9.3	16.4	9.7	9.5
1689	2.7	8.0	14.3	8.7	8.5	1729	1.7	6.7	15.9	11.6	9.3
1690	4.3	7.3	14.7	9.3	8.9	1730	4.6	9.1	15.2	11.8	10.0
1691	2.2	7.2	15.2	8.3	8.1	1731	2.5	8.3	16.2	11.8	9.9
1692	1.8	6.8	14.5	7.7	7.7	1732	4.7	8.8	15.7	10.6	9.7
1693	3.8	6.2	14.8	9.2	8.5	1733	5.0	9.0	16.5	9.5	10.5
1694	2.7	6.7	13.7	7.8	7.7	1734	6.1	9.5	15.5	9.3	9.8
1695	0.7	6.0	13.2	8.7	7.3	1735	4.1	8.5	14.8	10.3	9.5
1696	4.7	6.5	14.7	8.7	8.5	1736	5.0	8.7	16.6	10.6	10.3
1697	1.3	8.0	14.3	8.3	8.0	1737	5.6	9.1	15.7	9.7	9.9
1698	1.0	6.5	14.0	8.7	7.6	1738	4.7	8.9	15.5	9.7	9.8
1699	3.4	6.8	15.5	9.4	8.8	1739	5.6	8.0	15.3	8.8	9.2

Table D.1 (cont.)

Year	Win	Spr	Sum	Aut	Ann	Year	Win	Spr	Sum	Aut	Ann
1740	-0.4	6.3	14.3	7.5	6.8	1780	1.4	9.0	16.2	9.7	9.1
1741	2.8	6.9	15.8	11.2	9.3	1781	3.4	9.3	17.0	10.4	10.2
1742	3.1	7.1	15.5	8.6	8.4	1782	4.2	6.1	14.9	7.7	8.0
1743	3.4	8.0	15.8	10.8	9.8	1783	3.2	7.9	16.5	9.6	9.3
1744	3.1	7.4	15.4	9.7	8.8	1784	1.2	7.3	14.3	9.4	7.8
1745	3.2	7.8	14.4	10.1	8.8	1785	1.4	7.3	15.4	9.3	8.5
1746	2.2	7.6	15.3	8.4	8.6	1786	3.0	7.1	15.4	7.5	8.3
1747	4.8	7.6	16.6	10.2	9.8	1787	4.1	8.6	15.1	9.0	9.3
1748	3.2	6.2	15.3	10.2	8.8	1788	3.8	8.9	15.7	9.8	9.2
1749	5.0	8.1	14.9	10.2	9.4	1789	2.1	7.3	15.3	8.8	8.9
1750	5.1	8.9	15.5	9.5	9.7	1790	5.7	8.1	15.0	9.5	9.4
1751	3.2	7.5	14.9	8.4	8.4	1791	4.4	8.9	15.3	9.5	9.3
1752	3.1	7.6	15.4	10.3	9.2	1792	2.6	8.7	15.1	9.2	9.2
1753	3.3	8.5	15.2	9.2	9.1	1793	3.9	7.1	15.4	9.7	9.1
1754	3.5	7.4	14.7	10.0	8.8	1794	4.8	9.5	16.4	9.4	9.9
1755	2.3	7.8	15.1	8.9	8.5	1795	0.5	7.5	15.0	10.7	8.7
1756	4.3	7.3	14.9	9.0	8.8	1796	6.2	8.2	14.8	9.1	9.0
1757	2.4	7.9	15.9	9.5	8.9	1797	2.6	7.7	15.6	8.5	9.0
1758	3.2	8.7	15.1	8.6	8.9	1798	4.1	9.5	16.5	9.4	9.6
1759	5.2	8.9	16.5	9.8	10.0	1799	2.0	6.1	14.6	8.9	7.9
1760	2.7	9.2	16.0	10.2	9.8	1800	2.1	8.5	16.1	9.5	9.2
1761	5.8	9.4	15.5	9.9	10.0	1801	4.2	9.0	16.0	9.7	9.6
1762	4.4	8.9	16.7	8.7	9.6	1802	2.3	8.4	14.8	9.7	8.9
1763	2.6	8.2	15.1	9.1	8.9	1803	2.9	8.6	15.8	8.6	9.1
1764	4.6	7.8	15.1	8.6	8.7	1804	4.4	8.3	15.9	10.5	9.6
1765	2.7	8.0	14.9	8.8	8.5	1805	2.8	8.2	15.2	9.2	9.0
1766	1.4	7.3	15.3	9.9	8.6	1806	4.0	8.0	15.5	10.6	9.8
1767	2.9	7.3	14.4	10.1	8.7	1807	4.4	7.5	16.1	8.3	8.6
1768	3.0	8.3	15.2	8.8	8.9	1808	2.4	7.6	16.6	8.6	8.8
1769	3.3	8.0	14.8	8.9	8.8	1809	3.3	8.1	14.5	9.2	8.9
1770	4.4	6.0	14.7	9.4	8.5	1810	3.3	7.4	14.8	9.7	8.8
1771	2.6	6.9	14.8	9.2	8.6	1811	3.1	9.6	14.9	11.2	9.7
1772	2.9	7.0	16.4	10.6	9.1	1812	3.7	6.6	13.8	9.1	8.2
1773	3.8	8.4	15.9	9.2	9.2	1813	3.1	8.7	14.4	8.3	8.7
1774	2.9	8.6	15.6	9.2	9.1	1814	0.4	7.2	14.3	8.5	7.8
1775	4.7	9.5	16.4	9.5	10.1	1815	3.7	9.3	14.8	9.0	9.1
1776	2.2	8.9	15.2	9.8	9.0	1816	2.4	6.8	13.4	8.7	7.9
1777	2.9	8.5	14.9	10.6	9.1	1817	4.7	7.3	14.3	9.6	8.9
1778	2.6	7.8	16.5	8.8	9.2	1818	3.2	7.6	16.6	11.6	9.8
1779	5.6	9.7	16.6	10.6	10.4	1819	4.1	9.0	15.7	8.9	9.2

Table D.1 (cont.)

Year	Win	Spr	Sum	Aut	Ann	Year	Win	Spr	Sum	Aut	Ann
1820	1.4	8.3	14.7	8.7	8.6	1860	2.3	7.4	13.5	8.5	7.9
1821	3.5	8.2	14.5	11.3	9.5	1861	2.7	8.1	15.2	9.7	9.1
1822	5.8	9.6	16.0	10.4	10.1	1862	4.3	8.9	13.8	8.8	9.2
1823	1.5	8.0	13.6	9.3	8.4	1863	5.7	8.6	14.8	9.6	9.7
1824	4.6	7.6	14.8	10.1	9.3	1864	3.7	8.6	14.4	9.5	8.9
1825	4.3	8.6	15.9	10.4	9.7	1865	2.7	8.7	15.8	10.9	9.7
1826	3.8	8.8	17.6	9.7	10.1	1866	5.3	7.8	15.2	10.1	9.6
1827	2.7	8.9	15.2	10.7	9.5	1867	4.7	7.9	15.1	9.2	9.0
1828	5.7	9.1	15.6	10.6	10.3	1868	4.5	9.7	16.9	9.2	10.4
1829	3.9	7.8	14.8	8.0	8.2	1869	6.8	7.8	15.3	10.0	9.6
1830	1.1	9.5	14.2	9.7	8.7	1870	3.0	8.5	16.1	9.0	9.0
1831	2.7	9.3	16.3	10.7	10.1	1871	2.4	9.1	15.1	8.6	9.1
1832	4.1	8.4	15.5	10.1	9.5	1872	5.2	8.2	15.5	9.5	9.8
1833	4.0	8.9	14.9	9.6	9.5	1873	4.1	7.7	15.3	8.9	9.0
1834	6.5	9.3	16.2	10.4	10.5	1874	4.9	8.8	15.4	9.9	9.3
1835	4.7	8.6	16.1	9.6	9.6	1875	2.8	8.7	15.0	9.7	9.4
1836	3.4	8.0	15.1	8.5	8.9	1876	4.1	7.3	16.0	10.0	9.5
1837	3.8	5.6	16.0	9.4	8.8	1877	5.9	7.0	15.0	9.2	9.2
1838	1.4	7.2	15.0	9.0	8.1	1878	5.0	8.7	16.0	9.0	9.2
1839	3.6	6.9	14.6	9.7	8.7	1879	0.7	6.4	13.7	8.5	7.4
1840	3.8	8.3	14.6	8.1	8.5	1880	2.5	8.2	15.2	9.0	9.1
1841	1.6	9.3	13.8	9.1	8.7	1881	2.3	8.1	14.6	9.6	8.6
1842	3.1	8.5	15.7	8.6	9.2	1882	5.1	9.1	14.4	9.2	9.4
1843	4.4	8.2	14.3	9.3	9.1	1883	4.8	6.9	14.6	9.6	9.0
1844	4.3	8.4	14.5	9.5	8.6	1884	5.5	8.3	15.9	9.7	9.8
1845	1.5	6.7	14.2	9.2	8.3	1885	4.4	7.0	14.6	8.5	8.6
1846	5.8	8.7	17.1	10.4	10.1	1886	2.4	7.4	15.1	10.5	8.7
1847	1.7	8.2	15.5	10.0	9.2	1887	2.7	6.3	16.1	7.8	8.3
1848	4.1	9.3	14.6	9.4	9.4	1888	2.5	6.7	13.7	9.3	8.2
1849	5.1	8.2	15.0	9.7	9.3	1889	3.7	8.2	15.1	9.4	9.0
1850	3.5	7.9	15.4	9.2	9.1	1890	4.0	8.3	14.0	9.9	8.7
1851	5.0	7.9	14.8	8.9	9.1	1891	1.5	6.5	14.6	9.7	8.5
1852	4.8	8.0	15.9	9.5	9.8	1892	3.3	7.2	14.3	8.6	8.2
1853	4.5	7.3	14.6	9.2	8.4	1893	2.9	10.2	16.5	9.3	10.0
1854	3.1	8.7	14.6	9.6	9.3	1894	4.4	8.5	14.5	9.6	9.3
1855	1.9	6.4	15.3	9.4	8.0	1895	1.2	8.6	15.3	10.0	8.6
1856	3.8	7.3	15.2	9.3	9.1	1896	4.4	9.2	15.6	8.1	9.3
1857	3.8	8.0	16.5	11.0	10.1	1897	3.8	7.9	15.9	9.9	9.4
1858	4.2	7.8	15.8	9.5	9.1	1898	5.4	7.7	15.1	11.2	10.1
1859	5.1	8.8	16.4	9.1	9.6	1899	5.8	7.6	16.9	10.2	9.7

Table D.1 (cont.)

Year	Win	Spr	Sum	Aut	Ann	Year	Win	Spr	Sum	Aut	Ann
1900	3.1	7.4	15.8	10.2	9.6	1940	1.5	9.1	15.7	9.8	9.1
1901	4.3	8.1	15.8	9.5	9.1	1941	2.6	7.0	15.7	10.5	9.1
1902	3.2	7.7	14.3	9.7	8.8	1942	2.2	8.5	15.5	9.6	9.1
1903	5.3	8.2	14.2	10.0	9.3	1943	5.9	9.6	15.6	10.1	10.0
1904	3.6	8.0	15.2	9.1	9.0	1944	4.3	8.9	15.7	9.3	9.6
1905	4.2	8.3	15.5	8.1	9.1	1945	3.7	10.1	15.7	11.2	10.3
1906	4.4	7.6	15.6	10.7	9.4	1946	4.5	8.6	14.7	10.6	9.4
1907	3.1	8.1	13.6	10.0	8.8	1947	1.1	8.6	17.0	10.9	9.6
1908	4.1	7.6	14.9	10.7	9.3	1948	5.1	9.6	14.8	10.4	10.0
1909	3.4	7.8	13.9	9.0	8.6	1949	5.6	8.8	16.5	11.5	10.6
1910	4.2	8.2	14.7	8.8	9.2	1950	5.1	8.8	15.9	9.4	9.4
1911	5.0	8.5	17.0	9.8	10.1	1951	2.9	7.0	15.0	10.7	9.3
1912	5.1	9.4	14.3	8.5	9.4	1952	3.9	9.9	15.7	7.9	9.1
1913	5.3	8.6	14.7	11.1	9.8	1953	3.5	8.5	15.4	10.7	9.8
1914	5.2	8.9	15.5	10.1	9.9	1954	4.1	8.2	14.1	10.5	9.2
1915	4.3	8.0	14.8	8.4	8.9	1955	3.5	7.4	16.5	10.1	9.3
1916	5.5	7.7	14.5	10.1	9.2	1956	2.9	8.3	14.1	9.9	8.8
1917	1.5	7.1	15.5	9.8	8.5	1957	5.5	9.5	15.6	9.9	10.0
1918	4.2	8.5	14.9	8.9	9.5	1958	4.2	7.4	15.3	10.8	9.4
1919	3.9	8.1	14.6	7.8	8.5	1959	3.6	9.8	16.6	11.5	10.5
1920	5.6	9.1	14.0	10.1	9.6	1960	4.6	9.4	15.4	10.2	9.7
1921	5.4	9.0	16.2	10.5	10.5	1961	4.9	9.7	15.0	10.7	9.9
1922	4.9	7.6	13.7	8.8	8.7	1962	3.6	6.9	14.4	9.5	8.6
1923	5.7	7.8	15.1	8.5	9.1	1963	−0.3	8.4	14.8	10.7	8.5
1924	3.9	7.5	14.4	10.2	9.3	1964	3.5	8.8	15.1	10.1	9.5
1925	5.8	8.0	15.7	8.5	9.2	1965	3.4	8.3	14.6	9.2	8.9
1926	4.7	8.6	15.6	9.5	9.7	1966	4.4	8.2	15.0	9.8	9.4
1927	4.2	8.8	14.7	9.7	9.2	1967	5.1	8.4	15.5	9.9	9.6
1928	4.4	8.6	14.8	10.2	9.6	1968	3.5	8.1	15.1	11.0	9.3
1929	1.7	8.1	14.9	10.5	9.0	1969	3.2	7.3	15.7	10.8	9.3
1930	4.6	8.1	15.4	10.1	9.4	1970	3.3	7.8	15.9	11.0	9.6
1931	3.8	7.8	14.7	9.4	9.0	1971	4.4	8.1	15.0	10.4	9.7
1932	4.8	7.4	15.8	9.4	9.4	1972	4.9	8.4	14.2	9.5	9.2
1933	4.1	9.4	17.0	10.2	9.8	1973	4.8	8.2	15.7	9.7	9.5
1934	3.2	8.0	16.2	10.4	10.0	1974	5.4	8.3	14.8	8.9	9.6
1935	6.1	8.2	16.3	10.0	9.7	1975	6.4	7.7	16.9	9.9	10.0
1936	3.0	8.3	15.4	9.8	9.3	1976	5.2	8.3	17.8	10.1	10.1
1937	5.4	8.3	15.7	9.6	9.6	1977	3.3	8.2	14.4	10.6	9.5
1938	4.6	9.1	15.3	11.2	10.2	1978	4.1	8.3	14.5	11.5	9.4
1939	4.7	8.7	15.4	10.4	9.7	1979	1.6	7.5	15.0	10.5	8.8

Table D.1 (cont.)

Year	Win	Spr	Sum	Aut	Ann	Year	Win	Spr	Sum	Aut	Ann
1980	4.6	8.2	14.8	10.1	9.4	1990	6.2	9.6	16.2	10.7	10.6
1981	4.5	9.0	15.0	10.3	9.2	1991	3.0	8.9	15.5	10.6	9.5
1982	2.6	8.8	15.9	10.8	9.8	1992	4.6	9.9	15.7	9.5	9.9
1983	4.3	7.8	17.1	10.6	10.0	1993	4.7	9.2	14.9	8.5	9.5
1984	4.2	7.6	16.3	10.9	9.7	1994	4.7	8.8	16.2	11.0	10.2
1985	2.7	8.0	14.5	9.9	8.9	1995	5.9	8.8	17.4	11.5	10.5
1986	2.9	7.3	14.8	10.0	8.7	1996	3.0	7.4	15.8		
1987	3.5	8.2	14.8	9.9	9.1						
1988	5.3	8.8	14.8	9.6	9.8						
1989	6.5	9.0	16.5	10.9	10.5						

Table D.2 'England and Wales' precipitation (mm)

Year	Win	Spr	Sum	Aut	Ann	Year	Win	Spr	Sum	Aut	Ann
1766	200	217	249	201	806						
1767	275	192	271	241	941						
1768	268	145	362	399	1247						
1769	259	160	279	276	951						
1770	190	217	234	402	1079	1810	173	204	254	300	973
1771	218	129	207	252	793	1811	273	233	227	281	950
1772	277	200	215	401	1032	1812	235	241	225	268	923
1773	183	226	186	379	1034	1813	166	193	175	289	828
1774	309	215	257	250	995	1814	115	175	227	219	829
1775	240	116	347	335	1012	1815	194	211	204	222	799
1776	203	119	294	226	847	1816	216	206	312	261	1007
1777	127	220	291	229	860	1817	238	162	346	132	883
1778	145	138	225	309	888	1818	272	326	103	303	943
1779	137	164	279	284	901	1819	243	186	190	236	902
1780	213	203	124	293	699	1820	208	170	204	218	776
1781	157	101	197	217	734	1821	149	248	200	346	1038
1782	199	363	338	244	1109	1822	261	192	236	329	905
1783	231	155	226	237	834	1823	252	166	261	245	991
1784	132	209	294	147	811	1824	215	196	218	388	1027
1785	138	55	227	272	719	1825	219	197	160	293	840
1786	231	144	184	313	889	1826	240	124	122	267	730
1787	182	180	268	276	918	1827	172	226	173	310	935
1788	226	113	229	146	614	1828	315	210	356	224	1073
1789	207	200	296	308	1109	1829	180	166	396	245	913
1790	201	165	233	239	856	1830	133	223	321	283	1006
1791	320	137	216	294	928	1831	225	189	229	323	1005
1792	207	270	315	298	1117	1832	180	188	270	251	865
1793	241	166	173	235	777	1833	275	151	215	232	914
1794	177	182	169	378	895	1834	331	111	316	170	835
1795	168	164	237	293	878	1835	185	199	157	368	900
1796	269	155	232	215	863	1836	165	210	225	362	1015
1797	158	219	329	339	1090	1837	262	118	216	237	828
1798	236	137	259	306	869	1838	181	169	262	298	894
1799	209	179	324	397	1079	1839	184	142	351	358	1076
1800	157	229	92	352	902	1840	270	114	213	283	801
1801	248	194	193	308	948	1841	156	187	269	385	1064
1802	219	102	252	203	757	1842	208	155	198	300	820
1803	196	137	167	184	751	1843	204	258	256	269	955
1804	313	216	210	243	883	1844	167	102	189	258	718
1805	184	161	222	139	735	1845	129	157	272	234	874
1806	259	123	230	219	892	1846	242	227	256	262	927
1807	227	157	171	271	746	1847	157	200	152	255	848
1808	127	175	229	282	809	1848	307	212	344	299	1130
1809	275	179	267	180	941	1849	225	192	179	267	867

Table D.2 (cont.)

Year	Win	Spr	Sum	Aut	Ann	Year	Win	Spr	Sum	Aut	Ann
1850	218	177	216	229	801	1890	194	174	264	219	826
1851	211	191	237	168	781	1891	105	187	288	319	997
1852	232	99	337	456	1213	1892	244	128	248	288	829
1853	267	159	289	247	874	1893	214	72	205	231	756
1854	151	117	185	185	673	1894	263	177	264	285	980
1855	132	139	269	233	745	1895	191	147	247	259	856
1856	194	208	224	225	901	1896	151	142	185	314	839
1857	217	197	242	253	845	1897	290	218	232	201	916
1858	92	188	190	217	739	1898	205	187	166	233	770
1859	193	201	228	311	939	1899	287	186	133	254	840
1860	243	215	371	247	1083	1900	323	121	245	236	975
1861	210	153	254	273	848	1901	222	179	197	195	805
1862	154	276	223	260	929	1902	225	173	222	202	758
1863	212	120	230	308	862	1903	227	252	310	391	1180
1864	164	177	118	250	703	1904	284	171	190	165	815
1865	229	152	257	268	903	1905	153	197	224	236	765
1866	280	151	261	320	1049	1906	246	171	170	290	926
1867	270	246	209	191	891	1907	165	216	232	253	899
1868	230	169	143	247	907	1908	223	228	216	174	807
1869	381	226	99	299	943	1909	151	217	255	252	933
1870	253	117	131	276	733	1910	329	160	260	244	997
1871	203	167	251	271	869	1911	240	142	150	276	839
1872	287	220	304	394	1285	1912	338	184	410	225	1099
1873	299	161	246	241	834	1913	270	274	114	268	868
1874	167	127	180	308	852	1914	185	205	218	227	·971
1875	266	118	303	399	1037	1915	423	141	238	195	990
1876	192	208	186	324	1057	1916	374	226	215	303	1028
1877	418	249	291	302	1144	1917	186	183	303	245	863
1878	197	226	271	295	984	1918	194	159	204	325	963
1879	250	191	409	168	983	1919	324	212	173	193	926
1880	138	147	279	379	1028	1920	298	273	250	187	953
1881	268	169	288	261	967	1921	210	138	133	164	629
1882	235	217	303	370	1146	1922	275	184	267	162	928
1883	332	132	228	356	974	1923	333	193	192	312	1013
1884	231	153	187	169	792	1924	222	233	284	320	1083
1885	265	193	148	373	918	1925	330	204	177	282	967
1886	192	260	189	307	1051	1926	293	159	223	304	905
1887	241	140	112	252	669	1927	200	186	337	327	1108
1888	147	183	317	228	878	1928	320	158	253	286	1023
1889	172	253	209	230	849	1929	171	106	185	344	894

Table D.2 (cont.)

Year	Win	Spr	Sum	Aut	Ann	Year	Win	Spr	Sum	Aut	Ann
1930	330	198	247	331	1025	1970	288	177	195	308	934
1931	256	213	318	246	982	1971	201	176	267	197	824
1932	145	272	184	317	923	1972	214	220	160	183	853
1933	213	168	154	205	718	1973	205	177	219	184	740
1934	119	187	173	216	850	1974	295	102	243	378	1028
1935	328	164	189	424	1015	1975	236	200	136	218	759
1936	290	144	266	259	956	1976	146	129	74	397	792
1937	362	248	149	195	962	1977	340	179	216	221	959
1938	246	95	226	315	892	1978	297	187	210	129	897
1939	319	159	260	318	1008	1979	335	316	164	222	1023
1940	210	183	148	354	907	1980	327	152	288	284	967
1941	265	197	238	184	867	1981	185	310	141	337	992
1942	193	215	183	209	840	1982	216	175	251	331	989
1943	292	148	199	239	832	1983	253	290	111	234	885
1944	174	111	213	378	894	1984	318	140	128	373	930
1945	233	159	235	188	847	1985	186	203	293	166	893
1946	277	162	298	323	1052	1986	272	258	218	247	1013
1947	235	314	146	150	833	1987	236	211	251	329	945
1948	308	161	259	206	963	1988	293	218	268	216	982
1949	193	174	122	321	789	1989	186	205	156	207	857
1950	272	168	273	320	1015	1990	421	85	157	231	842
1951	285	269	221	295	1095	1991	258	159	202	230	802
1952	209	204	194	303	901	1992	143	210	262	336	980
1953	170	165	250	224	755	1993	208	211	212	290	1013
1954	179	173	307	378	1093	1994	388	241	157	298	1050
1955	239	190	138	187	773	1995	415	147	67	258	841
1956	259	96	331	182	865						
1957	282	127	265	260	904						
1958	293	166	310	262	1057						
1959	218	177	154	219	828						
1960	374	147	283	439	1195						
1961	310	154	192	261	905						
1962	257	184	186	215	814						
1963	145	231	251	299	878						
1964	89	226	182	159	725						
1965	221	205	247	287	1032						
1966	363	216	274	264	1061						
1967	272	244	185	348	1010						
1968	212	207	274	317	1015						
1969	255	252	198	197	905						

Table D.3 North Atlantic Oscillation index (hPa units)

Year	Win	Sum	Ann	Year	Win	Sum	Ann
1865	13.2	15.1	14.4				
1866	25.9	14.7	15.2				
1867	15.2	13.9	10.4				
1868	29.4	20.0	22.4				
1869	31.3	10.8	15.5				
1870	13.7	11.6	10.8	1910	30.6	10.6	16.9
1871	17.2	14.1	13.6	1911	24.0	10.0	16.2
1872	23.4	10.1	13.9	1912	16.3	10.1	14.9
1873	20.5	19.1	15.5	1913	28.3	13.8	19.6
1874	28.2	15.7	19.0	1914	24.6	14.9	18.7
1875	14.1	15.1	12.2	1915	28.3	7.5	10.1
1876	20.3	20.6	12.6	1916	25.3	9.8	13.6
1877	22.4	10.5	15.3	1917	4.1	10.4	10.0
1878	24.5	10.4	8.7	1918	11.6	12.9	15.3
1879	8.5	12.1	14.1	1919	14.9	15.0	12.5
1880	18.6	10.7	11.0	1920	33.5	14.3	19.9
1881	-2.0	16.2	11.2	1921	22.1	11.4	18.5
1882	30.1	16.0	18.2	1922	35.3	19.0	18.0
1883	26.4	11.0	18.4	1923	28.4	16.1	18.5
1884	26.1	14.3	18.1	1924	21.6	15.7	15.1
1885	19.8	11.7	13.0	1925	35.3	17.3	15.0
1886	16.9	19.1	15.6	1926	16.1	13.3	14.6
1887	27.4	10.1	10.7	1927	23.5	12.7	13.8
1888	5.9	9.9	10.3	1928	26.1	10.0	17.3
1889	25.3	14.8	18.6	1929	10.6	11.9	14.1
1890	30.8	17.6	20.0	1930	30.1	16.3	17.2
1891	21.0	8.1	15.0	1931	28.1	10.7	14.0
1892	19.3	11.6	10.8	1932	14.0	14.3	13.4
1893	14.0	9.4	13.5	1933	20.7	18.7	14.1
1894	35.9	17.1	19.0	1934	23.9	13.3	17.9
1895	4.7	12.3	9.8	1935	24.2	19.2	17.3
1896	20.3	13.8	17.1	1936	3.1	16.0	13.0
1897	23.2	10.8	15.3	1937	27.2	14.9	14.3
1898	28.8	12.3	17.3	1938	21.5	17.0	21.1
1899	16.9	13.7	11.4	1939	23.4	12.7	13.3
1900	15.3	11.5	14.8	1940	5.4	15.2	11.8
1901	16.4	16.7	13.5	1941	9.9	14.7	10.9
1902	12.6	4.6	10.6	1942	18.5	13.6	13.7
1903	28.8	9.1	17.7	1943	27.3	11.8	19.8
1904	28.6	9.3	17.5	1944	24.4	11.6	15.7
1905	22.0	13.4	16.4	1945	22.8	14.6	12.3
1906	30.8	12.1	18.1	1946	19.8	19.3	17.1
1907	28.0	10.4	18.2	1947	12.6	13.9	10.4
1908	28.7	13.0	18.4	1948	17.5	11.3	16.1
1909	27.5	16.4	13.3	1949	27.8	9.1	17.1

Table D.3 (cont.)

Year	Win	Sum	Ann		Year	Win	Sum	Ann
1950	26.1	15.1	16.7		1980	19.7	11.6	12.2
1951	23.4	12.0	15.6		1981	27.0	14.0	13.3
1952	27.3	12.3	11.8		1982	17.0	12.1	19.4
1953	17.0	17.7	17.5		1983	28.1	16.5	15.6
1954	27.0	15.7	19.4		1984	32.6	13.4	18.8
1955	13.3	15.3	10.1		1985	12.1	13.0	9.6
1956	11.0	10.7	15.1		1986	14.2	12.9	20.9
1957	26.7	8.3	13.8		1987	15.6	9.6	11.1
1958	16.0	7.7	11.0		1988	18.4	17.4	15.0
1959	14.7	15.7	17.9		1989	36.2	16.0	20.3
1960	12.0	11.0	11.6		1990	31.4	16.2	22.4
1961	29.3	19.8	16.9		1991	26.3	14.9	17.4
1962	17.7	13.0	12.9		1992	25.4	17.0	20.6
1963	0.7	11.0	12.3		1993	30.2	11.6	17.5
1964	7.0	13.0	13.8		1994	28.0	17.2	20.3
1965	8.7	15.3	12.3		1995	34.7	9.7	13.2
1966	9.3	13.3	12.1					
1967	23.3	16.7	17.9					
1968	13.7	10.3	10.8					
1969	2.0	16.3	12.0					
1970	19.0	15.0	14.2					
1971	16.8	12.2	15.4					
1972	24.6	17.8	19.3					
1973	31.0	18.0	15.3					
1974	30.1	15.6	19.7					
1975	26.9	13.1	15.5					
1976	22.8	11.0	16.6					
1977	7.8	12.3	12.7					
1978	13.7	11.3	14.7					
1979	5.5	15.3	15.0					

Table D.4 Northern Hemisphere (land areas only) surface air temperature anomaly (°C)

Year	Win	Spr	Sum	Aut	Ann	Year	Win	Spr	Sum	Aut	Ann
						1890	-0.02	-0.34	-0.33	-0.50	-0.35
1851		-0.83	0.09	-0.09	-0.23	1891	-0.81	-0.68	-0.39	-0.49	-0.49
1852	-0.28	-1.07	0.14	-0.54	-0.29	1892	-0.10	-0.78	-0.23	-0.42	-0.56
1853	0.44	-0.99	0.29	-0.31	-0.40	1893	-1.98	-0.64	-0.20	-0.34	-0.69
1854	-1.05	-0.21	0.16	-0.23	-0.13	1894	-0.44	-0.08	-0.20	-0.41	-0.28
1855	-0.60	-0.28	0.18	-0.46	-0.57	1895	-1.22	-0.43	-0.36	-0.23	-0.58
1856	-0.92	-0.83	0.02	-1.17	-0.51	1896	-0.33	-0.50	-0.02	-0.50	-0.31
1857	-0.48	-0.91	0.04	-0.51	-0.39	1897	-0.35	-0.26	0.04	-0.20	-0.25
1858	-0.44	-0.56	0.27	-0.55	-0.42	1898	-0.23	-0.74	-0.08	-0.25	-0.27
1859	0.27	0.23	0.36	-0.30	0.07	1899	-0.25	-0.44	-0.16	0.29	-0.20
1860	-0.69	-0.88	-0.13	-0.73	-0.68	1900	-0.56	-0.27	0.02	0.08	-0.07
1861	-1.51	-0.64	0.39	-0.30	-0.37	1901	-0.18	0.00	0.22	-0.31	-0.15
1862	-1.26	-0.13	-0.38	-0.73	-0.73	1902	-0.13	-0.43	-0.40	-0.55	-0.42
1863	0.57	0.16	-0.05	-0.04	0.30	1903	-0.22	-0.35	-0.52	-0.47	-0.37
1864	-0.64	-0.47	0.04	-1.22	-0.72	1904	-0.72	-0.65	-0.46	-0.32	-0.49
1865	-0.94	-0.46	0.09	0.13	-0.18	1905	-0.86	-0.50	-0.12	-0.08	-0.37
1866	0.24	-0.59	0.04	-0.52	-0.17	1906	-0.20	-0.15	-0.09	-0.24	-0.16
1867	0.14	-1.07	-0.33	-0.45	-0.55	1907	-0.49	-0.86	-0.59	-0.36	-0.62
1868	-1.05	-0.04	0.58	-0.54	-0.11	1908	-0.33	-0.55	-0.37	-0.55	-0.45
1869	0.75	-0.47	0.12	-0.47	-0.10	1909	-0.61	-0.82	-0.31	0.02	-0.45
1870	-0.75	-0.28	0.32	-0.41	-0.46	1910	-0.40	-0.13	-0.30	-0.43	-0.31
1871	-1.91	-0.16	0.05	-0.92	-0.67	1911	-0.70	-0.57	-0.13	-0.21	-0.35
1872	-0.75	0.24	0.24	-0.13	0.01	1912	-0.23	-0.46	-0.51	-0.90	-0.55
1873	-0.12	-0.68	0.26	-0.62	-0.24	1913	-0.42	-0.49	-0.45	-0.15	-0.32
1874	0.32	-0.64	0.08	-0.25	-0.16	1914	0.43	-0.20	-0.18	-0.25	-0.12
1875	-1.35	-1.06	0.02	-1.05	-0.90	1915	-0.02	-0.36	-0.20	-0.11	-0.13
1876	-0.24	-0.56	0.31	-0.73	-0.38	1916	0.06	-0.51	-0.28	-0.37	-0.38
1877	-0.48	-0.85	0.28	-0.23	-0.13	1917	-1.07	-1.11	-0.15	-0.30	-0.70
1878	0.46	0.50	0.29	0.15	0.23	1918	-0.95	-0.52	-0.33	-0.10	-0.36
1879	-0.44	-0.27	-0.23	-0.50	-0.41	1919	-0.25	-0.49	-0.12	-0.39	-0.35
1880	-0.37	-0.09	-0.15	-0.88	-0.31	1920	-0.41	-0.09	-0.09	-0.36	-0.22
1881	-0.96	-0.37	-0.07	-0.55	-0.40	1921	0.09	0.24	0.07	-0.32	0.07
1882	0.68	-0.14	-0.34	-0.60	-0.20	1922	-0.39	-0.07	-0.03	-0.19	-0.18
1883	-1.06	-0.83	-0.18	-0.51	-0.58	1923	-0.20	-0.53	-0.16	0.25	-0.12
1884	-0.24	-0.83	-0.59	-0.69	-0.62	1924	-0.14	-0.37	-0.07	-0.04	-0.24
1885	-0.92	-0.70	-0.40	-0.63	-0.60	1925	-0.23	-0.08	-0.04	-0.08	-0.03
1886	-0.80	-0.53	-0.29	-0.35	-0.53	1926	0.57	-0.18	-0.09	0.10	0.04
1887	-0.72	-0.31	-0.25	-0.59	-0.51	1927	-0.19	-0.36	0.05	0.22	-0.12
1888	-1.18	-0.85	-0.43	-0.53	-0.71	1928	-0.14	-0.44	-0.15	0.07	-0.07
1889	-0.52	-0.03	-0.22	-0.76	-0.35	1929	-0.90	-0.39	-0.11	0.06	-0.40

Table D.4 (cont.)

Year	Win	Spr	Sum	Aut	Ann	Year	Win	Spr	Sum	Aut	Ann
1930	−0.34	−0.02	0.16	0.05	0.02	1970	0.20	−0.11	0.03	−0.14	−0.07
1931	−0.03	−0.28	0.29	0.32	0.09	1971	−0.27	−0.37	−0.18	0.07	−0.15
1932	0.28	−0.24	0.12	0.06	0.03	1972	−0.63	−0.19	−0.09	−0.32	−0.33
1933	−0.45	−0.42	0.06	−0.02	−0.28	1973	0.23	0.28	0.14	−0.05	0.18
1934	−0.19	−0.10	0.10	0.26	0.13	1974	−0.26	−0.10	−0.09	−0.21	−0.21
1935	0.39	−0.30	0.09	−0.04	−0.02	1975	0.11	0.19	0.02	−0.06	0.06
1936	−0.48	−0.20	0.32	0.14	0.01	1976	−0.07	−0.39	−0.30	−0.66	−0.37
1937	0.19	−0.26	0.33	0.42	0.10	1977	−0.44	0.46	0.07	0.20	0.12
1938	0.14	0.36	0.25	0.54	0.31	1978	−0.03	0.07	−0.21	0.07	−0.03
1939	0.00	−0.12	0.24	0.04	0.19	1979	−0.31	−0.05	0.04	0.20	0.07
1940	0.04	0.00	0.10	0.17	0.01	1980	0.36	−0.01	0.13	0.20	0.09
1941	0.29	−0.08	0.17	−0.02	0.04	1981	0.74	0.52	0.27	0.15	0.47
1942	−0.33	−0.11	0.05	0.20	−0.03	1982	0.13	−0.05	−0.06	0.01	0.00
1943	−0.06	−0.01	0.03	0.27	0.10	1983	0.83	0.25	0.32	0.54	0.44
1944	0.70	0.08	0.12	0.22	0.18	1984	0.08	0.06	0.13	−0.19	−0.06
1945	−0.55	−0.09	0.02	0.07	−0.15	1985	−0.61	0.02	−0.07	−0.10	−0.13
1946	−0.02	0.09	0.01	0.07	0.06	1986	0.24	0.28	0.07	−0.06	0.15
1947	−0.53	0.14	0.08	0.39	0.07	1987	0.41	−0.04	0.24	0.15	0.25
1948	0.24	0.02	0.12	0.21	0.13	1988	0.54	0.39	0.49	0.26	0.42
1949	0.00	−0.07	−0.03	0.20	0.01	1989	0.67	0.45	0.31	0.25	0.39
1950	−0.58	−0.10	−0.17	−0.27	−0.27	1990	0.69	0.90	0.46	0.63	0.67
1951	−0.56	−0.12	0.10	0.16	−0.04	1991	0.62	0.42	0.50	0.36	0.46
1952	0.33	−0.13	0.20	−0.17	0.00	1992	0.74	0.24	−0.13	−0.20	0.17
1953	0.18	0.27	0.28	0.12	0.25	1993	0.56	0.33	0.11	−0.21	0.21
1954	−0.21	−0.33	0.00	0.25	−0.12	1994	0.21	0.57	0.50	0.61	0.48
1955	0.07	−0.42	0.04	−0.06	−0.10	1995	1.17	0.54	0.62	0.57	0.70
1956	−0.55	−0.49	−0.28	−0.43	−0.45	1996	0.38	0.06	(0.25)*	(0.05)	(0.18)
1957	−0.36	−0.30	0.03	0.11	−0.04						
1958	0.56	0.04	−0.01	0.08	0.14						
1959	0.20	0.17	0.10	−0.18	0.05						
1960	0.29	−0.43	0.11	0.01	0.03						
1961	0.38	0.21	0.11	−0.09	0.10						
1962	0.18	0.07	−0.06	0.08	0.11						
1963	0.24	−0.16	0.05	0.48	0.13						
1964	−0.19	−0.33	−0.17	−0.26	−0.26						
1965	−0.33	−0.29	−0.27	−0.17	−0.23						
1966	−0.03	−0.15	0.13	−0.03	−0.07						
1967	−0.46	0.04	−0.06	0.09	−0.05						
1968	−0.33	0.17	−0.22	−0.15	−0.16						
1969	−0.92	−0.15	−0.06	0.01	−0.22						

* Parentheses indicate provisional values.

Table D.5 Global average (land and marine areas) surface air temperature anomaly (°C)

Year	Win	Spr	Sum	Aut	Ann	Year	Win	Spr	Sum	Aut	Ann
						1890	−0.25	−0.35	−0.36	−0.50	−0.38
						1891	−0.46	−0.34	−0.28	−0.34	−0.33
						1892	−0.20	−0.42	−0.41	−0.40	−0.41
						1893	−0.83	−0.48	−0.31	−0.33	−0.45
						1894	−0.34	−0.37	−0.35	−0.41	−0.37
						1895	−0.51	−0.42	−0.31	−0.22	−0.36
1856		−0.42	−0.26	−0.47	−0.36	1896	−0.25	−0.28	−0.08	−0.11	−0.16
1857	−0.32	−0.52	−0.38	−0.62	−0.46	1897	−0.13	−0.07	−0.10	−0.22	−0.15
1858	−0.61	−0.46	−0.26	−0.34	−0.42	1898	−0.26	−0.52	−0.24	−0.35	−0.34
1859	−0.23	−0.17	−0.22	−0.30	−0.23	1899	−0.30	−0.33	−0.20	−0.04	−0.23
1860	−0.31	−0.47	−0.17	−0.45	−0.39	1900	−0.27	−0.17	−0.09	−0.14	−0.14
1861	−0.75	−0.54	−0.12	−0.39	−0.41	1901	−0.16	−0.15	−0.15	−0.35	−0.23
1862	−0.63	−0.27	−0.45	−0.56	−0.53	1902	−0.28	−0.40	−0.34	−0.42	−0.37
1863	−0.25	−0.28	−0.39	−0.31	−0.25	1903	−0.27	−0.39	−0.48	−0.56	−0.44
1864	−0.53	−0.46	−0.20	−0.50	−0.45	1904	−0.60	−0.56	−0.48	−0.40	−0.48
1865	−0.38	−0.30	−0.20	−0.19	−0.24	1905	−0.49	−0.43	−0.30	−0.31	−0.37
1866	−0.07	−0.39	−0.06	−0.29	−0.21	1906	−0.23	−0.23	−0.30	−0.42	−0.30
1867	−0.19	−0.38	−0.28	−0.24	−0.30	1907	−0.45	−0.51	−0.49	−0.47	−0.49
1868	−0.55	−0.12	−0.04	−0.31	−0.20	1908	−0.46	−0.55	−0.48	−0.57	−0.52
1869	−0.02	−0.31	−0.30	−0.37	−0.29	1909	−0.56	−0.62	−0.45	−0.34	−0.50
1870	−0.36	−0.32	−0.20	−0.30	−0.32	1910	−0.48	−0.42	−0.43	−0.49	−0.46
1871	−0.65	−0.18	−0.19	−0.48	−0.36	1911	−0.60	−0.60	−0.45	−0.40	−0.48
1872	−0.39	−0.19	−0.09	−0.21	−0.20	1912	−0.27	−0.34	−0.42	−0.53	−0.40
1873	−0.24	−0.35	−0.20	−0.38	−0.29	1913	−0.42	−0.52	−0.43	−0.37	−0.41
1874	−0.19	−0.58	−0.31	−0.40	−0.39	1914	−0.12	−0.27	−0.26	−0.27	−0.24
1875	−0.59	−0.39	−0.27	−0.46	−0.42	1915	−0.13	−0.15	−0.13	−0.17	−0.15
1876	−0.37	−0.42	−0.26	−0.51	−0.41	1916	−0.26	−0.37	−0.38	−0.38	−0.37
1877	−0.34	−0.38	−0.05	−0.03	−0.13	1917	−0.64	−0.68	−0.26	−0.31	−0.49
1878	0.16	0.15	−0.04	−0.13	−0.00	1918	−0.57	−0.51	−0.41	−0.23	−0.38
1879	−0.23	−0.24	−0.26	−0.32	−0.29	1919	−0.17	−0.21	−0.28	−0.39	−0.28
1880	−0.33	−0.22	−0.26	−0.40	−0.27	1920	−0.36	−0.13	−0.19	−0.25	−0.23
1881	−0.34	−0.17	−0.14	−0.36	−0.24	1921	−0.28	−0.21	−0.18	−0.26	−0.21
1882	−0.06	−0.20	−0.25	−0.31	−0.23	1922	−0.27	−0.33	−0.28	−0.32	−0.31
1883	−0.43	−0.34	−0.17	−0.35	−0.31	1923	−0.28	−0.34	−0.34	−0.22	−0.28
1884	−0.30	−0.40	−0.33	−0.39	−0.36	1924	−0.20	−0.33	−0.27	−0.38	−0.34
1885	−0.47	−0.40	−0.33	−0.23	−0.33	1925	−0.44	−0.26	−0.22	−0.23	−0.24
1886	−0.28	−0.22	−0.19	−0.29	−0.26	1926	0.05	−0.12	−0.14	−0.12	−0.09
1887	−0.41	−0.34	−0.31	−0.36	−0.37	1927	−0.17	−0.26	−0.14	−0.12	−0.20
1888	−0.55	−0.37	−0.25	−0.15	−0.31	1928	−0.26	−0.33	−0.23	−0.18	−0.23
1889	−0.14	−0.05	−0.18	−0.33	−0.17	1929	−0.47	−0.42	−0.32	−0.19	−0.38

Table D.5 (cont.)

Year	Win	Spr	Sum	Aut	Ann	Year	Win	Spr	Sum	Aut	Ann
1930	-0.38	-0.20	-0.14	-0.06	-0.16	1970	0.15	0.01	-0.05	-0.08	-0.03
1931	-0.11	-0.12	0.02	-0.11	-0.08	1971	-0.22	-0.25	-0.18	-0.13	-0.19
1932	-0.06	-0.17	-0.13	-0.09	-0.12	1972	-0.28	-0.07	0.03	0.02	-0.04
1933	-0.25	-0.24	-0.14	-0.24	-0.24	1973	0.21	0.16	0.10	-0.03	0.09
1934	-0.31	-0.23	-0.03	-0.07	-0.12	1974	-0.24	-0.16	-0.08	-0.14	-0.17
1935	-0.07	-0.28	-0.14	-0.17	-0.17	1975	-0.13	-0.05	-0.10	-0.17	-0.12
1936	-0.26	-0.20	-0.00	-0.08	-0.12	1976	-0.26	-0.28	-0.16	-0.20	-0.21
1937	-0.05	-0.14	0.05	0.05	-0.04	1977	-0.05	0.12	0.07	0.08	0.06
1938	-0.04	0.10	0.08	0.12	0.06	1978	0.00	-0.01	-0.10	-0.02	-0.03
1939	-0.07	-0.09	0.09	-0.17	-0.02	1979	-0.06	-0.03	0.06	0.14	0.06
1940	-0.03	-0.03	0.05	-0.04	-0.02	1980	0.24	0.13	0.10	0.08	0.11
1941	0.07	0.01	0.04	0.11	0.06	1981	0.20	0.15	0.08	0.03	0.13
1942	0.16	0.06	0.03	0.00	0.04	1982	0.12	0.04	0.01	0.07	0.06
1943	-0.06	-0.05	0.03	0.13	0.04	1983	0.39	0.22	0.21	0.21	0.24
1944	0.23	0.12	0.27	0.19	0.19	1984	0.13	0.08	0.05	-0.02	0.03
1945	-0.01	0.01	0.13	0.15	0.06	1985	-0.11	0.03	0.00	0.01	0.01
1946	0.03	-0.09	-0.22	-0.10	-0.12	1986	0.12	0.12	0.09	0.07	0.10
1947	-0.28	-0.07	-0.09	-0.07	-0.11	1987	0.20	0.14	0.26	0.26	0.24
1948	-0.13	-0.12	-0.06	-0.12	-0.11	1988	0.36	0.30	0.24	0.17	0.25
1949	-0.09	-0.13	-0.12	-0.09	-0.11	1989	0.18	0.15	0.19	0.19	0.18
1950	-0.32	-0.18	-0.12	-0.23	-0.21	1990	0.30	0.41	0.30	0.34	0.34
1951	-0.38	-0.14	0.05	0.02	-0.08	1991	0.34	0.32	0.34	0.21	0.29
1952	0.12	-0.01	0.05	-0.06	0.01	1992	0.29	0.21	0.07	0.00	0.14
1953	0.05	0.13	0.07	-0.00	0.07	1993	0.28	0.26	0.14	0.08	0.19
1954	-0.11	-0.23	-0.18	-0.08	-0.18	1994	0.11	0.27	0.24	0.34	0.26
1955	-0.12	-0.30	-0.13	-0.17	-0.18	1995	0.53	0.33	0.39	0.37	0.39
1956	-0.31	-0.31	-0.21	-0.25	-0.26	1996	0.26	0.21	(0.25)*	(0.15)	(0.22)
1957	-0.15	-0.02	0.11	0.08	0.04						
1958	0.25	0.09	0.06	0.03	0.10						
1959	0.09	0.06	0.05	-0.03	0.03						
1960	0.03	-0.17	0.03	-0.00	-0.01						
1961	0.12	0.09	0.05	-0.04	0.03						
1962	0.01	0.01	-0.00	0.04	0.02						
1963	0.05	-0.05	0.05	0.15	0.05						
1964	-0.08	-0.24	-0.19	-0.30	-0.24						
1965	-0.27	-0.22	-0.13	-0.10	-0.16						
1966	-0.09	-0.10	0.02	-0.06	-0.07						
1967	-0.22	-0.04	-0.07	-0.02	-0.08						
1968	-0.20	-0.11	-0.06	-0.03	-0.10						
1969	-0.15	0.09	0.05	0.06	0.03						

* Parentheses indicate provisional values.

GLOSSARY

$\delta^{18}O$ **record**: the past record of the oxygen isotope ratio ($^{18}O/^{16}O$) found in carbonate and carbon dioxide samples expressed as a departure from the current ratio found in Standard Mean Ocean Water (SMOW). A $\delta^{18}O$ value of -10, for example, indicates a sample with an $^{18}O/^{16}O$ ratio 1 per cent (or ten parts in a thousand) less than SMOW.

500 hPa level (or geopotential height): a surface of constant pressure in the atmosphere, where the pressure equals 500 hectopascals (or millibars). Commonly used as a measure of the state of the total **tropospheric** column of air, the 500 hPa level is usually at around 5,500 to 6,000 metres above sea level at mid-latitudes.

acid rain: a colloquial term used to describe precipitation in a polluted environment, where the raindrops become acidified by either sulphur dioxide or by a combination of sulphur dioxide and nitrogen oxides emitted from e.g. motor vehicle exhausts.

advection: the process by which the property of a mass of air is transferred by movement, usually in the horizontal direction.

aerosols: suspended minute particles (solid or liquid) of dust, sea salt, sulphates and, in urban environments, carbon, lead and aluminium compounds, produced by combustion of fuels.

air mass: a body of air in which the horizontal gradients of temperature and humidity are relatively slight and which is separated from an adjacent body of air by a transition zone (front) in which these gradients are relatively large.

albedo: the proportion of solar radiation which is diffusely reflected from a nonluminous body.

allergens: a substance capable of provoking an allergy.

ambient temperature: the temperature of the environment surrounding a subject or object.

anemograph trace: the record of wind speed by a pen on paper trace. Now largely superseded by electronic loggers.

anemometer: an instrument that measures wind speed, usually by means of three rotating cups.

angular momentum: the angular momentum of a particle rotating about a fixed axis is the product of the particle's momentum (its mass $\times$ velocity) and its perpendicular distance from the axis of rotation.

anthropogenic: of human origin.

anticyclone: an area of high pressure and subsiding air, generally accompanied by dry, settled conditions.

anti-oxidant: a substance which 'mops up' oxidising (potentially harmful) compounds in the human body.

Atlantic period: part of the **Holocene** according to the Blytt/Sernander climate sub-division, referring to a period of supposedly warm and moist conditions in north-western Europe, roughly dated between about 7,500 and 5,200 BP.

atmospheric sounder: a satellite-based multi-spectral **radiometer** used to make indirect measurements of the vertical distribution of some atmospheric property (e.g., temperature, humidity, etc.).

atmospheric thickness: the vertical distance, expressed in metres, between two pressure surfaces in the atmosphere, often 1,000 hPa and 500 hPa.

barotropic: the state of a gas or fluid whereby surfaces of constant density lie parallel to or coincide with surfaces of constant pressure.

baroclinic: the state of a gas or fluid whereby surfaces of constant density intersect surfaces of constant pressure, implying major atmospheric instability, probable around the frontal zone.

baroclinic wave: a wave depression which forms in a strongly **baroclinic** region of the atmosphere.

Beaufort scale: a numerical wind force scale ranging from 0 (calm) to 12 (hurricane) devised by Admiral Beaufort in 1805.

benthic foraminifera: unicellular microscopic organisms that live on or in the sediments of the sea-floor and which commonly, but not always, secrete a skeleton of calcium carbonate.

blocking: the occurrence of a persistent anticyclone (or cyclone) at the surface, and pressure ridge (or **trough**) at height in the atmosphere, which disrupts the dominant **zonal flow** in mid-latitudes. Over the British Isles, blocking is usually associated with weather extremes (heat or cold, drought or flood).

Boreal: a climate zone characterised by long snowy winters and short summers. It is a term also applied specifically to the coniferous forests of the Northern Hemisphere (or more generally to refer to the Northern Hemisphere). The Boreal period may also refer to a part of the **Holocene** according to the Blytt/Sernander climate sub-division, referring to a period of cold winters, warm summers and dry conditions in north-western Europe, roughly dated between about 9,500 and 7,500 BP. It gave way to the **Atlantic** period.

boundary layer: used in the sense of the planetary boundary layer (i.e., the layer of air from the surface to the level where the frictional influence is absent). Thus, the average wind speed within the boundary layer is lower than the free-stream value. A typical height of the boundary layer is 600 m, but this varies depending on atmospheric conditions, such as temperature and humidity.

bubbler: a bottle containing a gas absorbent which can be analysed after a volume of air is bubbled through.

carbon dioxide equivalent: refers to the increase in the concentration of carbon dioxide which would give the same increase in **radiative forcing** as the combined effect of specified increases in the concentration of an ensemble of greenhouse gases.

carcinogenic: capable of causing cancer.

circulation types: a **synoptic** configuration of the atmospheric circulation which is similar to one of several pre-defined configurations. These types form the basis of synoptic weather classifications.

circumpolar vortex: the circumpolar flow of winds circuiting from west to east around the Earth over each hemisphere, mainly over the middle latitudes, and carrying most of the momentum of the atmosphere. The depth of this flow through the **troposphere** may vary from about 2 km to 15 to 20 km above the surface.

climate analogue regions: regions which are climatically similar to a source region even though they may be geographically distant.

climate response surface: a three-dimensional representation of the geographical distribution of, for example, some plant taxon graphically expressed in terms of different temperature and moisture axes (e.g., mean July temperature on one axis, annual temperature range on the second, and precipitation minus evaporation on the third).

climate sensitivity: can be defined as the eventual change in global average surface air temperature which occurs following a doubling of the concentration of carbon dioxide in the atmosphere.

coccolithophores: planktonic (floating) microscopic algae, which live in ocean surface waters and which contribute their minute calcium carbonate skeletons to the ocean floor in vast quantities.

cold front: the boundary line between advancing cold air and a mass of warmer air. The passage of a cold front is normally marked at the surface by a rise in pressure, a fall in temperature and **dew point** and a veer in wind.

continental drift: the concept, originating in the nineteenth century, which suggests that continents have moved relative to one another and around the Earth's surface.

convection: a type of heat transfer which occurs in

a fluid by the vertical movement of large volumes of the heated material by differential heating (at the bottom of the atmosphere) thus creating, locally, a less dense, more buoyant fluid.

convergence: negative **divergence.**

cooling degree days: the number of degrees above a specific threshold temperature, accumulated over all days in the year or season on which the temperature is above the same threshold value. This is a useful measure of energy use for space cooling applications.

Coriolis parameter: the factor $2\Omega \sin\phi$, where Ω is the Earth's angular velocity and ϕ is the angle of latitude.

cryosphere: that part of the Earth's surface, including the oceans, which comprises ice or snow.

data quality control: the application of a systematic set of procedures whereby errors of whatever type are identified and eliminated from a climate dataset.

degree days: the number of degrees above (or below) a specific threshold temperature, accumulated over all days in the year or season on which the temperature is above (or below) the same threshold value.

deposit gauges: essentially a large rain-gauge with a screen to prevent birds from contaminating the collection. Every month the bottle below the funnel is replaced with a new one and the contents of the previous month's sample analysed (typically for chloride, sulphate, ammonia, calcium, total undissolved matter, tarry matter, ash and acidity or alkalinity (pH)).

depression: an area of low pressure and rising air, accompanied by frontal systems and unsettled weather.

Devensian: British terminology for the last glacial stage of the Pleistocene between about 11,700 and 10,000 BP, named after the many glacial deposits situated around the Roman city of Deva (modern-day Chester).

dew-point temperature: the temperature to which the air must be cooled to produce saturation with respect to water at its existing atmospheric pressure and humidity.

diagenetic changes: chemical changes which are experienced by sediments when they are at or near to the Earth's surface as a result of compaction and cementation at relatively low temperatures and pressures.

diatoms: floating, or bottom-dwelling, microscopic single-celled marine or freshwater algae that secrete silica/siliceous skeletons (cf. **Radiolaria**).

diurnal temperature range: the difference between daily maximum and minimum temperatures recorded at a site.

divergence: in meteorology this normally refers to the stretching of the air in the local horizontal dimension. This is comprised of the two components ($\partial u/\partial x + \partial v/\partial y$).

drag coefficient: a measure of the drag on the air passing over a surface, and hence related to the roughness length. However, the drag coefficient is also dependent on the stability of the air and the wind speed.

emissions scenario: an estimate of future emissions of greenhouse gases into the atmosphere. Such an estimate has to make a large number of assumptions about future demography, technology, energy consumption patterns, etc.

European monsoon: a summer period of generally higher precipitation, often commencing in late June, and heralded by the return of westerly, **progressive** conditions over north-west Europe following a period of more frequent **blocking** in spring/early summer.

Föhn effect: a warm and relatively dry wind which descends on the leeward side of a mountain range, when moist air is forced to rise over the topographic barrier during the passage of a depression.

frontogenesis: the development or marked intensification of a front.

gas cycle model: a numerical model which converts gas emissions to concentrations over a specified time period by simulating the fluxes and chemical reactions within the atmosphere–ocean–biosphere–lithosphere system.

geostationary satellite: a weather satellite that orbits the Earth in an equatorial plane at a fixed altitude of approximately 36,000 km.

geostrophic wind: the horizontal equilibrium wind

which blows parallel to the isobars and represents an exact balance between the horizontal pressure gradient force and the horizontal component of the Coriolis force.

glaze: a generally homogeneous and transparent deposit of ice formed by the freezing of super-cooled drizzle droplets or raindrops on objects the surface temperature of which is below or slightly below 0°C.

global radiation: the amount of direct and diffuse solar radiation which reaches the Earth's surface at a given location; equivalent to **insolation**.

graben: a German term for a structurally defined block of land downthrown between parallel faults.

gravity waves: a type of wave in the atmosphere whose existence is dependent upon the restoring force provided by buoyancy.

greenhouse effect: the description of the process whereby a certain group of atmospheric gases (predominantly water vapour and carbon dioxide) absorb outgoing long-wave radiation whilst allowing incoming short-wave radiation to pass through the atmosphere. The net effect is to warm the lower atmosphere.

Grosswetterlagen: the **synoptic** situations corresponding to the Grosswetter – a German term which illustrates, with special reference to long-range weather forecasting, the main features of weather over a specified region and period of time.

heating degree days: the number of degrees below a specific threshold temperature, accumulated over all days in the year or season on which the temperature is below the same threshold value. This is a useful measure of energy use for space heating applications.

Heinrich events: episodes of very high rates of iceberg discharge into the North Atlantic, mainly from the edge of the North American ice-sheet, during the Last Glacial period. This led to extreme surface water cooling and a reduction in salinity. Named after the investigator who first recognised the significance of tell-tale layers of ice-rafted material in deep-sea sediment cores.

Holocene: a term which means 'wholly recent life' and referring to all of the time which has elapsed since the end of the last glacial period about 10,000 years BP.

homogeneous: when used in relation to climate data it refers to a time series in which the variations are related only to climatic factors and not to changes in the instrumentation, the measurement practice or other non-climatic changes in the environment of the recording site.

Horse latitudes: the belts of high pressure, characterised by calm and light winds, which occur in the sub-tropics at 35°N to 30°N and at 35°S to 30°S.

hydrometer: a general term for the items produced when water vapour condenses in the atmosphere. Hydrometers include ensembles of falling particles which may either reach the Earth's surface (rain, drizzle, snow and hail), or evaporate during their fall (**virga**); ensembles of particles suspended in the air (cloud, fog and mist); particles lifted from the Earth's surface (drifting or blowing snow, and spray); particles deposited on the earth or exposed objects (dew, frost and glaze).

hygroscopic: a hygroscopic nuclei is one which absorbs moisture by accelerating the condensation of water vapour.

hypsithermal: a term used to describe the warmer conditions of the **Holocene** which are thought to have existed between various dates between about 9,000 and 2,600 BP.

Ice Age: any period in the Earth's history during which high latitude ice sheets expand considerably and surface temperatures in the temperate latitudes are lowered.

induction: a system of reasoning which attempts to explain phenomena by drawing on empirical rather than theoretical evidence.

insolation: the amount of diffuse and direct solar radiation which reaches the Earth's surface at a given location; equivalent to global radiation.

interstadial: a short phase of climatic amelioration with a glacial period. Generally shorter and colder than an interglacial period.

Ipswichian: the last interglacial stage of the Pleistocene in the British Isles between about 130,000 and 117,000 BP.

jet dust counter: an early device, invented by J.S. Owens, for determining the relative concentration of fine particles in the atmosphere.

katabatic wind: a night-time wind created when very cold air is formed in exposed upland areas due to heat loss by radiation and becomes sufficiently dense to drain downhill.

kinematic: of motion considered abstractly, without reference to force or mass.

Last Glacial Maximum: the period of maximum ice sheet extent during the last (**Devensian**) glaciation. This was reached about 21,000 years ago.

latent heat: the quantity of heat absorbed or emitted, without change of temperature, during a change of state of unit mass of a material.

Laurentide ice sheet: a massive extent of ice that covered most of north-east North America during the late **Quaternary**.

lead candle: a tube coated with lead peroxide which absorbs sulphur dioxide and which is a method of determining the rate of deposition of gaseous sulphur dioxide.

Little Ice Age: a cool period, variously dated between the fourteenth and nineteenth centuries, during which glaciers re-advanced or were re-established in some of the world's mountain regions. Latest research suggests that such a cool period may not have been very coherent over time, nor as geographically extensive as was earlier thought.

long wave: a smooth wave-shaped undulation in the airflow of the middle and upper troposphere with a wavelength of about 2,000 km. Also known as a Rossby wave.

Lusitanian: a term used to define a west European marine domain, which extends from the Straits of Gibraltar to Lands End (in the United Kingdom) and into the western English Channel.

Markovian: a term used to describe a data series in which the conditional probability of the next term in the series, given the entire past, depends on the previous observations. A first-order Markovian process uses only the previous observation; second-, third-, or n-order models can be defined. Named after the Russian statistician Andrei Markov who

developed the theory of describing data series in this way.

Medieval Warm Epoch/Period: a period in the Middle Ages, variously dated between the ninth and fourteenth centuries, when north-west Europe was traditionally regarded as experiencing a relatively mild climate. Latest research suggests that such a warm period may not have been very coherent over time, nor as geographically extensive as was earlier thought.

meridional flow: airflow oriented along lines of longitude, meridional flow is from the general direction of the north or the south. In mid-latitudes, the **zonal flow** to the west of a **blocking** situation is transferred into meridional flow, branching polewards or equatorwards.

meteorograms: the reading from a **meteorograph**, usually in the form of a continuous paper trace.

meteorograph: an instrument which gives an automatic record of two or more standard meteorological elements. The term is usually used today for an instrument, now obsolete, which was attached to a kite or balloon to measure the pressure, temperature and humidity of the upper atmosphere.

miasmas: a historical term used to describe badly smelling air which was thought to act as the medium for disease transmission.

Model Output Statistics: a statistical method of post-processing numerical weather prediction model output to provide forecasts of variables such as maximum and minimum temperature used by weather forecasters.

moraines: an accumulation of heterogeneous rubbly material, including angular blocks of rock, boulders, pebble and clay, that has been transported and deposited by a glacier or ice-sheet.

natural forcing: radiative forcing changes which are caused by natural factors, i.e., factors unrelated to human activity. Examples include changes in the strength of solar radiation or changes induced by volcanic eruptions.

natural seasons: seasons of the year defined by H.H. Lamb, among others, on the basis of the regular timing through the year of shifts in the frequencies of certain weather types. They differ from the standard three-month calendar seasons.

North Atlantic Drift: a current of anomalously warm ocean water, originating from the Gulf Stream, which moves northwards and reaches the vicinity of the British Isles at around 50°N.

North Atlantic Oscillation: a measure of the gradient of sea level pressure between the Azores and Iceland, where a high value indicates strong **zonal flow** and hence strengthened mid-latitude westerly winds in the North Atlantic sector, and vice versa.

obliquity cycle: the regular change in the orbit of the Earth in which the angle between the Earth's rotational axis and the plane of the ellipse in which it travels around the Sun varies over a 41,000-year cycle. See **orbital variations**.

occluding: the process by which an **occlusion** is formed.

occlusion: an atmospheric process in which the cold front overtakes the **warm front** of a depression, thereby lifting the warm sector off the ground to form an occluded front.

orbital forcing: the **radiative forcing** changes caused at the top of the atmosphere by the **orbital variations** of the Earth.

orbital variations: cyclical variations in the orbit of the Earth which relate to its distance from the Sun and the axis and tilt of its rotation. These variations – also known as Milankovitch cycles – cause changes in the amount of radiation received from the Sun by the top of the atmosphere.

Ozone Hole: a colloquial term used to describe the depletion of stratospheric ozone in recent decades, especially at high latitudes.

Palaeolithic: an archaeological term describing the human culture when Mankind fashioned the earliest known artefacts.

partial thin-plate splines: a statistical technique which fits a series of mathematical functions (or splines) to describe an n-dimensional dataset. These are called 'partial' when the splines include sub-models which account for variations in the target variable caused by higher order dimensions.

particulate phase: usually used in the context of chemical reactions which occur on the surface of small particles.

periglacial: a term which describes the type of climate, and the climatically controlled surface conditions and features, in a broad zone around the Pleistocene ice-sheets. Under periglacial conditions, the landscape is modified by various freeze–thaw processes.

permafrost: permanently frozen ground – only a superficial surface layer may thaw out during the summer season. Typically the mean annual air temperature of such regions is below about −1° or −2°C.

phenology: the study of the sequence of seasonal changes in nature. All natural phenomena are included but the observations are usually limited to the dates at which certain trees and flowering plants come into leaf and flower each year, and to the first and last appearances of certain birds and insects.

photochemical reaction: a chemical reaction in the atmosphere which is triggered by sunlight.

planktonic foraminifera: drifting or floating unicellular microscopic organisms found at various depths in seas, lakes and rivers and which commonly, but not always, secrete a skeleton of calcium carbonate.

PM_{10}: a shorthand description for particulate matter (PM) below a given size. The subscript defines the upper size of the particulate in microns or millionths of a metre. Thus PM_{10} is particulate matter with a diameter smaller than 10 microns.

polar front: a front which divides polar and tropical **air masses** and on which most of the travelling weather disturbances of mid-latitudes develop. The term may also be applied to thermal differences in ocean water masses.

polar low: a secondary depression of a non-frontal character which forms, more especially in winter, within an apparently homogeneous polar air mass.

pressure tendency: the local time rate of change of atmospheric pressure. On synoptic charts, the pressure tendency refers to the pressure change during the last three hours.

Principal Component Analysis: a statistical technique which enables a three-dimensional dataset to be transformed into a new set of uncorrelated variables which are derived in decreasing order of importance, allowing the effective dimensionality of the original dataset to be reduced.

progressive: a term used to describe the dominant westerly airflow at mid-latitudes. See also **zonal flow.**

proxy climate data/sources: a biological, glaciological or geological process, such as tree-rings, ice cores or sedimentation rates, which has been influenced by past climate and can be used to estimate a climate variable, (e.g., temperature).

Quaternary: the younger of the two periods of the Cenozoic era. It comprises two epochs, the Pleistocene and the **Holocene,** and has so far lasted some 1.8 to 2 million years.

radiative forcing: is defined here as a change in average net radiation at the top of the atmosphere. A radiative forcing change is usually expressed in terms of watts per square metre.

radiatively active: a term usually referring to gases in the atmosphere which absorb incoming solar radiation or outgoing long-wave radiation, thus affecting the vertical temperature profile of the atmosphere.

radioactive decay: the gradual breakdown of an element by the emission of charged particles from the nuclei of its atoms, leading to the formation of stable isotopes.

Radiolaria: microscopic organisms characterised by internal siliceous skeletons, which are important constituents of the marine zooplankton.

radiometer: an instrument capable of measuring the radiant energy emitted or reflected by the Earth and its atmosphere. Sensitivity is normally confined to a specified range of wavelengths in the electromagnetic spectrum.

radiosonde: a balloon-borne instrument that measures and transmits pressure, temperature and humidity to a ground-based radio-receiving station. Wind velocity may also be obtained by tracking the radiosonde, by radar, from reflections of a radar target also carried by the balloon.

rain-shadow: an area with a relatively small average precipitation due to sheltering by a range of hills from the prevailing rain-bearing winds.

rate constant: the constant of proportionality used in the determination of the rate of a chemical reaction.

return period: the average length of time separating events of a similar magnitude.

ridge: an extension of an **anticyclone** or high-pressure area.

rock glaciers: accumulations of ice in which there are very high concentrations of debris as a result of rockfall.

sea-surface temperature: temperature of the surface layer of ocean water, down to a depth of between 5 centimetres and 5 metres. Sea-surface temperatures in the open ocean closely follow the surface air temperature.

sensible heat: the heat of a body which can be measured (i.e., its temperature).

shear vorticity: a measure of the degree of rotational flow in a fluid (e.g., air) caused by horizontal shear (e.g., in the wind field). In the Northern Hemisphere, the horizontal shear vorticity associated with the westerly jet stream is cyclonic poleward of the jet and anticyclonic equatorward of the jet.

singularities: periods of distinctive weather, or changes in the weather, regularly occurring at the same time of the year. For example, the 'European Monsoon' signals a rise in the westerly or north-westerly winds in the middle of June, often accompanied by an increase in precipitation.

slantwise convection: a type of motion considered significant in the general circulation of the atmosphere, in which exchange of air between different levels is effected despite the absence of a superadiabatic lapse rate.

specific heat capacity: the quantity of heat needed to raise the temperature of a unit mass of a substance by one degree Kelvin.

speleothems: a general term for mineral deposits (usually calcite) found in caves and formed through the re-precipitation of carbonate contained in dripping water. Includes stalactites, stalagmites, flowstones and other cave forms.

spherical harmonic functions: are defined for every point on the surface of the Earth in a spectral numerical weather prediction model and are composed of pairs of sine and cosine waves. Complex spatial fluctuations in, say, a field of atmospheric pressure can be described by combining a series of these spherical harmonic functions, each having a different wavelength and amplitude. Thus, the more complex the field, the greater the number of harmonics will be required for an accurate representation.

static instability: when a system is subjected to a disturbing influence and moves farther from its original position when the disturbing impulse is removed.

Stevenson screen: a standard housing for meteorological thermometers consisting of a white, wooden ventilated box allowing thermometers to measure air temperature without radiative influence. Designed by Thomas Stevenson.

stochastic: a term used for a method of obtaining a solution to a problem in which a random variable is present and in which, because of the presence of this element of randomness, various degrees of probability need to be attached.

stomatal conductivity: the ability of the epidermis of a leaf or stem to exchange a gas (e.g., water vapour) between the plant and the external environment. The stomata regulate this exchange process and can alter the conductivity.

stratosphere: the layer of the atmosphere above the **troposphere** and below the **mesosphere** (between 10 km and 50 km), generally characterised by an increase in temperature with height.

sub-grid scale: the geographical scale smaller than the size of the grid or cell which is represented in the dataset or the model. Physical processes which operate at sub-grid scales are not explicitly captured or resolved by the gridded dataset or model.

sun-synchronous satellite: a meteorological satellite whose plane of motion is inclined at a small angle to the Earth's axis of rotation and whose orbit processes slowly in space so that the orbital orientation is fixed relative to the Sun. Also referred to as polar-orbiting.

surface roughness length: a measure of the roughness of the surface, proportional to the average height of the roughness elements of the surface, such as vegetation, infrastructure, etc. A typical value of roughness length for a short grass cover is 0.03 m.

synoptic map: a type of map in which all known meteorological data are entered for a specific time in order to give an indication of weather conditions at a given time over a large area.

synoptic system: a large-scale system of weather and atmospheric circulation which describes the configuration of fronts, **troughs, anticyclones** and **depressions** at a given moment.

tectonic uplift: the uplift of the Earth's surface caused by internal pressures which deform the Earth's crust. When such processes act regionally they contribute to mountain-building.

teleconnection: a term used to describe events which are related but separated by large distances.

thermal expansion: the process whereby a fluid (e.g., ocean water) expands in volume as it warms. Thus a warmer ocean will lead to a rise in sea level. The relationship between the amount of heat applied at the ocean surface and the resultant rise in sea level depends on the rate of vertical diffusion of that heat and on the circulation of the oceans.

thermal inertia: or alternatively the 'specific heat capacity' of an object. It is a measure of the amount of heat necessary to raise the temperature of one gram of a given substance by 1°C. Water, for example, has a substantially greater thermal inertia or specific heat capacity than air.

thermohaline circulation: an ocean-water circulation resulting from differences of temperature and salinity. Thermohaline convection results in the slow process of the overturning of the deep water of the oceans.

thermoluminescence: the light emitted by many minerals when heated below the temperature of incandescence. By monitoring the amount emitted it is possible to estimate the value of a constant rock temperature imposed by a former climate and therefore to calculate the length of time since the change of temperature occurred in the rock's environment.

tropopause: the boundary between the **troposphere** and the **stratosphere**.

troposphere: the layer of the atmosphere extending from the Earth's surface up to the tropopause (about 10 km). The **troposphere** is characterised, in general, by a decrease in temperature with height and is the layer to which precipitation and clouds (apart from some rather rare types) are confined.

trough: an elongated zone of low atmospheric pressure at a horizontal surface. Not all troughs are frontal, but all fronts are marked by troughs.

urban warming: the effect of an urban area on the regional temperatures such that the city or built-up area is warmer than its surroundings.

virga: precipitation that falls from a cloud but evaporates before reaching the ground.

volatile organic compounds: organic compounds (i.e., containing chains of carbon atoms) which evaporate readily into the atmosphere, e.g., benzene and 1,3 butadiene.

vorticity: a measure of rotation of an air mass.

warm front: the boundary line between advancing warm air and a mass of colder air over which it rises. In this clash of **air masses**, a warm front is accompanied by a rise in temperature, rainfall or snowfall and a shift in wind direction as it passes.

Weibull frequency distribution: a distribution describing a common (but not universal) shape of the frequency distribution of wind speeds. Whereas the better-known normal distribution is symmetrical about the mean, the Weibull distribution has a positive skew.

wind chill equivalent temperature: an index, using formulae derived from wind speeds and air temperatures, devised by the US Army to assist in the design of cold weather clothing. In general, the stronger the wind the greater the heat loss from exposed skin, so that absolute air temperature is not a true guide to potential exposure.

Younger Dryas: the youngest of the five phases of the Late Glacial period and a 500-year period of climatic deterioration around the Atlantic Basin. The Younger Dryas terminated at about 10,300 years BP.

zonal flow: airflow oriented along lines of latitude, i.e., zonal flow is from the general direction of the west (positive sign) or east (negative sign). See also **progressive**.

NAME INDEX

INDEX OF PLACE NAMES

GENERAL INDEX

ablation 84, 95
abnormal weather events 265
absolute severity time-series 240
absolute vorticity 20
absolute vorticity, conservation of 302
acceleration due to gravity 307
account roll evidence 115
account rolls 115, 118–19, 122
accounts 115, 119
accumulated mean temperature 337
acid pollution 165
acid rain 246, 329
adiabatic changes 303
adiabatic cooling 301
adiabatic energy 307
adiabatic equation 301
adiabatic processes, dry and saturated 302
advection 13; of the humidity field 307; of tropical air
    272; of warm air 265
aerodynamic radius 259
aerosol forcing uncertainties 331
aerosol(s) 107; anthropogenic emissions of 329; effects
    333, 335; forcing 329, 331, 333–4
agricultural production 335
agricultural yields 197
agriculture 54, 335–7
air, malodorous 243; saturated 48
air flow 29, 34, 154, 161, 287, 303
air frost 40, 337; average annual frequencies of 337
air mass 29–30, 41, 98, 147, 155, 165, 265, 305;
    characteristics 165; concept 155; trajectory 165
air pollutants 24–8, 257–258
air pollution 6, 243–6, 253, 257–9; exposure to 257;
    legislation 6; monitoring equipment 245; monitoring
    network 246; policy 259; records 246; traffic-related
    258
air quality 243-4, 246, 258, 335; ambient 259; change
    251; standard 253

air temperature 40, 335; average annual 80
air trajectory analysis 302
airborne lead 255
airborne particulate matter 253
AIRMET 320
albedo 27, 31, 331
aldehydes 253
alert thresholds 246, 259
alkenes 253
All Ireland series 202–3, 206–7
allergens 255, 258
almanacs 139
amateur observations 150
amateur observer 137, 150–2
ambient temperature 53
analogue region 347
analogue techniques 319
ancient Egyptian civilisations 299
ancient Greeks 300
anemograph 225
anemograph trace 225
anemometer 223, 225, 227, 289; network 223; record
    225; sites 223, 227; wind speed 227; wind speed,
    monthly 227
Anglian glacial deposits 74
Anglian Glaciation 66, 74
angular momentum 17–18; conservation 16; transport
    17–18
annals 123; Icelandic 130
annual correlations 203
annual cycle 41, 166, 198
annual growth rings 103, 112
annual mean wind speeds, reconstructed 227, 231–2
annual range in temperature 15
annual series 202
annual standard deviation 198
annual temperature 129, 179, 183, 193; cycle 341;
    indices 120, 129; range in 175